Cyclophane Chemistry

Cyclophane Chemistry

SYNTHESIS, STRUCTURES AND REACTIONS

FRITZ VÖGTLE
Institut für Organische Chemie und Biochemie der Universität Bonn, Germany

Translated by

P. R. Jones
Department of Chemistry,
University of New Hampshire, USA

JOHN WILEY & SONS

Chichester · New York · Brisbane · Toronto · Singapore

Copyright © 1993 by John Wiley & Sons Ltd,
Baffins Lane, Chichester,
West Sussex PO19 1UD, England

Originally published as
Cyclophan-Chemie
© 1990 B. G. Teubner, Stuttgart

Other Wiley Editorial Offices

John Wiley & Sons, Inc., 605 Third Avenue,
New York, NY 10158-0012, USA

Jacaranda Wiley Ltd, G.P.O. Box 859, Brisbane,
Queensland 4001, Australia

John Wiley & Sons (Canada) Ltd, 22 Worcester Road,
Rexdale, Ontario M9W 1L1, Canada

John Wiley & Sons (SEA) Pte Ltd, 37 Jalan Pemimpin #05-04,
Block B, Union Industrial Building, Singapore 2057

Library of Congress Cataloging-in-Publication Data

Vögtle, F. (Fritz), 1939–
    [Cyclophan-Chemie. English]
    Cyclophane chemistry : synthesis, structures, and reactions /
Fritz Vögtle ; translated by P. R. Jones.
        p.   cm.
    Includes bibliographical references and indexes.
    ISBN 0 471 93199 3
    1. Cyclophanes. I. Title.
QD400.V6413   1993   93-10572
547'.59—dc20   CIP

British Library Cataloguing in Publication Data

A catalogue record for this book is available
from the British Library

ISBN 0 471 93199 3

Typeset in Times 10/12 pt by Dobbie Typesetting Limited, Tavistock, Devon
Printed and bound in Great Britain by Biddles, Guildford, Surrey

Contents

Preface to English Edition

Besides many corrections, new formulas and new text have been added to the English translation. Many dozens of new, more recent, literature references have also been introduced. The formulas have been completely redrawn, for which I am grateful to the following co-workers: G. Brodesser, M. Frank, R. Güther, W. Josten, K. Kadei, D. Karbach, H.-B. Mekelburger, H.-P. Michels, W. Schmidt, J. Schmitz, M. Schönberg and A. Schröder.

I would also like to thank Prof. Drs L. Ernst, H. Irngartinger, H. Lehner, E. Weber and Drs P. M. Windscheif and J. Schulz for valuable suggestions.

Bonn, Spring 1993   Fritz Vögtle

Introduction

RINGS IN THE COURSE OF HISTORY

The empty circle, the *Ring*, appeared in mythology as the serpent eating its own tail (Figure 1), as 'Ouroboros'. It symbolizes, essentially, unity, which knows no separation into periods of time, no rhythm, no beginning and no end, and no extremes.

Figure 1. 'Ouroboros', a serpent biting its own tail. It served as the symbol of eternity, the harmony achieved between consciousness and unconsciousness[1]

Goethe interpreted the circle of the closed serpent as a symbol of man's great longing. Man's most ardent wish would be to 'connect the beginning to the end'.

Is it any wonder that a symbol associated with such ideal qualities has elicited a magical fascination for the human mind? How else is it explained that in every age women and men have worn rings of every description, and not by any means only as jewelry. Examples are the picture of the two Countesses Rietberg painted by Hermann tom Ring, from the year 1564 (Figure 2) and the ear of Wolfgang Amadeus Mozart adorned with a ring (Figure 3).

A small collection of finger rings (Figures 4, 5) should serve as convincing examples of the extent to which the simple ring has again and again inspired human imagination to produce such impressive works of decorative art.

It is thus small wonder that molecules containing rings have held a similar fascination for scientists in chemistry from early times up until the present.

The following statement is attributed to the famous mathematician and astronomer Carl Friedrich Gauss: '*A peculiar magic surrounds the perception of measure and harmony*'.

2

Figure 2. The Countesses Rietberg by H. tom Ring

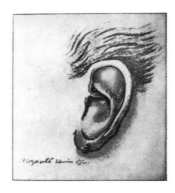

Figure 3. The Ear of Mozart. Anonymous Water Colour, *ca* 1791. Mozarteum, Salzburg

RINGS IN CHEMISTRY

By all accounts one of the most important rings in organic chemistry is benzene
(1), which was discovered by Faraday in 1825 in illuminating gas. The exact
geometric form of benzene was settled by Kekulé in 1865, but discussion about
the electronic structure continues to the present day. Just recently new proposals
for the basis of the special stability ('aromaticity') of this ring have been
advanced.[2]

Because of its stereochemistry, cyclohexane likewise was a high point of
interest for many years (Sachse, 1890; Hasse, 1950s). *Saturated aliphatic rings*
are subdivided today into *small rings* (cyclopropane, cyclobutane), *normal rings*
(cyclopentane, cyclohexane, cycloheptane), *medium rings* (C_8 to C_{12} rings), and

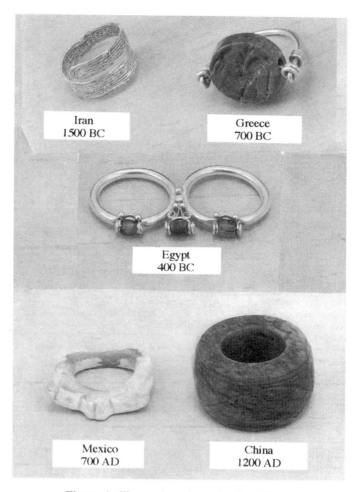

Figure 4. Finger rings from Ancient Times

large (many-membered) *rings* (larger than cyclododecane). The small and medium ring compounds have continually elicited interest because of their internal strain (Baeyer-, Pitzer-, transannular strain) (Figures 6, 7).[3]

The small, medium, and large conjugated *unsaturated* rings served as 'highlights' in organic chemistry in the course of research on non-benzenoid

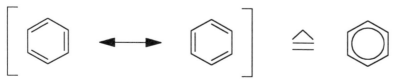

ca. 1700

ca. 1800   ca. 1700

ca. 1600   ca. 1700

Sigurd Persson
1961

Figure 5. Old and Contemporary Finger Rings (Selected Examples)

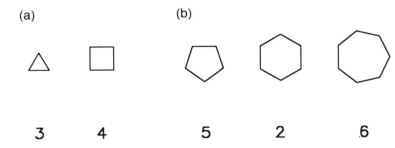

(a)   (b)

3   4   5   2   6

Figure 6. (a) 'Small rings'; (b) 'normal rings'

5

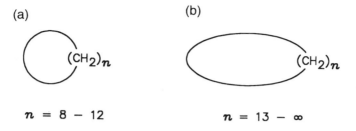

Figure 7. (a) 'Medium rings'; (b) 'large rings'

aromatic compounds and annulenes, and they were of particular significance in the development of *Hückel* molecular orbital theory. The non- or anti-Hückel aromatic ring compounds cyclobutadiene (**8**), cyclooctatetraene (**12**), and others have also been intensively studied.

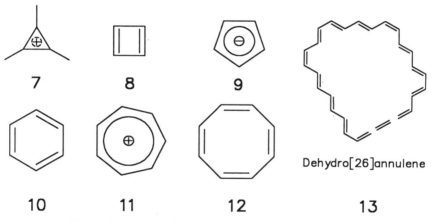

Figure 8. Some conjugated, unsaturated ring compounds

Other rings which have been extensively examined are methanoannulenes, fluxional hydrocarbons such as homotropilidene (**15**), and bullvalene (**16**),[4] as well as cyclic benzene isomers such as benzvalene (**17**; Figure 9); also *condensed* '*aromatics*'/'antiaromatics' and related molecules such as naphthalene, azulene, pentalene, heptalene, octalene, fulvene, calicene, and kekulene (Figure 10);[5]

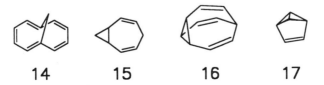

Figure 9. Several well studied ring compounds with interesting π-electron systems

6

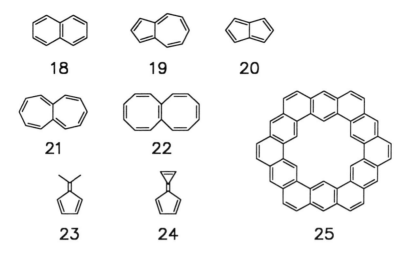

Figure 10. Some condensed 'aromatic'/'antiaromatic' and related ring systems

the first belt-like molecule, 'kohnkene' **(26)**[6] and, finally, 'Buckminster-fullerene' **(27**;[7] Figure 11) and further fullerenes.

With just these few examples one can glean a recent trend toward taking greater advantage of two-[8] and three-dimensionality, particularly in larger molecules. Among these are molecular belts[6] or triply bridged host molecules (see Figure 12) where planar structural units can be extended in all directions. Additional examples are to be found in Chapter 10 ('Exotic Phanes').

The first *cyclophane* {[2.2]metacyclophane} was discovered in 1899 (see below),[9] while the first direct synthesis of a three-dimensional (threefold) bridged cyclophane {2,11,20-trithia[3.3.3](1,3,5)cyclophane} **29** from two 'halves' was carried out in 1970.[10] Large-ring crown ethers such as **30** were first described in 1967.[11]

Also to be noted here are *heterocyclic compounds, organometallic compounds*, such as ferrocene and dibenzene chromium,[12] and other unusual synthetic molecules, such as the *catenanes* (Figure 13; see Chapter 10),[13] those occurring as natural products and those in biochemical systems.

From consideration and comparison of the structures in Figures 8–14, readers can glean somewhat the magical fascination for macroscopic and molecular rings. Thus it is understandable that a considerable number of chemists, over a period of several years, have initiated a search for exotic ring compounds. Many approaches to hypotheses and theories can be imagined by considering benzene, the cycloalkanes, the non-benzenoid aromatics, and the large saturated and unsaturated compounds. Thus, one can contend that ring structures in chemistry—including inorganic chemistry—were not only the starting point for research programmes but also widely useful in all areas of chemistry.

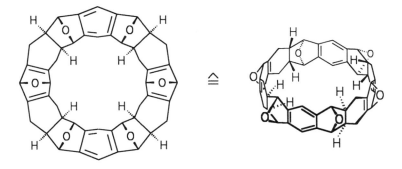

'Kohnkene' **26**

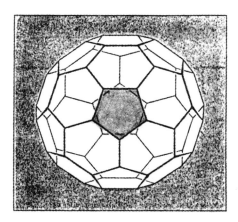

'Buckminsterfullerene' 27

Figure 11. Recent elaborate molecular bridging units and ring structures (in **27** the double bonds have been omitted)

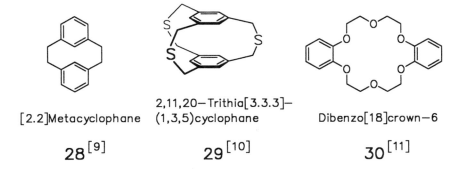

[2.2]Metacyclophane

28[9]

2,11,20−Trithia[3.3.3]−
(1,3,5)cyclophane

29[10]

Dibenzo[18]crown−6

30[11]

Figure 12. First cyclophanes and crown ethers {a hexaoxa[7.7]orthocyclophane}

8

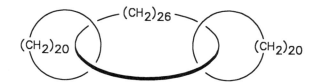

Figure 13. Exotic rings in organic chemistry: catenanes[13,14]

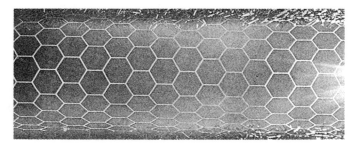

Figure 14. Section of 'hollow carbon fibre'.[15] These fibres are constructed from ordered, tubular carbon atoms, which are oriented in a six-membered ring pattern like that in graphite. Several layers of 'tubular graphite' form concentric tubes

CYCLOPHANES: HISTORICAL

Pellegrin succeeded in synthesizing the first cyclophane. In 1899 he prepared *[2.2]metacyclophane* (**2**, at the time designated 'di-*m*-xylylene') by a Wurtz coupling of 1,3-bis(bromomethyl)benzene (**1**, *m*-xylylene dibromide).[9] It should be noted that 'di-*o*-xylylene' {[2.2]orthocyclophane, **4**}—which was only later included in the cyclophane family—was also obtained in 1945 through Wurtz coupling:[16]

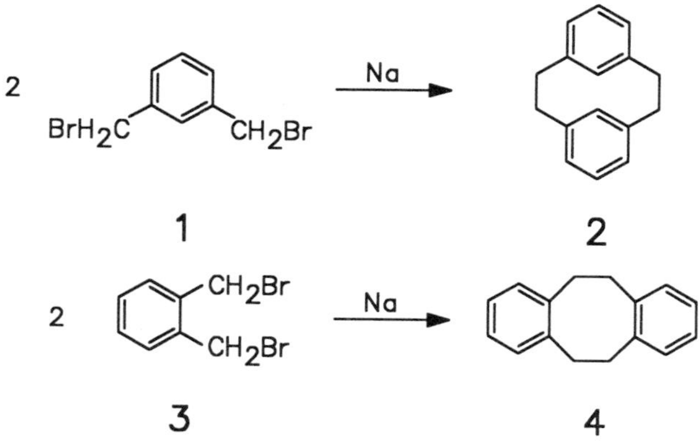

Figure 15. First syntheses of [2.2]metacyclophane and [2.2]orthocyclophane

In 1937 Lüttringhaus described molecules which he called *ansa compounds* (*ansa* (Lat.), handle).[17] Compounds **5** and **6**, which exist as atropisomers,[18] were separated into their stable enantiomers.[17] *Ansa* compounds of this type clearly fit into the modular system (construction unit) of *cyclophanes* (compare 'Nomenclature of cyclophanes and phanes').

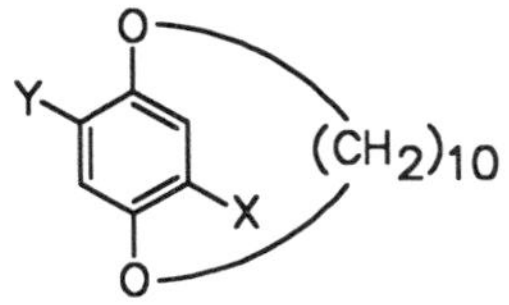

5 : X = CO_2H, Y = Br

6 : X = Y = Br

The 'Age of cyclophanes' actually began in 1949 with the first synthesis of 'di-*p*-xylylene' (**9**, later named [2.2]paracyclophane) by Brown and Farthing.[19] They obtained **9** by extraction from pyrolysis products of *p*-xylene (**7**), such pyrolyses first having been described by Szwarc.[20] Brown and Farthing published a low resolution X-ray structural analysis, which at that time was rare.

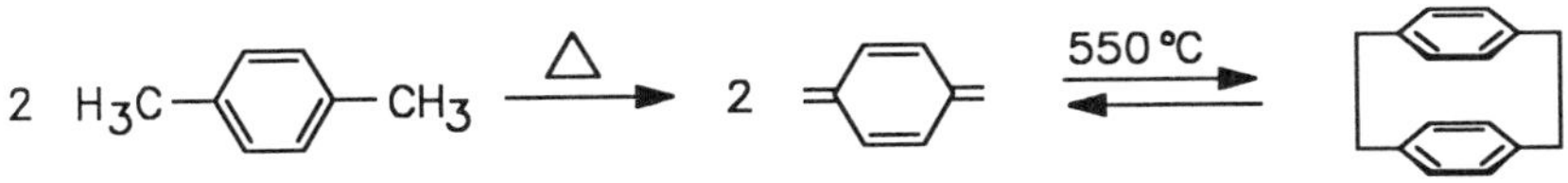

Figure 16. Preparation of [2.2]paracyclophane

The second major advance in cyclophane chemistry was introduced by Cram and Steinberg, when they published the first 'directed' synthesis of [2.2]paracyclophane:[21]

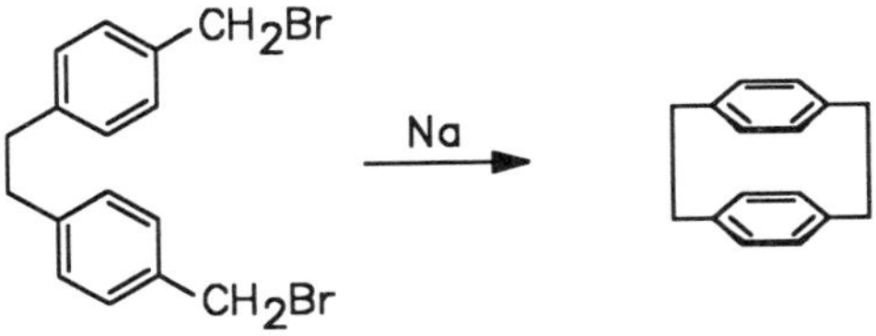

Figure 17. Wurtz synthesis of [2.2]paracyclophane

In his essay 'Cyclophanes: a personal account' in Chapter 1, Volume I of *Cyclophanes* (Keehn, P. M., Rosenfeld, S. M., editors)[22] Cram goes into considerable detail on the history of his concept of [2.2]paracyclophane. As a graduate student, he had already noted in his idea book that it should be possible to place two benzene rings, with their planes placed face-to-face, by means of methylene bridges. At the time this was in accord with Dewar's idea of the involvement of π-complexes in the benzidine rearrangement.[23] Cram's thought was that π-complexes, like those proposed for the benzidine rearrangement, could be stabilized by making them rigid. So in 1948 Cram assigned to his first graduate student, H. *Steinberg*, experiments in cyclophane chemistry. He was to synthesize the [2.2]-, [3.2]-, and [4.2]paracyclophanes **(9, 11, 12)**, which he succeeded in doing in 1950:

	m	n
9 :	2	2
11 :	2	3
12 :	2	4

[2.2]–, [3.2]– and [4.2]Paracyclophanes

These results were published in 1951.[21] Starting from bis(bromomethyl)-bibenzyl **(10)**, Steinberg prepared [2.2]paracyclophane—as outlined in Figure 17—by a high-dilution intramolecular Wurtz reaction in xylene (60 hours with molten sodium); the yield amounted to a mere 2.1% (Figure 17).

Among the German pioneers in constructing aliphatic bridged benzenes (=cyclophanes) are, first and foremost, A. *Lüttringhaus* (1937, *ansa* compounds **5, 6**; see above);[17] R. *Huisgen* (1952, high-dilution intramolecular Friedel–Crafts acylation):[24]

14   **15**   **16**

and H. *Stetter* (1955, 'bridged benzidines'):[25]

13: $n = 2 - 4$; R = Tos, H

Medium aliphatic ring compounds without an arene unit became generally known through the efforts of L. *Ruzicka* (1926), K. *Ziegler* (1933), V. *Prelog* and M. *Stoll* (1947).[26]

Nomenclature of Cyclophanes and Phanes

A unique system of nomenclature, independent of the complicated IUPAC rules, was introduced by *Cram*[21,22] and subsequently systematized by *Schubert*[27] and later *Smith*:[28]

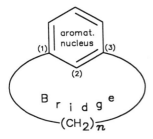

Figure 18. Components of cyclophanes (as well as phanes): aromatic nucleus (benzene or arene) and an aliphatic bridge

This *cyclophane nomenclature* was further systematized and extended by *Vögtle* and *Neumann*.[29] In place of the generic heading cyclophane they chose the class designation 'phane', which denotes all bridged aromatics (including condensed and heteroaromatics such as naphthalene and pyridine). The prefix 'cyclo' thus stands for the benzene ring, which is most commonly present in examples of phanes. In this way simple names were available for the simple bridged arenes, and designations which had already been introduced were maintained {e.g. [2.2]metacyclophane}. For more complex bridged benzenes, the prefix 'benzeno' can be inserted instead of 'cyclo'.[30a] In this way phanes can be subdivided into cyclophanes (more precisely, benzenophanes), hetero-, and heteraphanes. *Heterophanes* contain hetero atoms in the aromatic ring, while in *heteraphanes* they are present in the bridge.

[10]Metacyclophane   [7](1,8)Anthracenophane   [8](2,5)Pyrrolophane

The lengths of the bridges in a phane (the number of atoms between two aromatic points—bridgehead atoms)[30b] are placed in order of decreasing length in square brackets in front of the name; the syllable 'phane' serves as the suffix. Positions of the bridges can be designated by *ortho*, *meta*, *para*, or with numbers in parentheses (after the square brackets). Thus [2.2]paracyclophane can also be named [2.2](1,4)cyclophane.

In multiply-bridged cyclophanes, where all bridges are alike, the numbering can be simplified by means of a subscript to denote the number of bridges: thus, [2.2.2.2.2.2]cyclophane becomes [2_6]cyclophane (= superphane, **18**).

12

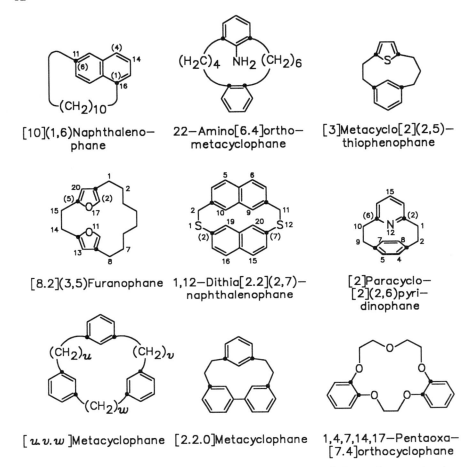

[10](1,6)Naphthaleno-
phane

22-Amino[6.4]ortho-
metacyclophane

[3]Metacyclo[2](2,5)-
thiophenophane

[8.2](3,5)Furanophane

1,12-Dithia[2.2](2,7)-
naphthalenophane

[2]Paracyclo-
[2](2,6)pyri-
dinophane

[u.v.w]Metacyclophane

[2.2.0]Metacyclophane

1,4,7,14,17-Pentaoxa-
[7.4]orthocyclophane

Figure 19. Examples of cyclophanes (benzene as the arene) and phanes (arene unit other than benzene, e.g. pyrrole, naphthalene, furan, pyridine). Bridging positions are designated with a heavy dot

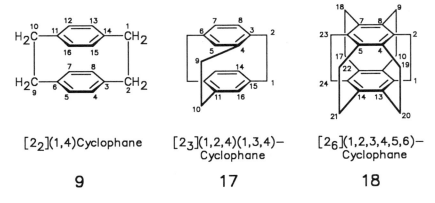

[2₂](1,4)Cyclophane

[2₃](1,2,4)(1,3,4)-
Cyclophane

[2₆](1,2,3,4,5,6)-
Cyclophane

9

17

18

Numbering in a molecule, according to *Boekelheide's* system, begins with the first bridging atom in the longest chain and then continues, as is illustrated with [2.2]cyclophane and its analogues.[33]

Some additional characteristic examples are provided in the following structures.*

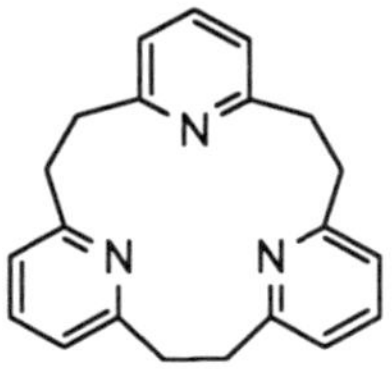

[2.2.2](2,6)Pyridinophane 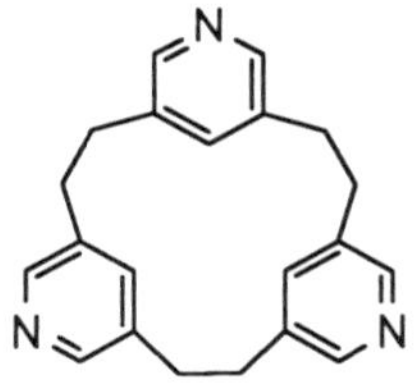   [2.2.2](3,5)Pyridinophane

The literature should be consulted for further details.[28d]

A variation in the above nomenclature system (aside from the IUPAC nomenclature) allows one to devise relatively clear names for particularly

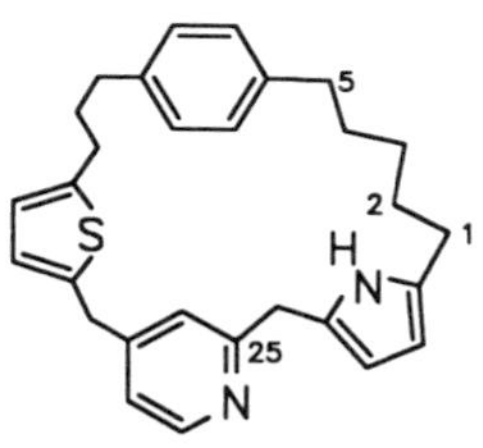

[5]Paracyclo[3](2,5)thiopheno−   [5](2,5)Pyrrolo[3](2,4)pyridino−
[1](4,6)pyridino[1](2,5)−   [1](2,5)thiopheno[2]−
pyrrolophane   paracyclophane

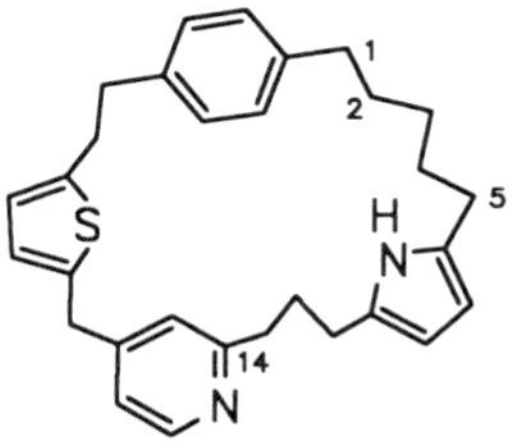

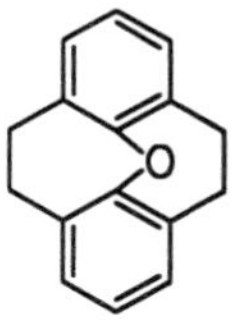

17−Oxa[2.2.1]− $(u > v > w)$ $(u > v > w)$
(1,2,3)cyclophane

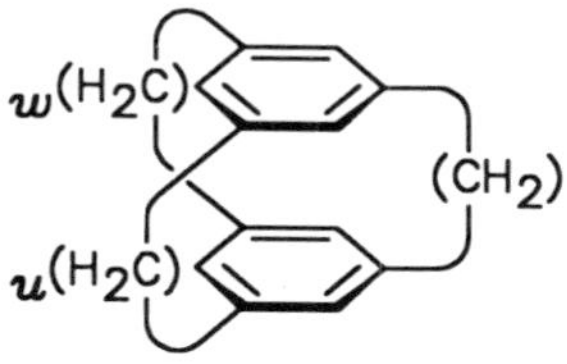

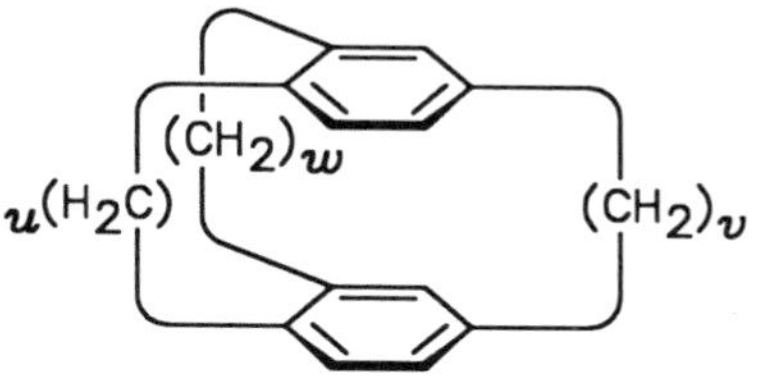

[$u.v.w$](1,3,5)Cyclophane   [$u.v.w$](1,4,2)Cyclophane

---

*For the use of inequality signs as brackets ($<\ >$) in designating 'ring ensembles' as 'aromatic' units, see Section 3.2.

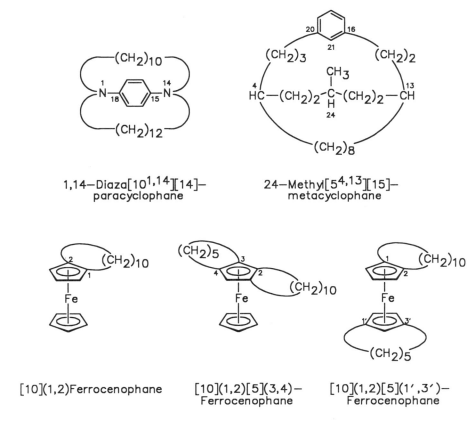

1,14–Diaza[10^{1,14}][14]–
paracyclophane

24–Methyl[5^{4,13}][15]–
metacyclophane

[10](1,2)Ferrocenophane

[10](1,2)[5](3,4)–
Ferrocenophane

[10](1,2)[5](1′,3′)–
Ferrocenophane

complicated phanes. First named 'arena nomenclature',[31] this system was later referred to as 'nodal nomenclature'.[32]

In the nodal nomenclature the aromatic rings 'collapse' to a point, whereupon the structure becomes a simplified framework. In the case of [2.2]meta-cyclophane one would draw a cyclohexane skeleton:

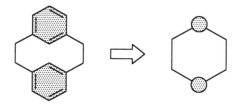

For a definitive name, the points must be translated into aromatic rings. For [2.2]metacyclophane the name becomes 1,4-di(1,3-benzena)cyclohexane or -cyclohexa-nodane. The suffix *nodane* specifies that an aromatic unit has been allowed to collapse according to the nodal nomenclature. The advantage over the IUPAC nomenclature is that one need not include aromatic double bonds and their locations, but yet the aromatic rings as such are part of the name. This simplifies somewhat the recognition of pyridino-, thiopheno-, naphthaleno-phanes, etc.:

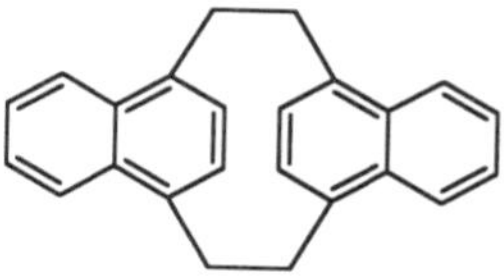

Phane Nomenclature: [2.2](1,4)Naphthalenophane
IUPAC Nomenclature: 6,7,14,15-Tetrahydro-5,16:
8,13-diethenodibenzo[a,g]cyclododecene
Nodal Nomenclature: 1,4-Di(1,4-naphthalena)-
cyclohexanodane

The phane nomenclature is easily applicable to relatively simple phanes and can be adapted to most complicated structures. From the name one easily recognizes the basic structure, connecting positions, and lengths of the bridges.

In spite of the simplicity of the phane nomenclature, trivial names for some old and new ring systems, which belong to the phanes, have persisted. Examples are the calixarenes, spherands, kekulene (see above), tri-o-thymotide (TOT), cyclotriveratrylene (CTV), etc. (for further details see following sections).

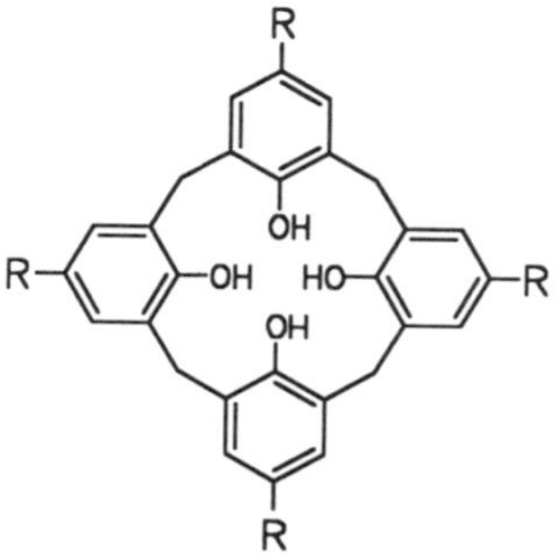

Calix[4]arene

19

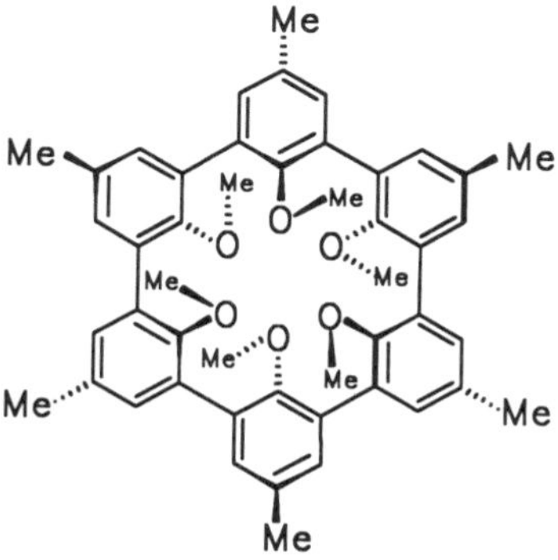

Spherand

20

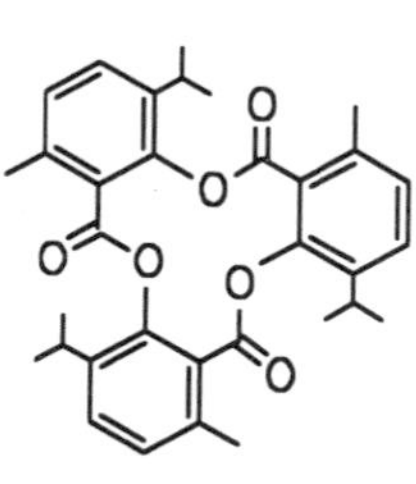

TOT

21

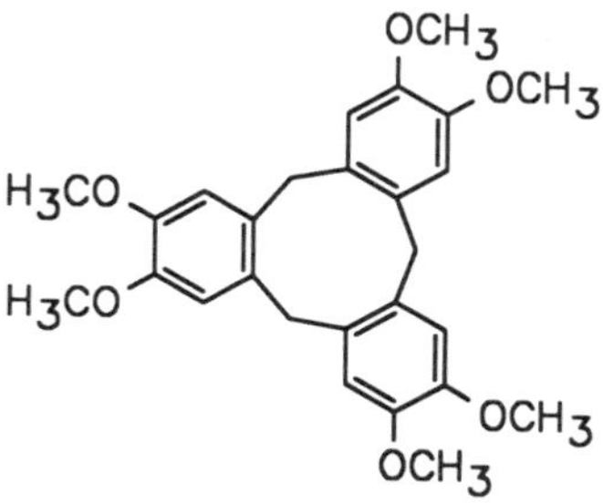

CTV

22

The reason for these trivial names is that the structural skeletons are fully retained in the course of conversion to derivatives, so that a simpler trivial name characterizes the basic structural unit and many of its derivatives, as a family of compounds.

Some additional remarks concerning questions and rules for less straightforward nomenclature—for example, the problem of purely aliphatic phane analogues—are to be found in Chapter 9.

WHY CYCLOPHANE CHEMISTRY?

In his first publication on cyclophanes Cram had already expressed his opinion about some peculiarities to be expected in cyclophanes, which he outlined as follows:

(a) electronic interaction between aromatic rings placed 'face-to-face';
(b) the resulting influence on substitution reactions in the aromatic rings; i.e., influence of substitution in one ring through transannular electronic effects, which are induced by the other ring;
(c) intramolecular charge-transfer complexes;
(d) ring strain; steric strain; transannular strain.

These predictions have been subsequently substantiated extensively. From studies it has been demonstrated that aromatic rings in many cyclophanes are not planar but distorted out of planarity through bending (*'bent and battered benzene rings'*).[34] It has been shown unequivocally by X-ray structure analysis that benzene rings can be distorted into *boat-*, *chair-*, and *twist-*forms by 'clamping' or bridging them in cyclophane structures. These stereochemical aberrations create spectroscopic and chemical consequences which are of interest to all chemists, particularly in connection with benzene, naphthalene, azulene, pyridine, etc.

Tightly 'clamped' cyclophanes are typically strained. The ring strain reaches an extreme in highly strained molecules such as [6]paracyclophane, available through a directed synthesis. The question arises as to what extent benzene rings can be bent and distorted and what are the resulting spectroscopic and chemical consequences? How closely can benzene rings be positioned face-to-face, and what ramifications with respect to spectroscopy and chemical reactivity result?

What influence does distortion of arene rings have on 'aromaticity'?

With the construction of molecules with several benzene rings stacked in a staircase fashion, it should be possible to study *transannular electronic and steric effects* transmitted through more than two rings.

Benzene rings can indeed be arranged into multi-layer cyclophanes, and transannular electronic effects through more than two aromatic rings can be detected spectroscopically.

Furthermore, phane molecules can orient themselves as planar chiral and helical arrangements, and the resulting enantiomers, once separated, have been examined with respect to their chiroptical properties.

Another consideration is the placement of 'intra-annular' substituents into the inner opening of the cyclophane. This raises the existing steric strain, attributed to van der Waals interactions between the protruding groups or substituents inside the cavity.

What conformations do cyclophanes assume; which of these are favoured? How are aliphatic chains arranged between the aromatic rings? How flexible are such chains and the aromatic rings? How is flexibility altered by the length of the bridge and substitution on the aromatic ring?

How is the strain energy in larger cyclophane molecules distributed among various groups?

Can strained cyclophanes serve as synthons for the preparation of other molecules?

In the case of the face-to-face arrangement of anthracenes, the question arises as to whether photochemical and topochemical (in the crystal) reactions might be stereospecific.

All of the above questions make it clear that phanes are ideal model compounds, in which benzene and other aromatic rings can be placed relative to one another in tailor-made fashion. Bridged cyclophanes are often rigidly symmetric and so convenient subjects for spectroscopic experiments.

With cyclophane chemistry one has the opportunity gradually and successively to increase the distortion. Consideration of extended π-systems such as anthracene, ferrocene, dibenzene chromium, etc., makes it clear that changes in the aromatic unit will affect the spectroscopic and chemical properties.

With the phane structure it is possible to place interesting functionalized bridging units close to the aromatic ring: for example to incorporate a threefold 'face-to-face' bridge over benzene rings. In this way one can achieve a particular arrangement of functional groups and thereby trigger their chemical interaction.

One can introduce into the rings of many cyclophanes donor substituents, which interact as ligands with cations and neutral molecules (crown ethers; host–guest chemistry; molecular recognition; further details in Chapter 12).

Although cyclophanes, at the time 20 years ago when this author began investigating them, were considered esoteric and unrealistic, there being no thought toward their industrial applications, the situation has changed completely. With the possibility of locating groups precisely in space, cyclophane chemistry has provided the building unit for nests, hollow cavities, 'multi-floor structures', helices, macropolycyclics, macro-hollow tubes, novel ligand systems, etc. (see later sections). Cyclophane chemistry has become a major component of *supramolecular chemistry*, of molecular recognition, of models for intercalation, of the building blocks for organic catalysts, receptor models, crown ethers, and cryptands. Apart from the varied research possibilities and their role as cyclo-oligomers for high molecular weight compounds ('polyphenylene sulphide', 'macrolone'[35]) the phanes find application in the 'high-tech' field.

Cyclophane chemistry has proved to represent a large variety of molecules and geometries, and it seems important to survey the current field, not only for cyclophane chemists. Cyclophane chemistry depends upon the three-dimensional geometry of its molecules. It is the wide variety of arrangements of partial structures, fixed or conformationally flexible, that affords such a rich field for research into molecular design, synthesis, structural analysis, physical investigations, chemical reactions, and all the way to supramolecular chemistry.

References

1. See, for example, *P. M.-Magazin*, 82 (1989).
2. I. Gutman and S. J. Cyvin (Eds.), *Advances in the Theory of Benzenoid Hydrocarbons. Top. Curr. Chem.*, **153**, Springer, Berlin, 1990.
3. A. de Meijere and S. Blechert (Eds.), *Strain and its Implications in Organic Chemistry*, NATO ASI Series C, Vol. 273, Kluwer Academic Publishers, Dordrecht, 1988.
4. Review: H. R. Christen and F. Vögtle, *Organische Chemie. Von den Grundlagen zur Forschung*, Vols. I, II, Salle-Sauerländer, Frankfurt a.M./Aarau, 1989.
5. H. A. Staab and F. Diederich, *Chem. Ber.*, **116**, 3487 (1983); H. A. Staab, F. Diederich, C. Krieger and D. Schweitzer, *Chem. Ber.*, **116**, 3504 (1983); F. Vögtle and H. A. Staab, *Chem. Ber.*, **101**, 2709 (1968); H. A. Staab and M. Sauer, *Liebigs Ann. Chem.*, 742 (1984).
6. J. F. Stoddart, *Nature*, **334**, 10 (1988); P. R. Ashton, N. S. Isaacs, F. H. Kohnke, G. Stagno-d'Alcontres and J. F. Stoddart, *Angew. Chem.*, **101**, 1269 (1989); *Angew. Chem. Int. Ed. Engl.*, **28**, 1261 (1989); F. H. Kohnke, J. P. Mathias and J. F. Stoddart, in E. Weber (Ed.) *Top. Curr. Chem.*, **165**, 1 (1993); F. Vögtle *et al.*, *Chem. Ber.*, **125**, 1881 (1992).
7. H. W. Kroto *et al.*, *Nature*, **318**, 162 (1985); Review: *Acc. Chem. Res.*, **25**, 106 (1992); H. W. Kroto and D. R. M. Walton, in *Carbocyclic Cage Compounds* (E. Osawa and O. Yonemitsu, Eds.), VCH, Weinheim, 1992; G. S. Hammond and V. J. Kuck (Eds), *Fullerenes*, ACS, Washington, 1992.
8. A.-D. Schlüter, *Nachr. Chem. Tech. Lab.*, **38**, 8 (1990).
9. M. M. Pellegrin, *Rec. Trav. Chim. Pays-Bas*, **18**, 457 (1899).
10. F. Vögtle, *Liebigs Ann. Chem.*, **735**, 193 (1970).
11. H. E. Schroeder and C. J. Pedersen, *Pure Appl. Chem.*, **60**, 445 (1988).
12. Ch. Elschenbroich and A. Salzer, *Organometallchemie*, Teubner, Stuttgart, 1988; *Organometallics—a Concise Introduction*, VCH Verlagsgesellschaft mbH, Weinheim, 1989.
13. (a) J. F. Stoddart *et al.*, *Angew. Chem.*, **101**, 1404 (1989); *Angew. Chem. Int. Ed. Engl.*, **28**, 1394 (1989).
    (b) A. M. Albrecht-Gary, C. O. Dietrich-Buchecker, Z. Saad and J.-P. Sauvage, *J. Am. Chem. Soc.*, **110**, 1467 (1988); C. O. Dietrich-Buchecker and J. P. Sauvage, *Angew. Chem.*, **101**, 192 (1989); *Angew. Chem. Int. Ed. Engl.*, **28**, 189 (1989).
    (c) G. Schill, E. Logemann and W. Littke, *Chemie in uns. Zeit*, **18**, 130 (1984).
14. Review: F. Vögtle, *Supramolecular Chemistry*, Wiley, Chichester, 1991.
15. G. Tibbetts, General Motors, in *Bild der Wissenschaft*, **26**, 44 (1989); P. M. Ajayam and S. Iijima, *Nature*, **361**, 333 (1993).
16. W. Baker, R. Banks, D. R. Lyon and F. G. Mann, *J. Chem. Soc.*, **1945**, 27.
17. A. Lüttringhaus and H. Gralheer, *Liebigs Ann. Chem.*, **550**, 67 (1942); **557**, 108 (1945); A. Lüttringhaus and G. Eyring, *Liebigs Ann. Chem.*, **604**, 111 (1957). A pure carbocyclic analogue was described by A. T. Blomquist, R. E. Stahl, Y. C. Meinwald and B. H. Smith, *J. Org. Chem.*, **26**, 1687 (1961).

18. Cf. 'Meilensteine der Chemie', *Nachr. Chem. Tech. Lab*, **38**, 36 (1990).
19. C. J. Brown and A. C. Farthing, *Nature (London)*, **164**, 915 (1949).
20. M. Szwarc, *J. Chem. Phys.*, **16**, 128 (1948).
21. D. J. Cram and H. Steinberg, *J. Am. Chem. Soc.*, **73**, 5691 (1951).
    (a) For new quantitative data and calculations of high dilution reactions see: G. Ercolani, L. Mandolini and P. Mencarelli, *Macromolecules*, **21**, 1241 (1988); *Gazz. Chim. Ital.*, **119**, 209 (1989); G. Ercolani and P. Mencarelli, *J. Chem. Soc. Perkin Trans 2*, 187 (1989); G. Ercolani, L. Mandolini and P. Mencarelli, *J. Chem. Soc. Perkin Trans 2*, 747 (1990); L. Mandolini *et al.*, *J. Chem. Soc., Chem. Commun.*, 538 (1993).
22. P. M. Keehn and S. M. Rosenfeld (Eds.), *Cyclophanes*, Vol. I, Academic Press, New York, London, 1983.
23. M. J. S. Dewar, *Nature (London)*, **156**, 784 (1945); *J. Chem. Soc.*, 406 (1946).
24. Review: R. Huisgen, *Angew. Chem.*, **69**, 341 (1957).
25. H. Stetter and E.-E. Roos, *Chem. Ber.*, **88**, 1390 (1955); cf. R. Hilgenfeld and W. Saenger, *Angew. Chem.*, **94**, 788 (1982); *Angew. Chem. Int. Ed. Engl.*, **21**, 781, Suppl. 1690 (1982).
26. Review: E. L. Eliel, *Stereochemistry of Carbon Compounds*, McGraw-Hill, New York, 1962. See also ref. 13c.
27. W. M. Schubert, W. A. Sweeney and H. K. Latourette, *J. Am. Chem. Soc.*, **76**, 5462 (1954).
28. Reviews:
    (a) B. H. Smith, *Bridged Aromatic Compounds*, Academic Press, New York, 1964.
    (b) D. J. Cram and J. M. Cram, *Acc. Chem. Res.*, **4**, 204 (1971).
    (c) S. Misumi and T. Otsubo, *Acc. Chem. Res.*, **11**, 251 (1978).
    (d) F. Vögtle and G. Hohner, *Top. Curr. Chem.*, **74**, 1 (1978); Computer program for generation of IUPAC notation for polycyclic molecules: G. Rücker and Ch. Rücker, *Chimia*, **44**, 116 (1990).
29. F. Vögtle, *Tetrahedron Lett.*, 3193 (1969); F. Vögtle and P. Neumann, *Tetrahedron Lett.*, 5847 (1970).
30. (a) Aliphatic bridges and aromatic nuclei merge at bridgehead positions in 'phanes'. The bridge consists of only one bond: zero bridges, [0]phane.
    (b) G. Hohner and F. Vögtle, *Chem. Ber.*, **110**, 3052 (1977).
31. Th. Kauffmann, *Tetrahedron*, **28**, 5183 (1972); K. Hirayama, *Tetrahedron Lett.*, 2109 (1972).
32. N. Lozac'h, A. L. Goodson and W. H. Powell, *Angew. Chem.*, **91**, 951 (1979); N. Lozac'h and A. L. Goodson, *Angew. Chem.*, **96**, 1 (1984); *Angew. Chem. Int. Ed. Engl.*, **23**, 33 (1984).
33. V. Boekelheide, in *Cyclophanes I* (F. Vögtle, Ed.), *Top. Curr. Chem.*, **113**, 87 (1983); H. Hopf, *Chem. in uns. Zeit*, **10**, 114 (1976).
34. D. J. Cram and J. M. Cram, *Acc. Chem. Res.*, **4**, 204 (1971).
35. D. J. Brunelle, E. P. Boden and Th. G. Shannon, *J. Am. Chem. Soc.*, **112**, 2399 (1990).

1 [n]Phanes

Notable progress has been made in the last few years in the area of compounds containing a benzene ring bridged at the *m*- or *p*-position { [*n*]metacyclophanes and [*n*]paracyclophanes}, in particular in their synthesis.[1]

1.1 [n]METACYCLOPHANES (AND HETERO-/HETERA-ANALOGUES)

1.1.1 SYNTHESIS

The coupling of Grignard reagents with dihalo-aromatic compounds in the presence of a nickel–phosphane complex as catalyst was achieved in 1975.

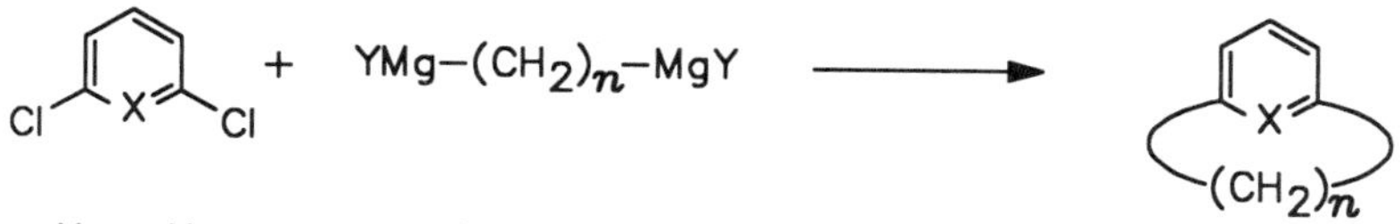

$$X = N; \quad n = 6 - 10, 12$$
$$X = CH; \quad n = 8 - 10, 12$$

In this way several metacyclophanes and 2,6-pyridinophanes, including the natural product *muscopyridine* (1) became available.[2]

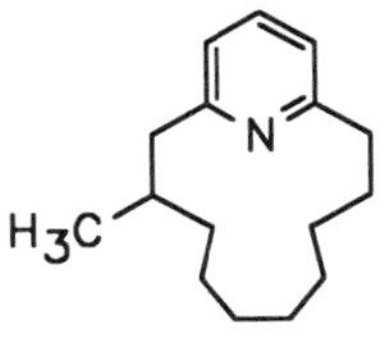

1

The short bridged *[5]metacyclophane* (3) was obtained by the following route (Bickelhaupt):

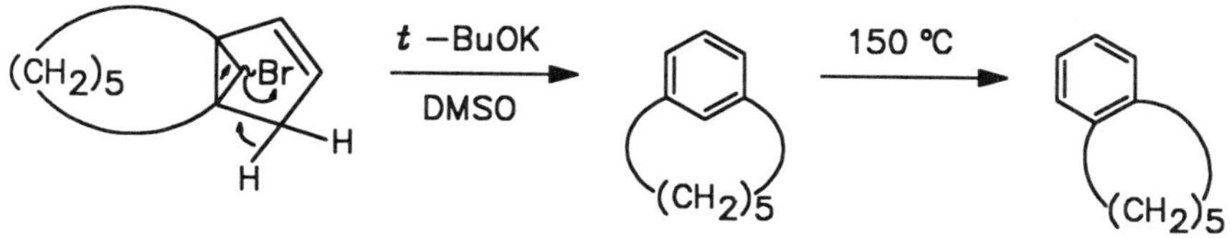

| 2 | 3 | 4 |

21

It undergoes facile thermal rearrangement to [5]orthocyclophane (**4**), possibly by way of a bridged Dewar benzene intermediate. The intermediate formation of *[4]metacyclophane* has been proposed in a synthesis analogous to that above; the corresponding Dewar isomer **5** could be isolated.[3]

5

[7]Metacyclophane (**7**) is generated by treatment of [7]paracyclophane (**6**) with fluorosulphonic acid and *p*-toluenesulphonic acid in benzene.[4]

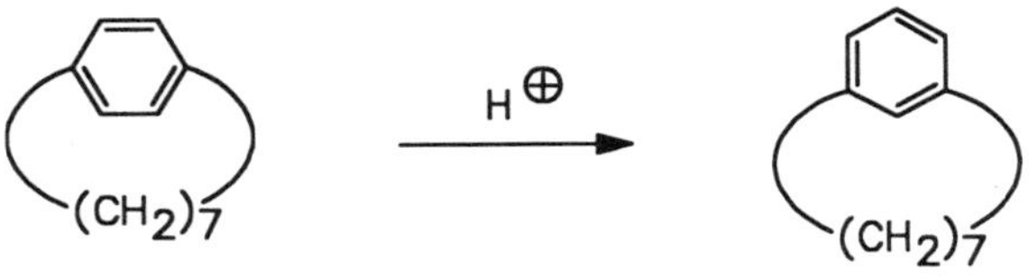

6   **7**

Rearrangement of [10]orthocyclophanes affords *[10]metacyclophanes*:[5]

R = H, CH$_3$

To date the shortest bridge in a 1,3-position is represented by [4]pyrrolophane **9**:[9]

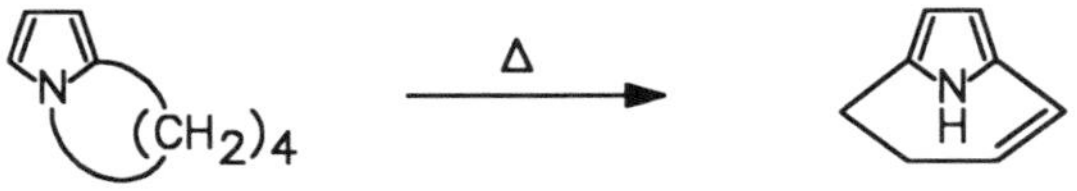

8   **9**

All members in the [*n*]metacyclophane series with a saturated hydrocarbon chain are known, where $n = 5–10$ and $n = 12$ and 13; thus the 5-carbon bridge appears to be the shortest possible. In the case of [*n*](1,3)naphthalenophanes the members with $n = 6$, 8, and 10 are known (see Section 1.3). The [n](2,5)thiophenophanes with $n = 8$–12 have likewise been prepared, as have some [n]furanophanes, [n]pyrrolophanes, etc.

In 1971 Fujita and Nozaki[6] developed a synthesis of [7](2,6)pyridinophane (13) and [7](2,6)pyryliophane perchlorate (12), as well as some substituted pyridinophanes starting with 9b-boraperhydrophenalene (10):

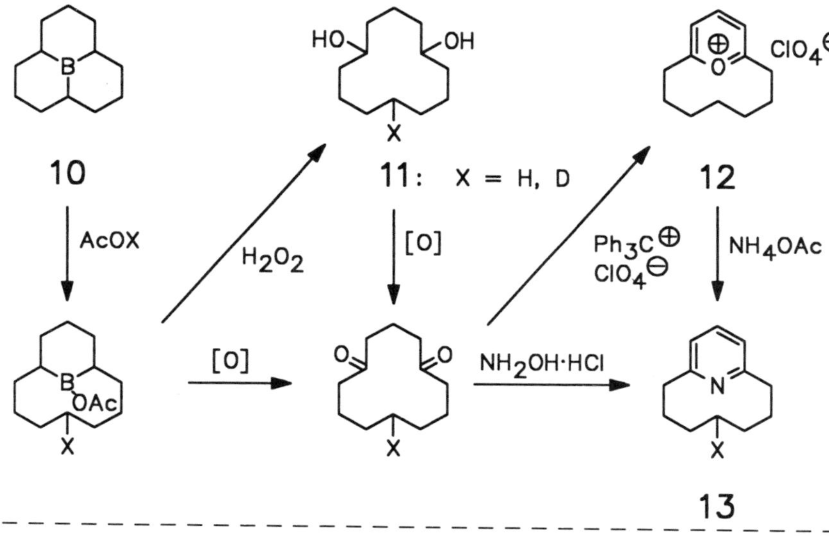

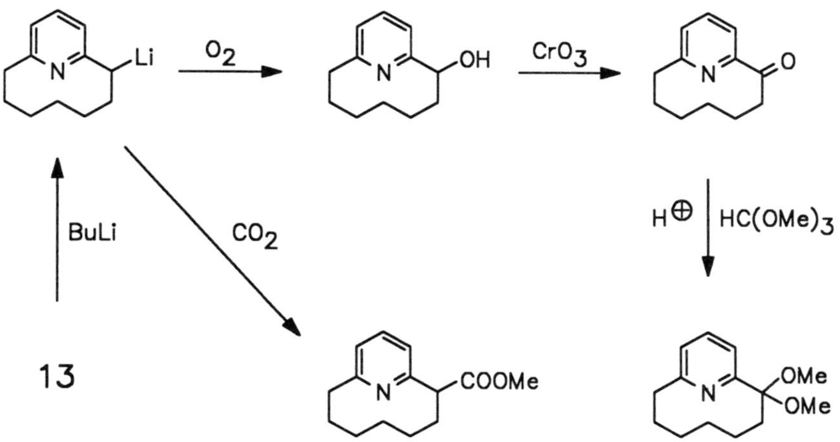

The dithia[n]metacyclophanes 14 ($n = 2$–9) and corresponding pyridino-, thiopheno- and other phanes are easily available by the method of *Vögtle et al.*:[7]

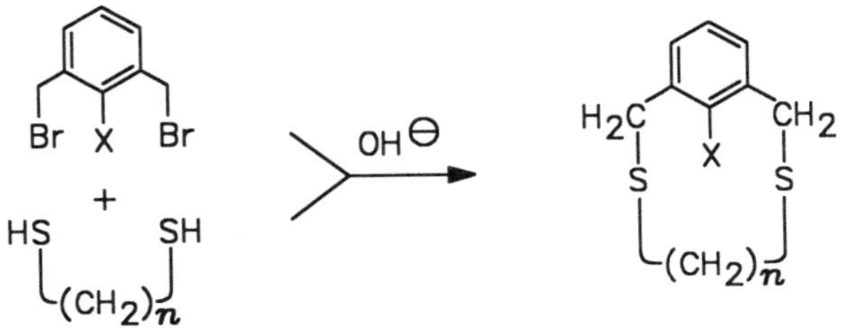

14: $n = 3 - 9$

Corresponding oxa-oxo[n]- (**15**) and diaza[n]metacyclophanes (**16**), like the dithia compounds, can be prepared efficiently:[8]

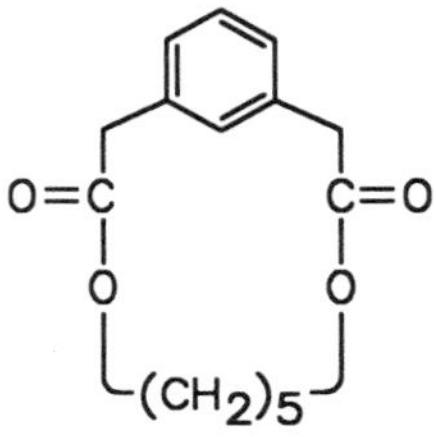

15

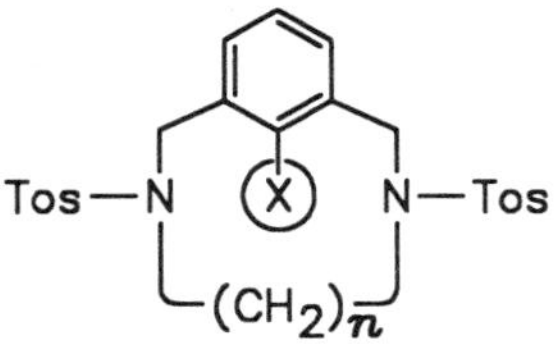

16

X = H, F, Cl; $n = 4$
X = F; $n = 5$
X = H, F, Cl, Br; $n = 6$

Nozaki described [n]furano-, thiopheno- and -pyrrolophanes of the type **17–19** via the Paal–Knorr cyclization:[9,10]

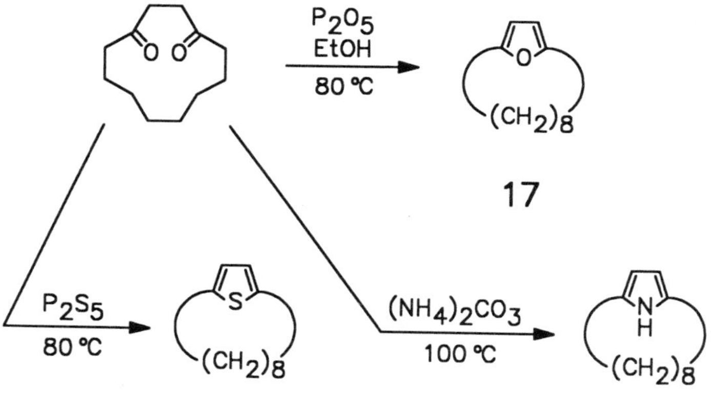

17

18   **19**

For additional routes to [n](1,3)phanes see reference 1.

1.1.2 SPECTROSCOPY

It was anticipated that in the [n]meta-, as in the [n]paracyclophane series, a dramatic loss of aromaticity would be associated with a decrease in n. This was, however, not the case. The UV spectra of [n]cyclophanes are typically characterized by a bathochromic shift and loss of fine structure, as the length of the methylene bridge decreases. This is attributed to a bending of the aromatic ring out of planarity. A qualitatively similar behaviour (bathochromic shift with 'maintenance of aromaticity') to that of the [n]paracyclophanes (compare Section 1.2) is exhibited by the [n]metacyclophanes, the [n](1,3)naphthaleno-phanes, [n](2,4)quinolinophanes, [n](2,6)pyridinophanes, [n](3,5)pyrazolo-phanes, and other [n](2,4)heterophanes.

In the NMR spectra of [n]cyclophanes one often observes upfield shifts for those aliphatic H_i-atoms of the bridge that lie above or below the plane of the aromatic ring. In *[7]metacyclophane*, for example, one sees a broad signal at $\delta = -0.18$ (at 25°C)—that is, 'beyond' the TMS standard; at -73°C the peak is at -1.33.[15]

The methylene chains in [n]metacyclophanes, as well as those in p-cyclophanes, with large chain lengths, give evidence of conformational mobility. Only in the case of the shortest *[5]metacyclophane* is the proton NMR spectrum not temperature-dependent. The bridge in *[6]metacyclophane* even at room temperature undergoes a pseudorotation and then 'flips' over and under the plane of the benzene ring (ring inversion). An intra-annular substituent R_i will impede the conformational mobility of the bridge. For example, while [7]- and [10]metacyclophane exhibit broad spectra at room temperature because of the ring inversion ('flipping') of the oligomethylene bridge, the benzyl protons in 2-bromo[7]- and 2-bromo[10]metacyclophane are not equivalent. Such conformational investigations have likewise been carried out with [n](1,3)naphthalenophanes, [8](2,5)pyrrolophanes, and other related phanes.[1]

In the case of [10]metacyclophane the ring inversion could not be 'frozen out' even at low temperatures, and for [7]metacyclophane only the ring inversion (coalescence temperature -28°C), but not the pseudorotation of the CH_2 groups, is observed. In contrast, the temperature-dependent proton NMR spectrum of [6]metacyclophane [signals at $\delta = -1.27!$ (-82°C) and at $\delta = +0.35$°C] provided the following details:[15] the pseudorotation of the methylene bridge (coalescence temperature $T_c = -31.5$°C, $\Delta G_c^{\ddagger} = 46 \, \text{kJ/mol}$) can be observed separately from the ring inversion, with a significantly higher energy barrier ($T_c = +76.5$°C).

The conformational mobility of the bridges will, of course, be influenced by the incorporation of hetero atoms in the bridge. In the dithia[n]meta-cyclophane series **14** (see above), for example, it was found the mobility of the sulphur compounds is higher than that of the hydrocarbon compounds; the reason is presumed to be the longer C—S bond, in comparison to the shorter C—C bond.

The concept of intra-annular substituents X in the rings **14** was introduced for the express purpose of creating '*steric interactions in the core of cyclic compounds*'. More than a dozen intra-annular groups (for example, H, F, Cl, Br, CH_3, OCH_3, SCH_3, OH, SH, NO_2, NH_2, CO_2CH_3, etc.) were appropriately incorporated into rings of varying size, such that the ring inversion process A ⇌ B is sterically hindered. Through dynamic ^{1}H NMR spectroscopy (diastereotopic arene-CH_2—S-methylene protons) the barriers to ring inversion could be quantified:[12]

14A   14B

As the size of the intra-annular substituent in such phanes (**14**) is increased, the interconversion barrier rises sharply.

14   20

Thus, by proper choice of an aliphatic chain length in the phane system **14** one can generate a ring with a diameter such that the intra-annular substituent X can just pass through the opening by the topological process **14A** ⇌ **14B**. The range of measurement in the dynamic NMR can be adjusted to the substituents.[12]

Because of a broad choice of synthetic approaches, it was possible to measure the activation enthalpies of the diastereotopic process A ⇌ B for a large number of substituents X (in **14**); the series could be expanded to include other substituents as well. The steric interaction for most substituents was investigated for two or three chain lengths ($n + 4$). The value of n where $\Delta G_c^{\ddagger} = 63$ kJ/mol could serve as a measure of the size of a substituent (the 'n-value').[12]

From this approach it follows that the relative size of substituents falls in the order:

$$N \rightarrow < CH \quad \text{and}$$

$$H < F < (\geq N)-O \approx OH < NH_2 < NO_2 \approx Cl$$

$$CH_3 < CN < Br < OCH_3 < I < SCH_3 < CO_2CH_3 < SO_2CH_3$$

In this way the size of the pyridine N-oxide group (in system **20**) could also be determined.

A change in the ring size in **14** strongly influences the barrier to ring inversion: shortening of the bridge by one methylene group where the ring size is critical raises $\Delta G_c^{\ddagger}$ by about 29 kJ/mol.

If one replaces methylene groups in the bridge with hetero atoms, the effect is that of changing the length of the bridging chain. Thus replacement of $-CH_2$ by $-S$ always facilitates ring inversion, as is evident in a comparison of **21** with **22** or **23** with **24**.

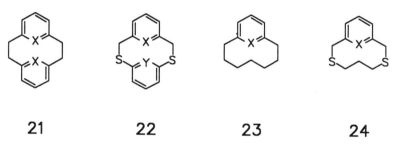

21   **22**   **23**   **24**

If one replaces the sulphur atom in phanes **25** or **14** with N(Tos), then the inversion barriers increase by about 20 kJ/mol:

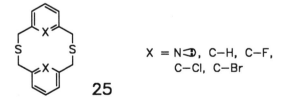

25

$$X = N \rightarrow, \ C-H, \ C-F,$$
$$C-Cl, \ C-Br$$

In the dioxaphane **26**, however, contrary to expectation, the topomerization barriers are found to be about 4 kJ/mol lower than that in **27**—an indication that the shortening of the chain by about 4×11 pm can be offset by the higher flexibility of the $C-O-C$ bond and the removal of some *gauche*-H–H interactions. Yet in **28** and **30** the value of $\Delta G_c^{\ddagger}$ is once again higher than that in the corresponding compounds **29** and **31**. The nature of the colliding atoms, which interact sterically, is thus of particular note.[12]

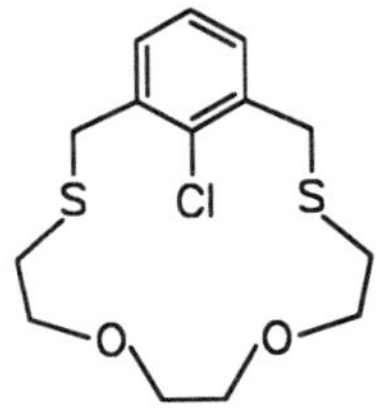

26 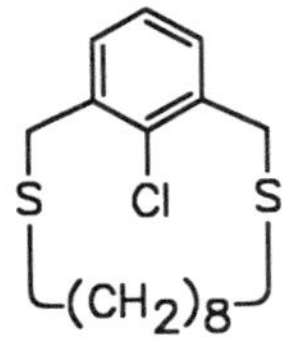   **27** 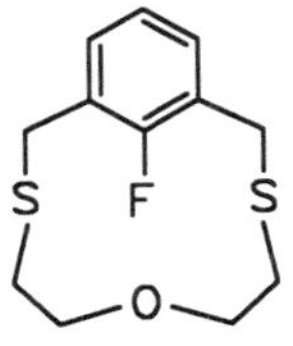   **28** 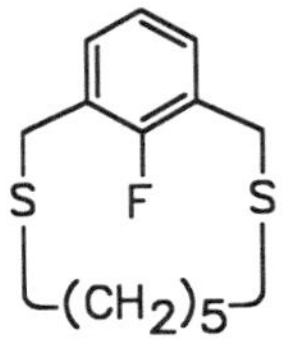   **29**

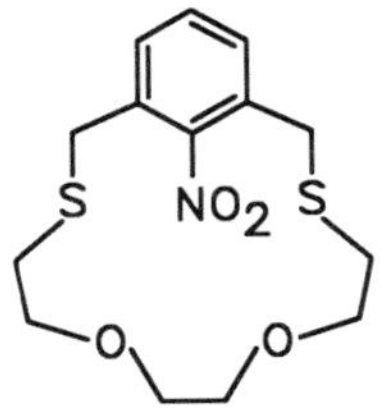

30 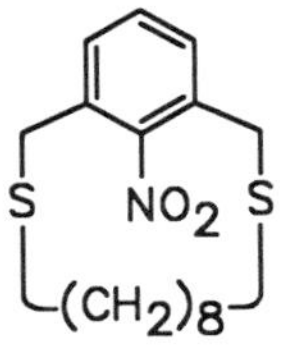   **31**

1.1.3 CHEMICAL REACTIONS

The 1,3-bridged naphthalenophane **32** undergoes a ring opening when treated with HBr.[13]

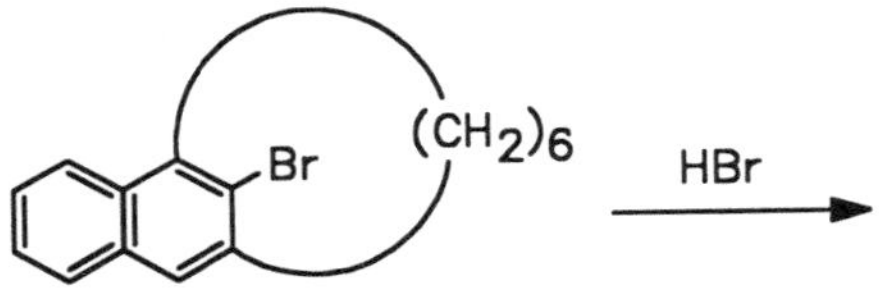

32

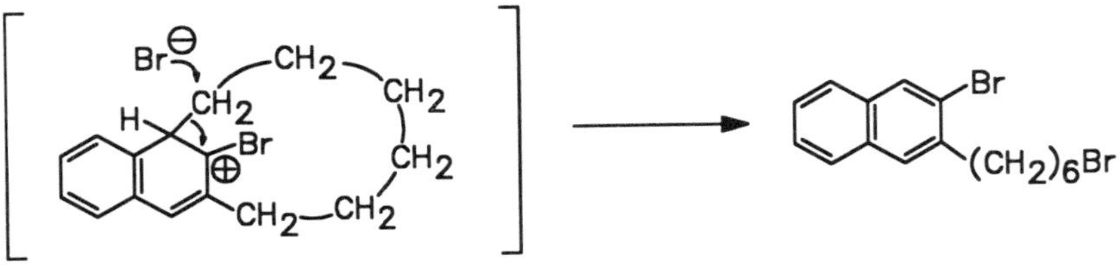

The Grignard reagent **33** derived from **32** leads to unexpected products on reaction with O_2; this demonstrates the ring strain and spatial proximity of certain hydrogen and carbon atoms in the bridge to the reaction site (transannular reaction):[14]

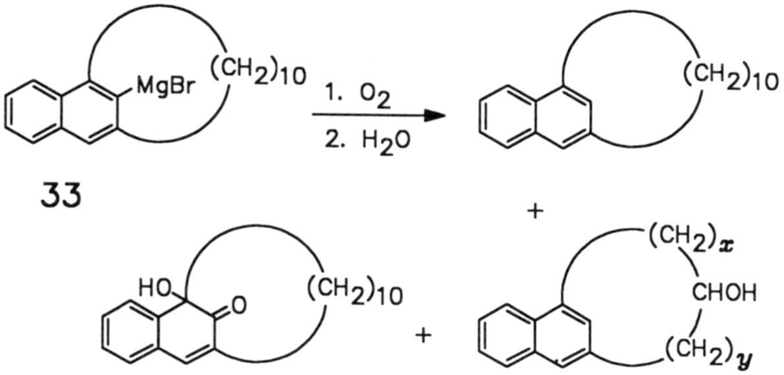

A large number of [*n*]metacyclophanes **35** with specific intra-annular functional groups can be prepared from the intra-annular lithiated [*n*]metacyclophane **34**.[15]

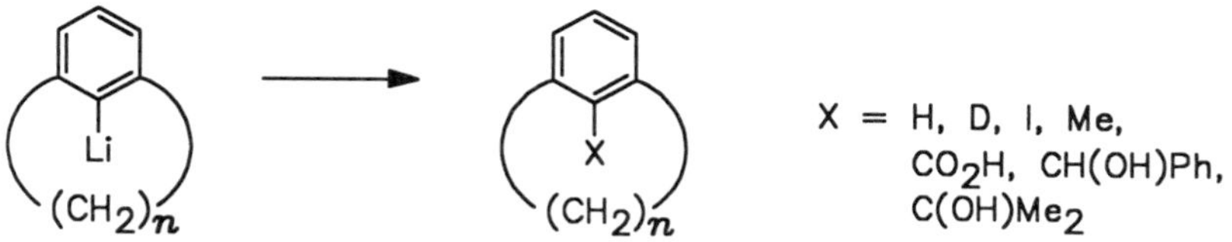

$$X = H, D, I, Me,$$
$$CO_2H, CH(OH)Ph,$$
$$C(OH)Me_2$$

34   **35**

[5]Metacyclophane readily undergoes Diels–Alder cycloadditions, the products with dienophiles being formed in quantitative yield.

Irradiation of 2-chloro[6](1,3)naphthalenophane (**36**) leads to transannular reactions:[16]

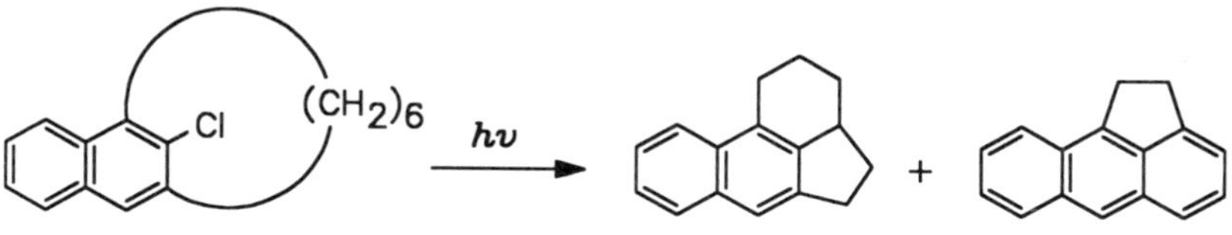

36   **37**   **38**

Similar ring closures have been observed in the photolysis of 2-bromo[6]- and -[7]metacyclophane.

According to Effenberger[17] the autoxidation of '[*n*](2,4)phloroglucino-phanes' **39** afford the hydroperoxides **40**, which are transformed to the more stable hydroxy molecules **41**. In alkaline solution the latter rearrange to the cyclopentenones **42**:

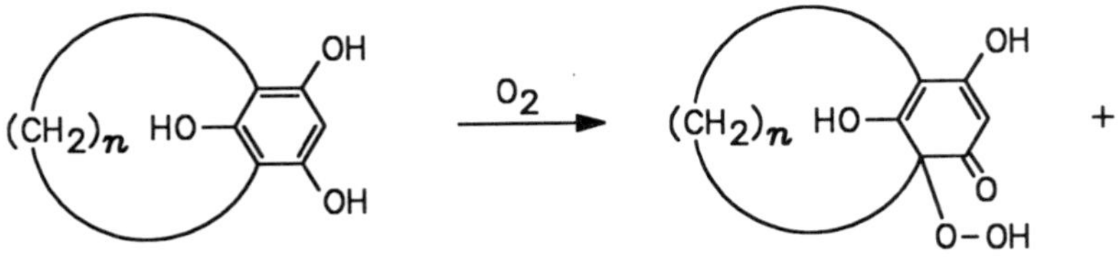

39 → **40** +

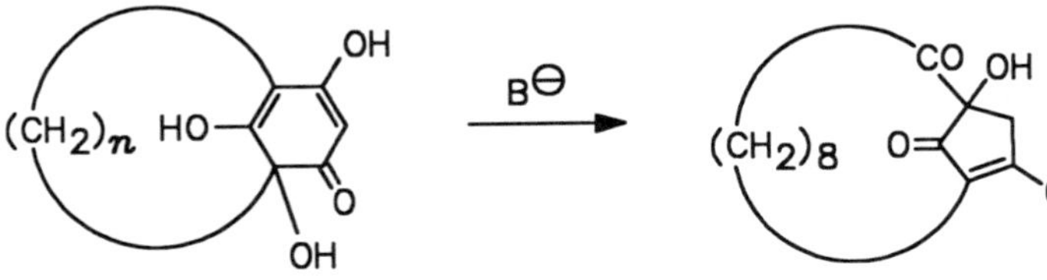

41 → **42**

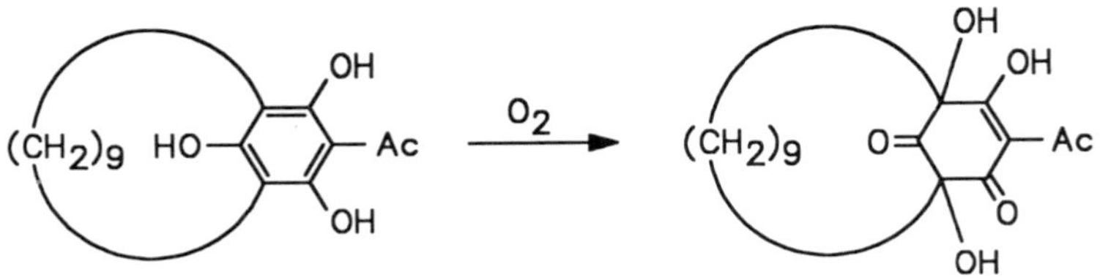

43 → **44**

By contrast the acyl derivative **43** reacts with oxygen to give the bis-hydroxylated compound **44**. 'Phloroglucinophanes' serve as models for the oxidation of deoxyhumulone to humulone and rearrangement of the latter to isohumulone, the bitter ingredient of hops.[17]

Crown ethers with the following general structures also belong to the family of [n]metacyclophanes:

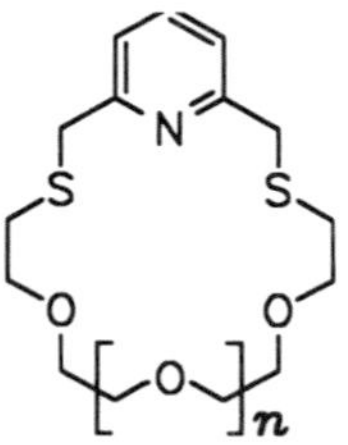

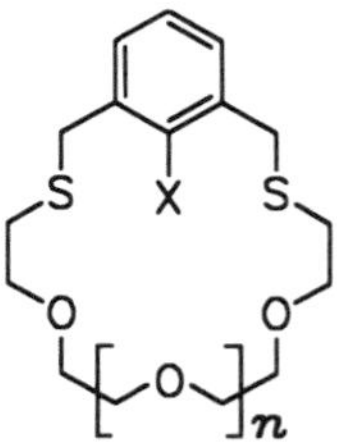

45   **46**

Such pyridine crowns and intra-annularly substituted benzene crowns[18] serve many uses, among them as neutral and acid chromoionophores[19] (chromoacerands[20]), which function effectively to distinguish between cations, amines, enantiomers, etc.

These crown ethers will not be discussed in any detail here, even though they constitute a major class of [n]metacyclophanes, the [2](2,6)- or -(2,5)heterophanes. Rather they will be included in the last chapter (Section 12.1), where the emphasis will centre on their role as crown ethers and host compounds. They are well covered in books on host–guest chemistry, crown ethers, and molecular recognition, etc. Further information can be found in the volume *Supramolecular Chemistry* (Wiley, Chichester, 1991).

References to Section 1.1

1. Review: S. M. Rosenfeld and K. A. Choe, in *Cyclophanes* (P. M. Keehn and S. M. Rosenfeld, Eds.), Vol. I, p. 311, Academic Press, New York, London, 1983.
2. K. Tamao, S. Kodama, T. Nakatsuka, Y. Kiso and M. Kumada, *J. Am. Chem. Soc.*, **97**, 4405 (1975).
3. L. A. M. Turkenburg, J. W. van Straten, W. H. de Wolf and F. Bickelhaupt, *J. Am. Chem. Soc.*, **102**, 3256 (1980); G. B. M. Kostermans, W. H. de Wolf and F. Bickelhaupt, in *Strain and its Implications in Organic Chemistry* (A. De Meijere and S. Blechert, Eds.), pp. 515, 517, Kluwer, Dordrecht, 1989; F. Bickelhaupt, *Pure Appl. Chem.*, **62**, 373 (1990); F. Bickelhaupt *et al.*, *J. Am. Chem. Soc.*, **112**, 6638 (1990).
4. K.-L. Noble, H. Hopf, M. Jones Jr and S. L. Kammula, *Angew. Chem.*, **90**, 629 (1978); *Angew. Chem. Int. Ed. Engl.*, **17**, 602 (1978).
5. G. Märkl and R. Fuchs, *Tetrahedron Lett.*, 4695 (1972).
6. S. Fujita and H. Nozaki, *Bull. Chem. Soc. Jpn.*, **44**, 2827 (1971).
7. F. Vögtle, *Chem. Ber.*, **102**, 1784 (1969); R. H. Mitchell and V. Boekelheide, *Tetrahedron Lett.*, 2013 (1969).
8. F. Vögtle and P. Neumann, *Tetrahedron Lett.*, 115 (1970).
9. J. M. Patterson, J. Brasch and P. Drenchko, *J. Org. Chem.*, **27**, 1652 (1962).
10. H. Nozaki, T. Koyama, T. Mori and R. Noyori, *Tetrahedron Lett.*, 2181 (1968).
11. H. Nozaki, T. Koyama and T. Mori, *Tetrahedron*, **25**, 5357 (1969).
12. H. Förster and F. Vögtle, *Angew. Chem.*, **89**, 443 (1977); *Angew. Chem. Int. Ed. Engl.*, **16**, 429 (1977); For pyrrolophanes from aziridinophanes, see F. Müller and J. Mattay, *Angew. Chem.*, **104**, 207 (1992); *Angew. Chem. Int. Ed. Engl.*, **31**, 209 (1992); Y. H. Lai, *Heterocycles*, **16**, 1739 (1981); R. H. Mitchell, in *Cyclophanes* (P. M. Keehn and S. M. Rosenfeld, Eds.), Vol. I, p. 239, Academic Press, New York, 1983; For [n]pyranophanes see F. W. Steuber *et al.*, *Chem. Ber.*, **123**, 1587 (1990).
13. W. E. Parham, D. R. Johnson, C. T. Hughes, M. K. Meilahn and J. K. Rinehart, *J. Org. Chem.*, **35**, 1048 (1970).
14. W. E. Parham, R. W. Davenport and J. K. Rinehart, *J. Org. Chem.*, **35**, 2662 (1970).
15. S. Hirano, H. Hara, T. Hiyama, S. Fujita and H. Nozaki, *Tetrahedron*, **31**, 2219 (1975).
16. D. J. Cram, C. S. Montgomery and G. R. Knox, *J. Am. Chem. Soc.*, **88**, 515 (1966).
17. F. Effenberger, B. Spachmann and K.-H. Schönwälder, *Chem. Ber.*, **122**, 1947 (1989).
18. F. Vögtle and E. Weber, *Angew. Chem.*, **86**, 126 (1974); *Angew. Chem. Int. Ed. Engl.*, **13**, 149 (1974).

19. H.-G. Löhr and F. Vögtle, *Acc. Chem. Res.*, **18**, 65 (1985); F. Vögtle and P. Knops, *Angew. Chem.*, **103**, 972 (1991); *Angew Chem. Int. Ed. Engl.*, **30**, 958 (1991).
20. T. Kaneda, S. Misuma *et al.*, *J. Am. Chem. Soc.*, **111**, 742, 1881 (1988).
21. M. T. Reetz, C. M. Niemeyer and K. Harms, *Angew. Chem.*, **103**, 1515, 1517 (1991); *Angew. Chem. Int. Ed. Engl.*, **30**, 1472, 1474 (1991).

1.2 [*n*]PARACYCLOPHANES (AND HETERO-/ HETERA-ANALOGUES)

1.2.1 HISTORICAL BACKGROUND

The first heterocyclic [*n*]paracyclophane {1,12-dioxa[12]paracyclophane, **1**} was obtained by Lüttringhaus in 1937.[1] Resolution of the racemate **2** was achieved in 1942 in the same research group.[2a]

1   **2** : R = CO_2H

In 1954 Huisgen and Cram independently synthesized the [9]- and [10]paracyclophanes, **3** and **4**, respectively.[3,4]

3   **4**

Cram and Allinger,[5] as well as Blomquist,[6] separated the enantiomers of **5**, while Huber resolved [9](2,5)pyridinophane (**6**).[7]

5   **6**

1.2.2 SYNTHESIS

The efficiency of practically every cyclization method has been tested in its application toward the synthesis of [n]paracyclophanes.[8] It is of historical interest that, in the book *Bridged Aromatic Compounds* by B. H. Smith published in 1964, the [8]paracyclophane was described as the smallest example. Yet Allinger had already speculated in 1963 that [7]paracyclophane, being about as strained as cyclopropane, should lend itself to synthesis. It was later demonstrated that this could be accomplished with the aid of new synthetic methods. In the following presentation only the most widely used methods are outlined. Further details are available in the book by B. H. Smith, mentioned above, and especially in the cited literature.[8]

Most syntheses of [n]paracyclophanes entail intramolecular ring-closure reactions; no specific problems are encountered in cases where n is relatively large. The usual methods fail, however, for [n]paracyclophanes with fewer than nine bridging atoms ($n<9$).

At the outset the acyloin condensation was applied with great success by Cram for the preparation of [9]-, [10]-, [12]-, and [14]paracyclophanes:

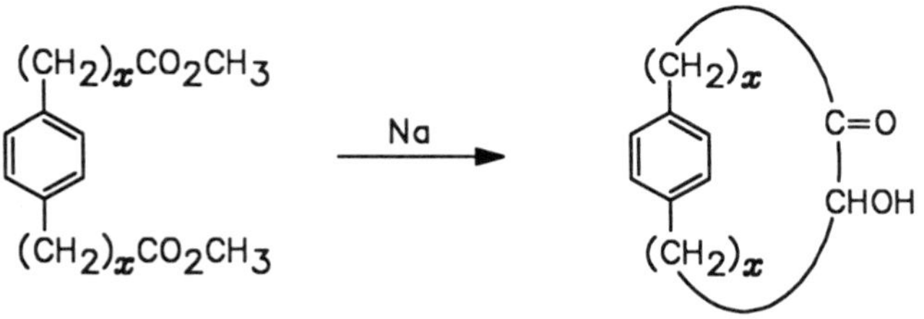

Although [n](1,4)naphthalenophanes, with $n=10, 14$, were accessible by way of the acyloin condensation, the method failed for [9](1,4)naphthalenophane.

The Friedel–Crafts acylation (under high dilution), introduced in 1954 by Huisgen and Schubert, afforded [n]paracyclophane compounds where $n=9$–14, 16:[3,8,9]

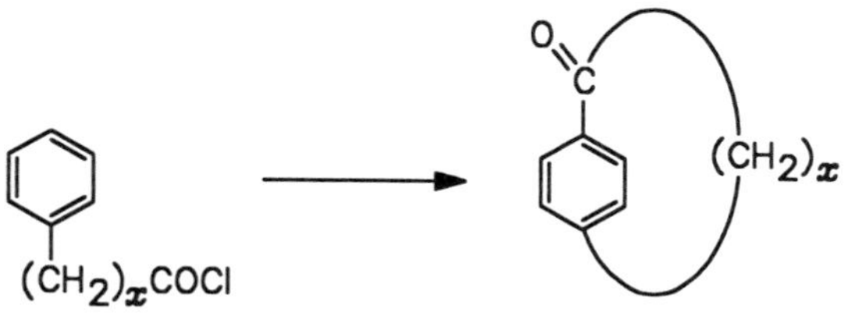

The intramolecular Eglinton coupling of bis-ethynyl compounds leads to [n]paracyclophandiynes:

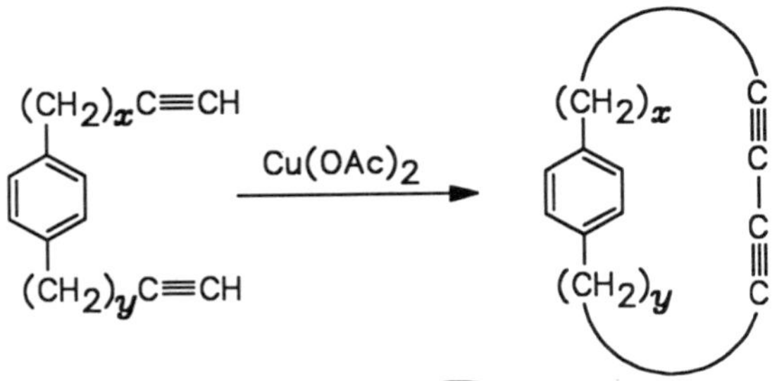

$$x = y = 3$$
$$x = 2; \; y = 4$$
$$x = 3; \; y = 4$$
$$x = y = 4$$
$$x = y = 5$$

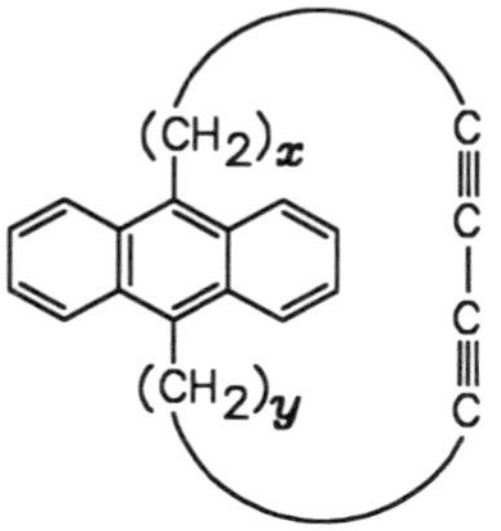

$$x = y = 3, 4$$

By this route Misumi succeeded in preparing [10]-, [11]-, [12]-, and [14]paracyclophandiynes, as well as [10]- and [12](9,10)anthracenophan-4,6-diynes.[10] By hydrogenation he also obtained [10](9,10)anthracenophane.

The intermolecular formation of dithiaparacyclophanes of type 7 proceeds easily and economically through the reaction of dithiols with dihalogen compounds, with yields of purified products ranging from 50 to 88% {dithia[8]- to [14]paracyclophanes}.[11,12] By subsequent pyrolysis of the corresponding sulphones Misumi[13] prepared [8]- to [12]paracyclophanes and [14]para-cyclophane (see 8) in 30–46% overall yields. Obtained in an analogous way were [10]- and [14]anthracenophanes[13,14] and [8](3,7)tropolonophane, as well as several dithia[n](2,5)pyridinophanes, where $n = 8$, 10, and 12.[8]

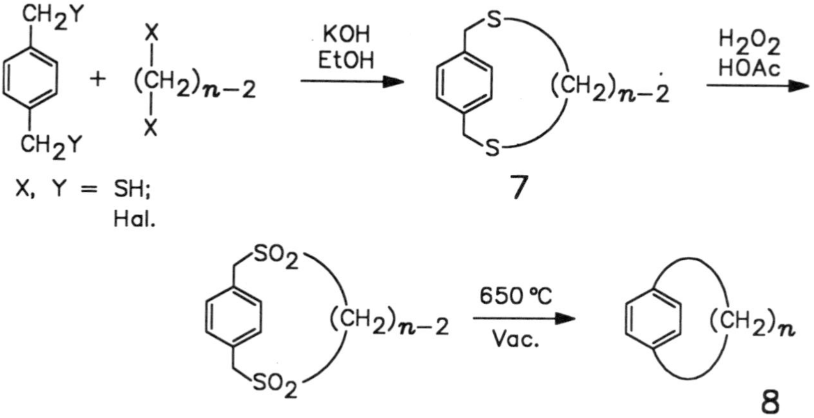

Cram was able to prepare [8]paracyclophane (**8d**) for the first time by opening of the furan ring in furanobenzenophanes, which had been synthesized through a crossed Hofmann elimination (see Chapter 6):[15,16]

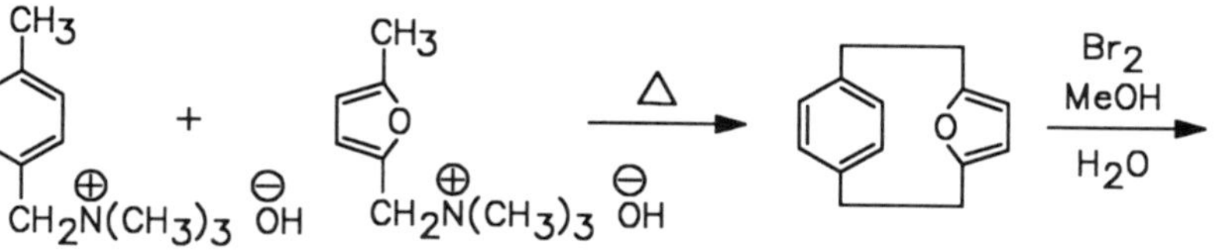

9

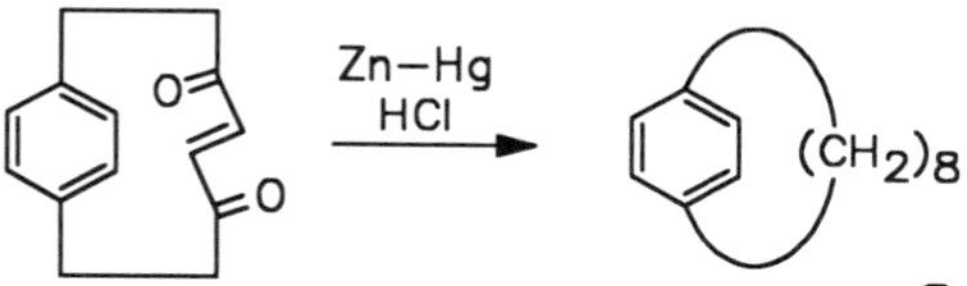

8d

In a similar way the syntheses of [8](1,4)naphthalenophane and the chiral [8][8]- and [8][10]paracyclophanes **10** were achieved:[18,19]

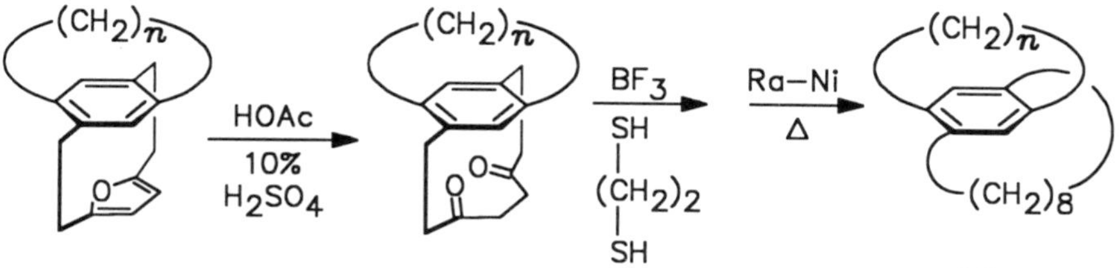

10

An attractive route, especially to obtain small paracyclophanes, is ring contraction of higher homologues. Allinger applied the Wolff rearrangement with success to the synthesis of [7]- and [8]paracyclophane derivatives:[20]

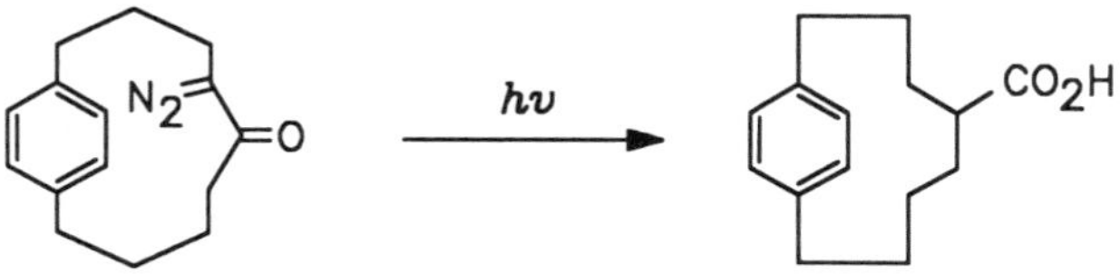

The [6]paracyclophane could not be obtained by this method because the required diazo ketone was not isolable.

Jones succeeded in preparing the smallest [n]paracyclophane which has been isolated so far, namely [6]paracyclophane,[21] as well as [7]paracyclophane[22] by way of the spiro dienone **11**:

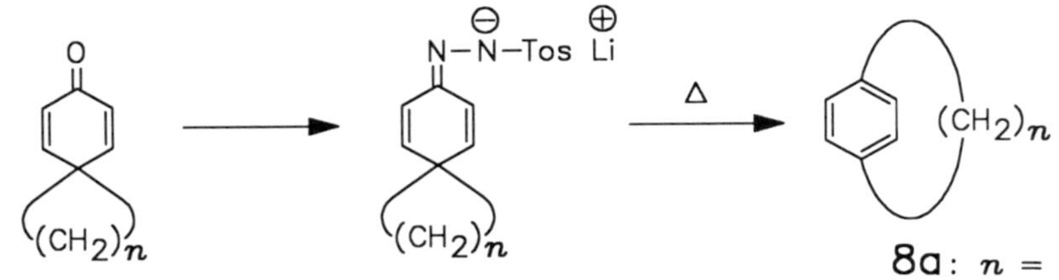

8a: $n = 5$
8b: $n = 6$
8c: $n = 7$
8d: $n = 8$

[6] Paracyclophane (**8b**) is formed in good yield by thermolysis of the corresponding Dewar isomer **13**, and the latter is likewise the sole product from photolysis of [6]paracyclophane.[23]

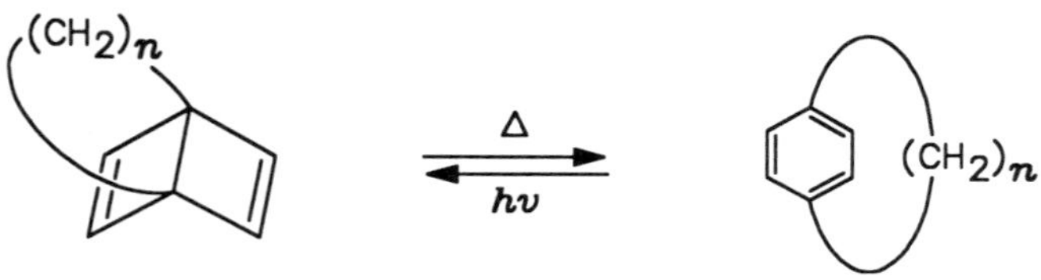

13

8a: $n = 5$
8b: $n = 6$

The smallest member of the [n]paracyclophanes, namely [5]paracyclophane, although unstable at room temperature, was postulated by Bickelhaupt[24] as an intermediate in the analogous thermal rearrangement of the corresponding Dewar isomer to a series of products. In the photostationary equilibrium **13** ⇌ **8** only about 6–7% of **8a** is present. Consideration of the mechanism of such a reaction led to a new synthetic route to [7]- and [8]paracyclophanes, albeit in low yields.

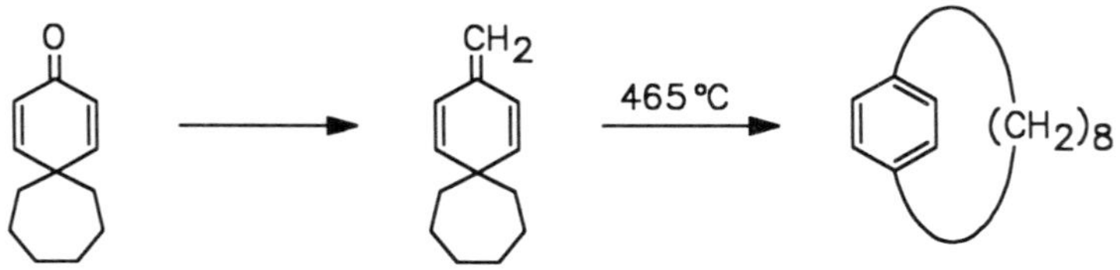

8d

The highly strained paracyclophane **8e** is even more elusive. More stringent conditions (300°C) are necessary for its production and it breaks down directly into ethylene and *p*-xylylene (**15**).[24a] When **8c** is produced from **13c** by irradiation in solution, it polymerizes at −60°C; however, it can be detected through a Matrix IV spectroscopic process. Moreover, it was demonstrated that it readily accepts protons to ease the ring strain. Thus it is unusually strongly basic.

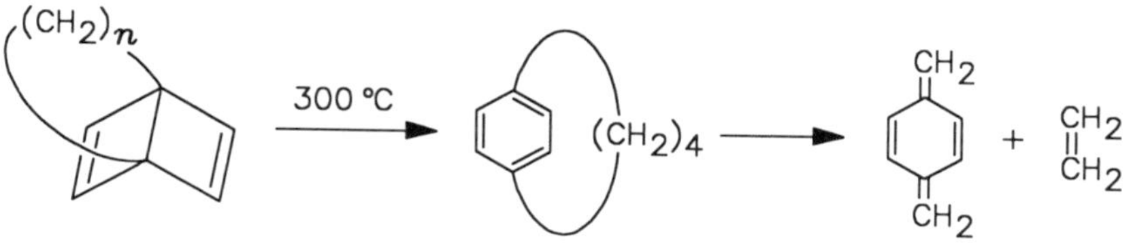

13

(n = 4)

8e

From the foregoing discussion it is clear that all members of the [n]paracyclophane series from n = 4–16 are known, with the exception of n = 15. Inasmuch as the paracyclophane **8b** (n = 6) is in all likelihood the smallest *isolable* member of the series, the full family of compounds is essentially complete. By contrast, only those in the [n](1,4)naphthalenophanes with n = 8, 10, 14 and those in the [n](9,10)anthracenophanes with n = 10, 14 are known.

Tochterman et al.[25] obtained crystalline [6]paracyclophanes **4** in good yield through the McMurry reaction with the epoxy dibromide **3**. The (+)- and (−)-enantiomers of [6]paracyclophane-8-carboxylic acid (**4**, R = CO$_2$H) were isolated by means of a stereospecific synthesis:[25b]

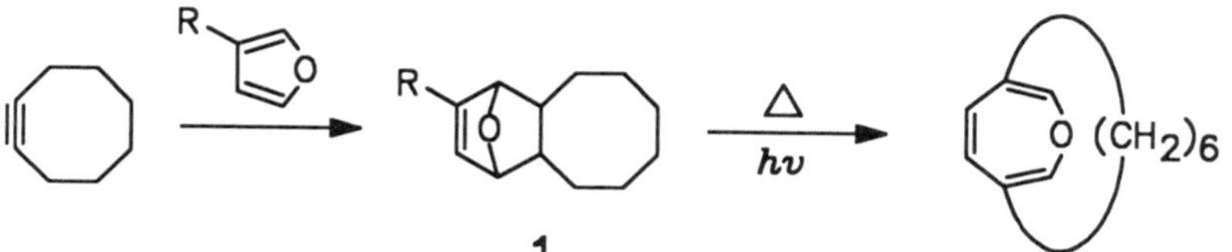

1

R = CH$_2$OAc

2

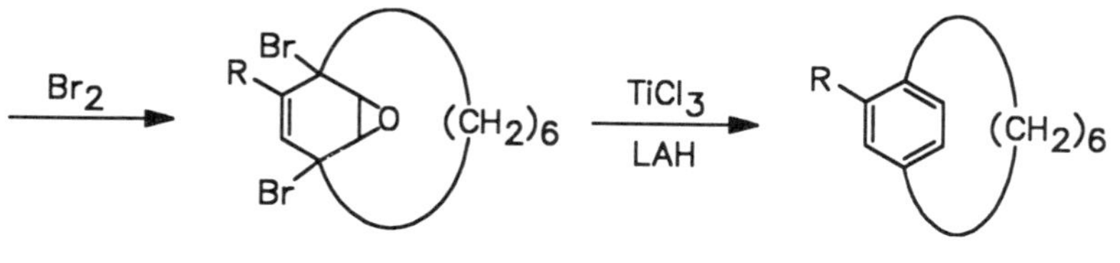

3

4

They possess a boat-like, distorted benzene ring: the distortion angles in the *ortho*-diester **5** were determined by X-ray crystal analysis to be 19.4° and 19.5°; the benzylic methylene groups are bent an additional 18.6° and 21.2° (Figure 1).[25b]

H_3CO_2C
H_3CO_2C $(CH_2)_8$

5

Through intermolecular ester or amide formation under high dilution, heterocyclic lactones and lactams can be constructed.[26,27]

CH$_2$COCl

R

R

CH$_2$COCl

$+$ $H_2N-(CH_2)_n-NH_2$ $\longrightarrow$

$$O=C$$
CH$_2$ NH
R
$(CH_2)_n$
CH$_2$ NH
C
O

R = H; n = 6 − 10
R = Me; n = 5 − 10
R = OMe; n = 5 − 10, 12
R = OEt; n = 12

1.2.3 SPECTROSCOPY

As mentioned above for [n]metacyclophanes, spectroscopic data indicate non-planarity of the aromatic ring as the value of n becomes small. The expectation of a drastic disappearance of aromaticity, as manifested in spectroscopic properties and chemical reactions, has, however, not been confirmed. MNDO calculations for the photoelectron spectra of [6]- and [7]paracyclophanes led to the conclusion that the observed splittings were mainly due to the electron-feeding effect of the bridging atoms rather than to a loss of aromatic 'character'.

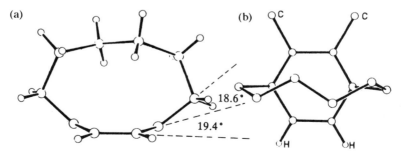

Figure 1. X-ray crystal structure of [6]paracyclophane-8,9-dicarboxylic acid dimethyl ester: (a) side view; (b) top view

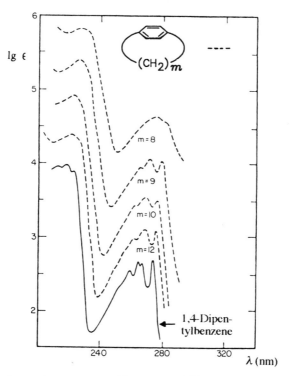

Figure 2. UV absorption spectra (Cram *et al.*[16]) of [8]-, [9]-, [10]- and [12]paracyclophane (in absolute ethanol). For comparison the model compound 1,4-dipentylbenzene (lowest, solid line; each curve is displaced 0.5 logarithmic unit along the ordinate from the one below it)

The UV spectra of the paracyclophanes, too,[8] are characterized by a bathochromic shift and loss of fine structure, which is attributed to deformation of the aromatic ring.

The EPR spectrum of the metastable triplet of [7]paracyclophane has been interpreted in terms of an altered π-molecular orbital, which is ascribed to a folding of the bridging carbon atoms out of the plane of the aromatic ring.

Figure 2 shows Cram's well-known UV spectra of some [n]paracyclophanes ($n = 8$–10, 12). The UV absorptions of [7]- and [6]paracyclophane were also recorded later ($\lambda_{max} = 216$, 245, 283; and 212, 253, 290 nm, respectively).

The UV spectra are used to deduce the extent of deformation of the benzene ring in the [n]paracyclophanes.

X-ray crystal structures for the carboxyl derivatives of [7]- and [8]paracyclophane gave values of $\Phi = 17°$ and 9.1°, respectively, for the deviation from planarity. These angles have been confirmed by molecular mechanics and force field calculations. On the basis of these values for Φ, the calculated electronic spectra are in good agreement with the experimental data.

Table 1. Calculated values for the distortion from planarity of the benzene ring (Φ) and strain energies in [n]paracyclophanes[8,29-31]

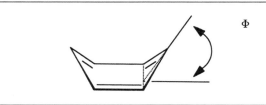

No.	n	Φ (°)	Strain energy (kJ/mol)		
			Ref. 8	Ref. 30	Ref. 31
8f	10	8.4	64.8		
8g	9	8.5	53.2		
8d	8	12.5	70.5	70.3	
8c	7	18.2	87.5	87.5	133.5
8b	6	22.4	120.3		
8a	5	26.5	163.5		

Analysis of calculated strain energies indicates that, for smaller values of n, the bending of the aromatic ring accounts for the major component of the strain energy. In contrast, the strain energy in [n]paracyclophanes, where $n = 9$, 10, is attributed mainly to torsional strain, which results from fewer stable conformations in the bridge (Table 1). In any case the strain energy is not attributed to a stretching of the C·C bridge bonds, as is verified by the X-ray structural analysis. There is evidence from the latter, however, for bond angle compression of the benzylic carbon atoms and C·C bond angle stretching of some methylene groups (up to a maximum of 117° in 3-carboxy[7]paracyclophane). In the most highly strained, isolable [n]paracyclophane, namely a ring substituted [6]paracyclophane, it was found by X-ray crystal structure analysis that the bond angle in the aliphatic bridge was expanded to 126.5° (ideal value in an unstrained molecule: 109.5°).[25]

1.2.3.1 Proton NMR spectra

As can be shown with molecular models, some of the bridging protons in [n]paracyclophanes with short bridges are located over the plane of the aromatic ring and are therefore shifted strongly upfield. Instances of the highest upfield shifts are listed in Table 2.

Peaks at even higher fields are observed in [10](1,4)naphthalenophane and [10](9,10)anthracenophane ($\delta = -0.40$ and 0.00) than those for [10]paracyclophane ($\delta = 0.48$). These resonances have been used as evidence for ring current theories, for example the 'free-electron model' and the Johnson–Bovey and Haigh–Mallion theories.

The ^{13}C-NMR spectra of [n]paracyclophanes have also been considered as

Table 2. Upfield shifts in the ^{1}H-NMR spectra of
[n]paracyclophanes[8]

No.	n	δ (p.p.m.)
8i	12	0.78[a]
8h	11	0.68[a]
8g	10	0.48[a]
8f	9	0.33[a]
8d	8	0.19[a]
8c	7	−0.3 to −0.9 (m, 2H)[b]
8b	6	−0.6 (m, 2H)[b]

[a]Decoupled; [b]centre of multiplet.

a measure of the influence of the aromatic ring current on the ^{13}C shifts in the aliphatic bridge.[32] Such a ring current seems to be in effect in the ^{13}C-NMR spectra of [8]- to [10]paracyclophanes.

The dynamic NMR spectra of [n]paracyclophanes have also been investigated.[33] In the [8]- to [10](1,4)naphthalenophanes the methylene bridges (at room temperature) are fixed at a position on one side or the other of the aromatic ring, whereas [14](1,4)naphthalenophane is conformationally flexible on the NMR time scale. Pseudorotation of the aliphatic bridge can be observed in the NMR spectrum of a ring-substituted [6]paracyclophane.[34]

The conformational behaviour of various hetera-substituted paracyclophanes and dioxa[n]paracyclophanes has been described as well.[35]

1.2.4 CHEMICAL REACTIONS OF [n]PARACYCLOPHANES

1.2.4.1 Transannular reactions

Acetolysis of optically active [9]paracyclophane-4-tosylate affords the corresponding acetate with 98 ± 2% retention of configuration. This suggests intermediate formation of the bridged ion **14**:[8]

14

The relative rates of acetolysis support the idea of phenyl ring participation, because placement of the tosylate group in the 3-, 4-, and 5-position causes an enhancement of the reaction rate. The drastic increase in the reactivity of [9]paracyclophane-4-tosylate is ascribed to a release of ring strain during ionization.

Acetolysis of [8]paracyclophane-3-tosylate, with formation of hydrocarbon products, also reflects the importance of neighbouring group participation of the benzene ring during ionization.[8,16]

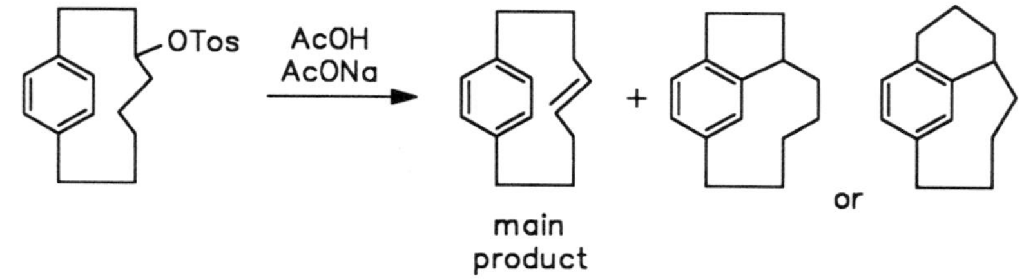

main
product

or

During the acetolysis of the ditosylate **15** one observes a twofold transannular substitution in the course of the thermally induced formation of **16**.[8]

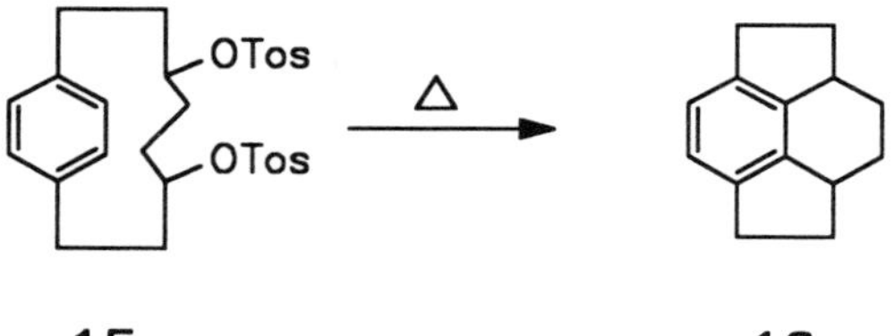

15   **16**

Bromination of [8]paracyclophane (**8d**) affords only isolable products resulting from rearrangement and cleavage of the bridge.[8]

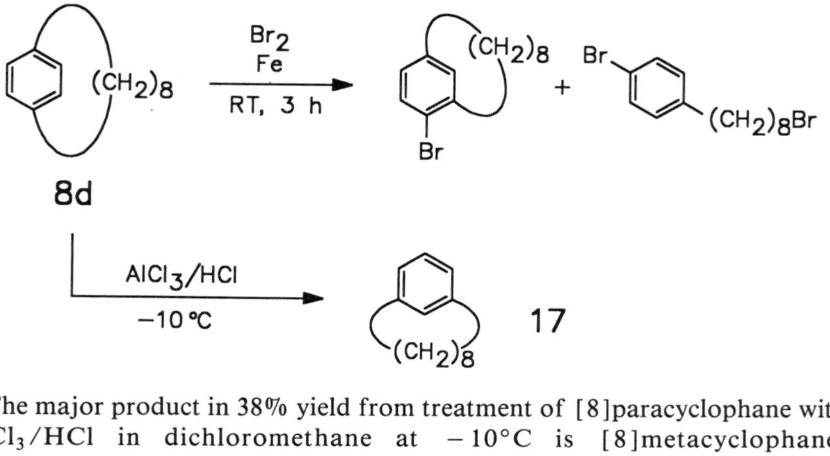

8d

17

The major product in 38% yield from treatment of [8]paracyclophane with $AlCl_3$/HCl in dichloromethane at $-10°C$ is [8]metacyclophane. [7]Paracyclophane (**8c**) can be converted to [7]metacyclophane under the influence of acid.

1.2.4.2 Diels–Alder reactions

The benzene ring is normally sluggish in Diels–Alder reactions, and this lack of reactivity is usually overcome only by drastic methods such as Lewis-acid catalysis or the choice of strong dienophiles. An alternative possibility for enhancing the reactivity is to incorporate a strongly sterically strained benzene or arene. [2.2]paracyclophane (strain energy approx. 130 kJ/mol; see Section 2.3), for example, reacts with dicyanoacetylene, as well as with tetrafluorobenzyne to give 1:1 and 1:2 adducts. Although the less reactive acetylene dicarboxylic acid dimethyl ester affords no cycloaddition, this dienophile does combine with the more highly strained [2.2.2](1,2,4)cyclophane (Section 5.1)[170°C, 1 h, 61% yield).

Because of their ready availability, [n]cyclophanes with smaller bridging atoms are convenient candidates for the study of [4 + 2]cycloadditions. Thus, for example, the Diels–Alder reactions of [7]- and [8]paracyclophane (8c,d), as well as of [7]metacyclophane, have been described.[30]

8c: $n = 7$
8d: $n = 8$

19a: R = CF_3, $n = 7$
19b: R = CF_3, $n = 8$
19c: R = CN, $n = 8$

With an excess of perfluorobutyne (18, R = CF_3) at 160°C in 3 h, 8c,d afford 1:1 adducts (bridged barrelenes) of the type 19 in 52 and 12.5% yields, respectively.

If the reaction time with 8d is increased to 24 h, the yield of 19b rises to 76%. This tightly bridged [n]paracyclophane, in turn, is even more reactive toward cycloaddition, a fact attributed to its higher strain energy as compared to that of 8d.

Dicyanoacetylene (18, R = CN) is unreactive with 8d under similar conditions; only by addition of $AlCl_3$ in o-dichlorobenzene at 120°C (24 h) is the solid 19c formed in 25% yield. Even with this activation method acetylene dicarboxylic acid dimethyl ester (18, R = CO_2CH_3) is unreactive toward 8d, its dienophilic character being weaker than that of either of the aforementioned, triply bonded dienophiles 18 (R = CF_3; R = CN).

When 8c is treated with a mixture of fluorosulphonic acid and p-toluenesulphonic acid in benzene, it isomerizes to [7]metacyclophane. The latter hydrocarbon affords the analogous 1:1 adduct in 50% yield in the presence of perfluorobutyne at 150°C.

Through the addition of dienophiles (such as benzyne) to anthracenophanes **(20)** one can prepare '*paddlanes*' (sulphones **21** and hydrocarbons **22**) with the triptycene structure:[14]

20 X = S, SO$_2$ **21** n = 8, 12   **22**

1.2.4.3 Donor–acceptor complexes

The π-base strength of arene rings in [n]paracyclophanes has elicited continued interest for several years. The [9]- and [10]-, and [12]paracyclophanes form 1:1 complexes with tetracyanoethylene (TCNE).

The position of the highest wavelength UV absorption band follows the order $n = 9 > 12 > 10$. This has been interpreted as a combination of back donation through transannular delocalization (hyperconjugation), which is stabilized by the ring, and a diminution of transannular delocalization as a consequence of the non-planarity of the benzene rings.[36]

1.2.4.4 Photochemistry

Photolysis of [6]paracyclophane in perdeuteriocyclohexane leads to a quantitative rearrangement to the corresponding bridged *Dewar benzene* (see above).[23] By contrast, [7]- and [8]paracyclophane[8,23] slowly polymerize under the same conditions.

1.2.4.5 Reactions within the bridge of paracyclophanes

By treatment of **24** with methyllithium one can isolate, in a remarkable 65% yield, the unusual allene **25**.[8]

23   **24** : X = Cl, Br   **25**

Although the diacetylene **26** is stable, it nevertheless reacts with TCNE at room temperature to give the 1:1 TCNE adduct **27**. The proposed nature of the combination of the reaction partners is presented in **27a**.[37]

[10](9,10)Anthracenophan-4,6-diyne (**28**) is transformed quantitatively after 10 minutes in sunlight to the photo dimer **30**:[8,10]

In this instance the intermediate **29** represented in parentheses is proposed.

The general conclusion from synthesis and chemistry of the lower [n]paracyclophanes is that by these means the possibility exists to deform benzene rings.

Furthermore, 'small' cyclophanes undergo unexpected reactions and in particular exhibit a reactivity reminiscent of the behaviour of oligo alkenes (olefins); that is, they have a tendency to behave as cyclohexatrienes. In sharp contrast, however, as is amply demonstrated from earlier spectroscopic results, in particular NMR and X-ray structures, delocalization and ring current are scarcely affected. It must be concluded that strongly deformed cyclophanes are to be considered as molecules with aromatic residues, and that their unusual properties are attributed to manifold consequences of their strain (and not a loss of 'aromaticity').[37a]

The [n]heterophane *[8](3,6)pyridazinophane* (**32b**) was prepared in 1974 by Nozaki *et al.* by reaction of cyclododecan-1,4-dione (**31**) with hydrazine and subsequent air oxidation.[38]

$$N_2H_4 \cdot H_2O$$
$$air-O_2$$

31 **32b**

For the synthesis by Gassman and Boardman[39] of *[7](3,6)pyridazinophane* (**32a**) it was necessary to devise a roundabout route. They started with (*Z,E*)-1,3-cycloundecadiene (**33**) and 4-phenyl-1,2,4-triazolin-3,5-dione (**34**). The product **32a** was obtained in the form of colourless crystals with m.p. 79–80°C.

C₆H₅

34

33

MCPBA
CH₂Cl₂

KO*t*Bu
DMSO

nickel–peroxide
(C₂H₅)₃N
(C₂H₅)₂O

LDA
THF

32a

The [7]phane **32a** was confirmed by X-ray crystal structure analysis, which showed that the pyridazine ring exists in a boat form with an average deformation angle of 17.45°. Two resonance signals (one proton each) at $\delta = 0.41$ and -1.93 indicated that the central carbon atom of the seven-membered bridge lies over the centre of the aromatic ring, such that the two geminal hydrogens experience the anisotropic effect of the pyridazine ring. The X-ray crystal structure analysis confirms this fact (Figure 3).

(a) (b)

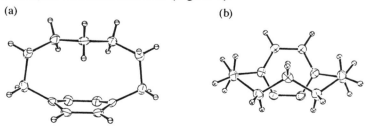

Figure 3. X-ray crystal structure of **32a**: (a) side view; (b) top view

When the NMR sample was cooled to $-80°C$, the resonance at $\delta = -1.93$ shifted to -2.33, while the same absorption changed to $\delta = -1.75$ by heating the sample to $100°C$. This relatively small change in the position of the proton with a temperature change of $180°C$ demonstrates that temperature exerts a relatively small effect on the equilibrium position of the C-10 hydrogen atoms.

References to Section 1.2

1. A. Lüttringhaus, *Liebigs Ann. Chem.*, **528**, 181 (1937).
2. (a) A. Lüttringhaus and H. Gralheer, *Liebigs Ann. Chem.*, **550**, 67 (1942).
   (b) A. Lüttringhaus and G. Eyring, *Angew. Chem.*, **69**, 137 (1957).
3. R. Huisgen, *Angew. Chem.*, **69**, 341 (1957).
4. D. J. Cram and H. U. Daeniker, *J. Am. Chem. Soc.*, **76**, 2743 (1954) cf. K. Wiesner, D. M. MacDonald, R. B. Ingraham and R. B. Kelly, *Can. J. Res. Sect. B*, **28**, 561 (1950).
5. D. J. Cram and N. L. Allinger, *J. Am. Chem. Soc.*, **77**, 6289 (1955).
6. A. T. Blomquist and B. H. Smith, *J. Am. Chem. Soc.*, **82**, 2073 (1960).
7. H. Gerlach and E. Huber, *Helv. Chim. Acta*, **51**, 2027 (1968).
8. Review: S. M. Rosenfeld and K. A. Choe, in *Cyclophanes* P. M. Keehn and S. M. Rosenfeld (Eds.), Vol. 1, pp. 311 ff, Academic Press, New York, 1983.
9. R. Huisgen, W. Rapp, I. Ugi, H. Walz and I. Glogger, *Liebigs Ann. Chem.*, **586**, 52 (1954).
10. T. Inoue, T. Kaneda and S. Misumi, *Tetrahedron Lett.*, 2969 (1974).
11. F. Vögtle, *Chemiker-Ztg.*, **94**, 313 (1970).
12. T. Otsubo and S. Misumi, *Synth. Commun.*, **8**, 285 (1978).
13. Review: F. Vögtle and L. Rossa, *Angew. Chem.*, **91**, 534 (1979); *Angew. Chem. Int. Ed. Engl.*, **18**, 514 (1979).
14. F. Vögtle and P. Koo Tze Mew, *Angew, Chem.*, **90**, 58 (1978); *Angew. Chem. Int. Ed. Engl.*, **17**, 60 (1978).
15. D. J. Cram and G. R. Knox, *J. Am. Chem. Soc.*, **83**, 2204 (1961).
16. D. J. Cram, C. S. Montgomery and G. R. Knox, *J. Am. Chem. Soc.*, **88**, 515 (1966).
17. K. B. Wiberg and M. J. O'Donnell, *J. Am. Chem. Soc.*, **101**, 6660 (1979).
18. M. Nakazaki, K. Yamamoto and S. Tanaka, *J. Org. Chem.*, **41**, 4081 (1976).
19. M. Nakazaki, K. Yamamoto, M. Ito and S. Tanaka, *J. Org. Chem.*, **42**, 3468 (1977).
20. N. L. Allinger, Th. J. Walter and M. G. Newton, *J. Am. Chem. Soc.*, **96**, 4588 (1974).
21. V. V. Kane, A. D. Wolf and M. Jones, Jr, *J. Am. Chem. Soc.*, **96**, 2643 (1974).
22. A. D. Wolf, V. V. Kane, R. H. Levin and M. Jones, Jr, *J. Am. Chem. Soc.*, **95**, 1680 (1973); L. W. Jenneskens, W. H. De Wolf and F. Bickelhaupt, *Tetrahedron*, **42**, 1571 (1986).
23. (a) S. L. Kammula, L. D. Iroff, M. Jones, Jr, J. W. van Straten, W. H. de Wolf and F. Bickelhaupt, *J. Am. Chem. Soc.*, **99**, 5815 (1977).
    (b) Y. Tobe, K. Ueda, K. Kakiuchi and Y. Odaira, *Tetrahedron*, **42**, 1851 (1986).
    (c) Cf. G. B. M. Kostermans, W. H. de Wolf and F. Bickelhaupt, in *Strain and its Implications in Organic Chemistry* A. de Meijere and S. Blechert (Eds.), p. 517, Kluwer, Dordrecht, 1989; F. Bickelhaupt, *Pure Appl. Chem.*, **62**, 373 (1990).
24. J. W. van Straten, W. H. de Wolf and F. Bickelhaupt, *Rec. Trav. Chim. Pays-Bas*, **96**, 88 (1977); see also F. Bickelhaupt *et al.*, *J. Chem. Soc.*, *Perkin Trans. 1*, 2119 (1985); J. E. Rice, T. J. Lee, R. B. Remington, W. D. Allen, D. A. Clabo, Jr and H. F. Schaefer III, *J. Am. Chem. Soc.*, **109**, 2902 (1987).
24a. F. Bickelhaupt, *Pure Appl. Chem.*, **62**, 373 (1990).
25. (a) J. Liebe, Ch. Wolff, C. Krieger, J. Weiss and W. Tochtermann, *Chem. Ber.*,

118, 4144 (1985); J. L. Jessen, G. Schröder and W. Tochtermann, *Chem. Ber.*, **118**, 3287 (1985); see also W. Tochtermann *et al.*, *Chem. Ber.*, **122**, 1653 (1989).
(b) W. Tochtermann, U. Vagt and G. Snatzke, *Chem. Ber.*, **118**, 1996 (1985); cf. W. Tochtermann, G. Olsson, A. Mannschreck, G. Stühler and G. Snatzke, *Chem. Ber.*, **123**, 1437 (1990).
26. K. Sakamoto and M. Oki, *Bull. Chem. Soc. Jpn.*, **46**, 270 (1973).
27. K. Sakamoto and M. Oki, *Bull. Chem. Soc. Jpn.*, **49**, 3159 (1976).
28. M. G. Newton, T. J. Walter and N. L. Allinger, *J. Am. Chem. Soc.*, **95**, 5652 (1973).
29. H. Hopf, *Chemie in uns. Zeit*, **10**, 114 (1976).
30. K.-L. Noble, H. Hopf, M. Jones, Jr and S. L. Kammula, *Angew. Chem.*, **90**, 629 (1978); *Angew. Chem. Int. Ed. Engl.*, **17**, 602 (1978).
31. H. Schmidt, A. Schweig, W. Thiel and M. Jones, Jr, *Chem. Ber.*, **111**, 1958 (1978).
32. T. Kaneda, T. Otsubo, H. Horita and S. Misumi, *Bull. Chem. Soc. Jpn.*, **53**, 1015 (1980).
33. M. Nakazaki, K. Yamamoto and S. Okamoto, *Tetrahedron Lett.*, **1969**, 4597.
34. J. Liebe, C. Wolff and W. Tochtermann, *Tetrahedron Lett.*, **23**, 2439 (1982).
35. K. Sakamoto and M. Oki, *Tetrahedron Lett.*, **1973**, 3989.
36. (a) R. E. Merrifield and W. D. Phillips, *J. Am. Chem. Soc.*, **80**, 2778 (1958).
(b) D. J. Cram and R. H. Bauer, *J. Am. Chem. Soc.*, **81**, 5971 (1959).
37. T. Kaneda, T. Ogawa and S. Misumi, *Tetrahedron Lett.*, **1973**, 3373.
(a) cf. C. W. Bird, *Tetrahedron*, **48**, 1675 (1992).
38. T. Hiyama, S. Hiramo and H. Nozaki, *J. Am. Chem. Soc.*, **96**, 5287 (1974).
39. P. G. Gassman and G. S. Boardman, *Tetrahedron Lett.*, **30**, 4649 (1989).

1.3 [*n*]NAPHTHALENOPHANES AND [*n*]QUINOLINOPHANES

1.3.1 [*n*](1,3)NAPHTHALENOPHANES

1.3.1.1 Synthesis

An interesting synthetic strategy was developed by Parham *et al.*[1] for *[n](1,3)naphthalenophanes* (**3**): bridged indenes (**1**) were converted into the corresponding dihalocyclopropanes (**2**), which with base afforded the desired phanes **3**:

The advantage of this methodology is that one has available intra-annular substituents, which can either be dehalogenated to give the phane hydrocarbon or can be derivatized.[1,2] The yields ranging between 73 and 89% in the last step are attributed to the relief of strain in opening of the cyclopropane ring and the accompanying aromatization of the naphthalene ring.

The synthetic method is broadly useful and was successful down to $n = 6$. The postulated next lower homologue 3 ($n = 5$) was later synthesized by Grice and Reese.[2] In the case of $n = 4$, however, H—X elimination could not be achieved.

The halogen-substituted naphthalenophanes 3 are reduced by way of the corresponding Grignard reagent or organolithium compound to the hydrocarbon. Analogous quinolinophanes such as 5 are available by this route as well.[3,4]

4 : $n = 6, 8, 10$ 5 : $n = 8, 10$

1.3.1.2 Chemical reactions

Such naphthalenophanes, in which the entire naphthalene ring is actually not bridged, but which are rather examples of benzo[n]metabenzenophanes, were studied from the point of view of reactivity and rearrangements.

Photolysis of 2-chloro[6](1,3)naphthalenophane (6) led to a transannular ring closure:[5]

6   7   8

The Grignard reagents in the [n](1,3)naphthalenophane series such as 9 react with oxygen (partially transannular reactions) to give anomalous products, which are attributed to ring strain and the spatial proximity of bridging hydrogen atoms (transannular neighbouring group effect):[6]

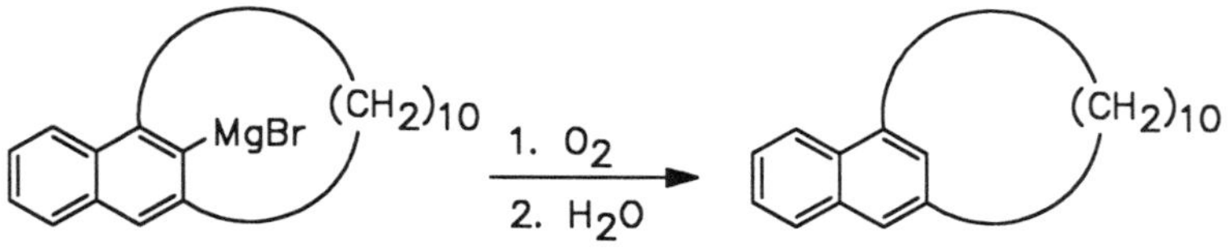

9

10

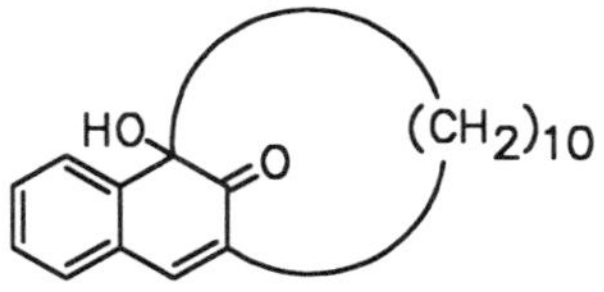

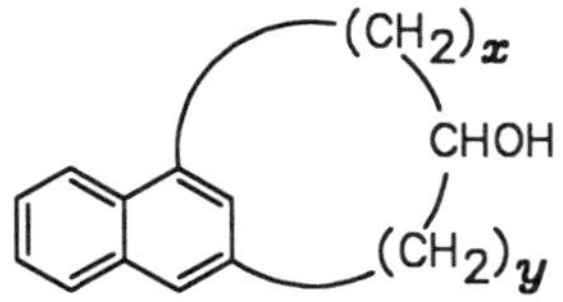

11

12: $x + y = 9$

1.3.2 [n](1,4)NAPHTHALENOPHANES

A series of [n]naphthalenophanes (15: $n = 8$–10, 14) have been prepared in the following way:

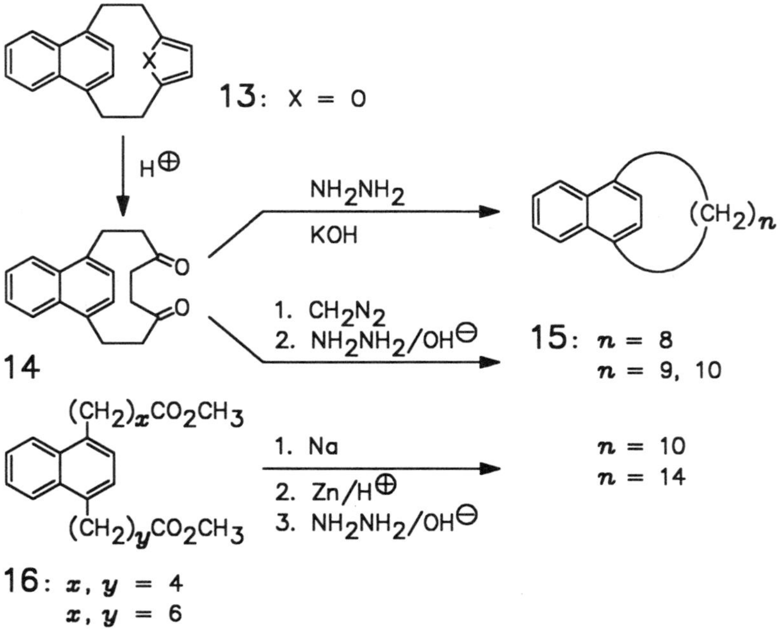

13: $X = 0$

14

15: $n = 8$
$n = 9, 10$

$n = 10$
$n = 14$

16: $x, y = 4$
$x, y = 6$

The starting material is the naphthalenofuranophane 13, which undergoes an acid-catalysed ring-opening, followed by homologization with diazomethane.

[10]- and [14](1,4)naphthalenophanes can be obtained by acyloin condensation of 16.[8]

The [12](1,4)naphthalenophane **18** was formed by cycloaddition of the bridged pentadienone **17** with benzyne.

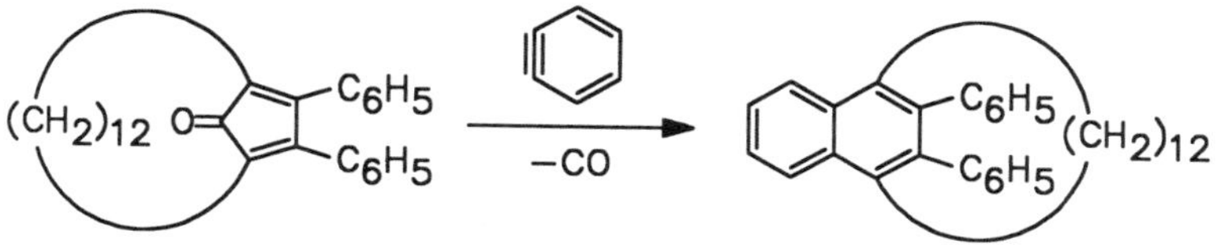

17   **18**

In 1989 Tobe *et al.*[9] reported the synthesis of the extremely short bridged and thus deformed [6](1,4)naphthalenophanes **19**, as well as of the [6](1,4)anthracenophanes **20**:

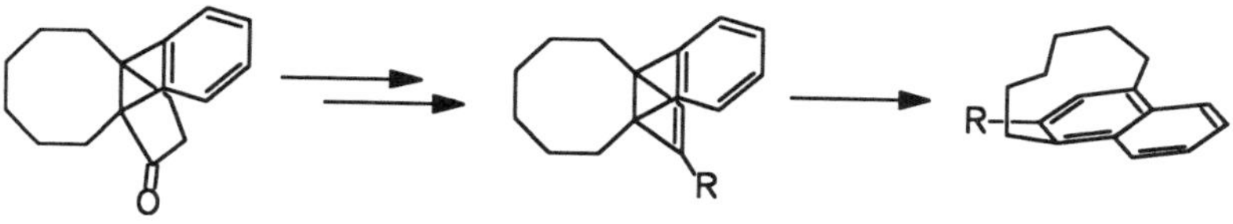

a : R = H
b : R = CO$_2$CH$_3$

19a,b

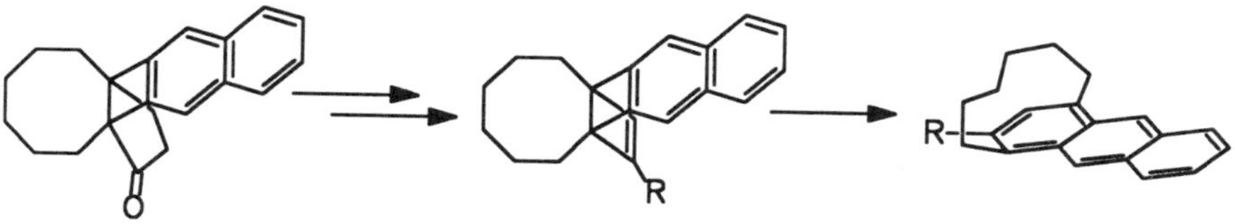

20a,b

Also to be mentioned is the synthesis by Sutherland *et al.*[10] of *[n](3,6)phenanthrenophanes* (**22**) by photochemical cyclization of the corresponding bridged stilbenes **21**. A [12](4,5)phenanthrenophane has been obtained by the acyloin condensation.

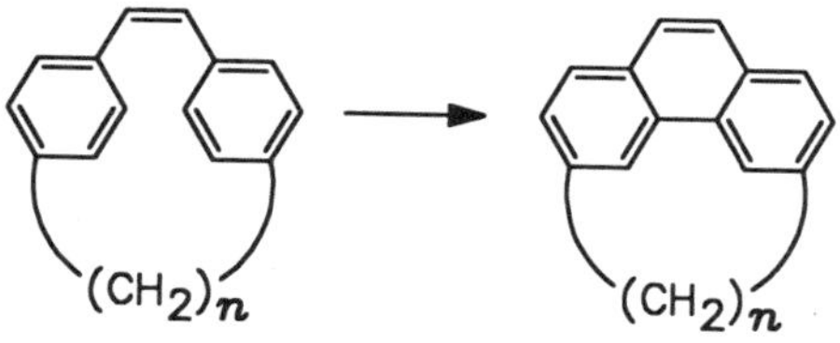

21 : n = 7 – 10   **22** : n = 7 – 10

From the ^{1}H NMR spectra of naphthalenophanes **15** it was shown that, only in the case of the [14]naphthalenophane where $n = 14$, is the molecule conformationally flexible on the NMR time scale at room temperature.

1.3.2.1 Chemical reactions of [n]naphthalenophanes

The naphthalenophanes **15** ($n = 8$–10) form Diels–Alder adducts **23** with dicyanoacetylene at 100°C, such that only the non-bridged aromatic ring reacts. In the case of [14]naphthalenophane the p-phenylene portion of the [n]naphthalenophane system reacts as well. The resulting products of type **24** were sought because of their paddlane character; for $n = 14$ the yields of **23** and **24** were 54% and 7%, respectively.[7]

23: n = 8, 9, 10, 14   **24**

References to Section 1.3

1. W. E. Parham, D. C. Egberg and W. C. Montgomery, *J. Org. Chem.*, **38**, 1207 (1973).
2. P. Grice and C. B. Reese, *Tetrahedron Lett.*, 2563 (1979); *J. Chem. Soc. Chem. Commun.*, 424 (1980); F. Bickelhaupt *et al.*, *J. Am. Chem. Soc.*, **112**, 8941 (1990).
3. W. E. Parham, R. W. Davenport and J. B. Biasotti, *J. Org. Chem.*, **35**, 3775 (1970).
4. W. E. Parham, D. C. Egberg and S. S. Salgar, *J. Org. Chem.*, **37**, 3248 (1972).
5. D. J. Cram, C. S. Montgomery and G. R. Knox, *J. Am. Chem. Soc.*, **88**, 515 (1966).
6. W. E. Parham, R. W. Davenport and J. K. Rinehart, *J. Org. Chem.*, **35**, 2662 (1970).
7. K. B. Wiberg and M. J. O'Donnell, *J. Am. Chem. Soc.*, **101**, 6660 (1979); Y. Tobe, T. Takahashi and K. Kobiro, *Tetrahedron Lett.*, **32**, 359 (1991).
8. D. J. Cram and H. Steinberg, *J. Am. Chem. Soc.*, **73**, 5691 (1951).
9. Y. Tobe, T. Takahashi and T. Ishikawa, *ISNA (International Symposium on Novel Aromatic Compounds)*, Osaka, August 1989; Y. Tobe, R. Gleiter *et al.*, *J. Am. Chem. Soc.*, **112**, 8889 (1990); Y. Tobe *et al.*, *J. Am. Chem. Soc.*, **114**, 9 (1992).
10. S. E. Potter and I. O. Sutherland, *J. Chem. Soc., Chem. Commun.*, 520 (1973).

1.4 OTHER [n]PHANES

X-ray crystal structure analysis of 2-thia[3]biphenylophane **1** showed that the central four-membered ring is contorted into a trapezoid by the CH_2—S—CH_2 'bracket'.[1]

1

'[*n*](1,5)Cyclooctatetraenophanes' (**2**) and 'semibullvalenophanes' (**3**) were described by Paquette *et al.*[3] (see also Chapters 9 and 10):

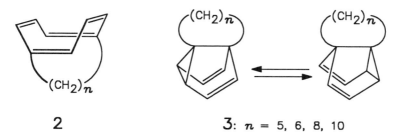

2   **3**: *n* = 5, 6, 8, 10

The first sandwich-phanes, *[n]ferrocenophanes* **4** (*n* = 3–5), were described as early as 1958 in a publication by Lüttringhaus *et al.*, where they were designated '*ansa-ferrocenes*'.[4]

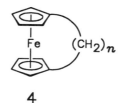

4

Since that time many other metallocenophanes have been described, among them ruthenocenophanes, bridged dibenzene chromium, as well as super-ferrocenophane (see also Chapter 10 as well as Ch. Elschenbroich, A. Salzer: *Organometallchemie* (Teubner, Stuttgart, 1988) and *Organometallics—A Concise Introduction* (VCH Publishers, Weinheim, 1989)).

The X-ray crystal structures of a number of [*n*]ferrocenophanes are known, including [3](1,1')ferrocenophane (TCNE complex), [3](1,1')ferrocenophan-1-one, and [15](1,1')ferrocenophane-8-one.[6]

References to Section 1.4

1. F. Vögtle, K. Saitmacher, S. Peyerimhoff, D. Hippe, H. Puff and P. Büllesbach, *Angew. Chem.*, **99**, 459 (1987); *Angew. Chem. Int. Ed. Engl.*, **26**, 470 (1987).
   (a) F. Vögtle, J. E. Schulz and K. Rissanen, *J. Chem. Soc., Chem. Commun.*, 120 (1992).
   (b) Ch. F. Wilcox, Jr, R. Gleiter *et al.*, *J. Am. Chem. Soc.*, **108**, 7693 (1986).
2. The 'phane' names are designated with quotation marks because COT and semibullvalene possess no 'aromatic' or 'arene' unit. For designation of such 'pseudophanes' containing non-aromatic or even purely aliphatic 'cores' see also Chapter 9 ('Protophanes' and 'Aliphanes').
3. L. A. Paquette, M. P. Trova, J. Luo, A. E. Clough and L. B. Anderson, *J. Am. Chem. Soc.*, **112**, 228 (1990).
4. A. Lüttringhaus and W. Kullick, *Angew. Chem.*, **70**, 438 (1958).
5. G. Oepen and F. Vögtle, *Liebigs Ann. Chem.*, **1979**, 1094; P. D. Beer, C. D. Bush

and T. A. Hamor, *J. Organomet. Chem.*, **339**, 133 (1988); M. C. Grossel, M. R. Goldspink, J. P. Knychala, A. K. Cheetham and J. R. Hriljac, *J. Organomet. Chem.*, **352**, C13 (1988).

6. Review: J. F. Liebman, in *Cyclophanes* (P. M. Keehn and S. M. Rosenfeld, Eds.), Vol. I, p. 23, Academic Press, New York 1983.

1.5 [*n*.1]PHANES

Some [*n*.1]phanes can be mentioned as close relatives of the [*n*]phanes.[1] Short bridged 'benzophenonophanes' and 'diphenylmethanophanes' such as **1–4** have been prepared as sterically hindered double-winged propeller molecules:

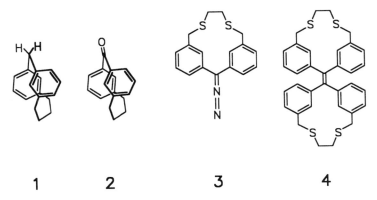

1   **2**   **3**   **4**

The [4.1]metabenzenophanes **1** and **2**, because of their very narrow bridging, exhibit a high barrier to topological isomerization for the interconversion of the propeller configurations: $\Delta G_c^{\ddagger} = 82 \text{ kJ/mol}$.

Carbocations (**6**), carbanions (**7**), and carbenes (**8**), as examples of crown ethers, have been prepared from the corresponding 'benzophenonophane' **5a**, and their interaction with cations studied:[3]

	X
a	O
b	$N_2(CH_3)_2$
c	NNH_2
d	NNHTos
e	NOH
f	NNHDNP
g	N_2

5

DNP = 2,4–Dinitrophenyl

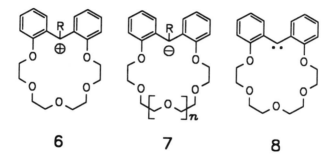

6   **7**   **8**

Turro and Staab *et al.*[4] have investigated the influence of molecular geometry on the spectroscopic and photochemical properties of α-oxo[*n*.1]paracyclophanes (**9**):

$$(CH_2)_{\overline{n}}$$

9: *n* = 8–12

References to Section 1.5

1. See also H. Hopf, *Chemie in uns. Zeit*, **10**, 114 (1976).
2. M. Atzmüller and F. Vögtle, *Chem. Ber.*, **111**, 2547 (1978); *Chem. Ber.*, **112**, 138 (1979).
3. F.-A. v. Itter and F. Vögtle, *Chem. Ber.*, **118**, 2300 (1985).
4. N. J. Turro, I. R. Gould, J. Liu, W. S. Jenks, H. A. Staab and R. Alt, *J. Am. Chem. Soc.*, **111**, 6378 (1989).

2 [2.2]Phanes

INTRODUCTION

Of the six possible [2.2]cyclophanes **1–6** all are known. The 'last' of these, [2.2]orthoparacyclophane (**6**), was first synthesized in 1992 by Tobe.[1a]

[2.2]Orthocyclophane (**1**) is of less interest as a cyclophane because of its *ortho*-bridging. Because of the commonly observed bridging between ortho positions, this type is, according to Lüttringhaus's definition, not an example of an *ansa* compound.

This discussion of the series **1,2,3,4,5** begins with a description of the synthesis and properties of the most interesting members of the cyclophane series, in particular those with simply bridged benzene rings. The compounds **2–5** all exhibit the transannular electronic and steric effects and spectral anomalies characteristic of cyclophanes with deformed rings; and they call for special synthetic strategies. These [2.2]phanes are central to cyclophane chemistry.

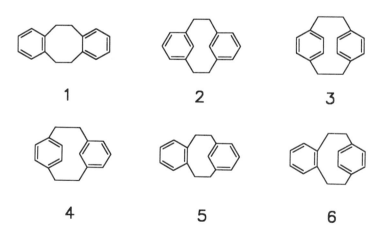

2.1 [2.2]ORTHOCYCLOPHANE

2.1.1 SYNTHESIS

In 1945 Baker obtained *[2.2]orthocyclophane* (1,2,5,6-dibenzocyclooctadiene, **1**), in 6% yield by treatment of 1,2-bis(bromomethyl)benzene (**7**) with an excess of a suspension of metallic sodium in boiling dioxane.[1] Along with acyclic

7   1

8   9

products, the next higher oligomer, [2.2.2]orthocyclophane (**8**), was also formed. Cope and Fenton in 1951 were able to isolate [2.2]orthocyclophane in 45% yield by the same reaction but with strict adherence to the high-dilution principle.[2] Even at relatively high concentrations, Müller and Röscheisen found they could obtain **1** and **8** in yields of 40% and 35%, respectively. [2.2.2.2]Orthocyclophane (**9**) can be obtained in 40% yield by treatment of 1,2-bis(2-bromomethylphenyl)ethane with phenyllithium[3] (see Section 7.1.1).

The *properties* of [2.2]orthocyclophane (**1**) warrant no particular comment, inasmuch as neither deformed benzene rings nor high-field inner protons are observed.[4] On the basis of its properties **1** is considered a dibenzocyclo-octadiene rather than a typical cyclophane. As already mentioned, it also does not belong to the family of '*ansa* compounds', because they are defined as those in which a benzene ring is bridged other than through the normal 1,2- or *o*-position.

References to Section 2.1

1. W. Baker, R. Banks, D. R. Lyon and F.G. Mann, *J. Chem. Soc.*, 27 (1945).
1a. Y. Tobe, M. Kawaguchi and K. Naemura, *ISNA-7*, July 1992, Poster No. 90; Y. Tobe *et al.*, *J. Am. Chem. Soc.*, **115**, 1173 (1993); for evidence for [1.1]para-cyclophane, see T. Tsuji, *J. Am. Chem. Soc.*, **115**, 5284 (1993).
2. A. C. Cope and S. W. Fenton, *J. Am. Chem. Soc.*, **73**, 1668 (1951); L. A. Paquette, M. A. Kesselmayer, G. E. Underiner, S. D. House, R. D. Rogers, K. Meerholz and J. Heinze, *J. Am. Chem. Soc.*, **114**, 2644 (1992).
3. E. D. Bergmann and Z. Pelchowicz, *J. Am. Chem. Soc.*, **75**, 4281 (1953).
4. Conformation of the eight-membered ring: D. J. Brickwood, W. D. Ollis, J. S. Stephanatou and J. F. Stoddart, *J. Chem. Soc., Perkin Trans. 1*, 1398 (1978); see also ref. 2.

2.2 [2.2]METACYCLOPHANES

As a model compound 'par excellence', *[2.2]metacyclophane* is well suited as a point of departure in the discussion of [2.2]phanes. The large assortment of its substitution products serves as ideal model systems to investigate benzene ring deformation, statistical and dynamic stereochemistry, and intramolecular and transannular steric and electronic effects. They have also served to answer questions concerning interactions with neighbouring groups not necessarily directly bonded to functional groups, questions about aromaticity of deformed benzene rings, transannular cyclizations, etc.

2.2.1 SYNTHESIS

Pellegrin prepared the hydrocarbon 2 in 1899 in varying yields by a Wurtz reaction as already described in the Introduction. 1,3-Bis(bromomethyl)benzene (1) was treated with an excess of sodium and a molar equivalent of bromobenzene in ether (phenylsodium).[1] Baker attempted to explain the facile formation of what were then considered unusual ring systems by the 'principle of rigid groups'.[1a]

In 1957 Müller and Röscheisen improved the yield of [2.2]metacyclophane from 11.7% (achieved by Baker in 1945) to 25%.[2] Allinger obtained [2.2]metacyclophane in 39% yield with phenyllithium in 1961,[3] while in the same year Boekelheide was able to realize a yield of 77% by the Müller–Röscheisen route.[4]

Beginning in 1966 Jenny achieved the synthesis of the higher homologues 3–10, along with [2.2]metacyclophane, by means of the Müller–Röscheisen reaction carried out in THF at $-80°C$ (TPE = tetraphenylethene). All members of the $[2_n]$metacyclophanes up to the 50-membered $[2_{10}]$metacyclophane (10), separated chromatographically, were characterized in this way.[5]

1

2 : $n = 2$

3–10 : $n = 3-10$

Yields for [2.2]metacyclophanes ranged between 33 and 35%; those of higher homologues, where $n = 3-10$, were approximately 8, 1–2, 5, 4, 0.5, 0.6, 0.5, and 0.3, respectively.

Sato *et al.* used the Müller–Röscheisen method to prepare [2.2]metacyclophanes with methyl substituents inside the ten-membered ring, which they could also accomplish with bis(*chloro*methyl) starting materials, which are inert toward phenyllithium.[6]

The bis-dithiane 11 of isophthalaldehyde can be alkylated by the Corey–Seebach method[7a] to the corresponding dithiane-substituted [2.2]metacyclophane 12. This, in turn, can be desulphurized by various methods and be converted via the diketone 13 to [2.2]metacyclophane 2.[7b,7c]

11   12   2 : X = CH$_2$
                                            13 : X = C=O

Sulphone pyrolysis of 2,11-dithia[3.3]metacyclophane tetroxide 15 from dithia[3.3]metacyclophane, itself easily obtained in high yield, affords the corresponding [2.2]metacyclophanes 16, usually in good yield and a high state of purity. Included are those containing inner, distorted substituents, which may be the same or different (X ≠ Y). This method, which can be applied to the synthesis of many other [2.2]phanes, has become one of the most widely applicable routes to cyclophane synthesis,[8] with several hundred successes having been recorded internationally.

14   15   16

The ketone pyrolysis described recently proceeds analogously; product selectivity is temperature-dependent:[8a]

The Stevens rearrangement of dithia[3.3]phanes provides a route to the [2.2]metacyclophanes, as well as to their corresponding dienes 20.[9] This reaction has also been modified with benzyne as a reagent.[10]

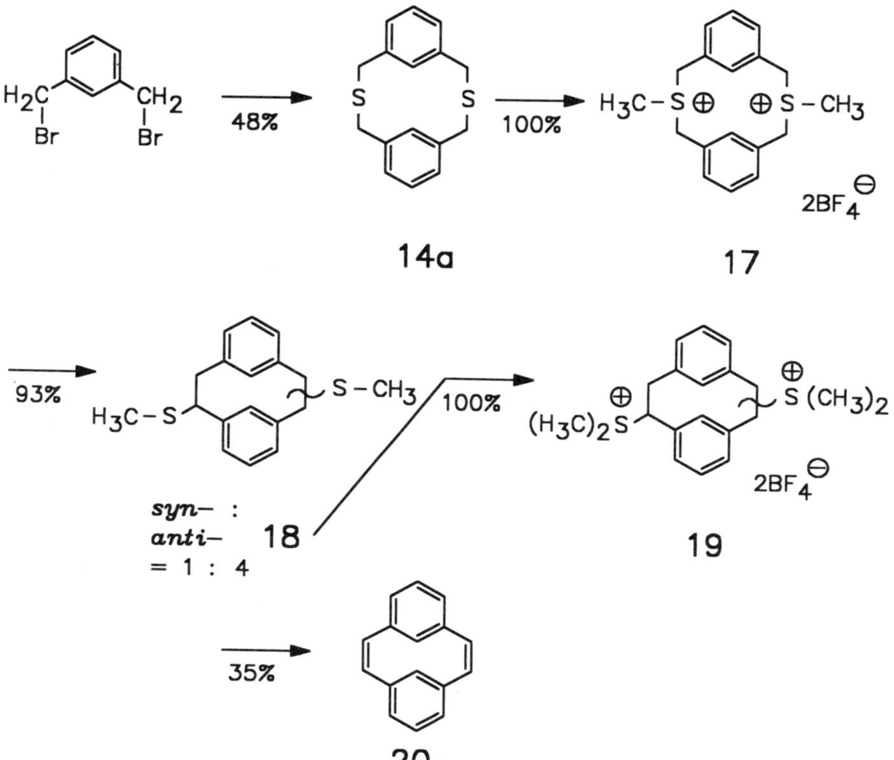

14a   17

syn- :
anti- 18
= 1 : 4

19

20

Sulphur can also be eliminated photochemically by the addition of a thiophilic phosphorus compound, presumably because neither the starting material nor the product is photo labile.

A novel method of ring contraction has recently been described by a group in Japan. N$_2$O is eliminated from the nitrosamines 23 of the corresponding 2,11-diaza[3.3]metacyclophanes 22 in high yield.[11]

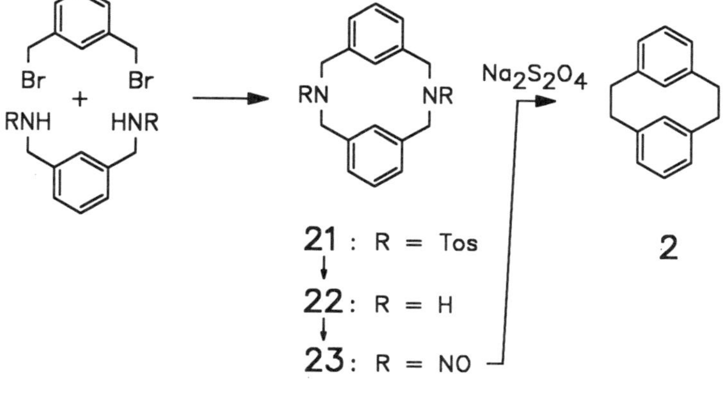

21 : R = Tos

22 : R = H

23 : R = NO

2

Likewise the appropriate 1,11-diseleno[3.3]phanes **25** undergo ring contraction to [2.2]phanes:[12]

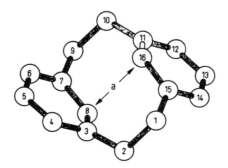

2.2.2 PROPERTIES[13]

[2.2]Metacyclophane forms colourless crystals with melting point 132–133°C. It easily sublimes and is transformed only at higher temperatures into pyrene. The molecule possesses a centre of symmetry, alternating axis, and a mirror plane (point group C_{2h}).

The X-ray crystal structure of **2** reveals molecules with a centre of symmetry. Both halves of the molecule are ordered in the form of steps. Because the benzene rings are not planar but distorted into boats, the steric overlap of the inner carbon atoms C(8) and C(16) and their attached hydrogens is diminished. The C(8)–C(16) distance (a in Figure 1) is 268.9 pm. The average C—C aromatic bond length is 138.6 pm, that of the aliphatic bonds 154.3 pm. It is remarkable that such a strong distortion of the normally planar hexagonal benzene ring is not accompanied by substantial changes in the interatomic distances.

The higher strain in the intra-annularly substituted *8,16-dimethyl[2.2]meta-cyclophane* **16a** (X = Y = CH₃), as compared to that in the parent hydrocarbon **2**, causes a lengthening of the C(1)–C(2) distance to 157.3 pm. It is also noteworthy that the two intra-annular methyl carbon atoms are coplanar with the C-atoms 8, 7 and 3 and 16, 11, and 15 (compare Figures 1 and 2).

Figure 1. Structure of [2.2]metacyclophane (**2**)

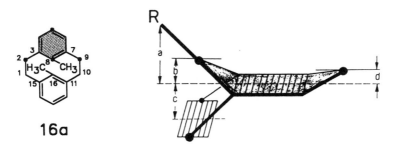

Figure 2. Deviation from planarity of the distorted boat-shaped benzene rings in the [2.2]metacyclophane systems **2** and **16** (only one of the two benzene rings is represented). For details see references 13a, 13c

The C(8)–C(16) distance increases from 268.9 pm (in **2**) to 281.9 pm (in **16a**). Rotation in the crystal of the methyl groups about their axes appears hindered if not blocked. Nevertheless proton resonance measurements in solution even at low temperatures do not suggest a rotational barrier for the methyl groups.

The ^{1}H-NMR spectrum of [2.2]metacyclophane in solution, depicted in Figure 3,[13a] shows clearly the consequence of the stair-like conformation. The bridging methyl protons are non-equivalent because they assume a staggered arrangement and thus behave as an AA'BB'-system. Especially noteworthy is the absorption of the intra-annular hydrogen atom H_i at unusually high field ($\delta = 4.25$), which signifies the location of the H_i-atoms over the opposite

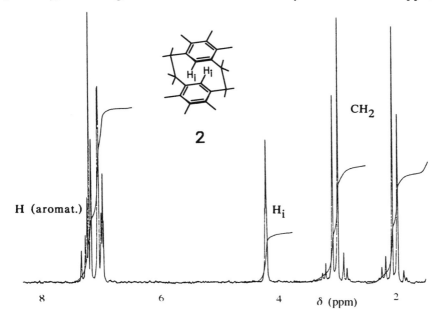

Figure 3. ^{1}H-NMR spectrum of [2.2]metacyclophane (**2**; 90 MHz, CDCl$_3$)

benzene rings with their corresponding anisotropy effect. The [2.2]metacyclophane molecule is completely fixed, as demonstrated by high- and low-temperature measurements; and a ring inversion is excluded. The methyl protons in 8-methyl[2.2]metacyclophane (**16b**) and in 8,16-dimethyl[2.2]meta-cyclophane (**16a**) are shifted to extremely high field ($\delta = 0.48$ and 0.56, respectively; compare toluene: $\delta = 2.32$).

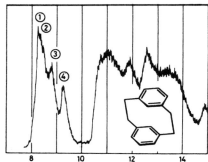

16a : R = CH_3
16b : R = H
16c : $R^1 = R^3$ = H, R^2 = CH_3
16d : $R^1 = R^3$ = H, R^2 = COOH

The rigid structural framework in [2.2]metacyclophane was conclusively demonstrated through resolution of the planar chiral compounds **16c** and **16d**, by means of TAPA [$(-)$-α-(2,4,5,7-tetranitro-9-fluorenylidenaminooxy)-propionic acid] and quinine.[14]

The UV spectrum of [2.2]metacyclophane shows a bathochromic shift (relative to the reference compound 3,3'-dimethylbibenzyl) at $\lambda_{max} = 272$ nm ($\log \epsilon = 2.64$).

Charge-transfer interactions between [2.2]metacyclophane and tetracyano-ethylene are indicated by the appearance of a band at 490 nm, which is shifted to longer wavelength by comparison to that of the complex with *m*-xylene. This suggests that the benzene ring in [2.2]metacyclophane which is not directly complexed with tetracyanoethylene exerts an electron-donating effect on the other, interacting benzene ring. The bathochromic shift, however, is not as pronounced as that observed in the [2.2]paracyclophane–tetracyanoethylene complex ($\lambda_{max} = 521$ nm) (see Section 2.3).

Bands appear in the IR spectrum of [2.2]metacyclophanes between 1400 and 400 cm^{-1}, which are either absent or weaker in reference compounds such as *m*-xylene, [2.2.2]metacyclophane (**3**; see above), and [2.2.2.2]metacyclophane (**4**; see above and Section 7.1.2). These are attributed to deformations in the benzene rings.

Figure 4. PE-spectrum of [2.2]metacyclophane (**2**)[15]

The photoelectron (PE) spectrum of [2.2]metacyclophane, shown in Figure 4, can be compared to those in Section 2.3 of [2.2]paracyclophane and other [2$_n$]phanes.

While all the discussion thus far concerns the *anti* conformation of [2.2]metacyclophane, Boekelheide succeeded a few years ago in synthesizing intra-annularly substituted *syn[2.2]metacyclophanes*.[16] Recently Mitchell was able to prepare the unsubstituted *syn*-[2.2]metacyclophane (see below).

The basis for both stereoselective syntheses was the rigid *syn* conformation in the larger ring 2, 11-dithia[3.3]metacyclophanes, whose *syn/anti* conformational isomerism had been discovered by Vögtle.[17] Boekelheide used the Stevens rearrangement to effect ring contraction in the *syn* conformer of the intra-annularly substituted dithia[3.3]metacyclophane **14** and thus obtained the intra-annularly substituted *syn*-[2.2]metacyclophane **16**. Mitchell, on the other hand, locked the conformation of the unsubstituted dithia[3.3]metacyclophane **14a** by complexation with chromiumtricarbonyl to form a transition metal–arene complex (see below).

anti : *syn* = 1 : 3

2.2.2.1 syn-[2.2]Metacyclophane

Synthesis and properties[18]

syn-[2.2]Metacyclophane (**2a**), first synthesized by Mitchell in 1985, has been characterized only in solution, not yet having been isolated in crystalline form. The starting material was the conformationally rigid bis(chromium carbonyl) complex **27** of the conformationally flexible *syn*-2,11-dithia[3.3]metacyclophane (*syn*-**14a**). Treatment of **14a** with chromium hexacarbonyl in refluxing di-*n*-butyl ether afforded *syn*-**26** (the monochromium carbonyl complex) in 70% yield.

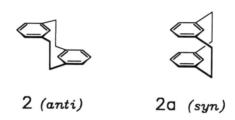

2 *(anti)*   **2a** *(syn)*

The *syn* conformation was confirmed from the intra-annular protons in the ¹H-NMR spectrum at δ = 7.23 and 4.83 p.p.m. Methylation of **26** and subsequent Stevens rearrangement afforded in 70% yield the monochromium carbonyl derivative **31D** of *syn*-[2.2]metacyclophane (**2a**) in the form of yellow crystals, m.p. 120–121°C. Assignment of the *syn* conformation to **31D** is based not only on the ¹H-NMR absorptions of the intra-annular protons at δ = 6.93 and 5.51 p.p.m. but also on the X-ray crystal structure. This *syn*-[2.2]metacyclophane chromium complex is transformed at 80°C within 1 hour into the corresponding *anti*-cyclophane (**33C**, m.p. 128°C, 'inner' protons at δ = 5.91 and 3.42 p.p.m.). The uncomplexed *syn*-[2.2]metacyclophane **30D** is generated by treatment of **31D** with cerium(IV) in acetonitrile at −35°C and then isolation and chromatography of the products, also at −35°C. The proton resonance spectrum of **30D** shows the inner protons at δ = 7.04 and 6.75, while the other aromatic protons are deshielded by the cofacial rings to δ = 7.0–6.3. When a solution of **30D** is warmed to 0°C, isomerization to the thiomethyl-substituted *anti*-cyclophane takes place.

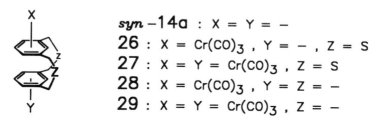

syn−**14a** : X = Y = −
26 : X = Cr(CO)$_3$, Y = − , Z = S
27 : X = Y = Cr(CO)$_3$, Z = S
28 : X = Cr(CO)$_3$, Y = Z = −
29 : X = Y = Cr(CO)$_3$, Z = −

As a route to uncomplexed, unsubstituted *syn*-[2.2]metacyclophane **2a**, the bis(chromium carbonyl) complex **27** of 2,11-dithia[3.3]metacyclophane was prepared with an excess of chromium hexacarbonyl. Stevens rearrangement followed by reduction with lithium in ammonia at −40°C afforded a mixture of **28** and **29**, the first chromium carbonyl derivatives of unsubstituted *syn*-[2.2]metacyclophane (**2a**). Removal of the chromium carbonyl groups with *m*-chloroperbenzoic acid or cerium(IV) at −45°C in acetonitrile gave, for the first time, the uncomplexed, unsubstituted *syn*-[2.2]metacyclophane **2a**, which rapidly isomerizes to the more stable *anti*-isomer **2** above 0°C. The ¹H-NMR spectrum of *syn*-[2.2]metacyclophane **2a** at −40°C exhibits the intra-annular protons at δ = 6.58, the external aromatic protons at 6.36 and 6.60, and the bridging protons at δ = 3.14 and 2.85 p.p.m.[18]

	−SMe
A:	1−eq, 3−eq
B:	1−eq, 4−eq
C:	1−ax, 3−ax
D:	1−ax, 3−eq

	X	Y		X	Y
30 :	−	−	**32** :	−	−
31 :	Cr(CO)$_3$	−	**33** :	Cr(CO)$_3$	−

2.2.3 CHEMICAL REACTIONS OF [2.2]METACYCLOPHANES

Dehydrogenation of [2.2]metacyclophane with Pd/carbon at 300°C affords pyrene (**34**) in 60% yield:

Electrophilic substitution of [2.2]metacyclophanes is typically accompanied by ring closure; in the case of nitration apparently tetrahydropyrene **35** is an intermediate in the formation of **36**.[18a]

In the case of the doubly methylated [2.2]metacyclophane **16a**, oxidation with FeCl$_3$ or with CrO$_3$/H$_2$SO$_4$ in acetone effects a *transannular ring closure* to afford in quantitative yield the bis-dienone **37**, which is stable towards acids.

Transannular ring closures are also observed in photochemical reactions of [2.2]metacyclophanes.[13a] Irradiation of [2.2]metacyclophane **2** in cyclohexane in the presence of iodine and sodium bicarbonate leads to 4,5,9,10-tetrahydropyrene (**35**) in 86% yield, whereas with iodine alone the starting material is recovered unchanged.

Metacyclophan-1-enes (**39**) undergo photoisomerization with formation of the coloured 4,5,10b,10c-tetrahydropyrenes (**40**). The reverse reaction can be accomplished by irradiation with visible light (λ approx. 500 nm) or thermally with the irradiated solution being allowed to stand in the dark. Compounds with structure **40** are oxygen-sensitive, being readily oxidized to 4,5-dihydropyrene **41**.

39 : R^1, R^2 = H, CH$_3$ **40**   **41**

It is remarkable that the unsubstituted *[2.2]metacyclophan-1,9-diene* (**42**) is considerably more stable than the intra-annularly substituted [2.2]metacyclophan-dienes **43**. The latter undergo spontaneous valence isomerization in the dark at room temperature, with formation of the more stable aromatic *trans*-10b,10c-dialkyldihydropyrene (**44**). By contrast the unsubstituted cyclophanediene **42** affords a mixture of *trans*-10b,10c-dihydropyrene (**45**) and pyrene, but only when it is irradiated.

42   **45**   **34**

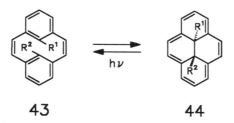

43   **44**

Reduction Catalytic hydrogenation of [2.2]metacyclophane (**2**) with Pt/HOAc gives perhydro[2.2]metacyclophane (**46**) in 91% yield. If the hydrogenation is interrupted after the consumption of 3 mol of hydrogen, however, it is possible to isolate the relatively strain-free 3,4,5,6,7,8-hexahydro-[2.2]metacyclophane (**47**).

46   **47**

Birch reduction at −75°C leads to 5,8,13,16-tetrahydro[2.2]metacyclophane (**48**), which presumably assumes the conformationally flexible *anti* conformation **48A**.[19]

48   **48A**

As is the case with other [2ₙ]cyclophanes, *anti*-[2.2](1,3)cyclophane (**2**) can be a route to double decker (**49**) and triple decker (**50**) ferrocene-like complexes.[13b,20] These are heat-and air-stable solids which can be stored indefinitely. Their ¹H- and ¹³C-NMR spectra provide insight into the nature of metal–arene binding as well as an assessment of ring current and electron-attracting effects resulting from metal complexation.

[2.2]meta–
cyclophane,
hν

49   50

References to Section 2.2

1. M. M. Pellegrin, *Rec. Trav. Chim. Pays-Bas*, **18**, 457 (1899).
1a. W. Baker, F. J. W. McOmie, and J. M. Norman, *J. Chem. Soc.*, 1114 (1951).
2. E. Müller and G. Röscheisen, *Chem. Ber.*, **90**, 543 (1957).
3. N. L. Allinger, M. A. DaRooge and R. B. Herrmann, *J. Am. Chem. Soc.*, **83**, 1974 (1961).
4. W. S. Lindsay, P. Stokes, L. G. Humber and V. Boekelheide, *J. Am. Chem. Soc.*, **83**, 943 (1961).
5. W. Jenny and R. Paioni, *Chimia*, **22**, 142 (1968); R. Paioni and W. Jenny, *Helv. Chim. Acta*, **52**, 2041 (1969); F. Vögtle, J. Schmitz and M. Nieger, *Chem. Ber.*, **125**, 2523 (1992).
6. Review: F. Vögtle and P. Neumann, *Synthesis*, 85 (1973).
7. (a) D. Seebach, *Angew. Chem.*, **81**, 690 (1969); *Angew. Chem. Int. Ed. Engl.*, **8**, 639 (1969); *Synthesis*, 17 (1969).
   (b) V. Boekelheide, P. H. Anderson and Th. A. Hylton, *J. Am. Chem. Soc.*, **96**, 1558 (1974).
   (c) Th. Hylton and V. Boekelheide, *J. Am. Chem. Soc.*, **90**, 6887 (1968); H. W. Gschwend, *J. Am. Chem. Soc.*, **94**, 8430 (1972).
   (d) Ab initio calculations: S. Tsuzuki and K. Tanabe, *J. Chem. Soc., Perkin Trans.* 2, 1687 (1990).
8. F. Vögtle, *Angew. Chem.*, **81**, 258 (1969); *Angew. Chem. Int. Ed. Engl.*, **8**, 274 (1969); F. Vögtle and L. Rossa, *Angew. Chem.*, **91**, 534 (1979); *Angew. Chem. Int. Ed. Engl.*, **18**, 514 (1979); J. Dohm and F. Vögtle, *Top. Curr. Chem.*, **161**, 69 (1991).
   (a) J. Breitenbach, F. Ott and F. Vögtle, *Angew. Chem.*, **104**, 360 (1992); *Angew. Chem. Int. Ed. Engl.*, **31**, 307 (1992); F. Ott, J. Breitenbach, M. Nieger and F. Vögtle, *Chem. Ber.*, **126**, 97 (1993).
9. R. H. Mitchell, T. Otsubo and V. Boekelheide, *J. Am. Chem. Soc.*, **96**, 1547 (1974); cf. R. H. Mitchell and V. Boekelheide, *Tetrahedron Lett.*, 219 (1975).
10. T. Otsubo and V. Boekelheide, *J. Org. Chem.*, **42**, 1085 (1977).
11. H. Takemura, T. Shinmyozu and T. Inazu, *Tetrahedron Lett.*, **29**, 1031 (1988).
12. H. Higuchi, K. Tani, T. Otsubo and Y. Sakata, *Bull. Chem. Soc. Jpn.*, **60**, 4027 (1987); cf. M. Hojjatie, S. Muralidharan and H. Freiser, *Tetrahedron*, **45**, 1611 (1989); cf. also: F. Vögtle, J. Breitenbach and M. Nieger, *J. Chem. Soc., Chem. Commun.*, 860 (1991).
13. Reviews:
    (a) F. Vögtle and P. Neumann, *Angew. Chem.*, **84**, 75 (1972); *Angew. Chem. Int. Ed. Engl.*, **11**, 73 (1972).

(b) V. Boekelheide, in *Cyclophanes, I* (F. Vögtle, Ed.), *Top. Curr. Chem.*, **113**, 87 (1983).

(c) For spectroscopic properties and X-ray crystal structure see: *Cyclophanes*, (P. M. Keehn and S. M. Rosenfeld, Eds.), Vol. I, Academic Press, New York 1983.

14. (a) T. Sato, S. Akabori, M. Kainosho and K. Hata, *Bull. Chem. Soc. Jpn.*, **39**, 856 (1966); *Bull Chem. Soc. Jpn.*, **41**, 218 (1968) and additional references therein. Cf. T. Sato *et al.*, *Tetrahedron*, **27**, 2737 (1971), and C. Glotzmann, E. Langer, H. Lehner and K. Schlögl, *Monatsh. Chem.*, **105**, 907 (1974).
    (b) Ch. Krieger, H. Lehner and K. Schlögl, *Monatsh. Chem.*, **107**, 195 (1976); H. Keller, Ch. Krieger, E. Langer, H. Lehner and G. Derflinger, *Liebigs Ann. Chem.*, 1296 (1977); H. Keller and H. Lehner, *Liebigs Ann. Chem.*, 595 (1978); H. Keller, Ch. Krieger, E. Langer, H. Lehner and G. Derflinger, *Tetrahedron*, **34**, 871 (1978); D. Krois, E. Langer and H. Lehner, *Tetrahedron*, **36**, 1345 (1980); H. Lehner, H. Paulus and K. Schlögl, *Monatsh. Chem.*, **112**, 511 (1981); See also K. Schlögl *et al.*, *Monatsh. Chem.*, **120**, 453 (1989) and K. Schlögl, *Top. Curr. Chem.*, **125**, 27; (1984); G. Derflinger, in *Chirality—From Weak Bosons to the α-Helix* (R. Janoschek, Ed.), Springer, Berlin, 1991.
    (c) F. Vögtle *et al.*, *Chem. Ber.*, **116**, 2630 (1983); *Chem. Ber.*, **121**, 823 (1988); *Chem. Ber.*, **122**, 343 (1989).
15. E. Heilbronner and Z. Yang, in *Cyclophanes, II* (F. Vögtle, Ed.), *Top Curr. Chem.*, **115**, 1 (1983).
16. R. H. Mitchell and V. Boekelheide, *Tetrahedron Lett.*, 1197 (1970); *J. Chem. Soc., Chem. Commun.*, 1555 (1970).
17. F. Vögtle and L. Schunder, *Chem. Ber.*, **102**, 2677 (1969); cf. F. Vögtle *et al.*, *Chem. Ber.*, **125**, 255 (1992).
18. R. H. Mitchell, T. K. Vinod and G. W. Bushnell, *J. Am. Chem. Soc.*, **107**, 3340 (1985); **112**, 3487 (1990); R. H. Mitchell *et al.*, *Pure Appl. Chem.*, **58**, 15 (1986); S. Ito *et al.*, *Tetrahedron Lett.*, **34**, 3787, 3789 (1993).
    (a) Nitration of [n.2]metacyclophanes: T. Yamato, M. Tashiro *et al.*, *Chem. Express* **6**, 663 (1991).
19. J. Reiner and W. Jenny, *Helv. Chim. Acta*, **52**, 1624 (1969).
20. H. Hopf, in *Cyclophanes*, (P. M. Keehn, S. M. Rosenfeld, Eds.) Vol. II, p. 512, Academic Press, New York 1983.

2.3 [2.2]PARACYCLOPHANES

2.3.1 SYNTHESIS[1]

Carbon–carbon coupling of 1,4-bis(bromomethyl)benzene (**1**) with sodium iodide in refluxing dioxane gives exclusively [2.2.2]paracyclophane (**2**). Nevertheless, [2.2]paracyclophane (**3**) can be obtained in 2.1% yield by intramolecular cyclization with sodium of 1,2-bis(4-bromomethylphenyl)ethane (**4a**).

2.3.1.1 Hofmann elimination

1,6-Hofmann elimination with *p*-methylbenzylammonium hydroxides has often been applied successfully, albeit in low yields, to the synthesis of [2.2]paracyclophanes. This method is especially advantageous for the preparation of multi-layer cyclophanes (see Chapter 6). The description in *Org. Syn.* starting with 4-methylbenzyltrimethylammonium bromide is still considered the standard procedure for the preparation of [2.2]paracyclophane.

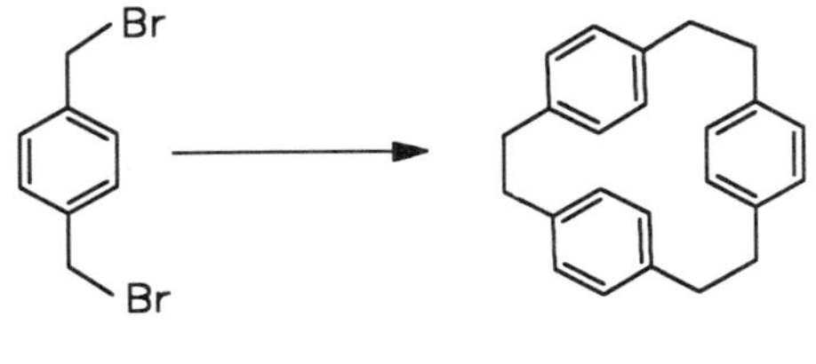

1   2

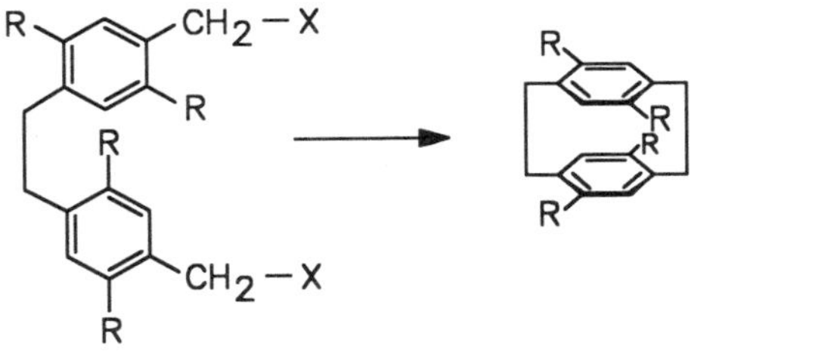

4a : R = H   3 : R = H

4b : R = CH$_3$ 5 : R = CH$_3$

6   7

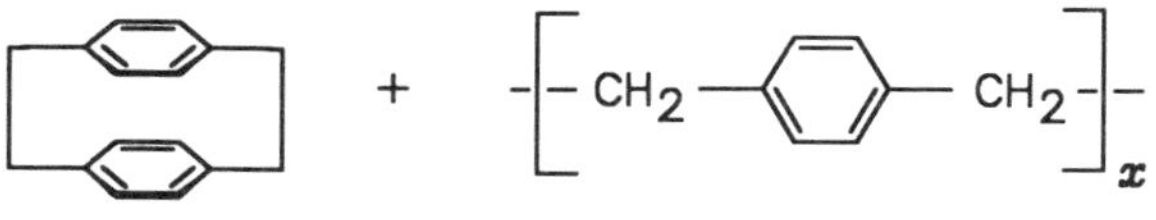

3   8

The water generated in the reaction is removed azeoptropically with toluene containing a small amount of phenothiazine (as inhibitor). The yield of [2.2]paracyclophane (3) is 17%, while that of the polymer 8 is 55%. In some instances the yield has been improved with newer modifications. In this latter

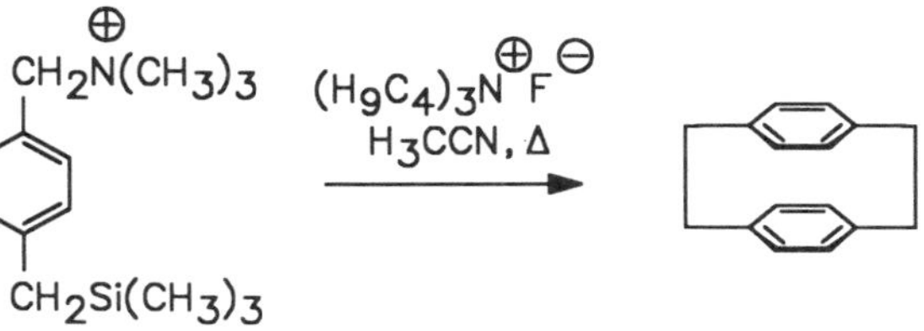

9   3

reaction fluoride attacks the silicon in the trimethylsilyl analogue **9**; the reported yield of 56% was achieved on a 25-mg scale.

Solvolysis of the ditosylate **10b** in refluxing pyridine affords [2.2]para-cyclophane (**3**) as the exclusive product in 40% yield.

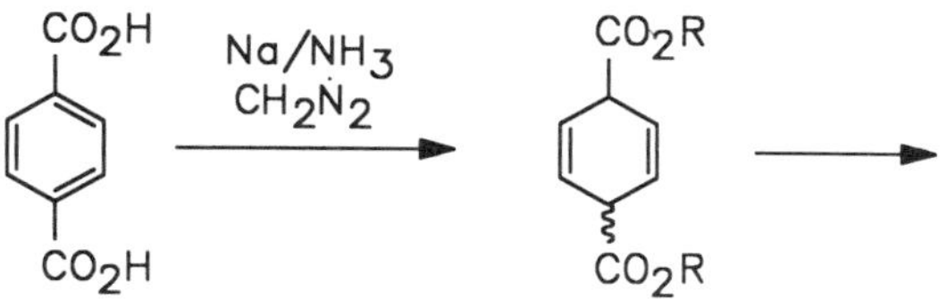

R = H, CH$_3$

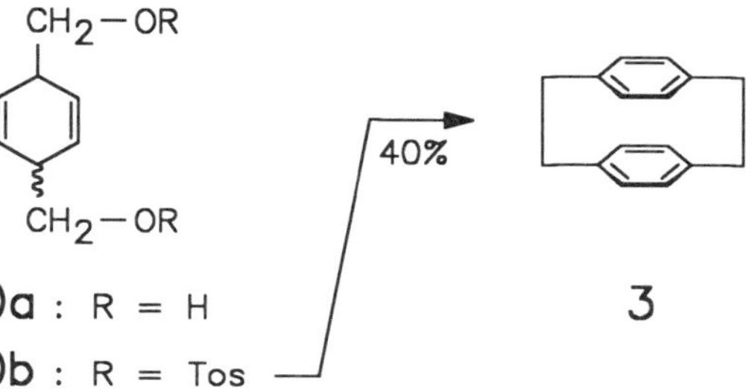

10a : R = H   **3**
10b : R = Tos

The tetrafluoro[2.2]paracyclophane **12** is formed in only 1% yield *via* the Wurtz reaction with **11**.

11   **12**

2.3.1.2 The dithiacyclophane route

In 1969 a new and general route to all types of phanes was developed. The strategy is to prepare sulphur-containing large rings (dithiaphanes) and effect a ring contraction by sulphur extrusion.[2] The method works in the [2.2]para-cyclophane series, for example, when the corresponding dithia[3.3]para-cyclophane is subjected to sulphur-photoextrusion or its sulphone is pyrolysed. Yields of [2.2]paracyclophane are 75–80%.[1b]

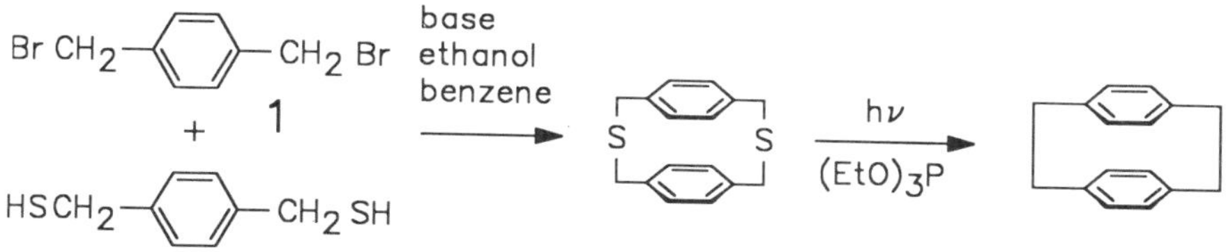

13   **3**

An alternative method of sulphur extrusion involves a modified Stevens rearrangement, whereby *in situ* generated benzyne affords a mixture of *cis/trans*-bisphenylsulphides **14**. Oxidation of **14** to the corresponding sulphoxide and pyrolytic elimination of benzenesulphinic acid in boiling xylene gives the [2.2]paracyclophandiene **15**, for which this route is the method of choice. Irradiation of the bis-sulphone of **13** and its analogues leads to [2.2]phanes.[1b] Photolysis of the dilactone **16** gives [2.2]paracyclophane in remarkably high yield.[3] (For ketone pyrolysis see Section 2.2.1.)

13   **14**   **15**

16 : $Z = CO_2$ **3**

2.3.1.3 Diels–Alder approach

According to Hopf[4,5] the reaction of 1,2,4,5-hexatetraene (**17**) with acetylenes (**18**) leads to substituted [2.2]paracyclophanes (**20**).

17   **18**   **19**   **20**

$$R = CO_2CH_3, \ CO_2C(CH_3)_3, \ CN, \ CF_3$$

This synthetic route can be carried out in concentrated solution, such that as much as 60 g, for example, (approx. 50% yield) of the desired [2.2]para-cyclophane is prepared in a single batch.[4c] In this way more highly substituted [2.2]paracyclophanes can be prepared; an example is 4,5,7,8,12,13,15,16-octamethyl[2.2]paracyclophane (**22**), formed in 22% yield from the cor-responding tetra ester **20a**.[6] The functional groups in **20** can subsequently be utilized for further bridging, so as to form [2n]phanes.

H3CO2C / H3CO2C — [structure] — CO2CH3 / CO2CH3

20a

1. LAH
2. PBr₃
3. LAH

H3C / H3C — [structure] — CH3 / CH3

20b

1. 'Rieche'
2. Pd/H₂

H3C / H3C — [structure] — CH3 / CH3 / CH3 / CH3

21

H3C / H3C — H3C — [structure] — CH3 / CH3 / CH3 / CH3

22

2.3.2 PROPERTIES

Because [2.2]paracyclophane in particular reflects the chemistry of cyclophanes to a significant extent,[1b,4b,7] it can be considered as the key member or the very essence of cyclophane chemistry. The two essential characteristics of [2.2]para-cyclophanes are the interaction between the π-electron systems of the two benzene-ring planes (the 'decks') and the deformation of the benzene rings, together with the consequences that result. As will be elaborated in the following sections, these characteristics are manifested in transannular directing effects in higher electrophilic substitution, neighbouring group effects of the [2.2]paracyclophane unit, *syn* additions at the aliphatic bridges, as well as isomerization, racemization, and photochemical reactions of [2.2]para-cyclophanes and their analogues. For this reason the [2.2]paracyclophanes have provided chemists with a challenging device with which to try to quantify these complicated electronic and steric effects.

The interaction between the π-systems of the two benzene rings in **3** leads to a novel, extended π-electron system spread over both rings, whose highest occupied molecular orbital (HOMO) is higher than that of the corresponding alkylbenzene as reference compound. Its lowest unoccupied molecular orbital (LUMO) lies at a lower level than that in the open chain reference molecule. This doubling of the molecular orbitals and narrowing of the HOMO/LUMO energy gap by comparison to benzene confer direct consequences on the chemical and physical properties of the [2.2]paracyclophanes. This same type of interaction, albeit weaker, exists in the [3.3]cyclophanes but is absent in the more widely separated [4.4]cyclophanes, where the individual benzene rings behave as separated, isolated π-electronic systems. With the availability of

X-ray crystallographic analysis for representation of concrete spatial relationships and of a variety of approaches to MO theory, the quantification of transannular interactions and the precise significance of their mechanism have become an ongoing problem in chemical research.

2.3.2.1 Molecular geometry

A number of X-ray crystal structure analyses of [2.2]paracyclophane (3) have been reported. They show that the four aromatic bridging atoms C(3), C(6), C(11), and C(14) are distorted out of the planes of the remaining benzene carbon atoms, with the result that the aromatic rings are deformed into boat-like shapes (see Figure 1). Figure 1 depicts the separation between the various carbon atoms as well as the angles. The intramolecular separation (h) between the central carbon atoms of the two benzene rings is shortened to 308.7–309.3 pm. (The normal van der Waals separation between parallel benzene rings is set at 340 pm as a minimum.)

This shortening is attributed to a considerable transannular π–π overlap, with the consequent reduction in the separation (g) between the bridging atoms C(3), C(14), and C(6), C(11) to only 275.1–277.8 pm (bold dots in Figure 1). As compensation for the effects of this transannular steric and electronic interaction, the CH_2-CH_2 bond length (d) in the bridge is unusually large: 163.0 pm (at 291K); 155.8–156.2 pm (X-ray crystal structure analysis at 93K).

Conformational equilibrium studies in multi-layer [2_n]phanes (see Chapter 6) have revealed that the boat conformation of a benzene ring is more stable than a chair by at least 16.7 kJ/mol. Hence the boat conformation is found to be predominant in [2_n]cyclophanes.

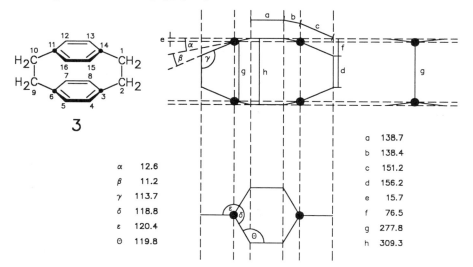

α	12.6
β	11.2
γ	113.7
δ	118.8
ε	120.4
Θ	119.8

a	138.7
b	138.4
c	151.2
d	156.2
e	15.7
f	76.5
g	277.8
h	309.3

Figure 1. Geometry of crystalline [2.2]paracyclophane (interatomic distances a–h in pm; bond angles α–θ in degrees

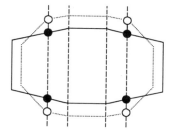

Figure 2. Oscillation in the [2.2]paracyclophane molecule (**3**), in the course of which the benzene rings approach and retreat from one another ('concertina movement')

Boyd estimated the strain energy (SE) in the [2.2]paracyclophane molecule at 129.8 kJ/mol. Although this strain is distributed throughout the molecule's framework, the largest strain resides in the distortion of the benzene rings from planarity.

From temperature-dependent X-ray data it has been shown that a 'concertina movement' is occurring in the [2.2]paracyclophane molecule, such that the benzene rings move simultaneously towards and away from each other in a direction perpendicular to the normal axis (see Figure 2).

In addition a simultaneous twisting of the benzene rings from their normal axis is occurring (see Figure 3).

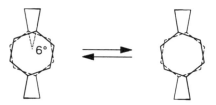

Figure 3. Twisting of the two benzene rings from the normal axis ('twisting vibration') in [2.2]paracyclophane (**3**)

2.3.2.2 NMR spectroscopy

The ^{1}H-NMR spectrum of [2.2]paracyclophane consists of two singlets of equal intensity for the methylene protons ($\delta = 3.04$) and the aromatic protons; because of the deshielding effect of the second benzene ring the latter appear at relatively high field ($\delta = 6.30$).

2.3.2.3 Electronic spectra

The electronic spectra of [2.2]paracyclophane provide insight into the extent and mechanism of transannular electronic interaction in strained π-electronic systems. The UV spectrum of [2.2]paracyclophane—by comparison with acyclic and larger ring model compounds—is characterized by bathochromic shifts, lack of fine structure, and diminished extinction coefficients. The absorption maxima appear at 225 nm ($\epsilon = 25\,000$), 244 (3163), 286 (324), and 302 (199) (see Figure 4). The absorption band at 302 nm, located beyond the long wavelength absorption in simple alkyl benzenes, is designated as the 'cyclophane band'; it is also characteristic of other $[2_n]$cyclophanes. The most important factors that contribute to the location and intensity of the cyclophane band are deduced by a comparison of the spectra in other $[2_n]$cyclophanes. From this it has been concluded that the major factors are the distance between the benzene rings and the extent and nature of their deformation.

Deeper insight into an understanding of the electronic and ESR spectra, as well as photochemical behaviour of [2.2]paracyclophane, has been provided by Gleiter.[8,15] He showed that separation of σ-and π-bonds is not justified, for the 1,2- and 9,10-bonds are parallel to the long axis of the p-orbitals of the benzene rings and thus are oriented for σ-π overlap.

Investigation of the ESR spectrum of the radical anion of [2.2]- and higher $[m.n]$paracyclophanes has shown that delocalization of the unpaired electron through both aromatic rings is impeded only when the bridging number $m > 3$. In non-bridged, open-chain model compounds $Ar(CH_2)_nAr$ the corresponding limit is $n > 1$. This finding supports the idea that two different mechanisms for the delocalization between the two benzene rings is operating for the bridged and non-bridged compounds.[8]

2.3.2.4 Chirality in substituted [2.2]paracyclophanes

$[m_n]$Paracyclophanes of type **23** are planar chiral molecules because the R-substituted benzene ring cannot rotate through the *para*-phenylene axis at ordinary temperatures. While the paracyclophane carboxylic acids **23a,b** can be separated into their torsionally stereoisomeric enantiomers, the phane **23c** with a larger bridging chain is not resolvable at room temperature.[9] Compound **23a** was assigned the absolute configuration S-4-carboxy[2.2]-paracyclophane by Falk and Schlögl in 1968.[11] Lüttringhaus established an analogous dependence between optical stability and bridging length much earlier in the *ansa* compounds {dioxa[n]paracyclophanes} of type **24**.[10]

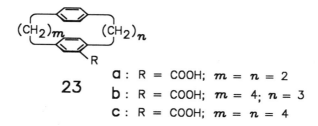

$$\text{(CH}_2)_m \qquad \text{(CH}_2)_n$$

23

a : R = COOH; $m = n = 2$
b : R = COOH; $m = 4$; $n = 3$
c : R = COOH; $m = n = 4$

24 ; (+)-(S)- 23a (-)-(R)- 23b

Both experimental measurements and calculations have been carried out concerning the circular dichroism in monosubstituted [2.2]paracyclophanes.[12] One approach has been to test whether the absolute configurations in the [2.2]paracyclophane system can be deduced by using the octant rule for the n–π* transition of the Ar—CO chromophore. In this case it is necessary to separate the n–π* effect of the 'cyclophane background' (the trough or peak from the 'naked' cyclophane skeleton itself). Tochtermann, Snatzke *et al.* have recently reported on a comparison of a variety of [n]- and [2.2]paracyclophanes and on the problems in interpretation of the CD spectra.[13]

2.3.2.5 ¹H-NMR spectroscopy

A strong shift in *all* aromatic protons is observed when a substituent in one of the aromatic rings of a [2.2]paracyclophane (3) is changed from electron-feeding to electron-withdrawing. The proton *ortho*- to the substituent experiences a high-field shift with electron-feeding substituents, and a comparable shift to lower field with electron-withdrawing substituents. Likewise the centre of absorption of the remaining aromatic protons is shifted in the same direction as that of the *o*-proton. It is of particular interest that the protons in the unsubstituted *p*-phenylene ring on the opposite side of the bridge are also shifted. This was termed the 'transannular electronic interaction'. The magnitude of the '*ortho*-shift' in [2.2]paracyclophanes lies between that for *Z*-protons in vinyl compounds and the smaller *ortho*-shift in normal benzene compounds. For this reason the *ortho*-'shifts' in [2.2]paracyclophanes were suggested as evidence for an enhanced double-bond character in the deformed benzene rings, with the possibility that depictions such as 25 are contributing structures.

In the IR spectrum of the pseudo-*geminal*-4-acetyl-13-bromo[2.2]para-cyclophane (26) the carbonyl stretching band appears at 1663 cm^{-1}, which lies somewhat out of the range found in other isomers (1666–1668 cm^{-1}). Reich

25   26

and Cram attributed this somewhat lower frequency to a transannular Br–CO interaction.

Of particular value for analysis of these phane types is the fact that low-ionization mass spectra of substituted [2.2]paracyclophanes generally display radical ions of both *p*-xylylene pieces. Thus mass spectrometry is well suited for determining the nature and number of substituents of both aromatic rings.

The UV spectra of 4,5,7,8-tetrafluoro[2.2]paracyclophane (**27**) and of the octafluoro compound (**28**) show a close relationship to the unsubstituted [2.2]paracyclophane (**3**). Bands in the region 286–291 nm are attributed to a distortion of the aromatic rings from planarity, by analogy to those in [2.2]paracyclophane. In **27** and **28** these are likewise shifted bathochromically by about 25 nm, in comparison to the fluoro-substituted open chain analogues.

Although [2.2]paracyclophane (**3**) shows only a weak shoulder at 302 nm, **27** has a band at 297 nm. The absence of absorption in this region for **28** suggests that the 297-nm band in **27** arises from a donor–acceptor transition; that is, with a transannular π–π interaction between the electron-withdrawing 'π-acid' tetrafluorophenylene- and the complementary benzene ring.

The proton resonance spectrum of **27** does not, as assumed formerly, point to transannular interactions. The fluorine atoms here are isochronous, but magnetically non-equivalent. They show a higher order spectrum (quintet at $\delta = 6.82$, $J_{H-F} = 0.8$ Hz).[13a]

The bathochromic and hypsochromic shifts in the long wavelength band of **28** are attributed both to the F–F repulsion and a higher extent of distortion in the benzene rings because of the larger size of the fluorine atoms.

2.3.2.6 [2.2]Paracyclophanes with substituents in the bridge

The [2.2]paracyclophanes containing substituents on the bridges have been less extensively investigated. In the course of synthesizing [2.2]para-cyclophandienes[1b,14] (see below) Cram prepared the bis-*geminal* dibromides **29a,b** and from them the diketones **30a,b**. The monoketone **31** is also known. The UV spectra of **29a,b** have maxima at 236 nm, which, by comparison with other bridge brominated [2.2]paracyclophanes, are assigned to the 225 band in the paracyclophane that has undergone a bathochromic shift through the inductive effect of the bromines.

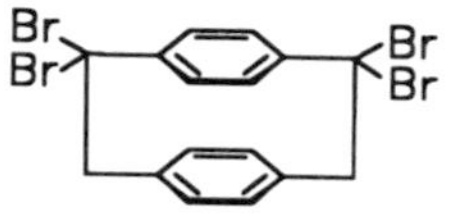

29a 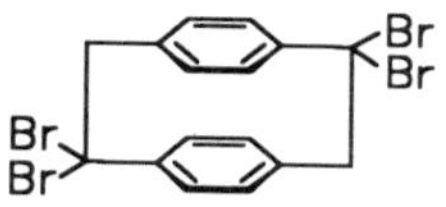   29b

In the ketones **30a,b,** with maxima at 235 nm, a similar effect by the carbonyl is plausible, when the position and intensity of these bands is attributed to the benzoyl chromophore.

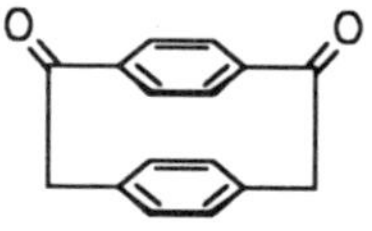

30a 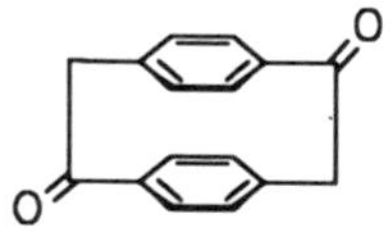   30b 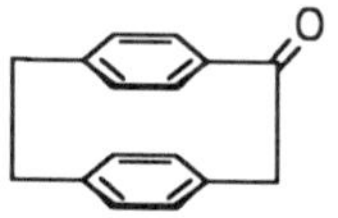   31

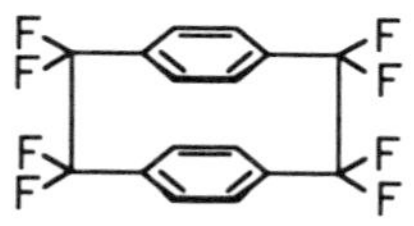

32 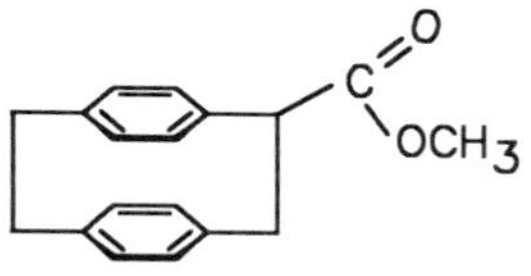   33

The bridge-fluorinated octafluoro[2.2]paracyclophane **32** resembles other [2.2]paracyclophanes in its spectroscopic properties (e.g. UV spectrum). The single absorption peak in the ^{1}H-NMR spectrum at $\delta = 7.3$ is substantially shifted from that in the unsubstituted [2.2]paracyclophane ($\delta = 6.3$). A strong band at $718\,cm^{-1}$ appears in the IR spectrum, as characteristic for other [2.2]paracyclophanes such as methoxycarbonyl[2.2]paracyclophanes **33** and **34.** In the parent hydrocarbon **3** the band appears at $725\,cm^{-1}$, but it is absent in the linear poly-*p*-xylylene **35.**

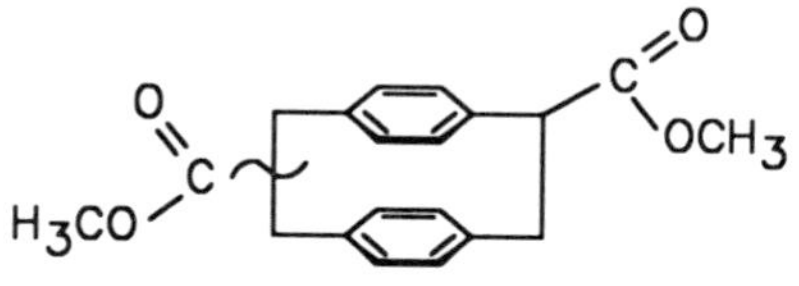

34

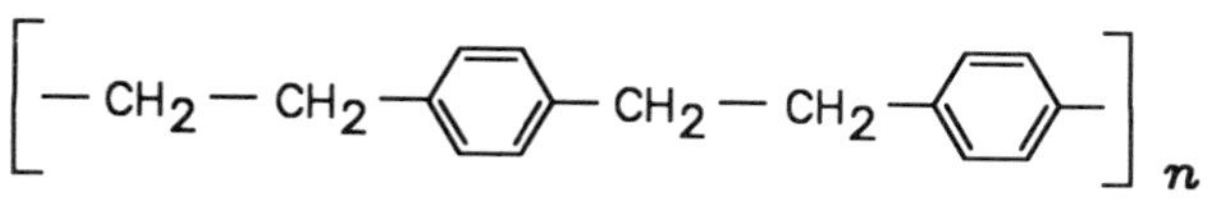

35

2.3.2.7 [2.2]Paracyclophan-1-ene and [2.2]paracyclophan-1,9-diene[16,14]

In a broad sense the ene **36** and the diene **37** can be grouped with the bridge-modified [2.2]paracyclophanes. Their preparation by way of dithiane alkylation or from dithia[3.3]paracyclophanes via the Stevens rearrangement was described in Section 2.3.1.

36   **37**

The UV spectra of [2.2]paracyclophan-1-ene (**36**) and [2.2]paracyclophan-1,9-diene (**37**) are very similar to that of [2.2]paracyclophane (**3**) itself (see Figure 4),[1b] although less definitive because the individual bands in the unsaturated paracyclophanes more closely overlap. They do not, however, exhibit the characteristic absorption of the open chain analogue Z-stilbene, because their fixed orientation between the benzene rings and C—C double bonds prevents substantial overlap according to the Hückel model.

Because of this orthogonal orientation of the π-electrons and its ring strain, [2.2]paracyclophan-1,9-diene (**37**) belongs to the interesting class of low molecular weight bridged aromatic compounds with a high order of symmetry. It has been shown from the X-ray crystal analysis that the benzene rings are

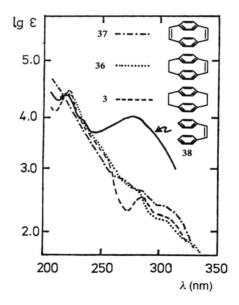

Figure 4. Comparison of the UV spectra of **3**, **36**, **37**, and Z-stilbene (**38**)[1b]

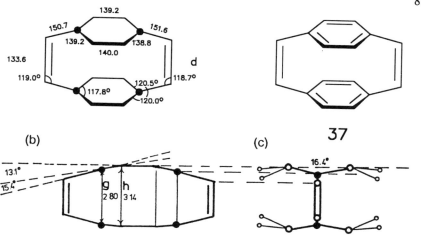

Figure 5. Geometry of [2.2]paracyclophane-1,9-diene (**37**)

distorted into boats, such that the bridging atoms lie between 16.6 and 17.8 pm out of the plane of the four unsubstituted aromatic carbons. The corresponding value in [2.2]paracyclophane (**3**) is considerably smaller (15.7 pm; see Figure 5).[1b,1c,7]

It is remarkable that the distances g and h between the two benzene nuclei in molecule **37**, 280 and 314 pm, respectively, are longer than those in [2.2]paracyclophane **3** (275 and 309 pm). The C=C double bond length of 133.6 pm is comparable to that in other alkenes (133.4–133.7 pm); and the

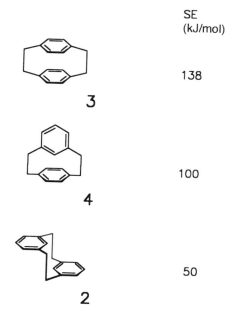

SE
(kJ/mol)

3   138

4   100

2   50

average C—C bond length in the benzene rings (139.9) is scarcely different from that in benzene (139.3–139.7 pm). Although the average C(aryl)–C(methine) bond length is comparable to that in toluene, it is nevertheless somewhat stretched by comparison with other similar systems. This can be attributed to decreased conjugation between the olefinic double bond and the benzene ring. Such a diminution in conjugation is expected because of the orthogonality between the π-orbitals of both unsaturated units, as is manifested in the UV spectrum of the diene (see above).

Total strain energy (SE) in the diene **37** is estimated at 163 kJ/mol. The ring strain appears not to be evenly distributed over all bonds and bond angles but rather concentrated mainly in the deformation of the benzene rings and somewhat less through lengthening and deformation of the C(aryl)–C(methine) bonds and torsional angles.

Strain energies of [2$_n$]phanes and their oligomethyl derivatives have been compared by Hopf:[4b] the experimentally determined strain energies on the previous page are probably those of the most important [2.2]phanes (see above).

Certain general observations can be made about the *photoelectron spectroscopy* of [2.2]phanes. In Figure 6 the PES spectra of [2.2]para-cyclophane **(3)**, its diene **37** and 'cyclopropa[2.2]phane' **39** are shown. The broad maximum present in all three at 8 eV is probably attributable to three closely lying π^{-1} ionizations.

Gleiter[15c] interpreted the photoelectron spectrum of **39**, the 'cyclopropano analogue' of [2.2]paracyclophane (see Figure 6), as demonstrating a strong σ–π interaction between the cyclopropano bridge and the benzene 'decks'. The longest wavelength band in **39** is not very different from that in [2.2]paracyclophane **(3)**. The only evidence for an unusual σ–π interaction is the enhanced intensity of the 'cyclophane band' at 304 nm ($\epsilon = 900$); compare [2.2]paracyclophane: $\lambda_{max} = 302$ nm ($\epsilon = 199$).

From a comparison in the [2.2]paracyclophane series it seems clear that the **39** 'through-bond interaction' is stronger than that in [2.2]paracyclophane **(3)**, as is also found for [2.2]paracyclophan-1-ene **(36)** and the diene **37**. In a comparative PES study of alkyl-substituted [2.2]paracyclophanes, furthermore, it is indicated that hyperconjugation plays a role but that the inductive effect of methyl groups is insignificant.[15]

2.3.2.8 Transannular interactions in the course of chemical reactions

Besides conventional chemical reactions leading to new bonds, there are also certain chemical transformations in the [2.2]paracyclophane series which manifest themselves in the formation of charge-transfer complexes.

2.3.3 REACTION IN THE AROMATIC RING[1b,4b,7]

2.3.3.1 π-Complex formation

Cram in particular posed the question as to whether and how electronic effects

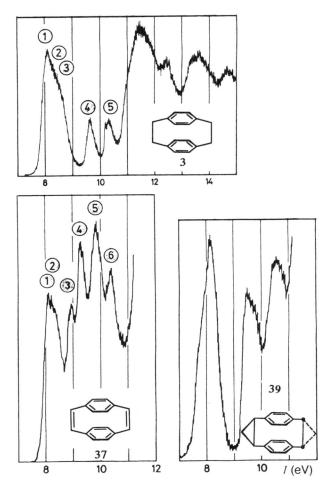

Figure 6. Comparison of the PES spectra of [2.2]paracyclophane (**3**), its diene **37** and [1:2,9:10]bis(methano)[2.2]paracyclophane (**39**)[15]

exerted by substituents in one ring could be transferred to the second ring. In a series of tetracyanoethylene (TCNE) π-complexes with [2.2]paracyclophanes the π-base strength was determined from the position of the longest wavelength 'charge-transfer' band in the UV spectrum and the association constants. From this it was shown that the complexes between tetracyanoethylene with [2.2]paracyclophanes containing electron-feeding groups (type A) were more stable than with those containing electron-withdrawing groups (type B). On the basis of the difference in π-basicity of the substituted and unsubstituted rings, the authors assume that, in complex A the substituted ring participates in the $\pi-\pi$ participation with TCNE, while in complex B the unsubstituted ring is more likely associated with the TCNE. This assumption is supported by the

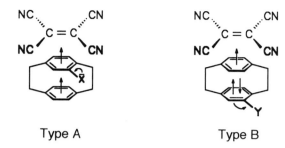

Type A   Type B

Figure 7. π-π Complexes with electron-donating substituents X (**A**) and with electron-acceptor substituents Y (**B**)

observation that the complex with 4-acetyl[2.2]paracyclophane is more stable than that with *p*-xylene, in spite of the electron-withdrawing substituent. The competing effects—electron donation by the benzene ring associated with TCNE on the one hand and electron withdrawal of substituents (acetyl, nitrile) on the other—are depicted in Figure 7 with appropriate arrows.

For both types of complexes **A** and **B** there is moreover a nearly linear relationship between the transition energy (E_T) of the 'charge-transfer' band and the Hammett substituent constants (σ_m), to which the authors ascribe the existence of *transannular substituent effects*.

2.3.3.2 Transannular directing effect in electrophilic aromatic substitution[7]

More direct evidence from chemical reactions for transannular electronic and steric effects has also been provided. Transannular effects in the [2.2]paracyclophane series are characterized by the steric or electronic effect exerted by one aromatic ring in the course of reactions taking place in the other aromatic ring. The strong dependence of transannular interactions on molecular geometry is evident.

Determination of the relative rates of acetylation in the [m_n]paracyclophane series reveals a remarkable situation. With the rate for [6.6]paracyclophane **23** (m = n = 6, R = H) set as 1, the relative rates for the following paracyclophanes are:

> [4.4]paracyclophane: 1.6
> [4.3]paracyclophane: 11
> [2.2]paracyclophane: >48

Thus, the closer the benzene rings are to each other, the more steeply the rate rises for the incorporation of the first acetyl group, while the rate for attack of the second acetyl group falls off. Both observations clearly support the involvement of an intermediate **40** for monoacetylation in which the partial positive charge is distributed over both benzene rings.

Reich and Cram studied the distribution of products from electrophilic substitution of monosubstituted [2.2]paracyclophanes. It was shown that

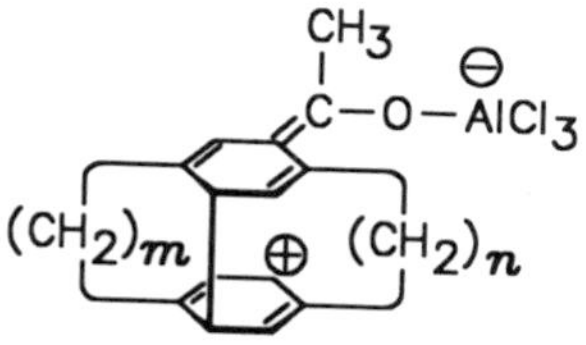

40

there was no correlation between the transannular directing effects of the substituents and their resonance effect in the ground states. Product distribution predicted from electrostatic ground-state models, such as canonical structures **41** for electron-feeding groups and **42** for electron-withdrawing groups, does not correspond to the experimental results.

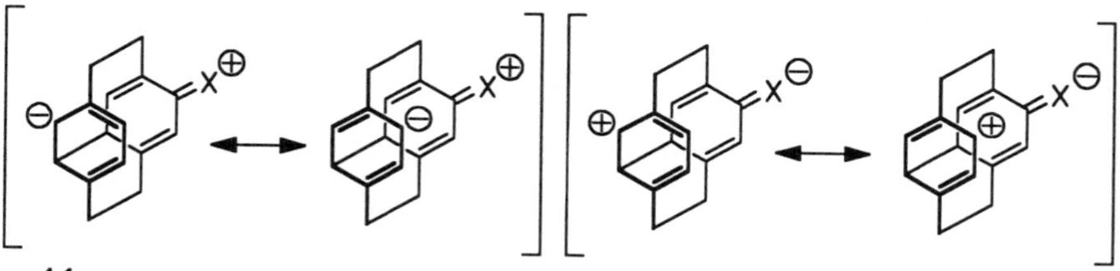

41 (X : electron−donating)   **42** (X : electron−accepting)

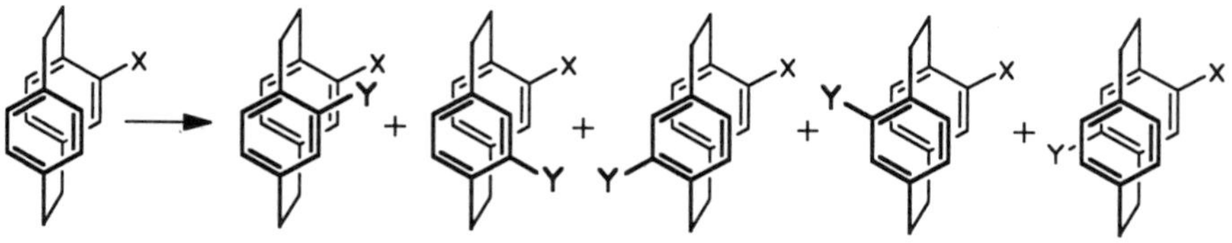

| pseudo− | pseudo− | pseudo− | pseudo− | para |
| geminal | ortho | para | meta | |

	X			Y	
43 :	Br		**44** :	Br	
43 :	Br		**45** :	COCH$_3$	
46 :	COOCH$_3$		**47** :	Br	
48 :	COCH$_3$		**49** :	Br	
50 :	NO$_2$		**51** :	Br	
52 :	CN		**53** :	Br	

Figure 8. Product distribution in the electrophilic substitution of monosubstituted [2.2]paracyclophanes

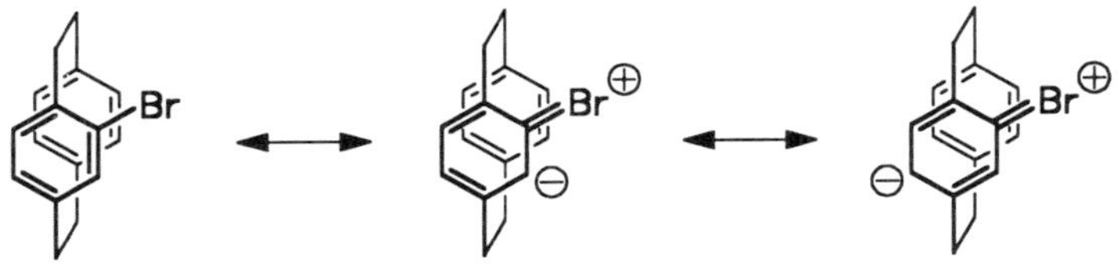

Figure 9. Resonance structures for 4-bromo[2.2]paracyclophane (**43**)

If one considers a limiting structure such as **41**, then attack of a second electrophile would be expected to occur in the pseudo-*m*- and pseudo-*geminal* position. What is observed, however, is predominant formation of pseudo-*o*- and pseudo-*p* products, as, for example, in the bromination and acetylation of 4-bromo[2.2]paracyclophane (**43**) (Figure 8).

The limiting structure **42** suggests the favoured locations for electrophilic attack to be the pseudo-*o*- and pseudo-*p*-positions. On the contrary, bromination of 4-methoxycarbonyl- (**46**) and 4-acetyl[2.2]paracyclophane (**48**) affords exclusively the pseudo-*geminal* product.

The experimental results can be rationalized by assuming that substitution in the [2.2]paracyclophane takes place predominantly pseudo-*geminal* to the most basic aromatic carbon or to the most basic substituent. The most basic positions in the bromo compound are those *o*- and *p*- to the bromine (see Figure 9); and, in fact, both bromination and acetylation lead predominantly to the pseudo-*o*- and pseudo-*p*- products.

Cram termed such polar formulas 'transannular resonance structures'. In nitro- and carbonyl-substituted [2.2]paracyclophanes, in which the most basic site is the oxygen of the substituent, the substitution proceeds at the pseudo-*geminal* position.

Correlation between the transannular directing effect and the basicity of the position which is pseudo-*geminal* to the original substituent suggests that a basic centre in the molecule can participate as a neighbouring group in the product-determining step of the substitution reaction. According to Cram the substituted ring can function in this way. The mechanism for pseudo-*p* substitution in electrophilic attack on 4-bromo[2.2]paracyclophane (**43**) or its monodeuterio analogue **54** serves as an example (Figure 10).

The electrophile E^+ attacks at the unhindered position of the unsubstituted aromatic ring. A proton (deuteron) is shifted in a transannular fashion (from one ring to the other), while the proton already located on the second ring shifts in the same direction. In this fashion the incoming electrophile and the departing proton follow the least hindered pathways in the [2.2]paracyclophane system. Deuterium migration could be confirmed experimentally in the bromination of 4-methyl[2.2]paracyclophane (**55**). The kinetic isotope effect k_H/k_D found for bromination of *p*-protio- and *p*-deuterio-4-methyl[2.2]paracyclophane in various solvents provides support for this mechanism, in particular that the C—H bond breaking in the σ-complex is the rate-limiting step.

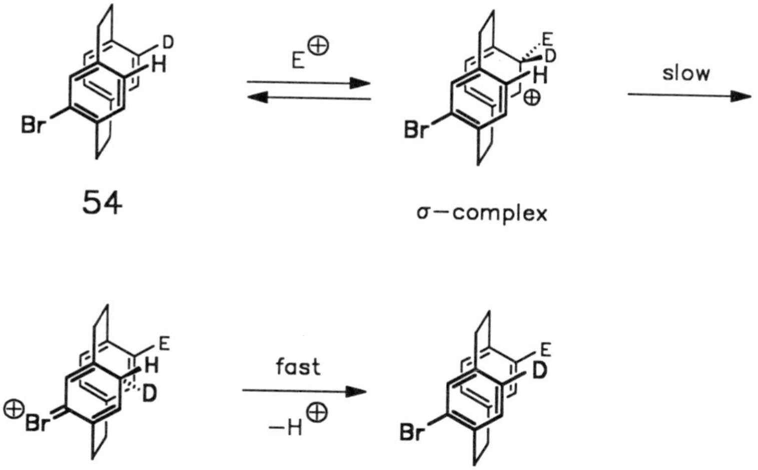

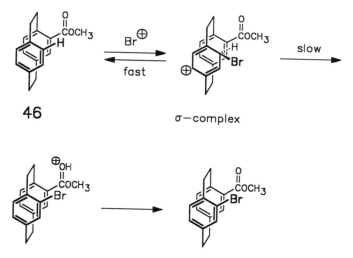

Figure 10. Mechanism for pseudo-*p*-electrophilic substitution

The most striking example of transannular directing effect is provided by the exclusive or predominant pseudo-*geminal* substitution in bromination of [2.2]paracyclophanes containing O-basic substituents (carboxy, methoxy, acetyl, nitro). The oxygen atom of the functional group is favourably situated to accept a proton from the pseudo-*geminal* position in the σ-complex. A variation of the above mechanism can be formulated for the example of bromination of the ester **46** (Figure 11). Here the oxygen atom of the electron-withdrawing substituent assumes the role of the intramolecular base, assisting in proton transfer. The linear cyano group, on the contrary, cannot orient itself

Figure 11. Mechanism of electrophilic substitution leading to pseudo-*geminal* product[7]

geometrically to accept a proton in a transannular fashion; indeed, bromination of **52** (see above) affords no pseudo-*geminal* disubstitution product.

2.3.3.3 Transannular reactions

The unique molecular geometry of the [2.2]paracyclophane system can be seen from the fact that no transannular ring closure takes place, although in the case of [4.4]paracyclophane **(56a)** the ring closure **56–57** proceeds in 28% yield.[7]

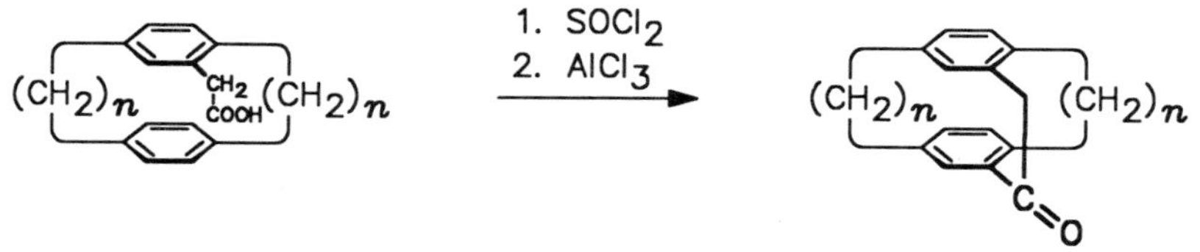

56a : $n = 4$
56b : $n = 2$

57a : $n = 4$
57b : $n = 2$

Forrester and Ramasseul have postulated a two-atom bridged intermediate as being involved in a [2.2]paracyclophane system. In the course of synthesizing the bis-hydroxylamine **58**, a precursor for the diradical **59**, they isolated as byproduct the pseudo-*p*-hydroxy compound **60**, which could be transformed in air or with silver oxide into the violet quinone **61**.

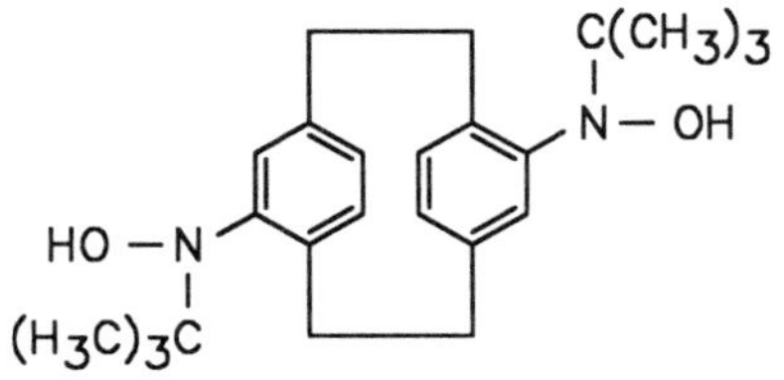

58

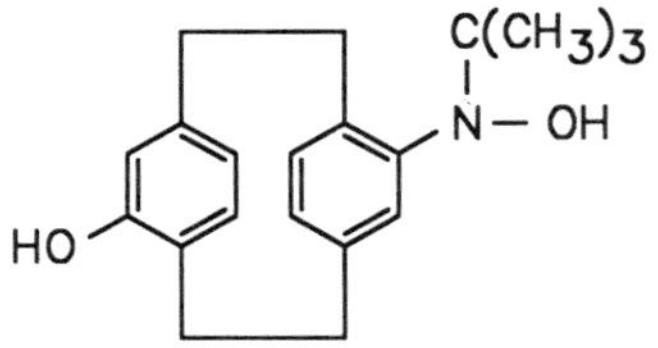

60

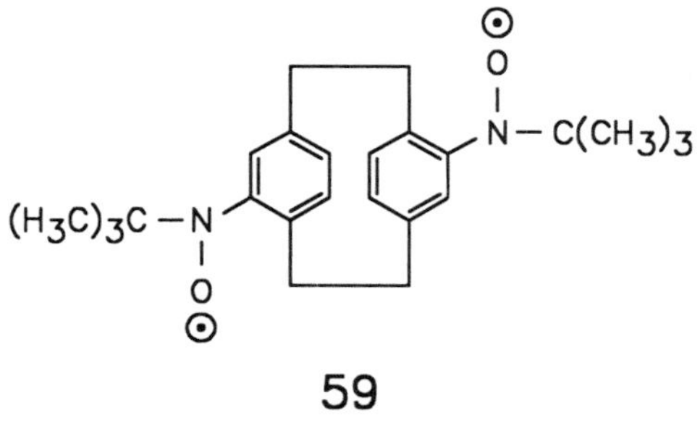

59

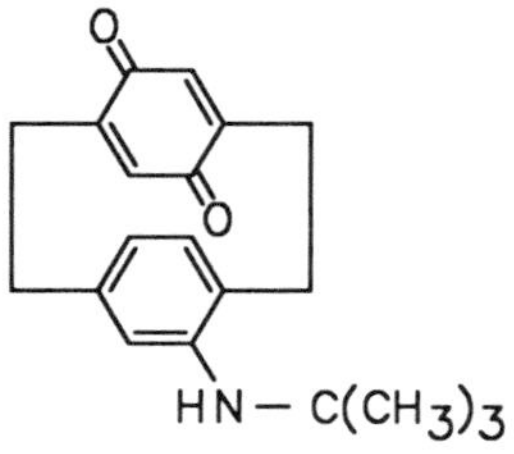

61

The following 'transannular' mechanistic pathway was proposed:

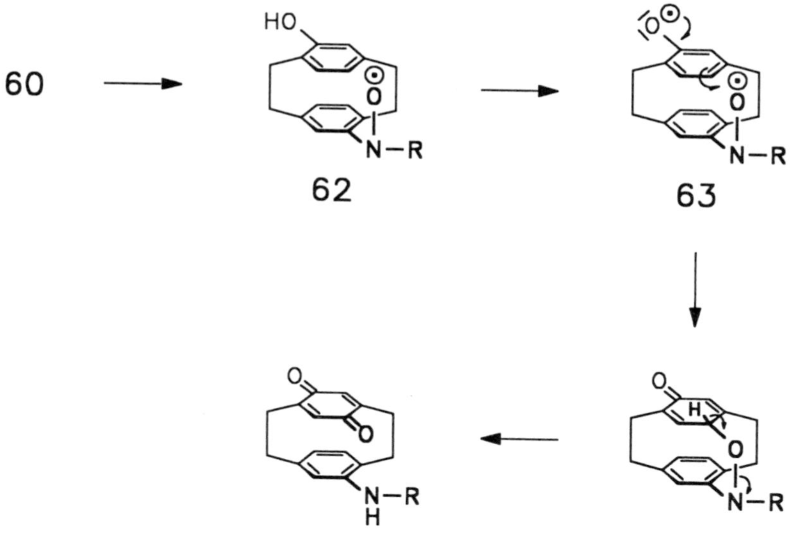

60 $\longrightarrow$

62   63

61   R = C(CH₃)₃   64

The crucial step is the transannular migration of the N-oxide oxygen (63→64). Such an intramolecular attack seems plausible because, as models show, the oxygen atom of the N–O group in 63 can approach the pseudo-*geminal* position in the other ring to within 260 pm. The analogous intermolecular oxidation of phenols to quinones by N-oxides is known.

The black *o*-quinone 66 very likely is formed in a similar fashion from the pseudo-*o*-substituted [2.2]paracyclophane 65.

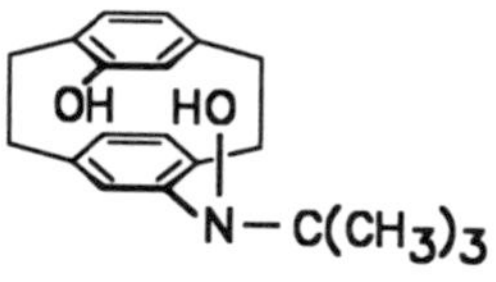

65 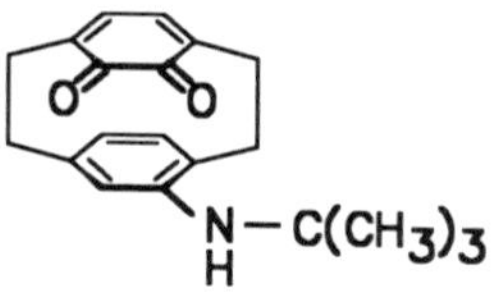   66

Hopf *et al* have investigated the epoxidation and de Meijere *et al.* the cyclopropanation of [2.2]paracyclophane and the diene (see below).[4b]

2.3.4 REACTIONS IN THE SIDE CHAIN[1b,4b,7]

2.3.4.1 Substitution, addition, and elimination

Cram *et al.* studied solvolysis reactions of optically active 1-tosyloxy[2.2]-paracyclophane (6)* and, as expected, found that all the transformations

---

*Compound numbers 1–6 in Sections 2.1–2.7 are otherwise reserved for the [2.2]phane hydrocarbons.

$7 \rightarrow 6$, $7 \rightarrow 8$, $7 \rightarrow 9$, and $7 \rightarrow 10$ proceeded with complete retention of configuration. Quite unexpected were the results of methanolysis, acetolysis, and trifluoroacetolysis of the tosylate **6**. Cram obtained the methyl ether **8**, the acetate **9**, and the trifluoracetate **10** with the same configuration and optical purity as those derived directly from the alcohol **7**. Thus, these solvolysis reactions at the bridge atom of [2.2]paracyclophane proceed with full retention. Furthermore, the rate of acetolysis of tosylate **6** is substantially faster than normal (aliphatic secondary tosylate); it is about 100-fold faster than that of 2-butyl tosylate and comparable to that of α-phenylneopentyl tosylate.

The high stereospecificity in the solvolyses of the tosylate **6,** together with the unexpectedly high rates, led Cram to propose the participation of the β-phenyl ring in the ionization step of the optically active carbon site, with

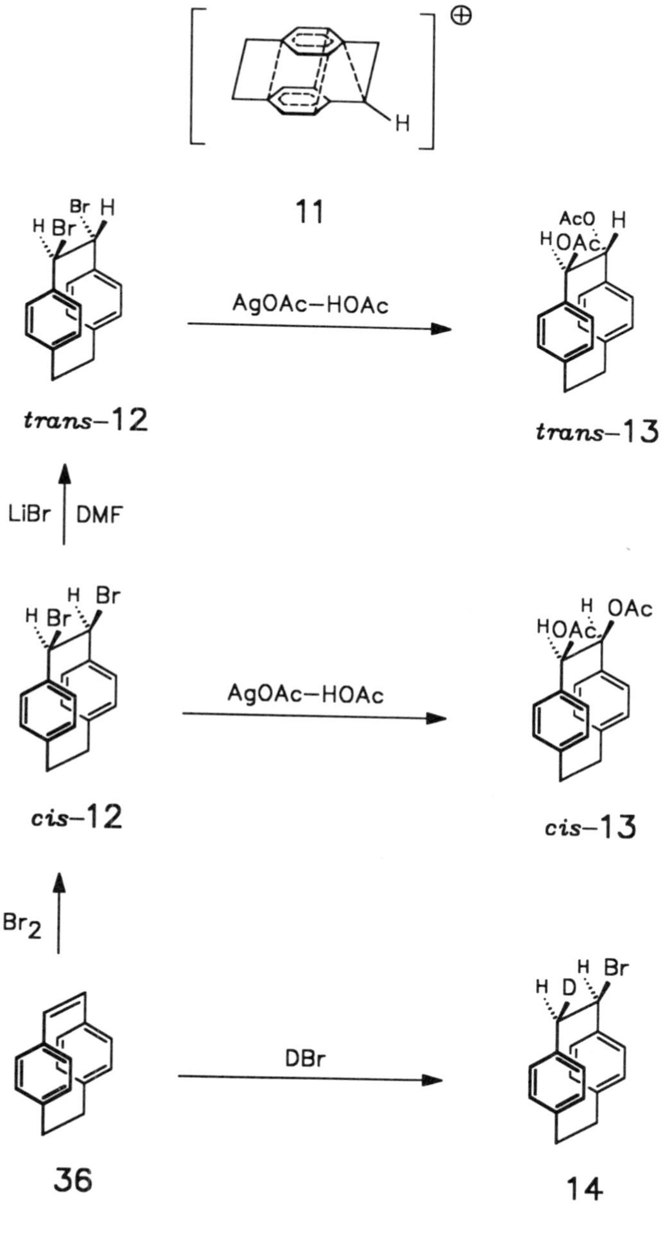

11

trans-12   AgOAc–HOAc   *trans*-13

LiBr | DMF

cis-12   AgOAc–HOAc   *cis*-13

Br₂

36   DBr   14

formation of the highly strained *bridged carbonium ion* **11**. This, in turn, would be opened in a second step leading to product. Both the formation of **11** and its ring opening must proceed with clean inversion, in order that the overall reaction is one of retention.

The positive charge in cation **11** can be distributed over both aromatic rings. Its formation reflects a compensation for the bond angle strain, which in the starting material offsets the π–π repulsions between the two benzene rings. The twisted skeleton in **11** corresponds to the somewhat distorted [2.2]para-cyclophane molecule as seen in the crystal structure.

According to Cram, the stereochemical course of some polar additions and substitutions in the bridge of the [2.2]paracyclophane system is best explained by involvement of structures analogous to **11**. For example, treatment of [2.2]paracyclophane-1-ene with bromine or deuterium bromide leads exclusively to the *cis* addition products **12** and **14**, which in turn undergo acetolysis with retention of configuration. *cis*-1,2-Dibromo[2.2]paracyclophane (*cis*-**12**) is converted with lithium bromide in DMF exclusively to *trans*-**12**, which, likewise affords only *trans*-diacetate (**13**) on acetolysis.

2.3.5 [2.2]PARACYLOPHANYL AS NEIGHBOURING GROUP[7]

Cram and Singer investigated the ability of a [2.2]paracyclophane to function as a neighbouring group in systems such as **15a,b** leading to carbenium ions.[7] Solvolysis rates and activation parameters demonstrated that the [2.2]paracyclophanyl group was a more effective neighbouring group than phenyl in the open-chain analogues **16a,b**. The paracyclophanyl bromide (**15a**) provides a stronger aryl participation in charge delocalization than is the case with **16a**.

a : R = CH_2Br
b : R = CH_2–CH_2–OTos
c : R = CH_2–CD_2–OTos

15   **16**

Hydrolysis of optically pure bromide **15a** in dioxane/water affords optically pure alcohol. This suggests that the transannular *p*-xylylene ring participates through π–σ charge delocalization (π–σ resonance) in carbenium ion formation, as depicted in **19**, and not through a direct displacement of the bromide to give a transannular bridged ion such as **17a**, for in the latter case the product would be racemic.

Cram and Singer also found a remarkable difference between substrates **15b** and **16b**. While the α-deuterium atoms in the model compound (**16c**) were extensively redistributed to the α-position (46 and 26%) during acetolysis and hydrolysis, the deuterium remained completely in the α-position during acetolysis and formolysis of **15c**.

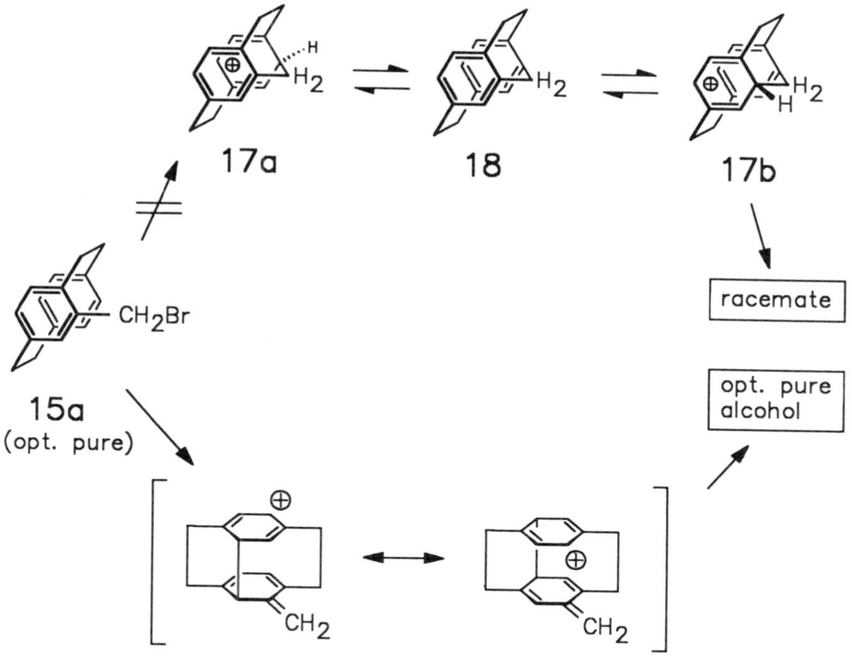

19

In a closer study of the stereochemistry of paracyclophanyl phenonium ions, Nugent et al.[12] raised the question whether the phenonium ion is formed and then converted to product *via* the *exo* or *endo* methylene (or deuteriomethylene) group. Determination of formolysis rates with 1-tosyloxymethyltetralin (**20**) and *exo*-**21a** and *endo*-17-tosyloxymethyl-4,5-tetramethylene[2.2]paracyclophane (**21b**) confirmed the neighbouring group participation of the paracyclophane: **21a,b** underwent solvolysis at faster rates than did **20**. It was also shown that carbocation formation is favoured with the *exo* configuration: *exo*-tosylate **21a** solvolysed seven times as fast as the *endo*-tosylate **21b**.

20

exo–**21a** : X = H; Y = CH_2OTos

endo–**21b** : X = CH_2OTos; Y = H

From the product analysis it was concluded that bridged ions of the type **22a,b** are involved, and that these are attacked by nucleophiles from the *exo* side because of steric and electronic factors.

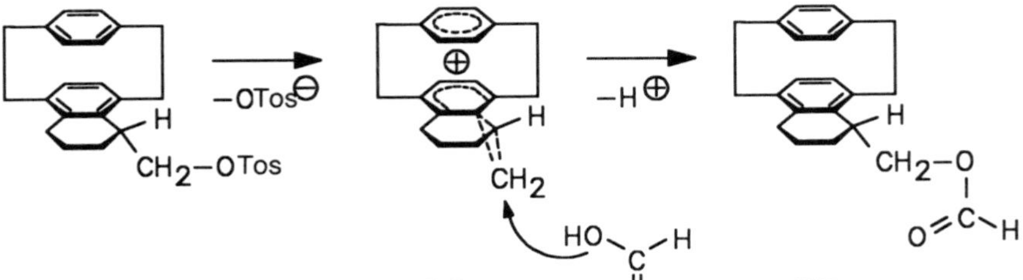

exo–21a   22a   23

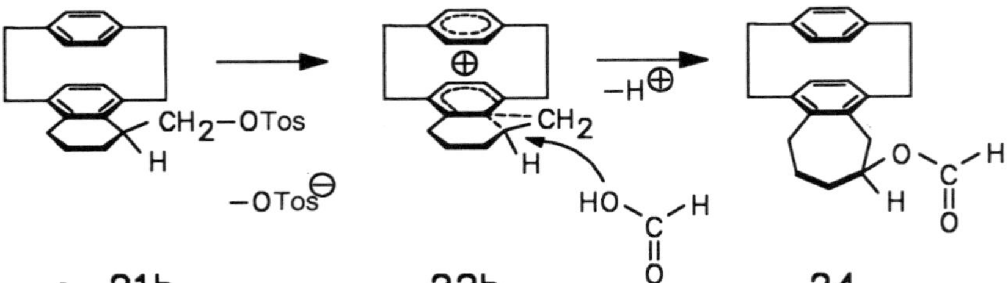

endo–21b   22b   24

2.3.6 PHOTOCHEMICAL REACTIONS OF [2.2]PARACYCLOPHANES

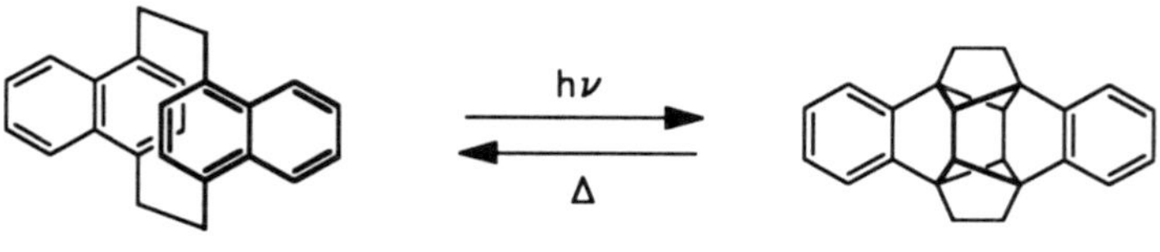

anti–25 (A)   26

Although the *anti*-isomer of [2.2](1,4)naphthalenophane (**25A**) affords dibenzoequinene (**26**) in 25% yield by photolysis (see Section 2.6), irradiation of [2.2]paracyclophane under a variety of conditions leads only to open-chain fragmentation products of **3**. Compound **27**, the analogue of equinene, was not found. Cram and Delton concluded that intermediates analogous to **27** are not intermediates in the photoracemization of various optically active ring- and bridge-substituted [2.2]paracyclophanes.

27

2.3.7 CYCLOADDITIONS OF [2.2]PARACYCLOPHANES AND THEIR ANALOGUES

2.3.7.1 Diels–Alder reactions[1b,1c,4b]

Ring strain in [2.2]paracyclophane lowers the activation energy barrier for the uncatalysed cycloaddition of dicyanoacetylene, so that 1:1 and 1:2 adducts are formed at relatively low temperatures. At 120°C a mixture of **28** and **29** is formed in 71% yield; at 170°C the only product is **29**, which can be sublimed at 210°C without decomposition.

28   **29**

2.3.7.2 Cycloaddition with benzynes

Brewer, Heaney, and Marples isolated the mono adduct **32** from the reaction of [2.2]paracyclophane (**3**) with tetrafluorobenzyne (**31**). With excess pentafluorophenylmagnesium bromide a mixture of adducts **33** and **34** was obtained.

The intermediate formation of 4,5-dehydro[2.2]paracyclophane (**35**) was demonstrated by Longone and Chipman through a trapping experiment with anthracene. *Steric overcrowding* in [2]paracyclo[2](1,4)triptycenophane (**36**) causes a deformation in the [2.2]paracyclophane portion.

30   **31**

32   **33** : X = F
                          34 : X = H

35   **36**

2.3.7.3 Reduction

Catalytic hydrogenation of [2.2]paracyclophane proceeds anomalously. The first four moles of hydrogen are taken up considerably more rapidly than the last two. Thus the reduction can be conducted so as to stop at octahydro[2.2]paracyclophane (**38**). This unstable, waxy substance, isolated in up to 91% yield, is, according to Cram, most likely **38a** or **38b**. The UV spectrum is consistent with an unconjugated diene; and, among all the conceivable isomers only these two can be constructed with molecular models. This 'cyclohexenophane' **38*** also appears to be the product from reduction of [2.2]paracyclophane with lithium in ethylamine.

Exhaustive hydrogenation (H_2/Pt, acetic acid) affords the perhydro compound **39**. Molecular models indicate that the inner cavity of this '*[2.2]cyclohexanophane*' is sterically highly strained. Van der Waals radii overlap considerably, and bond angles are likely deformed. Boyd *et al.* have calculated favourable conformations and strain energies for **39**.

Jenny and Reiner isolated a di- and tetrahydro compound (**40, 41**) by Birch reduction of [2.2]paracyclophane. On the basis of spectroscopic results the authors ascribe to **40** the spatial structure **40** (right), with a somewhat boat-like distorted benzene ring.

---

*For designation of 'aliphatic phane analogues' ('aliphane') see Chapter 9.

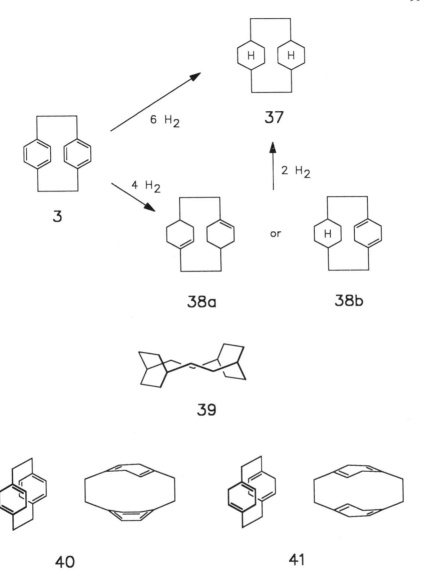

37

6 H₂

4 H₂

3

2 H₂

38a   or   38b

39

40   41

The tetrahydro product **41** is most likely depicted as the spatial arrangement shown above. Constitutional isomers can be excluded on the basis of ¹H-NMR spectroscopy. It is striking that **41**, when heated to 150–160°C, disproportionates to **40** and [2.2]paracyclophane.

At this point in the description of ring-substituted [2.2]paracyclophanes it is appropriate to mention the exceptional crown ether derivatives **42–45**, of which **43** as 'masthead formula' adorns the cover of the volumes *Cyclophanes*, often cited in this book.[1d-f]

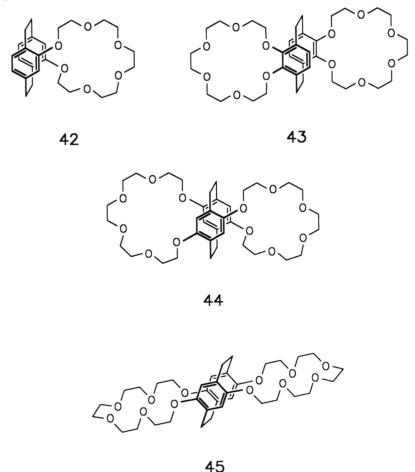

42

43

44

45

2.3.8 TRANSITION METAL COMPLEXES OF [2.2]PARACYCLOPHANES[1b,16,17]

Early work of Cram and Wilkinson[19] showed that [2.2]paracyclophane reacts rapidly with chromium hexacarbonyl, with formation of **46**; and Misumi[20] reported the synthesis of the double chromium carbonyl complex **47**. By means of the metal-vapour technique, Elschenbroich was able to prepare the unusual complex **48**, containing a chromium between the two phenylene rings, as well as the complex **49**, in which a chromium is bridged between two cyclophane molecules.[18]

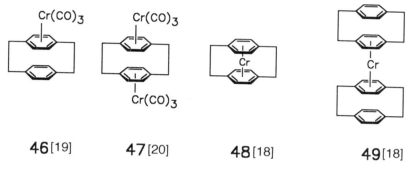

46 [19]   **47** [20]   **48** [18]   **49** [18]

Recently several dozen cyclophane ruthenium complexes with general structures **50a**, **51a**, **52**, and **53** have been prepared, as well as iron complexes **50b** and **51b**. The Fe(II)- and Ru(II)-complexes of [2_n]cyclophanes are stable to heat and light and can be handled without decomposition.

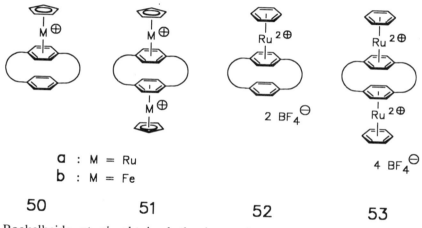

a : M = Ru
b : M = Fe

50   **51**   **52**   **53**

Boekelheide *et al.* obtained the interesting complex bis(η^6-hexamethyl-benzene)-{η^6,η^6-[2.2](1,4)cyclophane}diruthenium(II,II)tetrafluoroborate (**54**).[22] Through a two-electron reduction of this tetra cation, the benzene ring is transformed into a cyclohexadiene ring, while a bridging cyclobutane ring is generated. Figure 12 depicts the results of the X-ray crystal analysis of **55**.

The bond distance *d* of the 'inner' cyclobutane bond, formed between the former aromatic bridging carbons C(13) and C(23), is remarkably long. With a value of 196 pm, it can be considered the longest C—C bond found to date (normal value 154 pm).[22]

2.3.9 REACTIONS PROCEEDING WITH RING RUPTURE

2.3.9.1 Thermal isomerization and racemization[1a,1b,4b]

The high strain energy in [2.2]paracyclophane (**3**) facilitates ring opening by cleavage at the benzyl position. Pyrolysis at 400°C leads to

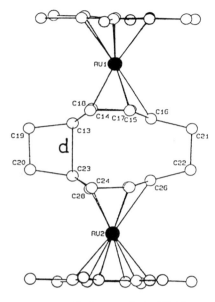

Figure 12. Molecular geometry of the product **55** {formed by reduction of the bis(ruthenium) complex **54** from [2.2]paracyclophane}[22]

4,4'-dimethylbibenzyl and 4,4'-dimethylstilbene. At 600°C p-xylylene (**58**) is generated, and it polymerizes spontaneously to poly-p-xylylene (**59**).

$$3 \xrightarrow{400°C} \left[H_2\overset{\odot}{C} \mathllap{-}\!\!\!\!\!\!\!\!\!\!\!\!\bigcirc\!\!\!\!-\!\!\!\!\!\bigcirc\!\!\!\!\!\overset{\odot}{C}H_2 \right] \longrightarrow H_3C \!-\!\bigcirc\!\!-\!\!\bigcirc\!\!-\!CH_3$$

56   **57**

$$3 \xrightarrow{600°C} H_2C{=}\!\!\bigcirc\!\!{=}CH_2 \longrightarrow {-}\!\left[{-}H_2C\!\!-\!\!\bigcirc\!\!-\!CH_2{-} \right]_n\!\!{-}$$

58   **59**

Reich and Cram carried out a series of experiments whose results point to the formation of the 4,4'-dimethylenebibenzyl radical (**56**) as an intermediate. Heating of [2.2]paracyclophane at 250°C in p-diisopropylbenzene afforded 4,4'-dimethylbibenzyl as the only isolated non-polymeric product in 21% yield.

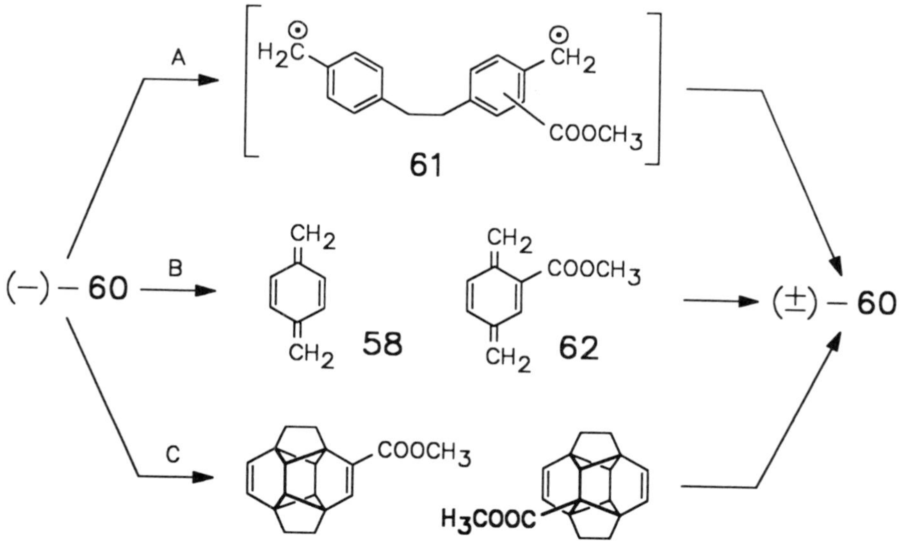

63 or **64**

When $(-)$-4-methoxycarbonyl[2.2]paracyclophane (**60**) is heated at 200°C it undergoes racemization without decomposition, the rate being relatively insensitive to solvent polarity (dimethyl sulphoxide, tridecane). Because of the rigid ring structure, racemization can only proceed by a ring opening followed by a recombination. The activation energy for the process was determined to be 159 kJ/mol. Besides the homolytic cleavage of a benzyl bond to give diradical **61** and its recombination after rotation of the benzene ring (path A), two other routes may be considered:

(B) cleavage to *p*-xylylenes **58** and **62** (the *p*-xylylene mechanism);
(C) by way of the polycyclic equinenes **63** and **64**, which could arise by intramolecular cyclization.

Investigation of the thermal isomerization of a variety of disubstituted [2.2]paracyclophanes at 200°C by Reich and Cram revealed that only the diradical mechanism (A) was consistent with the results. Thus, the isomerization of pure pseudo-*geminal* or pseudo-*m*-isomer gave the identical equilibrium

A:

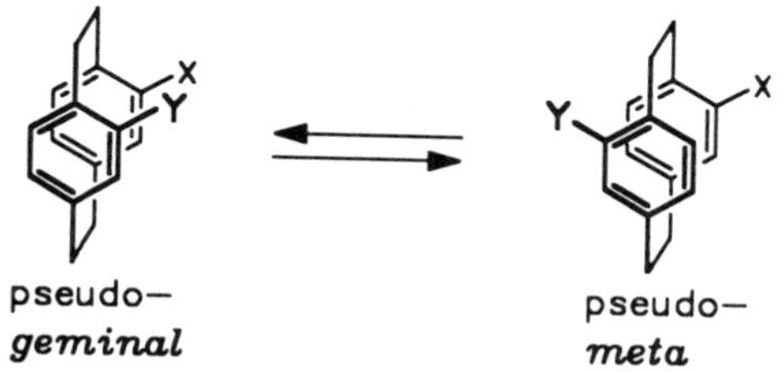

pseudo—
geminal

pseudo—
meta

$(X = Br, Y = COCH_3)$

B:

pseudo—
ortho

pseudo—
para

mixtures, as was the case for the pseudo-*o*- and pseudo-*p*-isomers, which can likewise be interconverted.

A subsequent isomerization of the system A to B—that is, from pseudo-*geminal* or pseudo-*m*- to pseudo-*o*- or pseudo-*p*-isomers—was not observed, a fact which is not consistent with the *p*-xylylene mechanism.

2.3.9.2 Ring expansion

Evidence pointing to the intermediacy of the diradical comes from Reich and Cram's experiments of heating [2.2]paracyclophane at 200°C for 40 h in the presence of dimethyl maleate or dimethyl fumarate. The result was the formation of *cis*- and *trans*-2,3-bis(methoxycarbonyl)[4.2]paracyclophane **66** and **67** in essentially equal amounts, the ratio being independent of the configuration of the 'olefin'. This reaction fits the pattern of other similar radical addition reactions, inasmuch as a concerted reaction with **3** or with the proposed diradical intermediate **56** is excluded as a possibility because of the lack of stereospecificity.

It should be possible to generate more stable radicals than **56** by thermolysis of 1-vinyl[2.2]paracyclophane (**68**) and (*E*)-beta-methoxycarbonyl-1-vinyl-[2.2]paracyclophane (**69**). When crystalline **68** was heated at 165°C or irradiated at room temperature with a 254-nm lamp, the ring-expanded product (*Z*)-[4.2]paracyclophan-1-ene (**72**) was formed in 90% and 13% yield, respectively.

When (*E*)-**69** is heated in a sealed tube in benzene it is converted to the (*Z*) isomer of 3-methoxycarbonyl[2.2]paracyclophan-1-ene (**73**) in 95% yield.

Thermal reaction at 100°C of **68** with dimethyl fumarate or dimethyl maleate led to several products. Besides **72** stereoisomeric mixtures of 1-vinyl-2,3-bis(methoxycarbonyl)[2.2]paracyclophan-1-ene (**74**) and *cis*-4,5-bis(methoxycarbonyl[2.2]paracyclophan-1-ene (**75**) were isolated. The thermal reaction of **69** with dimethyl maleate afforded a similar product mixture. Both the non-specificity of the insertion reaction and kinetic results are consistent with the previously outlined multi-step radical mechanism.

Ring-expansion reactions with [2.2]paracyclophanes afford a convenient entry to carbocyclic [*m*.*n*]paracyclophanes (**80**, see Chapter 4) which have hitherto been difficult to obtain; in particular those where *m* and *n* are small. By treatment of [2.2]paracyclophan-1-one (**76**) with diazomethane, Cram and Helgeson obtained the rearranged ketones **77** and **78**, which could be separated by gas chromatography. Wolff–Kischner reduction afforded the hydrocarbons **79** and **80**.

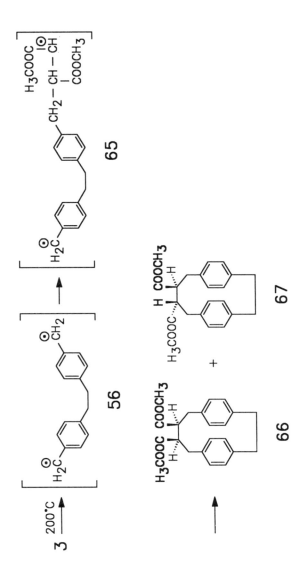

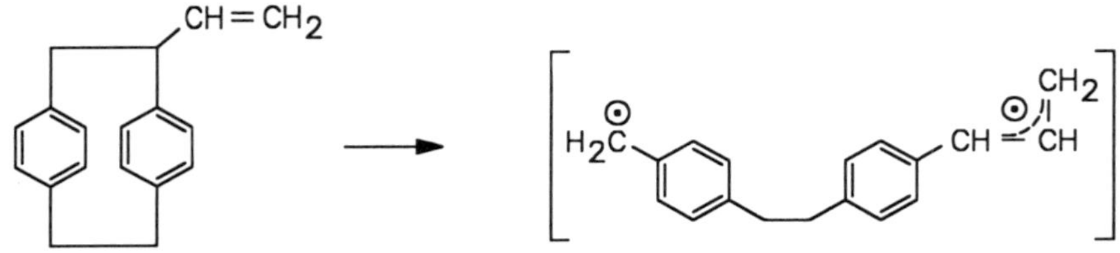

68

70

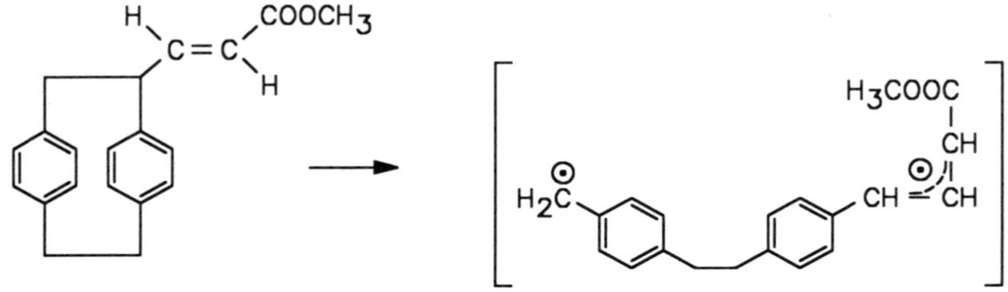

69

71

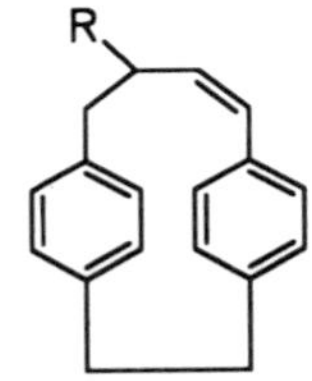

72 : R = H

73 : R = COOCH₃

68 + H₃COOC — CH = CH — COOCH₃ ⟶

72 +

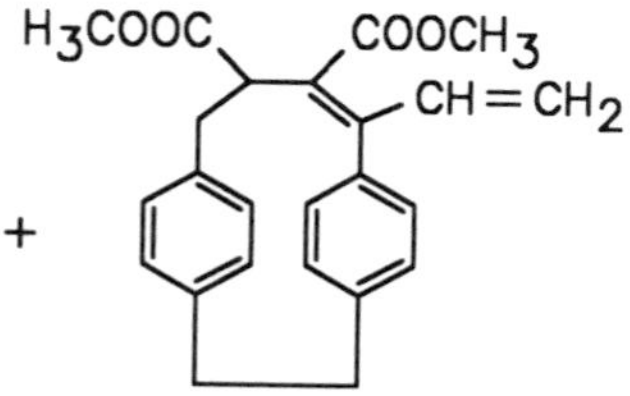

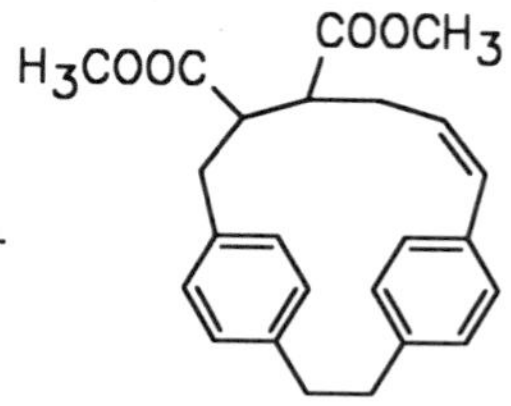

74

75

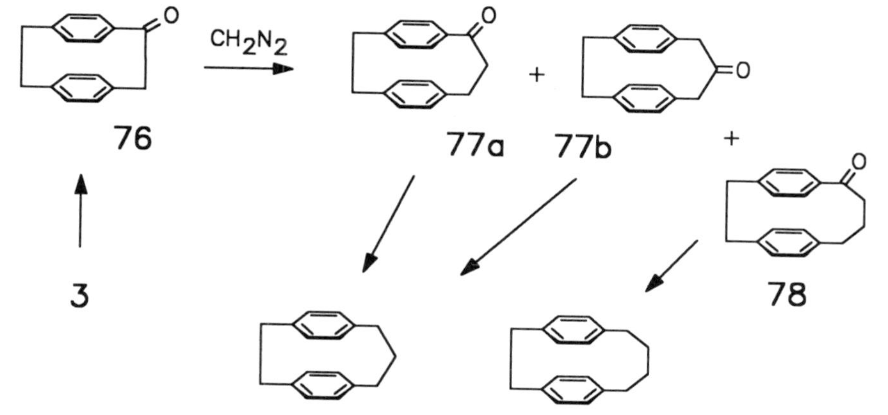

76   CH₂N₂ → **77a** + **77b** + **78**

3

79   **80**

The analogous treatment of [2.2]paracyclophane-1,9-dione (**81**) gave a complex mixture of [3.3]- and [4.3]paracyclophandiones **83** and **84**, whose reduction led to the carbocycles **82** and **85**. In the case of [3.3]paracyclophane (**82**) the route outlined is of preparative value, for previous routes afforded only

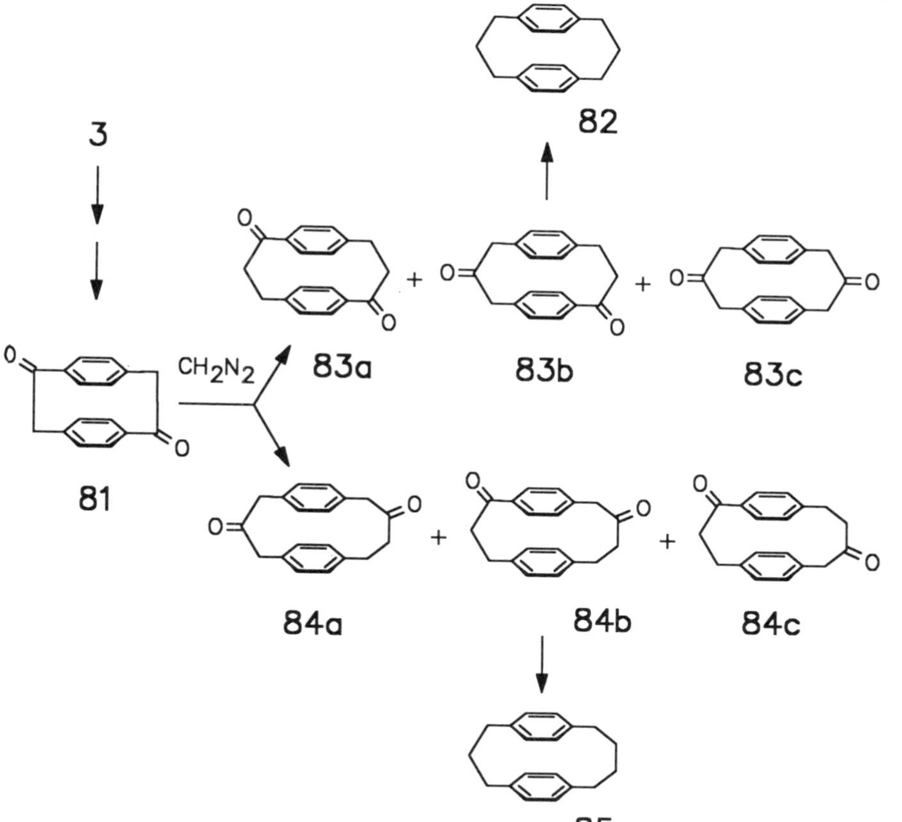

82

3

81   CH₂N₂

83a + **83b** + **83c**

84a + **84b** + **84c**

85

minimal amounts of this ring structure. By optimization of the reaction conditions, it was possible to achieve the synthesis of **82** in an overall yield of 19% from [2.2]paracyclophane (**3**).

The relatively fast reaction of diazomethane with small ring ketones, in comparison with open-chain analogues, and the absence of oxirane products in the reaction mixture are attributed to a relief of strain energy in the course of the reaction.

A ring-expansion reaction from the [2.2]- to the [3.3]paracyclophane skeletal structure described by Hedaya and Kyle is equally useful for synthesis and of interest mechanistically. From acetolysis of the tosylate **86** they isolated in 97% yield 2-acetoxyl[3.2]paracyclophane (**87a**), from which they could prepare [3.2]paracyclophane by way of **87b** and **77b**.

86   **87a** : R = OCOCH₃   **77b**
        87b : R = OH

Application of this reaction scheme to the ditosylate **88** leads to the rearranged ring **89**, as well as its elimination products **90** and **91**. Pyrolysis of **89** and **90** proceeds nearly quantitatively to the dienes **91**, which are smoothly hydrogenated to [3.3]paracyclophane (**82**).

2.3.9.3 Carbene addition

Reaction of [2.2]paracyclophane (3) with diazomethane in the presence of copper(I) chloride gives a mixture of methylated products 92 and 93 in small yield.[23,24] In the case of the tetrasubstituted [2.2]paracyclophanes 94 and 95, the higher methylated compounds 96 and 98 or 97 and 99 are isolated.[25]

$$3 \xrightarrow[\text{CuCl}]{\text{CH}_2\text{N}_2}$$

92   +   93

$$\xrightarrow[\text{CuCl}]{\text{CH}_2\text{N}_2}$$

+

94 : R = $-CO_2Me$ **96** : R = $-CO_2Me$ **98** : R = $-CO_2Me$

95 : R = $-Me$ **97** : R = $-Me$ **99** : R = $-Me$

Treatment of cycloheptatriene compounds such as 92 with trityl fluoroborate leads to cyclophanes where one of the 'decks' is a tropylium ion. These are of interest because of the charge-transfer interaction between the two decks.[27]

The reaction of [2.2]paracyclophane (3) with diazoacetic ester offers a new approach to cyclophanes containing a cycloheptatriene unit.[27]

The rapid transannular carbene insertion, a useful synthetic method for constructing additional bridges in $[2_n]$cyclophanes,[1b] has been applied in the synthesis of superphanes. A somewhat more difficult type of intramolecular carbene insertion was reported by Boxberger et al.[28] Pyrolysis of 100 led to 101 and 102, albeit in low yield.

A concluding example in this section is the addition of singlet oxygen to [2.2]paracyclophan-1,9-diene (37), observed by de Meijere, in which double bonds are indeed 'cleaved' but not in a simple way. The peroxide 103 is an intermediate. Epoxidation of 103 leads to 104, which is converted to the

$$\xrightarrow{270-300°C}$$

+

100   101   102

110

thermally stable, interesting compound trioxatris(σ-homo)benzene derivative **105** with a cobalt catalyst.

For photochemical properties of 4-acyl[2.2]paracyclophanes see reference 30. For the oxidation of hydroxy[2.2]paracyclophane with a molybdenum oxodiperoxo complex see reference 31.

37

$h\nu$ | O_2

103   [0]   **104** $Co^{2\oplus}$ **105**

References to Section 2.3

1. Reviews:
   (a) F. Vögtle and P. Neumann, *Synthesis*, 85 (1973).
   (b) V. Boekelheide, in *Cyclophanes, I* (F. Vögtle, Ed.), *Top. Curr. Chem.*, **113**, 87 (1983).
   (c) D. J. Cram, J. M. Cram, *Acc. Chem. Res.*, **4**, 204 (1971); D. J. Cram, R. B. Hornby, E. A. Truesdale, H. J. Reich, M. H. Delton and J. M. Cram, *Tetrahedron*, **30**, 1757 (1974).
   (d) P. M. Keehn and S. M. Rosenfeld (Eds.), *Cyclophanes*, Vols I, II, Academic Press, New York, 1983.
   (e) D. J. Cram, in ref. 1d, p. 1; see also 15.
   (f) J. F. Liebman, in ref. 1d, p. 23.
   (g) P. M. Keehn, in ref. 1d, p. 69.
   (h) R. H. Mitchell, ref. 1d, p. 239.
   (i) Formation through pyrolysis of xylene: T. Yamaguchi *et al.*, *Chem. Lett.*, 893 (1990).
2. F. Vögtle, *Angew. Chem.*, **81**, 258 (1969); *Angew. Chem. Int. Ed. Engl.*, **8**, 274 (1969); Review: F. Vögtle and L. Rossa, *Angew. Chem.*, **91**, 534 (1979); *Angew. Chem. Int. Ed. Engl.*, **11**, 419 (1972).
3. M. Hibert and G. Solladie, *J. Org. Chem.*, **45**, 4496 (1980).
4. (a) H. Hopf, *Angew. Chem.*, **84**, 471 (1972); *Angew. Chem. Int. Ed. Engl.*, **11**, 419 (1972). J. Kleinschroth and H. Hopf, *Angew. Chem.*, **91**, 336 (1979); *Angew. Chem. Int. Ed. Engl.*, **18**, 329 (1973).
   (b) Review: H. Hopf and C. Marquard, in *Strain and its Implications in Organic*

Chemistry (A. de Meijere, and S. Blechert, Eds.), p. 297, Kluwer, Dordrecht, 1989;
H. Hopf *et al.*, *Chem. Ber.*, **123**, 2015 (1990); cf. N. Mori *et al.*, *J. Chem. Soc.*,
Perkin Trans. 2, 2141 (1990).

(c) H. Hopf, I. Böhm and J. Kleinschroth, *Org. Synth.* (O. L. Chapman, Ed.),
Vol. 60, p. 41, Wiley, New York, 1981.
5. Cf. H. Hopf, *Chemie in uns. Zeit*, **10**, 114 (1976).
6. S. H. Eltamany and H. Hopf, *Tetrahedron Lett.*, **21**, 4901 (1980).
7. Review: F. Vögtle and P. Newmann, *Top. Curr. Chem.*, **48**, 67 (1974).
8. Review: F. Gerson, in *Cyclophanes, II* (F. Vögtle, Ed.), *Top. Curr. Chem.*, **115**,
   57 (1983).
9. D. J. Cram and N. L. Allinger, *J. Am. Chem. Soc.*, **77**, 6289 (1955); D. J. Cram,
   W. J. Wechter and R. W. Kierstead, *J. Am. Chem. Soc.*, **80**, 3126 (1958); D. J.
   Cram and R. A. Reeves, *J. Am. Chem. Soc.*, **80**, 3094 (1958).
10. A. Lüttringhaus and H. Gralheer, *Liebigs Ann. Chem.*, **557**, 112 (1945).
11. H. Falk and K. Schlögl, *Angew. Chem.*, **80**, 405 (1968); *Angew. Chem. Int. Ed.
    Engl.*, **7**, 383 (1968); K. Schlögl, *Top. Curr. Chem.*, **125**, 27 (1984).
12. M. J. Nugent and O. E. Weigang Jr, *J. Am. Chem. Soc.*, **91**, 4556 (1969).
13. W. Tochtermann, U. Vagt and G. Snatzke, *Chem. Ber.*, **118**, 1996 (1985).
    (a) R. Filler and J. E. Gadomski, *J. Fluorine Chem.*, **47**, 175 (1990); R. Filler *et
    al.*, *J. Org. Chem.*, **52**, 511 (1987).
14. R. H. Mitchell and V. Boekelheide, *J. Am. Chem. Soc.*, **92**, 3510 (1970).
15. (a) Review: E. Heilbronner and Z. Yang in *Cyclophanes, II* (F. Vögtle, Ed.), *Top.
    Curr. Chem.*, **115**, 1 (1983).
    (b) 'Data bank' of paracyclophane PE spectra: Z. Yang, B. Kovac, E. Heilbronner,
    J. Lecoultre, C. W. Chan, H. N. C. Wong, H. Hopf and F. Vögtle, *Helv. Chim.
    Acta*, **70**, 299 (1987).
    (c) R. Gleiter *et al.*, *Chem. Ber.*, **115**, 2009 (1982).
16. E. Langer and H. Lehner, *Tetrahedron*, **29**, 375 (1973).
17. A. F. Mourad and H. Hopf, *Tetrahedron Lett.*, **1979**, 1209.
18. C. Elschenbroich, R. Möckel and U. Zennek, *Angew. Chem.*, **90**, 560 (1978); *Angew.
    Chem. Int. Ed. Engl.*, **17**, 531 (1978).
19. D. J. Cram and D. I. Wilkinson, *J. Am. Chem. Soc.*, **82**, 5721 (1960).
20. H. Ohno, H. Horita, T. Otsubo, Y. Sakata and S. Misumi, *Tetrahedron Lett.*, 265
    (1977).
21. T. P. Gill and K. R. Mann, *Organometallics*, **1**, 485 (1982).
22. K.-D. Plitzko, B. Rapko, G. Wehrle and V. Boekelheide, *ISNA (International
    Symposium on Novel Aromatic Compounds)*, Osaka, Japan, August, 1989; V.
    Boekelheide *et al.*, *J. Am. Chem. Soc.*, **112**, 6545 (1990).
23. R. Näder and A. de Meijere, *Angew. Chem.*, **88**, 153 (1976); *Angew. Chem. Int.
    Ed. Engl.*, **15**, 166 (1976).
24. H. Horita, T. Otsubo, Y. Sakata and S. Misumi, *Tetrahedron Lett.*, 3899 (1976).
25. K. Menke and H. Hopf, *Angew. Chem.*, **88**, 152 (1976); *Angew. Chem. Int. Ed.
    Engl.*, **15**, 165 (1976).
26. J. G. O'Connor and P. M. Keehn, *J. Am. Chem. Soc.*, **98**, 8446 (1976).
27. Y. Sekine and V. Boekelheide, *J. Am. Chem. Soc.*, **103**, 1777 (1981).
28. M. Boxberger, L. Volbracht and M. Jones Jr, *Tetrahedron Lett.*, **21**, 3669 (1980).
29. I. Erden, P. Gölitz, R. Näder and A. de Meijere, *Angew. Chem.*, **93**, 605 (1981);
    Angew. Chem. Int. Ed. Engl., **20**, 583 (1981).
30. H. Hopf, T. Laue and M. Zander, *Z. Naturforsch. A: Phys. Sci.*, **46**, 815 (1991).
31. K. Krohn, H. Rieger, H. Hopf, D. Barrett, P. G. Jones and D. Döring, *Chem.
    Ber.*, **123**, 1729 (1990).

112

2.4 [2.2]METAPARACYCLOPHANES

2.4.1 SYNTHESIS

[2.2]Metaparacyclophane[1] (4)* was first obtained in small quantities by Cram in 1966 through a skeletal rearrangement of [2.2]paracyclophane.[1a,2] [2.2]Paracyclophane in finely powdered form was treated with HCl/AlCl₃ in dichloromethane for 30 minutes. Isolated products were [2.2]metapara-cyclophane (44%), hexahydropyrene (6; 10%), biphenyl (7; trace), and starting material (7%); the presence of [2.2]metacyclophane was also confirmed.

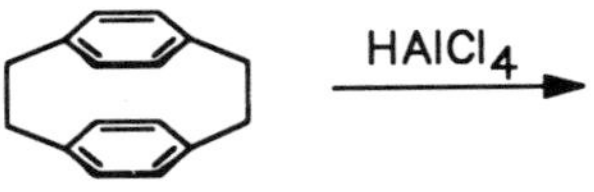

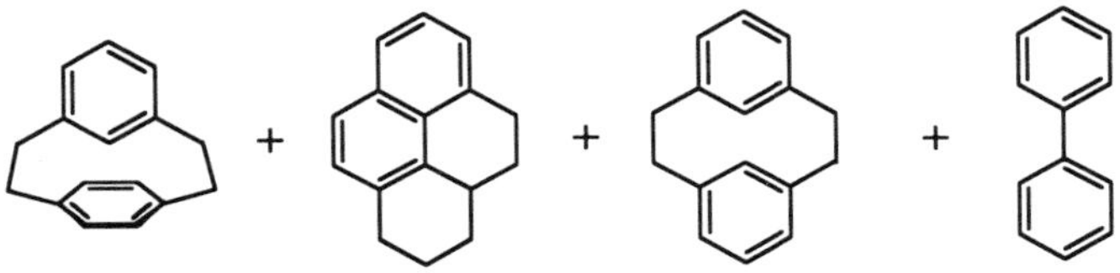

It is remarkable that this rearrangement is stereoselective; optically pure (+)-(S)-4-methyl[2.2]paracyclophane [(+)-8] affords optically pure (+)-(S)-12-methyl[2.2]metaparacyclophane [(+)-9] in 52% yield.[3] This result leads to the conclusion that, during the rearrangement, no ring-opening occurs and thus there is no possibility for the methyl-substituted ring to rotate.

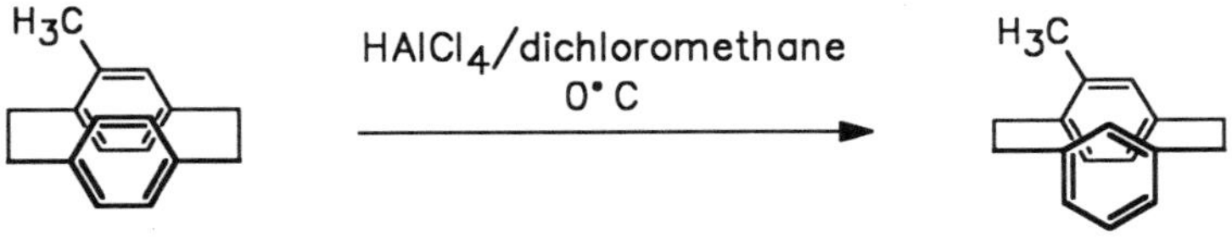

[2.2]Metaparacyclophane (4) can be conveniently prepared by the dithiaphane route, for example from sulphone pyrolysis.[4] In this way Vögtle et al. were able to prepare for the first time intra-annularly substituted [2.2]metapara-cyclophanes of the type 4 (Y = F).[4]

Another method which lends itself to the synthesis of a large number of [2.2]metaparacyclophanes is C—C coupling through nucleophilic alkylation.[1b,5,6]

---

*The compound numbers 1–6 are otherwise reserved for [2.2]phane hydrocarbons in Sections 2.1–2.7.

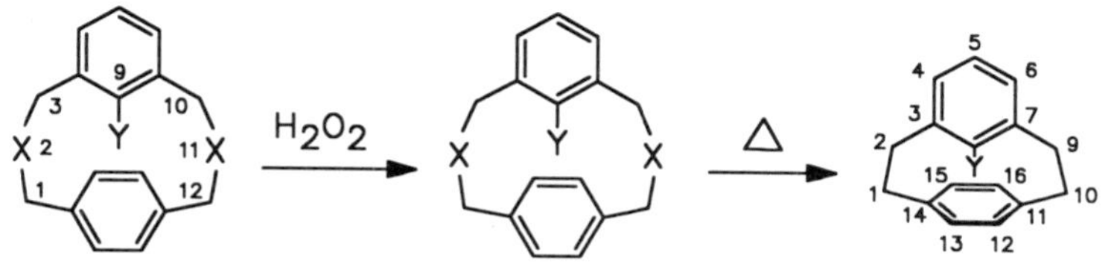

10: X = S **11**: X = SO$_2$ **4**: X = H, F
X = H

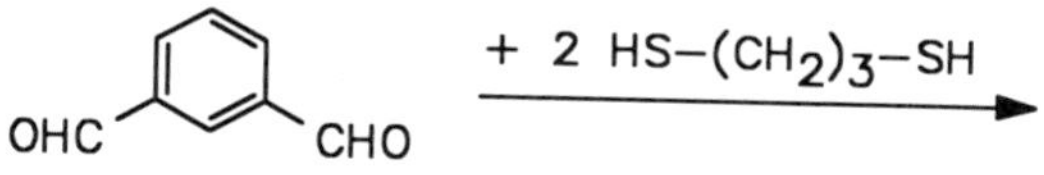

12

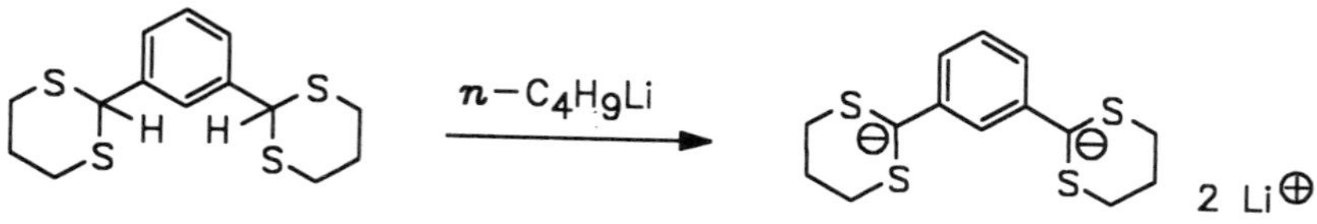

13 **14** 2 Li$^\oplus$

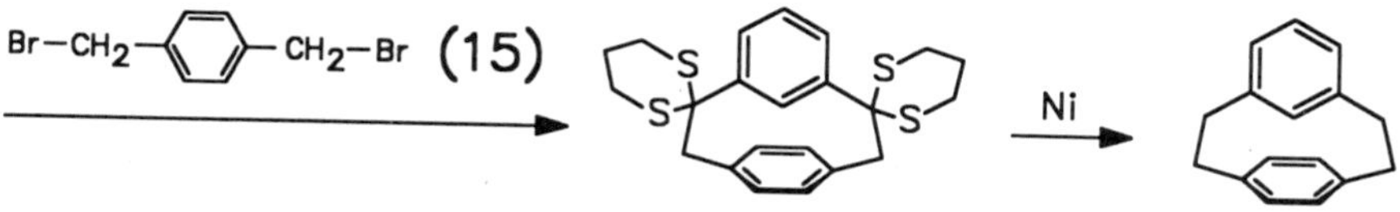

16 **4**

Cyclization of **14** with **15** leads to the dithiane derivative **16** in 36% yield, from which both the [2.2]metaparacyclophane-1,9-diene (**20**) and the corresponding alcohols **18** and the side chain ketone **17** can be obtained.

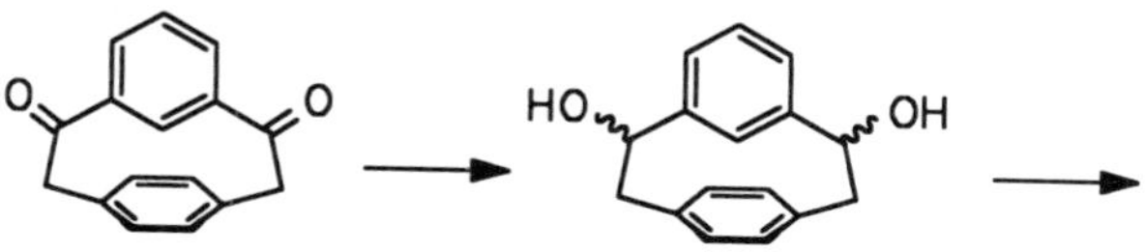

17 **18**

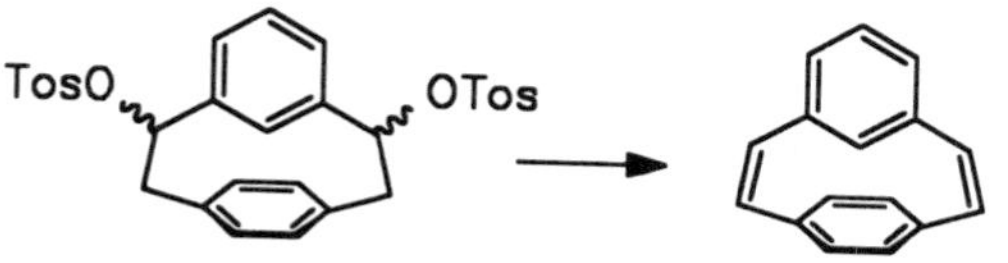

19 **20**

114

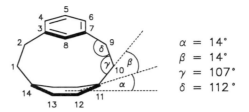

$\alpha = 14°$
$\beta = 14°$
$\gamma = 107°$
$\delta = 112°$

Figure 1. Geometry of crystalline [2.2]metaparacyclophane (**4**) (with bond angles)

2.4.2 PROPERTIES

X-ray crystal structure analysis of [2.2]metaparacyclophane (**4**; m.p. 81–81.5°C) shows that in this molecule the *para*-phenylene ring, with only seven bridging carbon atoms, is strongly distorted (Figure 1), more so than in [2.2]paracyclophane itself. In contrast, the less twisted *meta*-phenylene ring assumes a boat-like shape—quite unlike the situation in [2.2]metacyclophane— such that C(8) lies above the plane described by C(3), C(4), C(6), and C(7), while C(5) lies below the plane. As in the case of [2.2]paracyclophane, the van der Waals radii of the two benzene rings overlap.

Calculations by Boyd and others are in agreement with these findings.

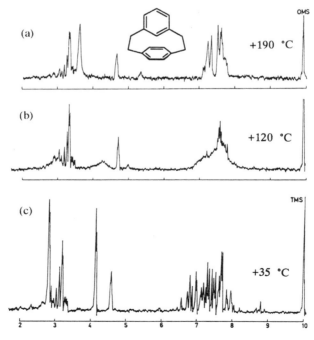

Figure 2. Proton resonance of [2.2]metaparacyclophane (**4**) at various temperatures: (a), (b) in DMSO-D$_6$; (C) in CDCl$_3$[7]

Temperature dependence of the NMR of **4** was first discovered by Vögtle.[7-9] Warming a solution of [2.2]metaparacyclophane in DMSO-D$_6$ induces a broadening not only of the H(A), H(A′), H(X), H(X′) resonances but also of the methylene signals (Figure 2). At 190°C the original signals of the *p*-phenylene protons disappear and are replaced by a new, somewhat broadened singlet at $\delta = 6.4$. The rotational barrier to the underlying conformational process, according to the Gutowsky–Holm equation, is calculated to be $\Delta G_c^{\ddagger} = 84–88$ kJ/mol (with $T_c = 140°C$ and $\Delta\nu = 85$ Hz).

Vögtle *et al.* clarified the nature of the conformational process by examining unsymmetrical, intra-annularly substituted [2.2]metaparacyclophanes such as **4** (Y = F), obtained by sulphone pyrolysis.[7] In this case the rotation of the *m*-phenylene unit is more strongly sterically hindered than in the unsubstituted [2.2]metaparacyclophane, for $\Delta G_c^{\ddagger}$ is found to be 95 kJ/mol in DMSO-D$_6$.

The observation that optically active ester **21c** fails to racemize when heated at 200°C for 25 hours suggests that rotation of the *p*-phenylene ring is most likely excluded as the mechanism for conformational change.

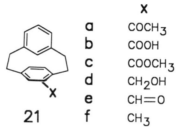

	X	
	a	COCH$_3$
	b	COOH
	c	COOCH$_3$
	d	CH$_2$OH
	e	CH=O
21	**f**	CH$_3$

Thus, the cause of the temperature dependence in the ^{1}H-NMR spectrum of [2.2]metaparacyclophane is attributed to a 'swinging process' A $\rightleftharpoons$ B, as depicted in Figure 3.

The intra-annular hydrogen atom H$_i$ of [2.2]metaparacyclophane (**4**, H$_B$ = H$_i$) absorbs in the region $\delta = 5.64–5.24$, whereas the corresponding signal in both [2.2]metacyclophane and [2.2]metaparacyclophandiene (**20**) is shifted to significantly higher field.[6,10] It has not been possible to distinguish between two explanations for this difference: a shorter distance between the H$_i$ atom and the *p*-phenylene ring or ring strain/steric effects.

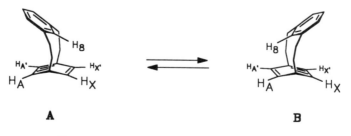

A   **B**

Figure 3. Swinging process A $\rightleftharpoons$ B (schematic)

116

An intense band at 240 nm in the UV spectrum of [2.2]metaparacyclophane is reminiscent of the 244-nm band in [2.2]paracyclophane (3). Both of these bands are attributed to charge-transfer interactions and normal excited states. The long wavelength bands at 283 and 291 nm, absent in the open-chain model compound 3,4'-dimethylbibenzyl, can be explained by the sharp deviation from planarity of the p-phenylene ring. The long wavelength charge-transfer band in the π-complex of [2.2]metaparacyclophane with tetracyanoethylene at 455 nm appears at a significantly shorter wavelength than those in the corresponding π-complexes of [2.2]metacyclophane (2; 486 and 490 nm) and of [2.2]para-cyclophane (3; 521 nm). From this one can conclude that [2.2]metaparacyclo-phane is a weaker π-*base* than [2.2]metacyclophane or [2.2]paracyclophane.

2.4.3 [2.2]METAPARACYCLOPHANE-1,9-DIENE

The X-ray crystal structure analysis[11] of [2.2]metaparacyclophan-1,9-diene (20) shows that the two aromatic rings are not perpendicular to each other; instead the planes formed by carbon atoms 3, 4, 6, 7 and 12, 13, 15, 16 are oriented at an angle of 41° to each other (Figure 4).

Unlike [2.2]metaparacyclophane (4), both benzene rings in the diene assume a boat conformation, the distortion being less pronounced in the m-phenylene ring. However, the distortion of the p-phenylene ring is more pronounced than that in [2.2]metaparacyclophane; the bridging atoms C(11), C(14) and the connecting methylene carbon atoms C(1), C(10) are displaced, respectively, 23.5 and 111 pm out of the plane described by the other four p-phenylene carbon atoms. The resulting angles are 18° (cf. Figure 4) as compared to 14° in [2.2]metacyclophane and 14 or 15° in [2.2]paracyclophan-1,9-diene (see Section 2.3, Figure 5).

It is surprising that the p-phenylene protons in the ¹H-NMR spectrum of [2.2]metaparacyclophan-1,9-diene[6,10] appear as a singlet (δ = 6.81), whereas the intra-annular hydrogen H_i absorbs as a broad singlet at δ = 4.29. The proton resonance spectrum of diene 20 is temperature-dependent, with a coalescence temperature at −96°C, corresponding to a $\Delta G^{\neq}$-value of only 35 kJ/mol. The barrier to conformational folding is thus significantly lower

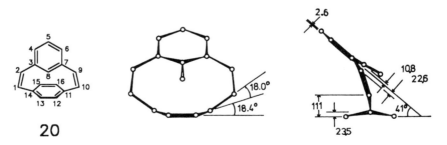

20

Figure 4. Some bond angles and bond lengths (pm) in the [2.2]metaparacyclo-phan-1,9-diene molecule[11]

than that in [2.2]metaparacyclophane (**4**). Reasons for the low energy barrier in **20** are to be found in alternative geometries of the skeletal structure, where there is no penetration of the C(8) proton into the π-electron density of the ring. Moreover, the transition state energy is reduced through conjugative stabilization, for the *m*-bridged aromatic ring aligns itself coplanar to the vinyl bridges.

The UV bands in **20** are intense, although the maxima appear at shorter wavelengths than those in [2.2]paracyclophan-1,9-diene and 8,16-dimethyl-[2.2]metacyclophan-1,9-diene (see section 2.2).[6]

The radical anion of **20** was characterized fully by means of hyperfine data of ESR-ENDOR and triple resonance spectra.[6a] The spin distribution is distinctly different from the one of [2.2]metaparacyclophane (**4**).

2.4.4 CHEMICAL PROPERTIES OF [2.2]METAPARACYCLOPHANE

2.4.4.1 Transannular ring closure

By acetylation of [2.2]metaparacyclophane (**4**) with acetyl chloride/AlCl$_3$ in dichloromethane at $-25°C$ Hefelfinger and Cram[9] obtained, along with the acetyl compound **21a** (see above; 20% yield), the transannularly bridged polycyclic compound **22**, which upon permanganate oxidation afforded benzene-1,2,3-tricarboxylic acid.

22

2.4.4.2 Skeletal rearrangement and isomerism

Delton, Gilman, and Cram[3] obtained [2.2]metacyclophane in 42% yield after irradiation of a cyclohexane solution of [2.2]metaparacyclophane (**4**) for 4 hours.

The same rearrangement of optically active $(-)$-12-methyl[2.2]meta-paracyclophane (**21f**) from an analogous irradiation in 8 hours leads to a mixture of methyl[2.2]metacyclophanes **23** along with 27% of extensively racemized starting material. The mechanism is envisioned as entailing a photochemical cleavage of the CH_2–CH_2 bond, whereupon a recombination as well as rotation of the benzene rings can ensue. Recombination of the benzylic carbons can lead via ring contraction to the less strained [2.2]metacyclophane.

$$(-) - 21f \qquad\qquad 23$$

2.4.5 SUBSTITUTED AND HETERO[2.2](1,3)(1,4)PHANES

2.4.5.1 Stereochemistry of substituted [2.2]metaparacyclophanes

4-Deuterio[2.2](1,3)(1,4)cyclophane (**24**) shows a k_D/k_H value of 1.20 ± 0.04 with regard to conformational flexibility, which qualifies as one of the strongest deuterium isotope effects ever observed.[12] More distantly located substituents also exert a significant influence. The rate of conformational 'flipping' of the 7-amino derivative **25** is only 0.189 times larger than that of the hydrocarbon. Surprisingly, however, the conformational mobility of the 7-nitro derivative **26** is slower, 0.705 that of **4**.[12] The pyridinophane **27**,[13] with the pyridine nitrogen in the 4-position, exhibits a coalescence temperature of $-43.5°C$ in the ^{1}H-NMR spectrum, which corresponds to a $\Delta G^{\neq}$ value of 45 kJ/mol. This observation demonstrates that a sterically smaller substituent, such as the lone pair on nitrogen, facilitates the ring inversion in this ring system.

The nitrogen analogue of diene **20**, 8-aza[2.2](1,3)(1,4)cyclophan-1,9-diene (**28**),[13] shows a symmetrical ^{1}H-NMR spectrum, which remains unchanged even at temperatures as low as $-110°C$. The X-ray crystal structure analysis of **28** revealed that, unlike the case of [2.2]metaparacyclophane, the two aromatic

4: R = H; X = H   **25:** R = H; X = NH_2 **27**

24: R = D; X = H   **26:** R = H; X = NO_2

rings are perpendicular to one another.[14] This geometry likely prevails in solution as well. In the case of **28** apparently a combination of reduced steric interaction and conjugative stabilization reduces the energy of the perpendicular orientation of the two benzene rings, such that this geometry becomes the ground state of the molecule.

20   **28**

References to Section 2.4

1. Reviews:
   (a) D. J. Cram and J. M. Cram, *Acc. Chem. Res.*, **4**, 204 (1971).
   (b) V. Boekelheide, in *Cyclophanes, I* (F. Vögtle, Ed.), *Top. Curr. Chem.*, **113**, 87 (1983).
   (c) F. Vögtle and P. Neumann, *Chimia*, **26**, 64 (1972).
   (d) F. Vögtle and P. Neumann, *Synthesis*, 85 (1973).
   (e) E. Heilbronner and Z. Yang, in *Cyclophanes, II* (F. Vögtle, Ed.), *Top. Curr. Chem.*, **115**, 1 (1983); F. Gerson, in *Cyclophanes, II* (F. Vögtle, Ed.), *Top. Curr. Chem.*, **115**, 57 (1983).
2. D. J. Cram, R. C. Helgeson, D. Lock and L. A. Singer, *J. Am. Chem. Soc.*, **88**, 1324 (1966).
3. M. H. Delton, R. E. Gilman and D. J. Cram, *J. Am. Chem. Soc.*, **93**, 2329 (1971).
4. F. Vögtle, *Chem. Ber.*, **102**, 3077 (1969); cf. F. Vögtle and L. Rossa, *Angew. Chem.*, **91**, 534 (1979); *Angew. Chem. Int. Ed. Engl.*, **8**, 274 (1979).
5. V. Boekelheide, P. H. Anderson and T. A. Hylton, *J. Am. Chem. Soc.*, **96**, 1558 (1974).
6. Th. Hylton and V. Boekelheide, *J. Am. Chem. Soc.*, **90**, 6887 (1968).
   (a) J. Bruhin, U. Buser, F. Gerson and T. Wellauer, *Helv. Chim. Acta*, **73**, 2058 (1990).
7. F. Vögtle, *Chem. Ber.*, **102**, 3077 (1969).
8. S. Akabori, S. Hayashi, M. Nawa and K. Shiomi, *Tetrahedron Lett.*, 3727 (1969).
9. D. T. Hefelfinger and D. J. Cram, *J. Am. Chem. Soc.*, **92**, 1073 (1970).
10. V. Boekelheide and P. H. Anderson, *Tetrahedron Lett.*, 1207 (1969).
11. A. W. Hanson, *Acta Crystallogr.*, **B27**, 197 (1971).
12. S. A. Sherrod, R. L. da Costa, R. L. Barnes and V. Boekelheide, *J. Am. Chem. Soc.*, **96**, 1565 (1974).
13. V. Boekelheide, K. Galuszko and K. S. Szeto, *J. Am. Chem. Soc.*, **96**, 1578 (1974); radical anions of intra-annularly substituted [2.2]metaparacyclophane-1,9-dienes (R = D, F, CN): J. Bruhin, U. Buser, F. Gerson and T. Wellauer, *Helv. Chim. Acta*, **73**, 2058 (1990).
14. L. H. Weaver and B. W. Mathews, *J. Am. Chem. Soc.*, **96**, 1581 (1974).

2.5 [2.2]ORTHOMETACYCLOPHANE

This 'unsymmetrical' cyclophane hydrocarbon 5 could only first be synthesized in 1989[1] with the aid of modern analysis, after intensive attempts at its preparation over nearly two decades had failed to lead to its isolation.

The successful synthesis of [2.2]orthometacyclophane (5) was achieved by preparation of the larger sulphur-containing ring 6 and desulphurization of its corresponding disulphone by 'sulphone pyrolysis'. This latter step afforded a yellow oil, which, according to its 400-MHz ^{1}H-NMR spectrum, contained at least three products. This mixture could be separated into two fractions by gas chromatography on an SE-30 stationary phase at 180°C. The first fraction (16%) was the desired [2.2](1,2)(1,3)cyclophane (5), in a *syn/anti* ratio of 4:1.

syn-5   *anti*-5

Both isomers show an AA′BB′ pattern for the *ortho*- and an A_2MX pattern for the *meta*-disubstituted benzene units, and the chemical shifts for *syn*- and *anti*-5 differ in a characteristic manner. All the aromatic protons in *syn*-5, with the exception of H(16) are more strongly deshielded than those in *anti*-5.[1] Although the intra-annular aromatic protons in *anti*-[2.2]metacyclophanes are typically strongly deshielded, this is not observed in *anti*-5 because its benzene rings do not lie in such close proximity as in the doubly *meta*-bridging arrangement.

The CH_2-protons of the bridges in *syn*-5 appear as a first-order spectrum, whereas those in *anti*-5 have not been analysed in detail because they have similar chemical shifts. The high-field shift of H(2a) is unexpected, for this hydrogen projects back into the shielding region of the *meta*-substituted ring.

Raising the temperature above 100°C broadens the ^{1}H-NMR spectrum of the isomers of **5** which, however, does not reach coalescence by 150°C, which means the energy barrier $\Delta G^{\pm}$ must lie between 84 and 100 kJ/mol. Thus separation of the isomers is not practical at room temperature.

The minimum distance between H(16) and H(2b) in *syn*-**5** is calculated to be 210 pm. In fact, H(2b) shows a strong nuclear Overhauser effect (16%) when the resonance at H(16) is saturated. Calculations indicate that the strain energy in *syn*-**5** should be less than that in *anti*-**5** ($\Delta SE = 7.3$ kJ/mol).

References to Section 2.5

1. G. Bodwell, L. Ernst, M. W. Haenel and H. Hopf, *Angew. Chem.*, **101**, 509 (1989); *Angew. Chem. Int. Ed. Engl.*, **28**, 455 (1989); cf. F. Vögtle and P. Neumann, *Tetrahedron*, **26**, 5229 (1970).

2.6 [2.2]NAPHTHALENOPHANES AND THEIR DIENES

2.6.1 INTRODUCTION

At the time of the publication of B. H. Smith's book *Bridged Aromatic Compounds* in 1964, only two naphthalenophanes **6** and **7** were known, the former (**6**)[1,2] having been described by Baker in 1951, the latter (**7**) by Cram[3] in 1954. Why should naphthalenophanes, especially [2.2]naphthalenophanes[4]

6   **7**

be of interest? Besides the normal features of cyclophanes with benzene rings, the naphthalene compounds possess a more extensive aromatic core. Interest centres around the nature and extent of deformation of the naphthalene ring, the strain energy, and static and dynamic stereochemistry, as well as charge-transfer effects between neighbouring condensed aromatic units. Thus naphthalenophanes serve as starting materials for topologically interesting aromatic compounds such as circulenes, propellicenes, and paddlanes. Optically active naphthalenophanes are of interest because of the structural–chiroptical correlations and their relationship to helicenes.

As with cyclophanes (benzenophanes) we do not include here the 1,1-binaphthol oligo ethers (e.g. **8**), better known for their crown ether units, because they belong more properly under 'molecular recognition' and 'crown ether complexation' (see *Supramolecular Chemistry*, Wiley, Chichester, 1991).

8

Because the naphthalenophanes to be discussed are quite varied in structure, the presentation of synthesis and properties does not lend itself to such well defined subsections, as in the case of the individual cyclophanes (benzenophanes). As a consequence only particularly important syntheses will be mentioned throughout the text. Synthetic methods are essentially those described for the benzene series, in particular the desulphurization of dithia[3.3]phanes, the crossed Hofmann elimination, acyloin condensation, and oxidative coupling of acetylenes. Novel approaches include the Paal–Knorr reaction for construction of heterocycles and succinylation for incorporation of a second condensed ring onto the benzene structure.

2.6.2 SYNTHESES

[2.2](2.7)Naphthalenophane (**7**) was obtained in 1951 by a Wurtz coupling of 2,7-bis(bromomethyl)naphthalene with sodium or phenyllithium (see above). Treatment with $AlCl_3/CS_2$ followed by dehydrogenation over Pd led to

9   **7**   **10**

coronene.[1] The first synthesis in low yield of [2.2](1,4)naphthalenopara-cyclophane (**13**) was achieved by Cram et al.[5] by attachment of a second benzene ring onto [2.2]paracyclophane through the Friedel–Crafts acylation with succinic anhydride.

Wasserman and Keehn[6] employed the 'crossed' Hofmann elimination with the quaternary ammonium salts **14** and **15** for the synthesis of **13**. The superior route to **13**, however, on the basis of high yield, is the dithia[3.3]phane route from **16**, which is subjected to photochemical sulphur extrusion, with triethyl phosphite serving as solvent and thiophilic trapping agent.[7]

3 → **11** → **12** → **13**

14 + **15** → **13** ← **16**

a: R = H
b: R = D

[2,2](1,4)Naphthalenophane exists in the two possible, stable isomers **19** and **20**. **19** was first synthesized (in 3% yield) in 1963 by dimerization of a benzo-1,4-xylylene intermediate **18**, which in turn had been generated from the ammonium hydroxide **17**. Its *anti* configuration was established by an alternate synthesis from [2.2]paracyclophane, in which naphthalene rings were built up on each benzene unit (overall yield of **19**: 0.07%).

Brown and Sondheimer developed an even better synthesis of [2.2](1,4)-naphthalenophane (**19**), starting with the dihydronaphthalene **21**, which was reduced to the diol and converted to tosylate **22**. Heating of the tosylate in boiling pyridine afforded cyclophane **19** in 90% yield! Inasmuch as the dimerization of **18** to **19** is formally a $[6\pi + 6\pi]$ cycloaddition and therefore thermally disallowed according to Woodward–Hoffmann rules, the reaction is assumed to be a multi-step process.

By working under high dilution in xylene in the presence of phenothiazine, Wasserman and Keehn were able to improve the elimination reaction of the quaternary bromide **14**—this time not a 'crossed' elimination with **15**. The result was a mixture of the *anti* isomer **19** in 40% yield and the *syn* isomer **20** in 4% yield. Apparently the higher reaction temperature is favourable to formation of the *syn* isomer **20**. The [2.2]paracyclophane tetracarboxylate **23** has served as starting material for another synthesis of naphthalenophane **19**. Although the overall yield of **19** was relatively small, the tetrabromide **24** is readily available (Hopf).[8]

The ^{1}H-NMR spectra of the *anti*- and *syn*-isomers **19** and **20** are distinctive. While the H_a and H_b protons in **19** are closely similar to those in 1,4-dimethylnaphthalene, these protons are shifted considerably upfield in **20**. ($\Delta\delta = 0.35$). In contrast, the H_c protons show the opposite effect; in **20** these are shifted downfield by 1.0 p.p.m. by comparison to those in **19**. The *syn* structure has also been confirmed by X-ray crystallographic analysis.[9]

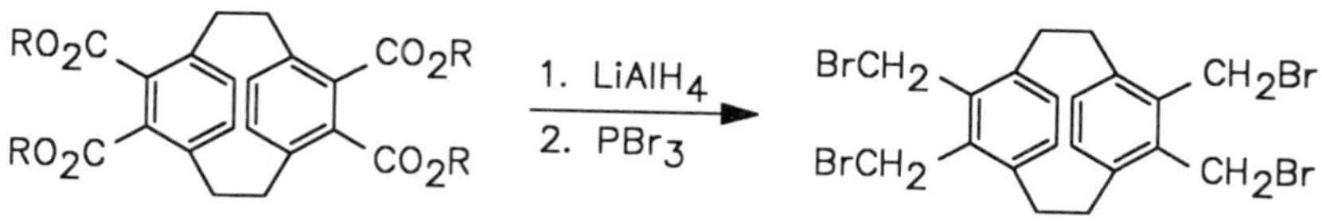

23 24

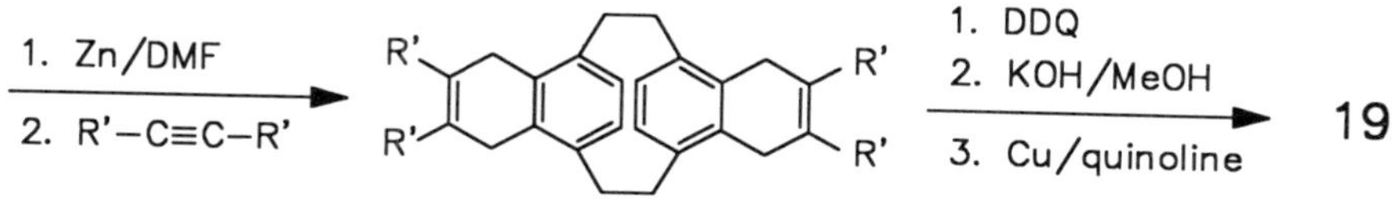

25: R' = CO_2Me

The 1,5-bridged naphthalenobenzenophane **26**, as well as the achiral (**27**) and chiral [2.2](1,5)naphthalenophane (**28**), has been synthesized by photolysis of the corresponding dithia[3.3]phanes by Haenel.[10] In the case of **27** an X-ray crystal structure has been done.[11]

26

27

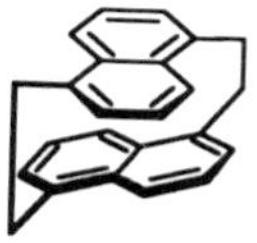

28

The (1,5)naphthalenopyridinophane **29** was likewise obtained by pyrolysis of the corresponding [3.3]disulphone, and the diene **30** was synthesized from the appropriate [3.3]dithianaphthalenopyridinophane. ^{1}H-NMR results indicate that the two aromatic rings are orthogonally positioned, as has been postulated for the corresponding benzenopyridinophane.[12]

29

30

Haenel and Staab synthesized the chiral [2.2](2,6)naphthalenophanes **31** and their dienes **32** via ring contraction and elimination from the corresponding dithia[3.3]-compounds.[13] The achiral isomers **33** and **34** were apparently not formed in the process. **31** was also obtained by sulphone pyrolysis, as well as by catalytic hydrogenation of the diene **32**. Resolution of **31** was achieved with Newman's reagent (TAPA), the laevorotatory (−)-**31** being assigned the *S*-absolute configuration.[14]

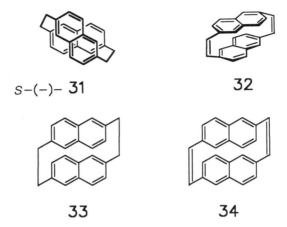

s-(-)- **31**

32

33   **34**

The two [2.2](2,6)(2,7)naphthalenophanes **35** and **36** were obtained by ring-contraction routes, the latter through pyrolysis of the bis-sulphoxide.[15] Both phanes **35** and **36** are conformationally flexible, exhibiting temperature-dependent ¹H-NMR spectra.

The *syn*- and *anti*-isomers of [2.2](1,4)naphthalenophan-1,13-diene (**37**, **38**) were prepared by Otsubo and Boekelheide.[16] As before, the easily obtainable *anti*- and *syn*-[3.3]dithiaphanes, formed in 64% overall yield in a ratio of 5:1, were subjected to ring contraction *via* benzyne, and the sulphoxides were subsequently pyrolysed (4 and 1% yield, respectively). The H_c-proton in the *anti*-isomer **37** absorbs at a higher field ($\Delta\delta = 1$ p.p.m.) than that in the *syn*-isomer **38**; the vinyl protons (H_d) in both dienes **37** and **38** absorb at $\delta = 7.45$ and 7.64. Thus the shielding effect of the naphthalene ring on the vinyl protons is even stronger than that in [2.2]paracyclophan-1,9-diene.

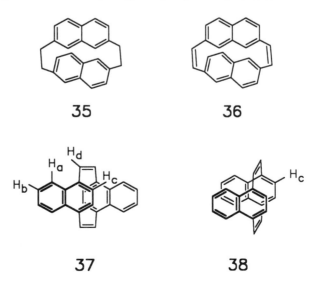

35   **36**

37   **38**

2.6.2.1 [2.2](1,3)(2,7)Naphthalenophanes

The dithia[3.3]phane route, with subsequent desulphurization and ring contraction led to [2.2](1,2)naphthalenometacyclophan-1,11-diene (**39a**). Condensed dienes of this type are of value because they can be easily converted photochemically or thermally into the corresponding dihydropyrene derivatives **40**. In an analogous way, the corresponding phane dienes **41** afford the

a: R = H
b: R = CH$_3$

39   **40**   **41**   **42**

trans-dihydropyrenes **42**, non-benzenoid aromatics with enhanced ring current effects and methyl groups shifted to very high fields.[4]

Although the dithia[3.3]phane route has already been depicted for benzenophanes, it is appropriate to reproduce the entire scheme here, in particular to outline the details of its execution.[17,18]

The ^{1}H-NMR spectra of dihydropyrenes obtained in this way provide good support for the hypothesis that equivalent Kekulé structures are associated with particularly strong diatropy (aromaticity) (Boekelheide and Mitchell).

The photoelectronic[19] and electron spin resonance spectra[20] of various [2.2]naphthalenophanes and their radical anions have already been summarized.

Sato showed by means of ^{1}H-NMR spectroscopy that both *anti*- and *syn*-isomers of **7** are apparently formed in the Wurtz synthesis (already described), in marked contrast to the situation for [2.2]metacyclophanes.[21] Thus **7** is the only [2.2](2,7)- or -(1,3)phane of this type which can be directly obtained as the *syn* conformer:

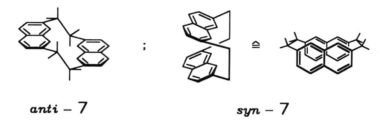

anti – 7   **syn – 7**

2.6.3 CHEMICAL REACTIONS

It is surprising that nitration of **7** affords the derivative **43a**, with an intra-annular nitro group as the major product in 55% yield,[22] a result in marked contrast to those of nitration of [2.2]metacyclophane and related compounds. This is in fact the only case of an intra-annular substitution reaction in the cyclophane series.

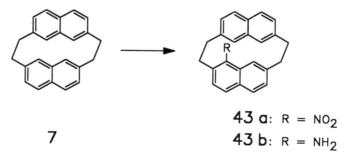

7

43 a: R = NO$_2$
43 b: R = NH$_2$

The [2.2](1,4)naphthalenophanes **19** and **20** undergo specific thermal and photochemical reactions because of their strained skeletal structures and the proximity of layered naphthalene rings. When the *syn*-isomer is heated above its melting point (243–245°C), it resolidifies and then melts again at 300–303°C.

This provides evidence for a complete isomerization to the *anti*-isomer **19**, which is visualized as proceeding by way of a diradical intermediate of the type **44**.

44

The *syn*- and *anti*-isomers **20** and **19** can be interconverted photochemically. Irradiation of the *syn*-isomer **20** in degassed benzene leads primarily to the *anti*-isomer, while continued irradiation of solutions of the *anti*-isomer **19** gives other products.[23] Irradiation of a cyclohexane solution of **19** above 290 nm effects rearrangement to the intermolecular addition product **48**,[24] which rearomatizes to **19** at room temperature with a half-life of 76 s at 20°C. The $[4\pi + 4\pi]$ addition product **48** appears to be the kinetic product, because extended irradiation for 10 days at room temperature leads to the thermodynamically more stable product, the dibenzoequinene **47**, in 25–50% yield. This remarkable hydrocarbon, whose skeletal structure has been confirmed by X-ray analysis,[25] presumably arises through two sequential $[2\pi + 2\pi]$ additions, such that the first intermediate is **46** and the second the end product **47**.

Irradiation of a solution of **19** in ether/ethanol at $-190°C$ effects a $[6\pi + 6\pi]$ addition to afford benzo-1,4-xylylene (**18**), which is sufficiently stable (approx. 30% loss after 20 h at 25°C) that it can be employed as synthon.[24] The naphthalene rings in **19** react under photosensitized autoxidation to the interesting polycyclic **50**.[26]

The naphthalenoparacyclophane **13a** also undergoes a photo-oxidation in the presence of methylene blue as sensitizer, whereby the oxidation and rearrangement products **51** and **52** are isolated in overall 15% yield. The following mechanism is proposed.[6]

References to Section 2.6

1. W. Baker, F. Glockling and J. F. W. McOmie, *J. Chem. Soc.*, 1118 (1951).
2. W. Baker, J. F. W. McOmie and W. K. Warbuton, *J. Chem. Soc.*, 2991 (1952).
3. J. Abell and D. J. Cram, *J. Am. Chem. Soc.*, **76**, 4406 (1954).
4. Review: J. A. Reiss, in *Cyclophanes* (P. M. Keehn and S. M. Rosenfeld, Eds.), Vol. II, Academic Press, New York, 1983.
5. D. J. Cram, C. K. Dalton and G. R. Knox, *J. Am. Chem. Soc.*, **85**, 1088 (1963).
6. H. H. Wasserman and P. M. Keehn, *J. Am. Chem. Soc.*, **94**, 298 (1972).
7. J. Bruhin, F. Gerson, W. B. Martin and C. Wydler, *Helv. Chim. Acta.*, **60**, 1915 (1977).
8. J. Kleinschroth and H. Hopf, *Tetrahedron Lett.*, 969 (1978).
9. A. V. Fratini, cited in a footnote in ref. 6.
10. M. W. Haenel, *Tetrahedron Lett.*, 4191 (1977).
11. M. W. Haenel, *Chem. Ber.*, **111**, 1789 (1978).
12. M. W. Haenel, *Tetrahedron Lett.*, 4007 (1978).
13. M. W. Haenel and H. A. Staab, *Chem. Ber.*, **106**, 2203 (1973).
14. P. Block, Jr and M. S. Newman, *Org. Synth. Coll.*, **5**, 103 (1973).
15. M. N. Iskander and J. A. Reiss, *Tetrahedron*, **34**, 2343 (1978).
16. T. Otsubo and V. Boekelheide, *J. Org. Chem.*, **42**, 1085 (1977).
17. R. H. Mitchell and J. S.-H. Yan, *Can. J. Chem.*, **55**, 3347 (1977).
18. R. H. Mitchell, R. J. Carruthers and L. Mazuch, *J. Am. Chem. Soc.*, **100**, 1007 (1978).
19. E. Heilbronner and Z. Yang, in *Cyclophanes, II* (F. Vögtle, Ed.), *Top. Curr. Chem.*, **115**, 1 (1983).
20. F. Gerson, in *Cyclophanes, II* (F. Vögtle, Ed.), *Top Curr. Chem.*, **115**, 57 (1983).
21. T. Sato, H. Matsui and R. Komaki, *J. Chem. Soc., Perkin Trans. 1*, 2051 (1976).
22. R. W. Griffin, Jr and N. Orr, *Tetrahedron Lett.*, 4567 (1969).
23. H. H. Wasserman and P. M. Keehn, *J. Am. Chem. Soc.*, **91**, 2374 (1969).
24. G. Kaupp and I. Zimmermann, *Angew. Chem.*, **88**, 482 (1976); *Angew. Chem. Int. Ed. Engl.*, **15**, 441 (1976).
25. A. V. Fratini, *J. Am. Chem. Soc.*, **90**, 1688 (1968).
26. H. H. Wassermann and P. M. Keehn, *J. Am. Chem. Soc.*, **88**, 4522 (1966).

2.7 NON-BENZENOID PHANES

In the last three decades cyclophane chemistry has provided important contributions to aromatic chemistry. Because non-benzenoid aromatic compounds occupy an important place in aromatic chemistry, the idea was to incorporate into phane structures other Hückel aromatic rings such as azulene or tropolone. It was of interest to examine possible deformation of the non-benzenoid Hückel aromatics caused by bridging. The π-interactions and charge-transfer effects could also be investigated here in multi-decked phanes; in such skeletal frameworks this would be of particular interest because of the high dipolar character of azulene and tropolone and the full electric charge in cyclopentadienide and tropylium ions. Furthermore, one could examine changes in resonance stabilization and resonance energy in bridged non-benzenoid aromatics, deformation perhaps leading eventually to disappearance of aromaticity altogether.[1]

2.7.1 SYNTHETIC METHODS

In some cases synthetic methodology for non-benzenoid phanes[1] is analogous to that for cyclophanes and heterophanes. The difference in approach lies in the fact that the non-benzenoid aromatics are often less stable than benzene, naphthalene, and pyridine and thus milder conditions must be employed. Because of the higher reactivity of non-benzenoid aromatics, however, the cyclization routes typically used for cyclophanes are less convenient than those in which the larger ring—the bridging unit—is already in place and then a five- or seven-membered ring is attached to a benzene nucleus. As a rule, method II is preferred over the cyclization via method I:

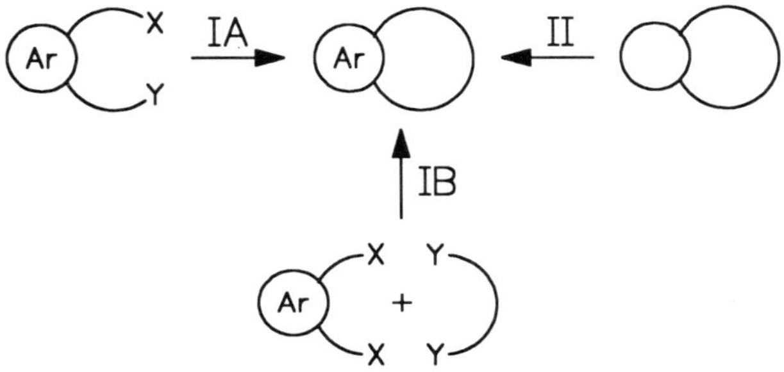

2.7.1.1 Examples of ring closures via method I

1. The dithiaphane route with subsequent sulphone pyrolysis (a) or photo-desulphurization (b):

(a)

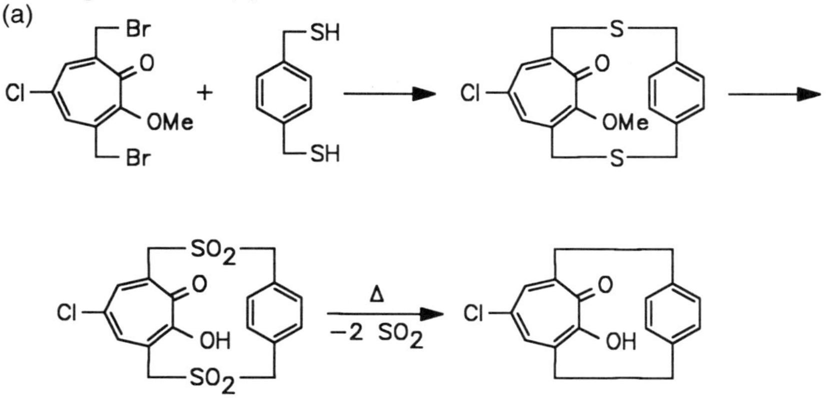

(b)

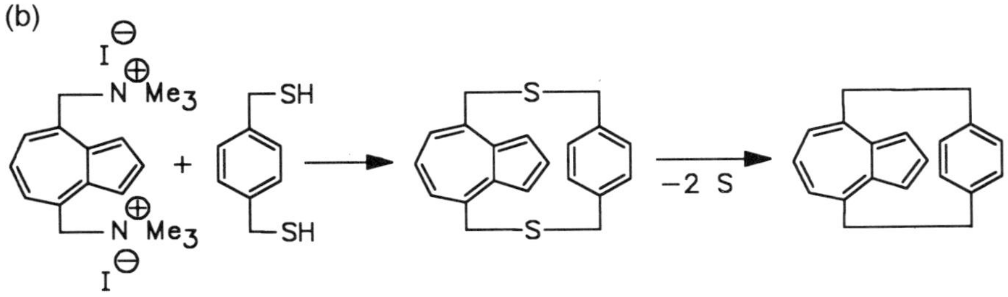

2. Coupling reactions: (a) Hofmann elimination, usually leading to isomers.

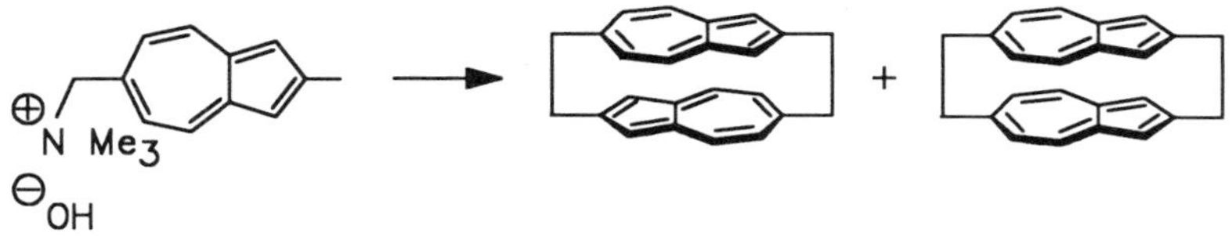

A variation of this route involves the dihalo compound **6**, which, in the presence of nickel tetracarbonyl in DMF leads to [2.2](2,6)-1,5(1,7)dihydro-*S*-indacenophane (**7**) in 10% yield.[2]

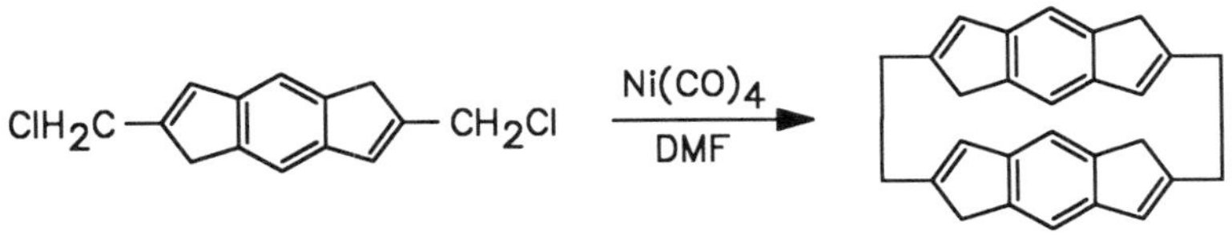

     6   **7**

3. Bridge cyclization *via* e.g. the Thorpe–Ziegler reaction, a method usually applicable only to large rings:

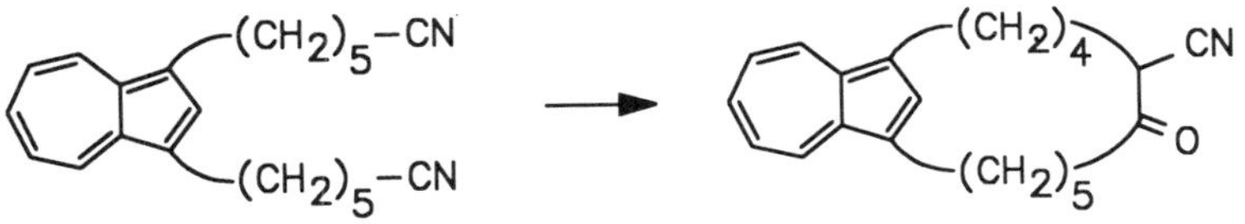

2.7.1.2 Method II: ring formation and aromatization

In this case the non-benzenoid aromatic ring is attached to an already existing hydrocarbon ring skeleton, effected, for example, by elimination or through direct ring formation by aldol condensation or the Hafner azulene synthesis.

(a)

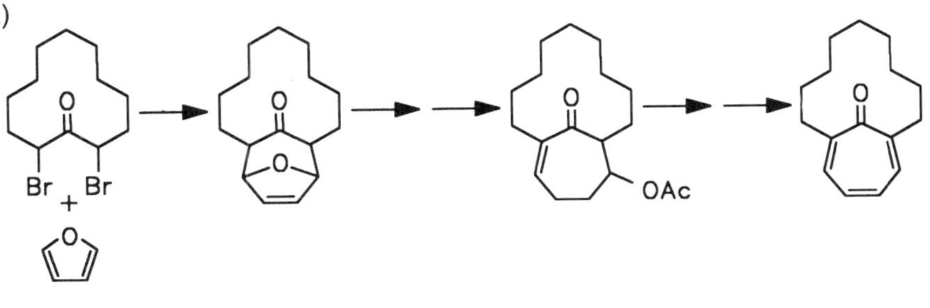

(b)

(c)

(CH₂)–COCHN₂

(d)

CHO
CHO

+ =O (CH₂)ₙ ⟶ =O (CH₂)ₙ

$n = 4-9, 10, 12, 13$

(e)

+ ⟶

2.7.2 PHANES CONTAINING NON-BENZENOID 6π SYSTEMS

2.7.2.1 Cyclopentadienidophanes

The only non-condensed [9](1,3)cyclopentadienidophane **8** known to date was described in 1971.[3] The diamagnetic ring current manifests itself in the ¹H-NMR spectrum. One observes high-field shifted methylene protons at δ = 0.5 (2H, m), the remaining protons absorbing at δ = 1.0–1.6. The protons in the five-membered ring appear at δ = 5.33 (1H, t) and 5.02 (2H, d), in the same region as those in cyclopentadienide itself (δ = 5.4).

8

2.7.2.2 *syn/anti*-[2.2](2,8)Fluorenidophanes

The molecules **9** and **10**[4] were obtained as dark red solutions by base-promoted deprotonation of the fluorene precursors, which in turn had been prepared by the dithiane route. An array of fluorene protons [H(1), H(8), H(9)] in **10** appear at substantially higher field than those in **9**, whereas H(3), H(6), H(4), and H(5) absorb at relatively lower field (fluorene numbering system).

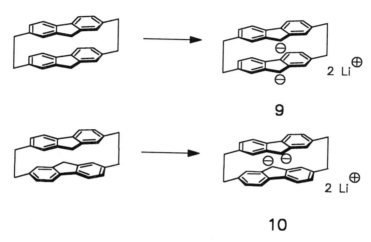

2.7.2.3 Troponophanes

The [n](2,7)troponophanes are usually synthesized by aromatization or ring fusion of a seven-membered ring onto an already existing ring skeleton:[5]

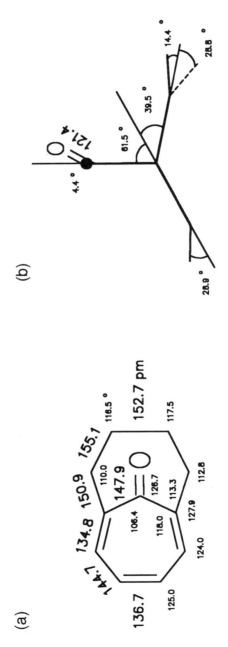

Figure 1. X-ray crystal structure of **13**.[5] (a) Bond lengths (pm) and bond angles (degrees); (b) side view of the molecule

Geometry and physical properties

The bridging atoms in troponophanes **11–13** and **14** are fixed in specific conformations, as is shown by the ^{1}H-NMR spectra from the non-equivalence of the 'benzylic' protons. Although the IR spectrum of **11** indicates the tropone ring is planar and of similar polarity to that in 2,7-dimethyltropone (**15**), the carbonyl band in **12** by comparison is shifted to higher frequency, suggesting a smaller carbonyl bond angle and reduced planarity.

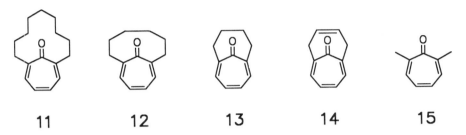

11   **12**   **13**   **14**   **15**

Troponophane **13** has been examined by X-ray crystallography.[5] The tropone ring is significantly deformed into a boat (Figure 1). In addition it exhibits pronounced variation in bond lengths and a shorter C=O bond distance than that in tropone itself, all of which points to reduced conjugation. ^{1}H- and ^{13}C-NMR spectra confirm this reduced conjugation. In Figure 2 are depicted the shifts in the ^{1}H- and ^{13}C-NMR spectra in the series from dimethyltropone (**15**) through tropones with varying spanning bridges to the bridged cycloheptatriene **16**.[1] In particular the carbonyl carbon atom and the C_β- and H_β- resonances are strongly dependent on the ring size. The carbonyl carbon is shifted downfield as far as that of a saturated, substituted carbonyl carbon atom, which is at least to some extent attributable to the partial positive charge on the carbon atom. Consistent with this picture is the shifting of H_β and C_β toward higher field to the region where tropilidenophane **16** absorbs.

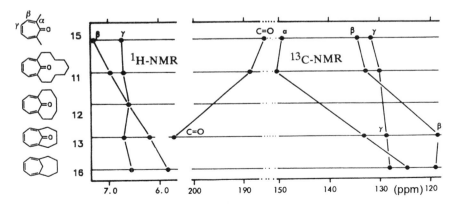

Figure 2. ^{1}H- and ^{13}C-NMR shifts in [n](2,7)troponophanes[1,5]

The UV spectra of troponophanes also point to diminished conjugation. The intense, broad band of the planar tropone at about 310 nm is reduced to a weak shoulder in **13**, and the general shape of the spectrum resembles that of **16**.

2.7.2.4 [n]Benzotroponophanes

Several [n]benzo[d]troponophanes (**17**, $n = 4$–10, 12, 13)[1] have been synthesized by a double aldol condensation between cycloalkanones and phthalaldehyde (see above, method IId).[6] The IR carbonyl frequency is significantly affected by the ring size. The half-wave reduction potential is likewise influenced by the bridging, as is the chemical shift (δ) of the tropone protons, when the bridge is shorter than $(CH_2)_7$. Noticeable changes are also seen in dipole moment and molecular refraction.

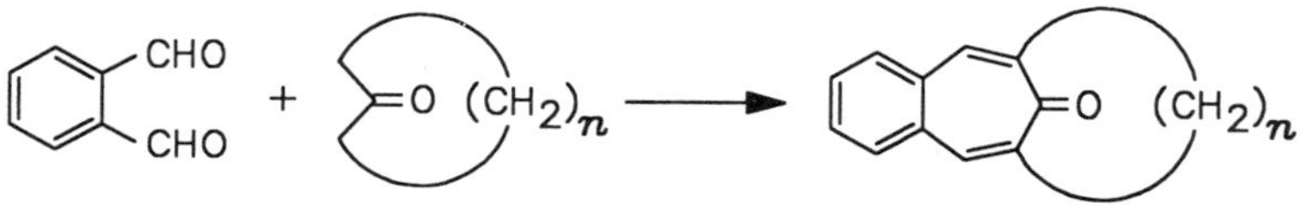

$n = 4$–10, 12, 13 $n = 4$–10, 12, 13

17

compare:

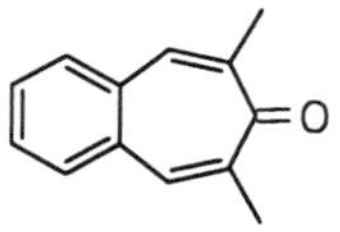

18

Although the UV spectra of dimethylbenzotropone and the long-chain bridged **17** ($n = 13$) exhibit four absorption bands and two shoulders, only the last two maxima (≈ 280 and 232 nm) and one shoulder (≈ 330 nm) are found in the spectra of [5]- and [6]benzotroponophane. The lower homologue with $n = 4$ shows only one maximum at 226 nm. All of these changes support the reduced angle of the carbonyl group and its suppressed conjugation with the remainder of the π-system. The reason lies in the forcing of the carbonyl carbon out of the plane of the ring, which is necessitated by the short bridging. This conjecture is supported by the X-ray crystal structure analysis of the two lowest homologues. The tropone ring in the [5]- and [4]-bridged compounds is a strongly distorted tub form, comparable to that in [4](2,7)troponophanes (see Figure 1). The delocalization energy for the homologue **17** ($n = 5$) is calculated to be 204 kJ/mol—146 kJ/mol less than that for **18**. In other words, [5]benzotroponophane is characterized by only slightly more delocalization than a benzene ring.

2.7.2.5 Tropyliophanes

When [9]troponophane **11** is protonated with one equivalent of trifluoroacetic acid (TFA), the tropone and 'benzylic protons' are shifted about 0.9–0.4 p.p.m. downfield, while some of the bridging protons experience an upfield shift. These changes are consistent with generation of the corresponding hydroxytropylium ion **19**.[7]

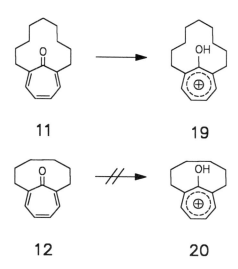

In contrast, [6](2,7)troponophane (**12**) shows no such change under similar acidifying conditions; and [4]troponophane (**13**, see above) in 100% TFA undergoes only a small low-field shift in the vinyl protons of about 0.1 p.p.m.[1] From this one can conclude the hydroxytropylium ion is only stable when the seven-membered ring can assume a nearly planar geometry. In 2-hydroxy[6](1,3)-tropyliophane (**20**) the molecule cannot approach planarity sufficiently so as to delocalize the positive charge.

Only benzo-analogues of [n]metatropyliophanes are known, examples being **23**, where n = 10, 12. The synthetic route is outlined in the following scheme:[8]

21: n = 9–12   **22**

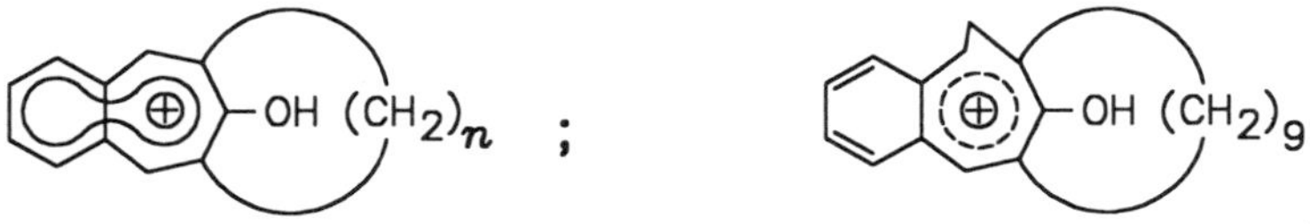

23: $n = 10, 12$ **24**

These phanes exhibit NMR spectra which resemble those of the hydroxy compounds **22**, in particular the tropylium protons at $\delta = 9.2$. The hydroxy[9]-benzohomotropyliophane (**24**, $n = 9$) was synthesized by homologation with carbene of **21a**.

There are a rather large number of double decker phanes (**25–30**)[9,10] which contain a tropylium ion unit and differ in the length, number, and position of the bridges. They were designed primarily to serve as models for the study of charge-transfer (CT) interactions. For more information on multiply bridged and related triple decker phanes containing tropylium units see Chapter 5.

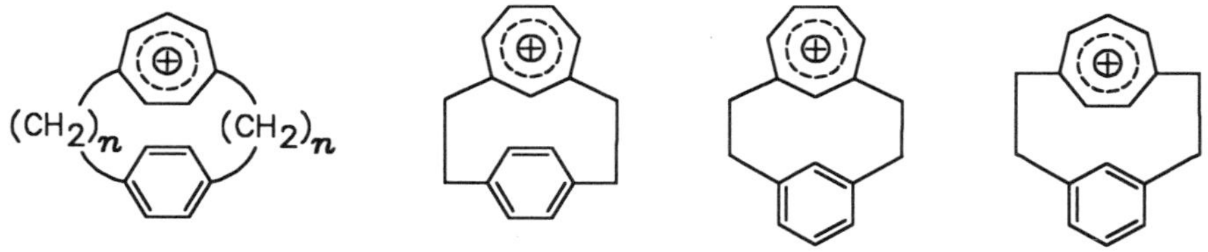

25: $n = 2$ **28**   **29**   **30**
26: $n = 3$
27: $n = 4$

In the following scheme the approach to the synthesis of these compounds from [2.2]paracyclophane is outlined.

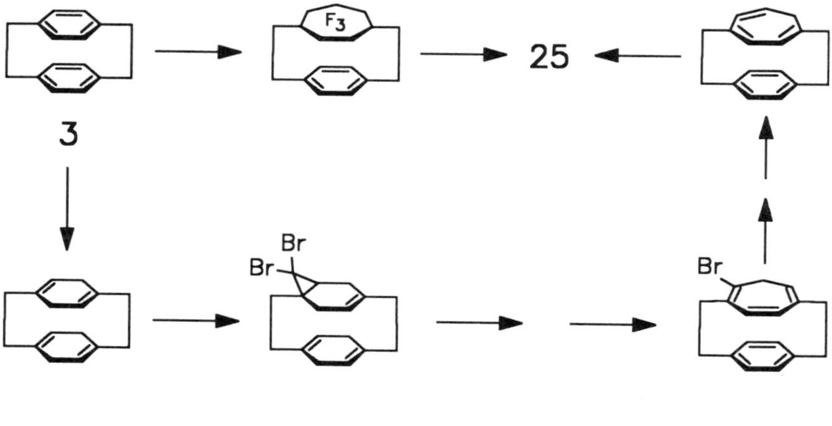

Molecular geometry in tropyliophanes

Although no X-ray structural analysis of tropyliophanes has been undertaken so far, one can make some conclusions about their molecular geometry by a comparison of the structures of doubly bridged tropyliophanes such as **23** and **28–30** with unsubstituted tropylium ion and quadruply bridged tropyliophanes such as **31**.[1] The tropylium protons in **31**, with their 'face-to-face' stereochemistry, appear at somewhat higher field ($\Delta\delta \approx 1.22$ p.p.m.) than those in unbridged tropylium ion ($\delta = 9.28$). This high-field shift is attributed to the dimagnetic ring current effect of the benzene ring and the enhanced electron density due to a charge-transfer effect.

25

28

29

30

31

The stair-like geometry of **29** and **30** can be deduced from the ^{1}H-NMR spectrum, as follows: the singlet absorptions for the benzene protons in **29** and **30** appear at much higher field than those in **31** ($\Delta\delta \approx 2.2$ p.p.m.), while the three remaining benzene protons appear at relatively lower field. The benzene protons in the ^{1}H-NMR spectra of paracyclophanes **25** and **28** are non-equivalent. This points to a fixed geometry (NMR time scale) at room temperature. Furthermore, in the case of **30**, there is no evidence of a conformational process (ring flip) up to a temperature of 120°C, whereas one observes an equilibration between the two equivalent conformers in **28** at higher temperatures. The energy barrier for this dynamic process is about 67 kJ/mol, that is, about 20 kJ/mol lower than that for [2.2]metacyclophane. This difference suggests a reduction in strain in the transition state as a result of a change in the size of the aromatic ring.[10]

Charge-transfer (CT) interactions

The extent of chemical shift difference (0.69 p.p.m.) between the benzene protons in **25** and those in [2.2]paracyclophane (**3**) can be used as an estimate of the charge-transfer interaction in **25**. The tropylium ring protons in **25** and **31** appear at somewhat higher field than those in unbridged tropylium fluoroborate, a consequence of a combination of ring current effect of the stacked aromatic rings and charge-transfer interaction. The higher field shift for the protons in **31**, as compared to those in **25**, is consistent with the results for the corresponding benzenophanes themselves.[11]

At the present time it is difficult to assess charge-transfer effects quantitatively because electronic spectra of various [2.2]phanes have been determined by different authors under diverse conditions. Boekelheide[12] has shown that a successive bathochromic shift in the charge-transfer bands accompanies a change from **25** to the corresponding triply and quadruply bridged benzeno-tropyliophanes. In this series both the distance between the two aromatic rings and their mobility are reduced. From this one can conclude that the charge-transfer interaction diminishes as the distance between the two aromatic units increases—as, for example, in going from [2.2]- to [3.3]- to [4.4]tropylio-phanes.

The influence of the relative positions of the two aromatic rings has been investigated by examining the spectra of the isomeric [2.2]tropyliobenzeno-phanes **25** and **28–30**. In those phanes in which the two aromatic rings more strongly overlap (**25, 28, 30**) one observes weak CT bands up to about 450 nm, whereas **29**, with *two* *m*-substituted rings, exhibits a strong band at 410 nm. In the triple decker compound **32** a broad band extends from 434 to 600 nm.[1]

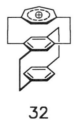

32

2.7.2.6 Tropyliocycloheptatrieno- and [2.2](1,4)tropyliophanes

The following scheme represents the synthesis of [2.2]tropyliophane bis(tetra-fluoroborate) (**33**) and of tropyliocycloheptatrienophane tetrafluoroborate (**34**).[13]

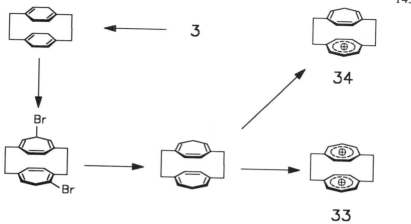

The electronic spectrum of **34** contains a band at 313 nm and shoulders at 360 and 423 nm, a pattern similar to that for [2.2](1,4)tropylioparacyclophane (**25**), which supports the electron-donor capacity of the cycloheptatriene ring.

The tropyliophane **33**, although unstable, has been characterized with the help of ^{1}H-NMR spectroscopy. By contrast the homologous benzologue [4.4](4,6)-benzotropyliophane (**35**) of **33** is stable.[1]

36: X = Cl

37: X = Br

38

144

2.7.2.7 Tropolonophanes

Only [2.2]paracyclo[2](3,7)tropolonophane (**38**) and its chloro and bromo derivatives **36** and **37** are known to date.[14] Their synthesis proceeds by the dithia[3.3]phane route with subsequent desulphurization by means of sulphone pyrolysis. Photochemical desulphurization led to less satisfactory results.

Tropolonophanes are of particular interest for the following reasons:

(a) Does the tropolone ring connected through a short bridge exist as the enolone form **A** or as one of the diketo forms **B** or **C**, which, because of their non-planarity, would result in diminished ring strain?

(b) If the enolone form predominates, what effect will ring strain or steric effects exert on the tendency for ketonization?

(c) What charge-transfer effects are exerted by the tropolone system?

(d) How is the overall strain distributed in the aromatic rings, in particular in the tropolone unit?

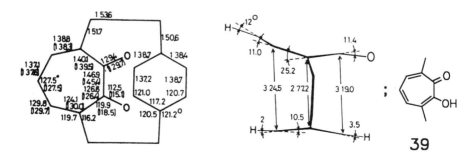

Geometry of tropolonophane

The X-ray crystal structure analysis[14] of **38** clearly showed that the molecule is a monomer; evidence for intermolecular hydrogen bonding is not observed in any compound except tropolone itself. The distance between the two aromatic rings (interplanar distance) is similar to that in [2.2]paracyclophane (**3**). Deformation of the tropolone ring from planarity is more pronounced than that of the benzene ring, a reflection of their relative delocalization. In spite of the distortion of the seven-membered ring, however, it exists in the tropolone form **A**, rather than as the keto form **B** or **C** (Figure 3).

Figure 3. X-ray structure of **38**. The comparative numbers in parentheses are those for the reference molecule dimethyltropolone (**39**). Bond lengths are given in pm

In solution compounds **38** and **36** also exist in the enolone form, showing no tendency for ketonization. Evidence for this is the positive colour reaction with $FeCl_3$ and formation of the methyl ether, both characteristic reactions of tropolone, as well as the failure of **38** to form the quinoxaline derivative.

In contrast to dimethyltropolone (**39**), **38** forms exclusively the bromo ketone **40** on bromination, the latter being converted on heating to **37**. Catalytic hydrogenation of **36** with Pd/C affords overhydrogenated product, also in contrast to the reaction of 3,7-dimethyltropolone (**39**). This behaviour points to the destabilization of the strained, bridged tropolone ring.[1]

Charge-transfer effects

Although the electronic spectra of **38** and **36** resemble that of dimethyltropolone (**39**), they exhibit broader bands and less extensive red shifts, all of which points to relatively less pronounced charge-transfer effects. In the spectrum of an acid solution, however, one notes a weak but significant shoulder at about 420 nm, corresponding to the charge-transfer absorption (400–430 nm) in tropyliopara-cyclophane (**25**). This points to generation of dihydroxytropylium ion **41**.[1]

In the rearrangement of the acetyl group (acetotropy) in the acetate **42**, stabilization of the intermediate through a charge-transfer effect seems to play a role. The activation energy of $\Delta G^{\ddagger} = 29 \, \text{kJ/mol}$ is considerably less than the value of 45 kJ/mol of the unbridged tropolone acetate. This difference is attributed to the stabilization of the intermediate **43** through charge-transfer interaction (arrow in formula **43**).

2.7.2.8 Tropoquinophanes

The interesting [2.2]paratropoquinophanes **44** and **45** were synthesized by the dithia[3.3]phane route.[15]

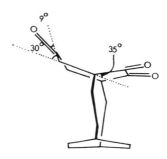

As is shown in the X-ray crystal structure of **44**, the strain imposed in the molecule through bridging is manifested in the deformation of the unsaturated ring. Deformation of the benzene ring is in fact less pronounced than in [2.2]paracyclophane (**3**), while the *p*-tropoquinone ring is distorted into a deep boat (Figure 4). Thus the diketone portion of the *p*-tropoquinone ring is

Figure 4. Molecular structure of **44** with bond angles[1]

oriented practically parallel to the centre plane of the benzene rings, but the dienone part is bent considerably out of the plane. These deformations are considerably reduced in [3.3]tropoquinophane (**46**).

Strain and deformation are also evident in the spectroscopic properties. The typical mass spectral fragmentation pattern in paratropoquinones involves the expulsion of three CO fragments, with formation of strong corresponding peaks, as is the case for the [3.3]phanes **46** and **47**. In the case of the strained [2.2]phanes **44** and **45**, however, one sees the loss of only two COs, which suggests that the loss of the third CO unit is a higher energy process.

46   **47**

The IR spectra of tropoquinones exhibit two groups of complex carbonyl bands of comparable frequency at about 1670 and 1605 cm^{-1}. Yet, by contrast to this behaviour for **46** and **47**, the latter bands diminish in intensity in **44** and **45**; in the case of **45**, presumably the most strained molecule, only absorption at 1670 cm^{-1} is observed.

From the ^{1}H-NMR spectra it is deduced that **45** exists in a stepladder shape, while **47** assumes a *syn* conformation, as indicated by the high-field shift of the quinone protons.[1]

Charge-transfer effects

The electronic spectra of **44–47** show a slight increase in intensity or a shoulder in the region 350–400 nm. Because of the bathochromic shift with an increase in solvent polarity, these are attributed to weak, broad charge-transfer bands, similar to those in benzoquinonophanes.

The first half-wave reduction potentials ($E_{1/2}$) for **44** (-0.64 V), **45** (-0.64 V), **46** (-0.53 V), and **47** (-0.52 V) can be attributed to a substantial increase in LUMO energy, by comparison to that for 3,7-dimethylpara-tropoquinone (-0.35 V). Because **46** and **47** clearly have the less distorted quinone rings, the energy difference of the LUMO energies [$(-0.52)-(-0.35) = -0.17$ V] can be attributed to charge-transfer effects.

2.7.3 [2.2]PHANES WITH 10π SYSTEMS

The [2.2]azulenophanes and [2]azuleno[2]phanes **48–50, 52** and **53** have all been synthesized from azulene bis(methyltrimethylammonium) diiodide by the sulphur route; that is, *via* desulphurization of dithia[3.3]phanes.[1] This starting material is more stable than the sensitive halomethylazulenes. **51** and **54** have been prepared only by intramolecular carbene insertion into [2.2]metacyclophane and [2.2]paracyclophane.[16]

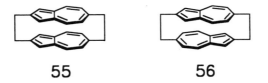

48   **49**

50a: R = H
50b: R = CH$_3$ **51**

52   **53**   **54**

The azulenophanes **55** and **56**, synthesized by Hofmann elimination, were subsequently separated by chromatography on H$_3$PO$_4$-impregnated silica gel.[17]

55   **56**

2.7.3.1 Geometry

As depicted in Figure 5, in all three phanes **48**,[18] **50a**,[19] and **53**[20] one sees deformation of the azulene rings (up to 9°), elongation of the C(9)–C(10) bond (azulene numbering) in **48** and **50a**, and elongation of the C(4)–C(5) and C(5)–C(6) bonds and bond angle deformation at C(4) and C(5) in **53**. This type of deformation is similar to that found in the [2.2]benzenophanes {[2.2]cyclophanes}. The distortion of the benzene rings in **48** and **53** closely resembles that in [2.2]metacyclophane.

^{1}H-NMR spectra show that this geometry is maintained in solutions and that it is conformationally fixed at room temperature. By comparison with open-chain reference compounds (substituted azulenes and xylene), the inner arene protons are strongly shifted to higher field while the outer protons are moderately shifted to lower field.

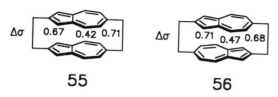

55   **56**

All molecules with two *m*-bridges (e.g. **49**, **50a,b**, **52**, **53**) exist in stepladder-like *anti* conformations.[1] At higher temperature **51** undergoes a flipping of the azulene ring, the energy barrier for this dynamic process being $\Delta G^{\neq} = 58$ kJ/mol ($T_c = 70°C$). The value for [2.2]metaparacyclophane, by comparison, is higher: 86 kJ/mol ($T_c = 146°C$). By contrast, there is no evidence for such a process in **48** up to 190°C; that is, the energy barrier is > 92 kJ/mol.

It is assumed that the differences in barriers to ring inversion stem from the extent of steric crowding in the intermediate between the inner H atom and the neighbouring benzene ring. The distance between the H atom and benzene ring in the intermediates for **48**, [2.2]metaparacyclophane (**4**), and **51** has been estimated to be 120, 150, and 175 pm, respectively.

The 2,6-bridged azulenophanes **55** and **56** (see above) appear to be conformationally flexible at room temperature, as shown by their IR and ^{1}H-NMR spectra.

2.7.3.2 Transannular interactions

The three major electronic absorption bands of *azulene* ($^1L_b \approx 670$ nm; $^1L_a \approx 400$ nm; $^1B_b \approx 270$ nm) undergo a modest bathochromic shift and are broadened, with loss of fine structure, when the azulene unit is incorporated into a double decker [2.2]phane system. In the [2]azuleno[2]benzenophane series, the ^1B_b band in the metacyclophanes **49** and **52** is shifted more significantly toward the red (≈ 2500 cm^{-1}) than is that in the paracyclophanes **48** and **51** (≈ 300 cm^{-1}). This difference can qualitatively be correlated with the molecular orbitals of azulene and xylene, as depicted in Figure 6. The HOMO of azulene, responsible for the ^1B_b band, does not interact effectively with the *p*-xylene unit in **48** and **51**, for reasons of geometry. Thus the next HOMO in these compounds differs little from that in azulene itself. On the contrary, the geometry in **49** and **52** allows substantial overlapping of the orbital coefficients of the inner carbon atoms, which causes the bathochromic shift. Azulenophane (**54**), which shows strong asymmetric overlap, exhibits bathochromic shifts in both ^1B_b and ^1L_a bands.

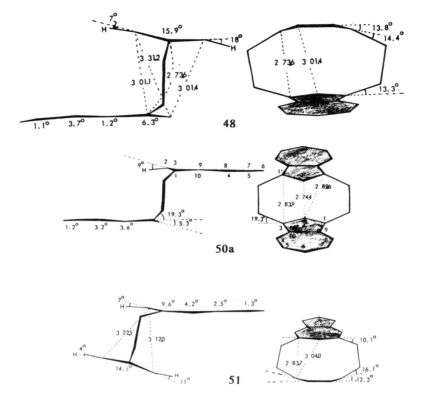

Figure 5. X-ray crystal structures for **48**, **50a**, and **51**.[18-20] Left: side view; right: front view. Bond lengths in pm; angles in degrees

The fact that the electronic spectra,[17] photoelectronic spectra,[21] and magnetic circular dichroism curves[22] of **55** resemble those of **56** points to a similar orbital interaction in the two azulene rings in these compounds. Lack of a strong orientation effect confirms the conception of azulene as a 10π annulene structure.

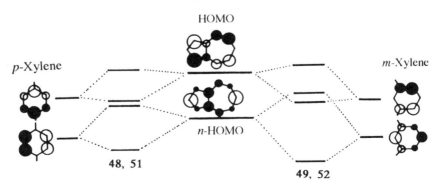

Figure 6. Orbital mixing between azulene and *p*- and *m*-xylene units in [2.2]phanes

The phanes **55** and **56** undergo fluorescence from the third excited singlet state. This anomaly comes about because of the large energy gap between the S_2- and S_3-states, as is the case with azulene itself, where the energy difference between S_1 and S_2 plays the major role.

Monoprotonation of [2.2]azulenophanes produces charge-transfer bands at 400 and 500 nm. On the basis of the acid concentration required in order to achieve maximum intensity of the charge-transfer bands, it was concluded that isomer **55** is a somewhat stronger base than isomer **56**. This points to a diminished orientation effect, which can be explained by a dipole–dipole interaction between the two azulene rings.

Transannular interactions in the radical anions of **48**, **50a**, **55**, and **56** have also been investigated.[23] The azulene ring in these compounds shows a marked tendency to form ion pairs with counter ions.

2.7.3.3 Transannular reactions

The azulenophanes **49** and **50a** can be subjected to photochemical transannular reactions, which are useful in the synthesis of azulenophenalene **58** and naphthodiazulene **60**.[24] The latter two structures are non-alternating isomers of benzpyrene and dibenzpyrene.

(a)

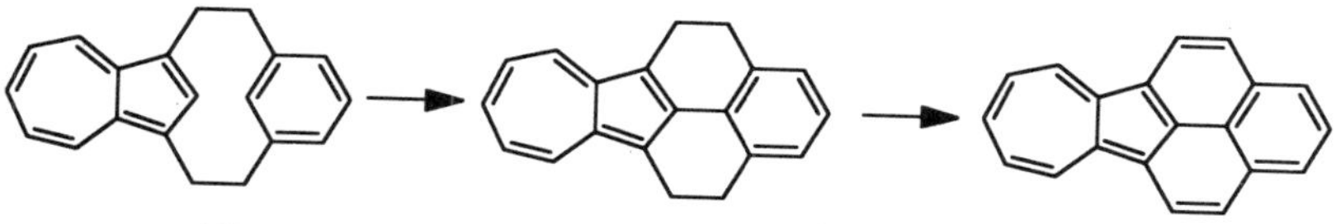

49   57   58

(b)

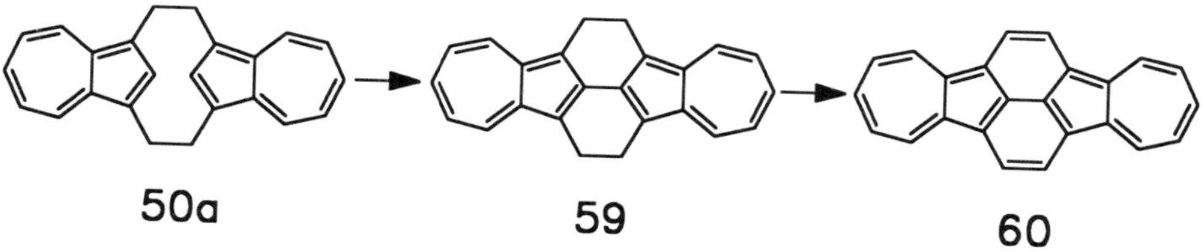

50a   59   60

49 ⟶

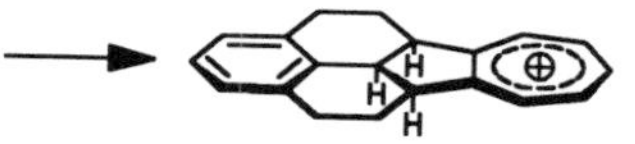

61

152

Acid treatment of **49** induces a transannular reaction, which leads to the tropylium ion **61**.[1]

References to Section 2.7

1. Review: S. Ito, Y. Fujise and Y. Fukazawa, in *Cyclophanes* (P. M. Keehn, S. M. Rosenfeld, Eds.), Vol. II, p. 485, Academic Press, New York 1983.
2. P. Bickert, V. Boekelheide and K. Hafner, *Angew. Chem.*, **94**, 308 (1982); *Angew. Chem. Int. Ed. Engl.*, **21**, 304 (1982).
3. S. Bradamante, A. Marchesini and G. Pagani, *Tetrahedron Lett.*, 4621 (1971).
4. M. W. Haenel, *Tetrahedron Lett.*, 1273 (1977).
5. Cf. Y. Fujise, T. Shiokawa, Y. Mazaki, Y. Fukazawa, M. Fujii and S. Ito, *Tetrahedron Lett.*, **23**, 1601 (1982).
6. R. E. Harmon, R. Suder and S. K. Gupta, *J. Chem. Soc., Perkin Trans. 1*, **1972**, 1746.
7. T. Hiyama, Y. Ozaki and H. Nozaki, *Chem. Lett.*, **1972**, 963; *Tetrahedron*, **30**, 2661 (1974).
8. R. E. Harmon, R. Suder and S. K. Gupta, *J. Chem. Soc., Chem. Commun.*, 472 (1972).
9. H. Horita, T. Otsubo, Y. Sakata and S. Misumi, *Tetrahedron Lett.*, 3899 (1976).
10. H. Horita, T. Otsubo and S. Misumi, *Chem. Lett.*, 1309 (1977).
11. R. Gray and V. Boekelheide, *J. Am. Chem. Soc.*, **101**, 2128 (1979).
12. Y. Sekine and V. Boekelheide, *J. Am. Chem. Soc.*, **103**, 1777 (1981).
13. J. G. O'Connor and P. M. Keehn, *Tetrahedron Lett.*, 3711 (1977).
14. N. Kato, Y. Fukazawa and S. Ito, *Tetrahedron Lett.*, 1113 (1979).
15. Cf. A. Kawamata, Y. Fukazawa, Y. Fujise and S. Ito, *Tetrahedron Lett.*, **23**, 4955 (1982).
16. T. Kawashima, T. Otsubo, Y. Sakata and S. Misumi, *Tetrahedron Lett.*, 1063 (1978).
17. N. Kato, H. Matsunaga, S. Oeda, Y. Fukazawa and S. Ito, *Tetrahedron Lett.*, 2419 (1979).
18. Y. Fukazawa, M. Aoyagi and S. Ito, *Tetrahedron Lett.*, 1067 (1978).
19. Y. Fukazawa, M. Aoyagi and S. Ito, *Tetrahedron Lett.*, 1055 (1979).
20. Y. Fukazawa, M. Sobukawa and S. Ito, *Tetrahedron Lett.*, **23**, 2129 (1982).
21. E. Heilbronner, in *Cyclophanes, II* (F. Vögtle, Ed.), *Top. Curr. Chem.*, **115**, 1 (1983).
22. H. Yamaguchi, K. Ninomiya, M. Fukuda, S. Ito, N. Kato and Y. Fukazawa, *Chem. Phys. Lett.*, **72**, 297 (1980).
23. M. Iwaizumi, Y. Fukazawa, N. Kato and S. Ito, *Bull. Chem. Soc. Jpn.*, **54**, 1299 (1981).
24. Y. Nesumi, T. Nakazawa and I. Murata, *Chem. Lett.*, 771 (1979); Y. Fukazawa, M. Aoyagi and S. Ito, *Tetrahedron Lett.*, **22**, 3879 (1981).

2.8 [2.2]ANTHRACENOPHANES

In 1961 Golden published an elegantly simple synthesis of [2.2]anthracenophane (**7**).[1,2] It consists of the reaction of sodium iodide in acetone with the easily obtainable bis(chloromethyl) compound of anthracene, which effects a 1,6-elimination to afford the corresponding bis-*exo*-methylene compound. Golden also discovered the smooth, photochemically induced cyclization of **7** to the cage compound **8**. By measuring the quantum yield of this photochemical reaction, Kaupp[3] could show that a diradical was being generated.

The [2.2]anthracenophanes below are of interest for the study of transannular π-electron interactions during excimer fluorescence. They are also fascinating compounds whose stereochemistry can be related to their photodimerization and ESR spectra. Both the *anti*- and *syn*-isomers 9 and 10 were prepared by dimerization of 1,4-anthraquinodimethane, which had been generated by a Hofmann elimination.[4] The *syn*-isomer 10 affords the cage structure 11 by way of a rapid light-induced cyclization, which is both thermally and photochemically reversible. The isomer 10 can also be rearranged thermally to the *anti* form 9. The static stereochemistry of 10 and 11 has been determined by X-ray crystallography.[1,5]

A crossed Hofmann elimination of quaternary ammonium hydroxides was used to obtain [2.2](1,4)(9,10)anthracenophane (12), as well as [2.2]para-cyclo(9,10)anthracenophane (13) and [2.2](1,4)naphthaleno(9,10)anthraceno-phane (14).[6] The UV bathochromic shifts in the long wavelength bands of these anthracenophanes, by comparison with their open-chain analogues, suggests significant transannular π-electron interaction.

Although the ESR spectrum of **12** appears complex, that of the radical anion of **7** suggests that the unpaired electron is delocalized into both anthracene rings. ENDOR studies of the radical anion of *anti*- and *syn*-isomers **9** and **10** showed that the spin density accumulates in the overlapping portions of the molecule.[7,8]

References to Section 2.8

1. Review: J. A. Reiss, in *Cyclophanes*, (P. M. Keehn, S. M. Rosenfeld, Eds.), Vol. II, p. 443, Academic Press, New York 1983.
2. J. H. Golden, *J. Chem. Soc.*, 3741 (1961).
3. G. Kaupp, *Liebigs Ann. Chem.*, 844 (1973).
4. T. Toyoda, I. Otsubo, T. Otsubo, Y. Sakata and S. Misumi, *Tetrahedron Lett.*, 1731 (1972).
5. T. Toyoda and S. Misumi, *Tetrahedron Lett.*, 1479 (1978); Radical ions: K. Müllen *et al.*, *J. Am. Chem. Soc.*, **113**, 1121 (1991).
6. A. Iwama, T. Toyoda, M. Yoshida, T. Otsubo, Y. Sakata and S. Misumi, *Bull. Chem. Soc. Jpn.*, **51**, 2988 (1978).
7. F. Nemoto, K. Ishizu, T. Toyoda, Y. Sakata and S. Misumi, *J. Am. Chem. Soc.*, **102**, 654 (1980).
8. F. Gerson, in *Cyclophanes, II* (F. Vögtle, Ed.), *Top. Curr. Chem.*, **115**, 57 (1983).

2.9 [2.2]PHENANTHRENOPHANES AND [2.2]PHENANTHRENOBENZENOPHANES

[2.2](3,6)Phenanthrenoparacyclophane (**15**) was obtained by sulphur extrusion of the corresponding dithiaphane. By treatment with $AlCl_3$ it could be converted to coronene.[1] The corresponding diene **16** was synthesized from the dithiaphane by a Wittig rearrangement and elimination of the bis-sulphoxide and also by a Stevens rearrangement by way of benzyne.[2,3] Irradiation of a degassed solution of diene **16** caused a change in colour to deep orange. The NMR spectrum of this solution fits the bridged annulene structure **17**. Oxidation with oxygen or iodine leads quantitatively to coronene. Boekelheide has also described [2.2]dihydrophenanthrenobenzenophandiene **18**.[4]

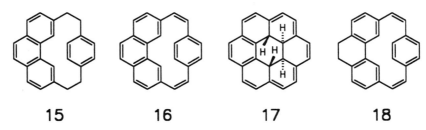

15   **16**   **17**   **18**

The phenanthrenonaphthaleno- and biphenylophanes **19** and **20**, also prepared by ring contraction from the dithia[3.3]phane precursors, exhibit NMR spectra which suggest *syn* conformations. Attempts to convert **19** and **20** or another phenanthrenophane **21** to polycyclic aromatics of the circulene type were unsuccessful.[6]

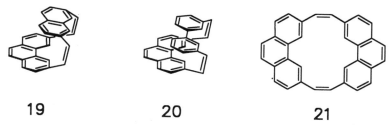

19　　　　　　**20**　　　　　　　　**21**

Haenel and Staab have described the synthesis of an isomeric mixture of [2.2](2,7)phenanthrenophane **22** and other similar unsymmetrical phenanthrenophanes containing biphenyl units.[7] During the formation of **24**, a potential intermediate in a synthesis of kekulene, there was also produced a larger ring [2₅](2,7)phenanthrenophane.[8]

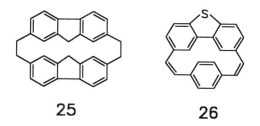

22　　　　　　**23**　　　　　　　**24**

Some examples of [2.2]phanes containing fluorene, fluorenone (Haenel), and dibenzothiophene units may also be mentioned; one such structure is [2.2](2,7)fluorenophane (**25**)[9] (see Section 2.7.2). Thiacoronene has been synthesized from **26**.[3,10]

25　　　　　　**26**

References to Section 2.9

1. J. T. Craig, B. Halton and S. F. Lo, *Aust. J. Chem.*, **28**, 913 (1975).
2. P. J. Jessup and J. A. Reiss, *Aust. J. Chem.*, **30**, 843 (1977).
3. R. B. DuVernet, O. Wennerström, J. Lawson, T. Otsubo and V. Boekelheide, *J. Am. Chem. Soc.*, **100**, 2457 (1978).
4. R. B. DuVernet, T. Otsubo, J. A. Lawson and V. Boekelheide, *J. Am. Chem. Soc.*, **97**, 1629 (1975).

5. D. N. Leach and J. A. Reiss, *Aust. J. Chem.*, **32**, 361 (1979).
6. Review: J. A. Reiss, in *Cyclophanes* (P. M. Keehn, S. M. Rosenfeld, Eds.), Vol. II, p. 443, Academic Press, New York 1983; [5]Circulene type: J. S. Siegel *et al.*, *J. Am. Chem. Soc.*, **114**, 1921 (1992); L. T. Scott *et al.*, *J. Am. Chem. Soc.*, **114**, 1920 (1992).
7. M. W. Haenel and H. A. Staab, *Tetrahedron Lett.*, 3585 (1970); H. A. Staab and M. W. Haenel, *Chem. Ber.*, **106**, 2190 (1973).
8. P. Baumgartner, R. Paioni and W. Jenny, *Helv. Chim. Acta*, **54**, 266 (1971).
9. M. W. Haenel, *Tetrahedron Lett.*, 3121 (1976).
10. J. Lawson, R. DuVernet and V. Boekelheide, *J. Am. Chem. Soc.*, **95**, 956 (1973).

2.10 PYRENOPHANES AND HIGHER CONDENSED PHANES

2.10.1 PYRENOPHANES

[2.2](2,7)Pyrenophane (**28**) was described as early as 1975 by Misumi,[1] Staab,[2] and Mitchell *et al.*,[3] in each case the synthesis involving a transannular ring closure of a metacyclophane. The X-ray crystal structure analysis of **28**[5] and its corresponding diene[6] confirm the bridged pyrene rings. Excimer formation is solvent dependent; ESR and ENDOR spectra have been reported for the radical anions of **28**.[7]

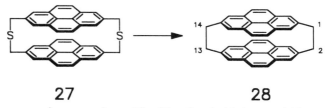

27   **28**

Several isomers of pyrenophane **28** with other bridging could be synthesized by the dithia[3.3]phane route. Some investigated in particular for their fluorescence have been [2.2](1,6)-, [2.2](1,6)(2,7)-, [2.2](1,8)-, and *[2.2](1,3)-pyrenophane* (**29**). A few 'mixed', 'unsymmetrical' pyrenonaphthaleno- and pyrenobenzenophanes, such as **30** and **31** are known, as well as the more highly condensed [2.2](1,3)phane **32**, reported by Misumi.[4]

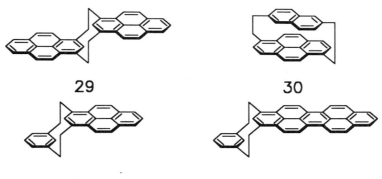

29   **30**

31   **32**

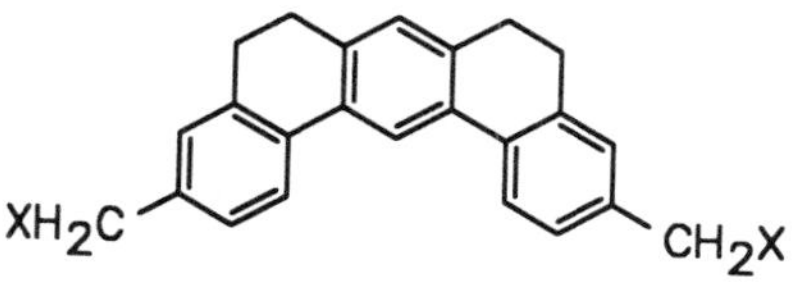

33

34

35

36

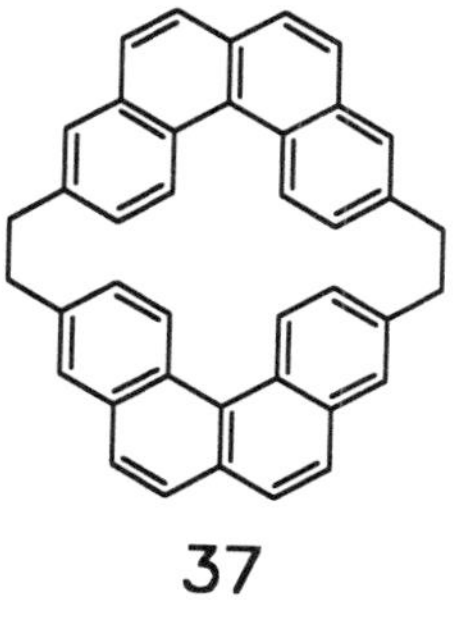

37

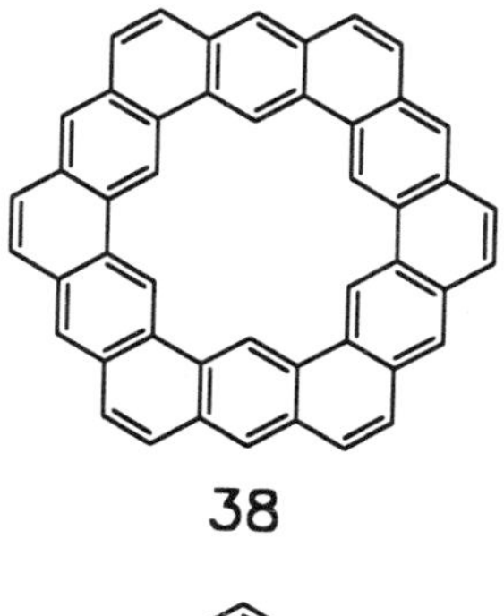

38

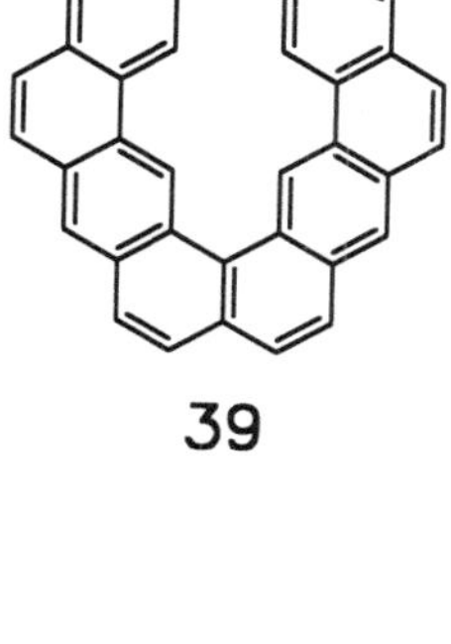

39

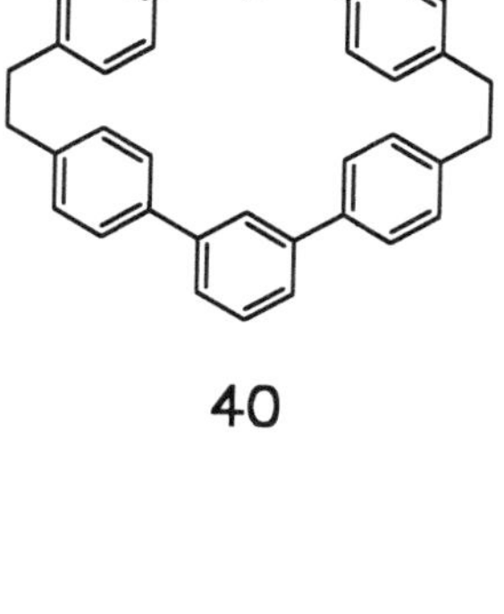

40

2.10.2 [2.2]PHANES WITH MORE HIGHLY CONDENSED RINGS AND OLIGOPHENYLENE UNITS

By starting with the tetrahydrodibenzanthracene system **33**, Staab and Vögtle[8] and Staab and Diederich[9] prepared the [2.2]phanes **34–36** and **37**.[10] These, in turn, could be converted by electrocyclization or dehydrogenation to kekulene **38**[11] and its smaller analogue **39**.[12] The *m*-terphenylophane **40** served as model compound for the macrocyclization.[8] [2.2](4,4′)Biphenylophane (**41**) was obtained in good yield by pyrolysis of the corresponding disulphone.[13]

41   **42**   **43**

[2.2](3,3′)(4,4′)Biphenylophane (**42**) and its corresponding 1,15-diene show temperature-dependent ¹H-NMR spectra; a ring inversion process (ring flipping), however, is ruled out as the cause.[14] By contrast, the triply bridged [2.2.2]biphenylophane **43** prepared by Vögtle *et al* is conformationally rigid.[15]

Among the phanes **44–48** stemming from quaterphenyl, quinquephenyl, and *p*-terphenyl, the 4,4‴-bridged **44**[16] is conformationally flexible, whereas **45**,[17] **46**,[18a] and the doubly bridged **47**[18b] are rigid. The latter consist of screw-shaped, oriented benzene rings and are thus helical chiral. Because no efficient chiral resolving agent for non-functionalized racemates was available just at the time these molecules were synthesized, the compounds have as yet not been separated into their enantiomers. Resolution today with 'Okamoto's resin' is very promising.

44   **45**   **46**

47   **48**

'Oligophenylenophanes' such as **44–48** should properly be designated as [2.0.0.0]cyclophanes (or benzenophanes), as dictated by the nature of the 'aromatic nucleus'. The abbreviation of the ring structure, however, greatly simplifies the names of these phanes. For a comparison, see Section 2.14 {'[2.0.0.0]Phanes'} and Chapter 7 {[m_n]Phanes}.

An energy barrier $\Delta G_c^{\neq} = 56 \, \text{kJ/mol}$ has been determined for the ring inversion process A $\rightleftharpoons$ B in [2.2]terphenylo(1,4)benzenophane **49**.[19]

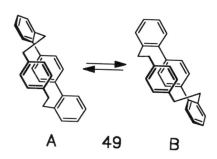

A   49   B

References to Section 2.10

1. T. Umemoto, S. Satani, Y. Sakata and S. Misumi, *Tetrahedron Lett.*, 3159 (1975).
2. (a) D. Schweitzer, K. H. Hausser, R. G. H. Kirrstetter and H. A. Staab, *Z. Naturforsch.*, **31A**, 1189 (1976).
   (b) H. Irngartinger, R. G. H. Kirrstetter, C. Krieger, H. Rodewald and H. A. Staab, *Tetrahedron Lett.*, 1425 (1977).
3. R. H. Mitchell, R. J. Carruthers and J. C. M. Zwinkels, *Tetrahedron Lett.*, 2585 (1976).
4. Review: J. A. Reiss, in *Cyclophanes* (P. M. Keehn and S. M. Rosenfeld, Eds.), Vol. II, p. 443, Academic Press, New York, 1983.
5. H. Irngartinger, R. G. H. Kirrstetter, C. Krieger, H. Rodewald and H. A. Staab, *Tetrahedron Lett.*, 1425 (1977).
6. Y. Kai, F. Hama, N. Yasuoka and N. Kasai, *Acta Crystallogr.*, **B34**, 1263 (1978).
7. Review: F. Gerson, in *Cyclophanes, II* (F. Vögtle, Ed.), *Top. Curr. Chem.*, **115**, 57 (1983).
8. F. Vögtle and H. A. Staab, *Chem. Ber.*, **101**, 2709 (1968).
9. F. Diederich and H. A. Staab, *Angew. Chem.*, **90**, 383 (1978); *Angew. Chem. Int. Ed. Engl.*, **17**, 372 (1978); C. Krieger, F. Diederich, D. Schweitzer and H. A. Staab, *Angew. Chem.*, **91**, 733 (1979); *Angew. Chem. Int. Ed. Engl.*, **18**, 699 (1979).
10. R. Peter and W. Jenny, *Chimia*, **19**, 45 (1965).
11. H. A. Staab and F. Diederich, *Chem. Ber.*, **116**, 3487 (1983); H. A. Staab, F. Diederich, C. Krieger and D. Schweitzer, *Chem. Ber.*, **116**, 3504 (1983); F. Vögtle and H. A. Staab, *Chem. Ber.*, **101**, 2709 (1968); H. A. Staab and M. Sauer, *Liebigs Ann. Chem.*, 742 (1984); S. E. Stein and R. L. Brown, *J. Am. Chem. Soc.*, **113**, 787 (1991).
12. D. J. H. Funhoff and H. A. Staab, *Angew. Chem.*, **98**, 757 (1986); *Angew. Chem. Int. Ed. Engl.*, **25**, 742 (1986).
13. M. W. Haenel and H. A. Staab, *Tetrahedron Lett.*, 3585 (1970); H. A. Staab and M. W. Haenel, *Chem. Ber.*, **106**, 2190 (1973).
14. D. N. Leach and J. A. Reiss, *Tetrahedron Lett.*, 4501 (1979); *Aust. J. Chem.*, **33**, 823 (1980).

160

15. F. Vögtle and G. Steinhagen, *Chem. Ber.*, **111**, 205 (1978).
16. F. Vögtle, M. Atzmüller, W. Wehner and J. Grütze, *Angew. Chem.*, **89**, 338 (1977); *Angew. Chem. Int. Ed. Engl.*, **16**, 325 (1977).
17. F. Vögtle and E. Hammerschmidt, *Angew. Chem.*, **90**, 293 (1978); *Angew. Chem. Int. Ed. Engl.*, **17**, 268 (1978).
18. (a) E. Hammerschmidt and F. Vögtle, *Chem. Ber.*, **112**, 1785 (1979).
    (b) E. Hammerschmidt and F. Vögtle, *Chem. Ber.*, **113**, 3550 (1980).
19. K. Böckmann and F. Vögtle, *Liebigs Ann. Chem.*, 467 (1981).

2.11 [2.2]- AND [m.n]DONOR–ACCEPTOR PHANES

Up until now we have discussed [2.2]phanes including some with more than two bridging units. Included in this section are those with longer bridging chains and those containing multiple layers.

In 1844 Wöhler discovered the dark green, crystalline *quinhydrone*, easily obtainable from hydroquinone and *p*-benzoquinone. This prompted Staab to design a cyclophane-like structure, in which these units were geometrically fixed. The binding between the donor (hydroquinone) and acceptor (quinone) is not dependent solely on hydrogen bonding, as is shown by the fact that *p*-dimethoxybenzene and several other π-donors besides hydroquinone, all incapable of hydrogen bonding, also form intensely coloured crystalline complexes with quinone. The properties of intramolecularly bridged CT complexes are consistent with this picture. Staab and Rebafka synthesized the parallel and crossed oriented [2.2](2,6)quinhydronophanes **6** and **7** and found a strong orientation effect for the charge-transfer interaction.[2] The extinction coefficient for the charge-transfer band in **6** is about tenfold that in **7**. Detailed studies of the zero field splitting parameter[8] and molecular orbital analysis[9] were carried out for such CT-paracyclophanes with [2.2]para-,[3,4] [2.2]meta-,[5] [2.2]metapara-,[6] and [m.n]phane structures.[7]

The analogous [2.2](1,3)quinhydronophane **8** could also be synthesized. Its geometry allows only weak intermolecular hydrogen bonding interaction, by comparison to the strong intramolecular hydrogen bonding found in **6** and **7**.

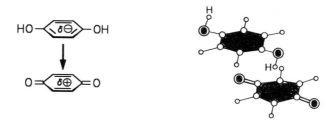

Figure 1. Quinhydrone. Left: donor–acceptor effect; right: 'face-to-face' orientation of hydroquinone and *p*-benzoquinone in the crystal[1]

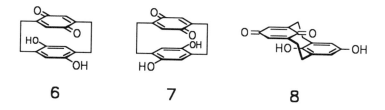

6   **7**   **8**

The tetrasubstituted [2₄](1,2,4,5)cyclophanes **9–11** were prepared by a combination of the dithiacyclophane-sulphur extrusion route and a stepwise construction of the additional bridges.[1]

9   **10**   **11**

One finds bands in the short wavelength region of the absorption spectrum of **11** which are scarcely shifted from those in the bis-quinone **15**.[11] The spectrum of **10** contains an intense long wavelength CT band extending from 330 to 650 nm, which, on the basis of its concentration independence, is attributed to an intramolecular CT complex. The CT absorption for **11** is surprisingly similar to that of the pseudo-*geminal* [2.2]paracyclophane quinhydrone **6**, even though a longer wavelength CT absorption would be predicted for the shorter donor–acceptor distance in **11**. Apparently the greater deviation from planar orientation of the donor and acceptor in **11**, because of more extensive ring strain, shifts the donor–acceptor overlapping to shorter wavelength. This is also the conclusion in a comparison between [2.2]- and [3.3]paracyclophane quinhydrones. A comparison of the CT bands in **11** with those in the pseudo-*geminal* [2.2]paracyclophane quinhydrone **6** also shows that the band width in **11**, in which the donor–acceptor orientation is rigidly held, is not diminished over that of the more flexible compound (Figure 2).

The CT absorption in **12** [$\lambda_{max} = 370$ ($\epsilon = 1350$), in dioxane; 383 (1470), in CHCl₃] appears about 100 nm towards shorter wavelength than that in **11**. This is attributed to the reduced donor ability because of the bending of the methoxy groups out of conjugation with the donor ring.[11]

12

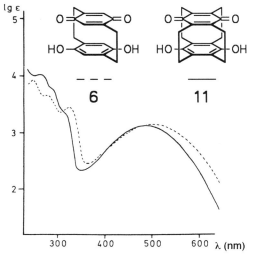

Figure 2. CT absorption bands for $[2_4](1,2,4,5)$cyclophane quinhydrone (**11**) (———) and for the pseudo-*geminal* [2.2]paracyclophane quinhydrone (**6**) (-----) in methanol[11]

The quadruply bridged quinonebenzenophane **16** was obtained from $[2_4](1,2,4,5)$cyclophane (**13**), which was subjected to singlet oxygen oxidation, such that the epidioxide **14** was first formed. This in turn was converted with base to the stable hydroxydienone **15**, a tautomer of the corresponding hydroquinone, which could be transformed into the quinone **16**.

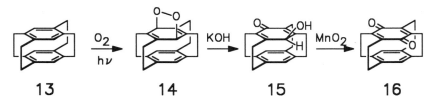

Staab *et al.* also studied charge-transfer effects in the naphthalenophane series.[13] The *anti-* and *syn-*isomers of [2.2](1,4)naphthalenophanes **17** and **18** were generated in a ratio of 17:1 (combined yield 26%) by pyrolysis of the quaternary ammonium hydroxide **1** in xylene, the isomers being separated by chromatography. Demethylation and oxidation without isomerization led to the corresponding naphthoquinonophanes **19** and **20**. At higher temperatures (230°C, 1 h, argon), however, **20** rearranges to **19**, the result being a 1:1 mixture of the two isomers.

In a similar way, [2](1,4)naphthaleno[2]paracyclophane (**21**) was obtained in 5–10% yield by crossed pyrolysis of I and II.[14] A comparison was made between the electronic spectra of the quinones **19**, **20**, and the three derived from **21**.[13]

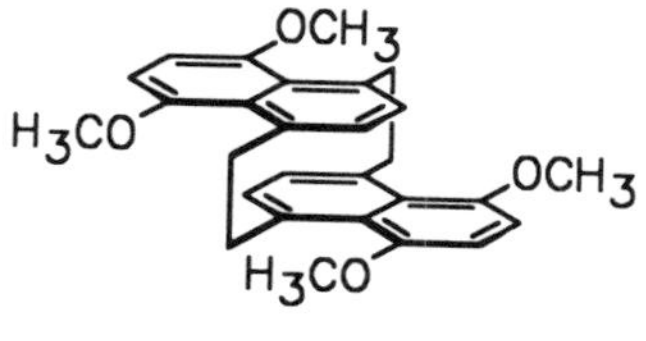

17

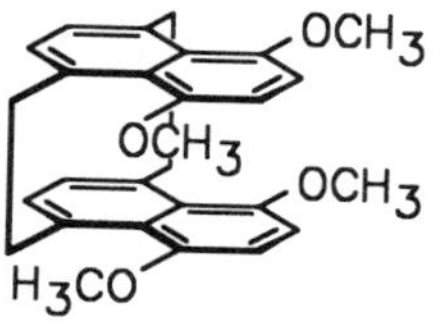

18

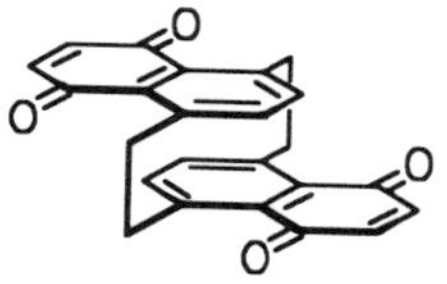

19

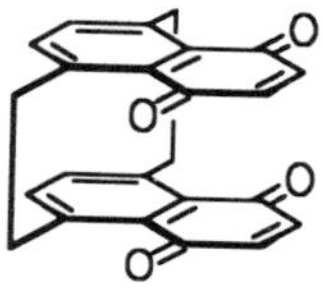

20

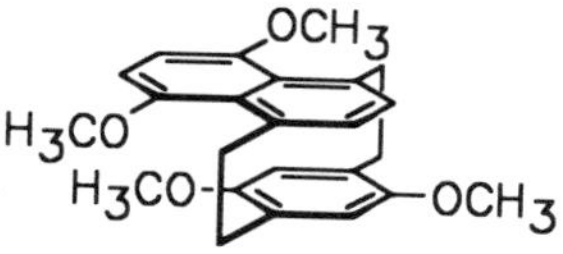

21

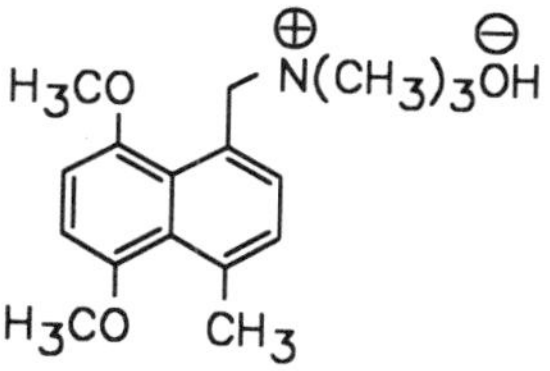

I

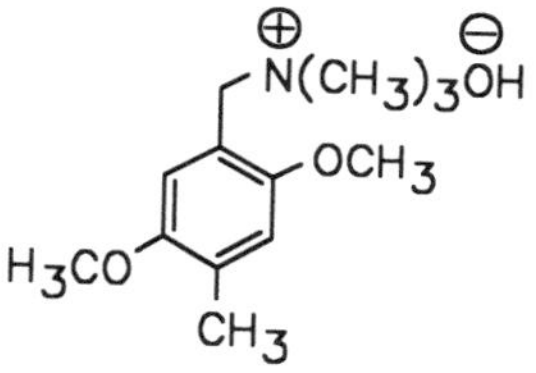

II

2.11.1 MULTI-LAYER DONOR–ACCEPTOR PHANES

Although the multi-layer phanes will be discussed in context in Chapter 6 ('Multi-layered Phanes'), some selected examples of this structural type are being included here because of their significance.[15] In the following multi-layer donor–acceptor phanes, those functioning as donors are benzene rings, methoxy-, hydroxy-, and tetrathiafulvalene (TTF) groups: acceptors are quinone-, tetracyanoquinodimethane (TCNQ)-, terephthalic ester-, and *p*-dicyanobenzene units.

2.11.1.1 Multi-layer paracyclophane quinones

Presented in the following structures are some of the multi-layer paracyclophane quinones synthesized by Staab[16] and Misumi *et al.*,[17] together with their parent double-layer compound **22**.[15]

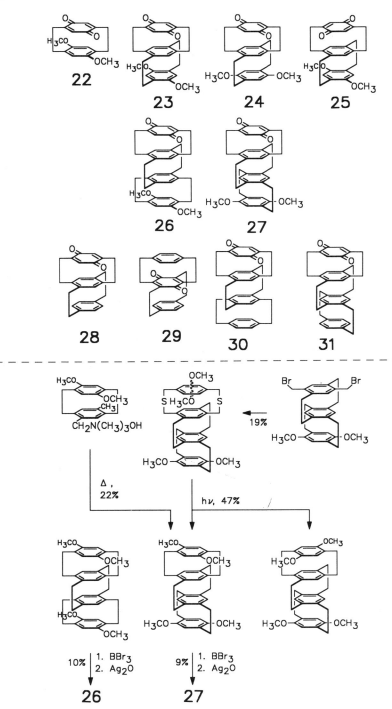

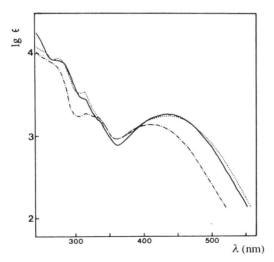

The synthetic methodology is summarized in the scheme on p. 164.[15]

The electronic spectra of all donor–acceptor cyclophanes **22–31** exhibit broad, featureless charge-transfer bands in the long wavelength region (Figure 3).[15] It is remarkable that the CT transitions in the multi-layer donor–acceptor phanes, contrary to the case for the double-layer phanes (**22**), show almost no dependence on the relative orientation of the donor and acceptor units. The isomeric pairs **23/25** and **26/27** are characterized by practically superimposable absorption spectra (Figure 3).

In the dimethoxy quinone series **22**, **23**, and **26**, the shifts in the charge-transfer band maxima do not parallel the increase in the number of aromatic layers. This CT band maximum in **22** is shifted to longer wavelength because the dimethoxybenzene unit is so closely situated to the quinone structure. Unlike

Figure 3. Electronic spectra of three isomeric, triple-layer quinonophanes **23** (········), **24** (-·-·-·-), and **25** (————) in dichloromethane[15]

166

22, the multi-layer compounds **23** and **26** possess two types of donors: the durene unit (the middle or inner benzene ring) and the dimethoxybenzene unit. In the case of **26** one can conclude that the donor unit is durene, because its absorption curve is practically superimposable with that of **30**. Thus, the methoxy groups confer essentially no influence on the CT transition, and so one observes no orientation dependence between the quinone and dimethoxybenzene units in the CT transitions in **23–27**.[15]

2.11.1.2 Multi-layer paracyclophanes containing tetracyanoquinodimethane units

The electronic spectra of double- and triple-layer 'TCNQ-phanes' **32**[18] and **33–36**[19] exhibit unusually broad CT bands up to a width of 1000 nm. As is

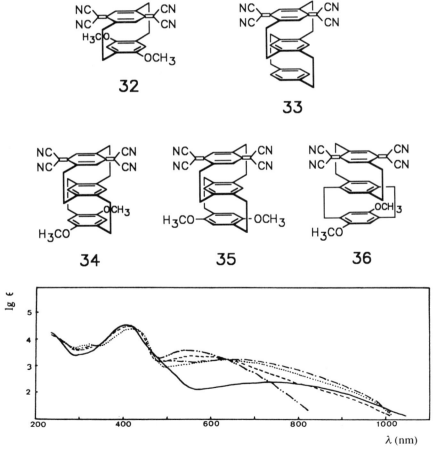

Figure 4. Electronic spectra of TCNQ-phanes **32** (———), **33** (·······), **34** (--------), **35** (-·-·-·-), and **36** (········) in dichloromethane[15]

shown in Figure 4, the absorption maximum in the double-layer parent structure **32** is shifted to longer wavelength, by comparison to the maxima in the triple-layer compounds **34–36**. The fact that the CT bands in **34–36** show a pronounced red shift, by comparison to **33**, suggests that the dimethoxybenzene unit, in the presence of strong acceptors, functions as a strong π-donor and transmits its electron density through the 'electronic cavity' of the sandwiched, middle ring.

Intramolecular exciplex formation

From fluorescence and nanosecond laser photolysis studies it was shown that the excited singlet state of the intramolecular complex for the three isomeric triple-layer dicyano[2.2]paracyclophanes **37–39** possesses the structure $(DD)^+A^-$ (D = donor, A = acceptor).[21]

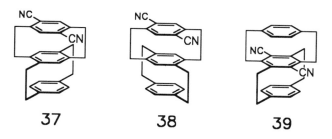

2.11.1.3 Other [2.2]- and [3.3]phanes as donor–acceptor systems

Additional compounds containing quinones, quinhydrones, and other donor–acceptor pairs have been prepared in the [2.2]- and [3.3]metaparacyclophane series (Staab *et al.*[7]). The donor–acceptor compound **40a** with a *meta/para*[2.2]phane structure has been resolved into enantiomers, whose circular dichroism has been measured.

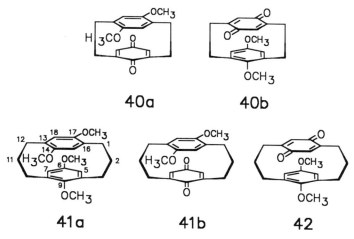

168

X-ray structures have been reported for the [2.2]- and [3.3]phanes **43–45** with TTF units and for the intermolecular charge-transfer complex **46** (Staab).[22]

43   **44**   **45**

:   (1:4)

46

References to Section 2.11

1. Cf. H. R. Christen and F. Vögtle, *Organische Chemie. Von den Grundlagen zur Forschung*, Vol. II, Salle/Sauerländer, Frankfurt/Aarau, 1990.
2. H. A. Staab and W. Rebafka, *Chem. Ber.*, **110**, 3333 (1977).
3. H. A. Staab and H. Haffner, *Chem. Ber.*, **110**, 3358 (1977); H. A. Staab and V. Taglieber, *Chem. Ber.*, **110**, 3366 (1977).
4. W. Rebafka and H. A. Staab, *Angew. Chem.*, **85**, 831 (1973); *Angew. Chem. Int. Ed. Engl.*, **12**, 776 (1973); *Angew. Chem.*, **86**, 234 (1974); *Angew. Chem. Int. Ed. Engl.*, **13**, 203 (1974); H. A. Staab and W. Rebafka, *Chem. Ber.*, **110**, 3333 (1977); H. A. Staab, C. P. Herz and H.-E. Henke, *Chem. Ber.*, **110**, 3351 (1977); H. A. Staab and C. P. Herz, *Angew. Chem.*, **89**, 839 (1977); *Angew. Chem. Int. Ed. Engl.*, **16**, 799 (1977); H. A. Staab, A. Döhling and C. Krieger, *Liebigs Ann. Chem.*, 1052 (1981); H. A. Staab, C. P. Herz, C. Krieger and M. Rentzea, *Chem. Ber.*, **116**, 3813 (1983); H. A. Staab, B. Starker and C. Krieger, *Chem. Ber.*, **116**, 3831 (1983); see also R. Gleiter, W. Schäfer and H. A. Staab, *Chem. Ber.*, **121**, 1257 (1988).
5. H. A. Staab, C. P. Herz and A. Döhling, *Tetrahedron Lett.*, 791 (1979); H. A. Staab, C. P. Herz, A. Döhling and C. Krieger, *Chem. Ber.*, **113**, 241 (1980); H. A. Staab, L. Schanne, C. Krieger and V. Taglieber, *Chem. Ber.*, **118**, 1204 (1985).
6. H. A. Staab, M. Jörns, C. Krieger and M. Rentzea, *Chem. Ber.*, **118**, 796 (1985); M. Tashiro, K. Koya and T. Yamato, *J. Am. Chem. Soc.*, **105**, 6650 (1983).
7. H. A. Staab, M. Jörns, C. Krieger and M. Rentzea, *Chem. Ber.*, **118**, 796 (1985); see also C. Krieger, J. Weiser and H. A. Staab, *Tetrahedron Lett.*, **26**, 6055 (1985); [3.3]Metacyclophane type: H. A. Staab, A. Döhling and C. Krieger, *Tetrahedron Lett.*, **32**, 2215 (1991).
8. D. Schweitzer, K. H. Hausser, V. Taglieber and H. A. Staab, *Chem. Phys.*, **14**, 183 (1976): H. Vogler, G. Ege and H. A. Staab, *Mol. Phys.*, **33**, 923 (1977).
9. H. Vogler, G. Ege and H. A. Staab, *Tetrahedron*, **31**, 2441 (1975).

10. M. Tashiro, K. Koya and T. Yamato, *J. Am. Chem. Soc.*, **104**, 3707 (1982).
11. H. A. Staab and V. M. Schwendemann, *Liebigs Ann. Chem.*, 1258 (1979).
12. I. Erden, P. Gölitz, R. Näder and A. de Meijere, *Angew. Chem.*, **93**, 605 (1981); *Angew. Chem. Int. Ed. Engl.*, **20**, 583 (1981).
13. W. Rebafka and H. A. Staab, *Angew. Chem.*, **85**, 831 (1973); *Angew. Chem. Int. Ed. Engl.*, **12**, 776 (1973).
14. R. Gray, L. G. Harruff, J. Krymowski, J. Peterson and V. Boekelheide, *J. Am. Chem. Soc.*, **100**, 2892 (1978).
15. Review: S. Misumi, in *Cyclophanes* (P. M. Keehn and S. M. Rosenfeld, Eds.), Vol. II, p. 573, Academic Press, New York, 1983.
16. H. A. Staab, U. Zapf and A. Gurke, *Angew. Chem.*, **89**, 841 (1977); *Angew. Chem. Int. Ed. Engl.*, **16**, 801 (1977); H. A. Staab and U. Zapf, *Angew. Chem.*, **90**, 807 (1978); *Angew. Chem. Int. Ed. Engl.*, **17**, 757 (1978).
17. H. Machida, H. Tatemitsu, Y. Sakata and S. Misumi, *Tetrahedron Lett.*, 915 (1978); H. Machida, H. Tatemitsu, T. Otsubo, Y. Sakata and S. Misumi, *Bull. Chem. Soc. Jpn.*, **53**, 2943 (1980).
18. H. A. Staab and H.-E. Henke, *Tetrahedron Lett.*, 1955 (1978).
19. H. Tatemitsu, B. Natsume, M. Yoshida, Y. Sakata and S. Misumi, *Tetrahedron Lett.*, 3459 (1978); M. Yoshida, Y. Tochiaki, H. Tatemitsu, Y. Sakata and S. Misumi, *Chem. Lett.*, 829 (1978).
20. M. Yoshida, H. Tatemitsu, Y. Sakata, S. Misumi, H. Masuhara and N. Mataga, *J. Chem. Soc. Chem. Commun.*, 587 (1976).
21. H. Masuhara, N. Mataga, H. Yoshida, Y. Tatemitsu, Y. Sakata and S. Misumi, *J. Phys. Chem.*, **81**, 879 (1977).
22. Review: P. M. Keehn, in *Cyclophanes* (P. M. Keehn and S. M. Rosenfeld, Eds.), Vol. I, Academic Press, New York, 1983.

2.12 [2.2]HETEROPHANES

2.12.1 SYNTHESIS

2.12.1.1 Dithia[3.3]phane route with subsequent ring contraction

[2.2](2,6)Pyridinophane (**10**) was first prepared as long ago as 1958 by Baker *et al.*, through a Wurtz coupling reaction (see below).[2] A preferable route from a preparative standpoint, however, is desulphurization of dithiapyridinophane worked out by Rasmussen and Martell in 1971[3] by starting with the 2,11-dithia[3.3](2,6)pyridinophane (**7**) prepared by Vögtle *et al.*[4] After deoxygenation of the pyridine *N*-oxide functional groups with iron/TFA, the sulphone pyrolysis proceeds in 46% yield.[3]

[2.2]Metacyclo(3,5)pyridinophane (**15**) was also obtained by the dithiaphane route, although the desulphurization step was carried out by photolysis of the sulphide. In the case of **15** the yield was 23%. The fact that the monosulphide **14**, a 2-thia[3.2]phane, could be isolated from the reaction mixture indicates that the removal of sulphur is a stepwise process.

Pyridinophanes with other forms of bridging—**16**, **17**, **18**—could be synthesized by analogous methodology.[1b] The [2.2]paracyclopyridinophane **16**, prepared by Hofmann elimination, can also be obtained by pyrolysis of the corresponding bis-sulphone. Irradiation of the appropriate dithiaheterophane

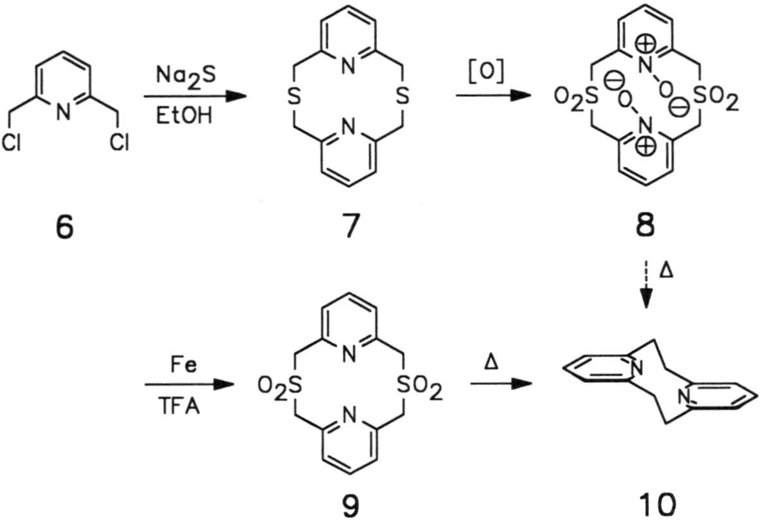

6 7 8

9 10

11 12 13

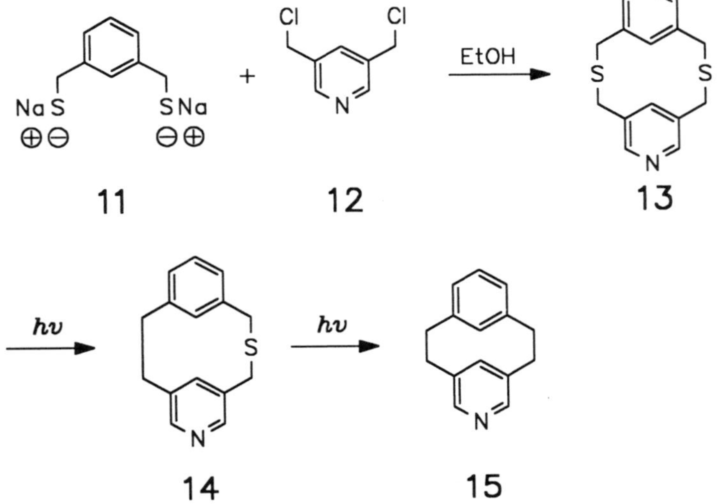

14 15

16 17 18

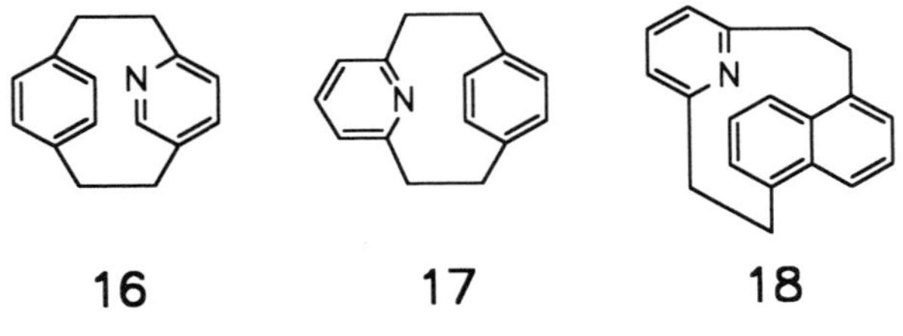

in triethyl phosphite affords [2.2]paracyclo(2,6)pyridinophane (**17**) in good yield. Whereas pyrolysis to **10** (see above) requires prior reduction of the N-oxides, the heterophane **17** can be obtained by direct pyrolysis of the sulphone N-oxide precursor.[7] The naphthalenophane **18**[18] is obtained in the substantial yield of 65% by pyrolysis of the corresponding bis-sulphone.

2.12.1.2 Stevens rearrangement

The identical starting materials, the dithia[3.3]phanes, can be used as for the sulphone pyrolysis route. Thus, the pyridinophane **22** was obtained from **7** by sequential methylation with Meerwein reagent (trimethyloxonium tetrafluoroborate) and treatment with potassium *tert*-butoxide and then a repetition of the same two steps. The intermediate **20**, one of a group of separable isomers, afforded the diene **22** in 20% yield.[9]

7   19   20

21   22

The Stevens rearrangement was also applied to the synthesis of [2.2]paracyclo(2,6)pyridinophan-1,9-dienes such as **25**. By Raney-nickel desulphurization of **23** [2.2]phanes with saturated bridges, such as **24**, were obtained. Sulphone pyrolysis was also successful.[10]

2.12.1.3 [2.2]Heterophanes *via* the Wurtz reaction

The first application of organometallic coupling reactions to the synthesis of [2.2]heterophanes was described by Baker *et al.* in 1958.[2] In this case the immediate precursor was **26**; but it was later shown that coupling with phenyllithium of **27** could also lead to the desired [2.2]pyridinophane **10** (Boekelheide, 1970[9]).

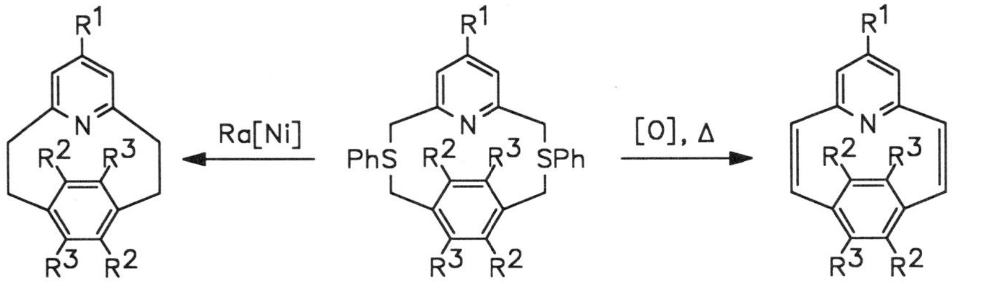

24a : R^1 = R^3 = H;
R^2 = CH$_3$

24b : R^1 = H;
R^2 = R^3 = CH$_3$

24c : R^1 = CH$_3$;
R^2 = R^3 = H

23

25a : R^1 = R^2 = R^3 = H

25b : R^1 = R^3 = H;
R^2 = CH$_3$

25c : R^1 = H;
R^2 = R^3 = CH$_3$

25d : R^1 = CH$_3$;
R^2 = R^3 = H

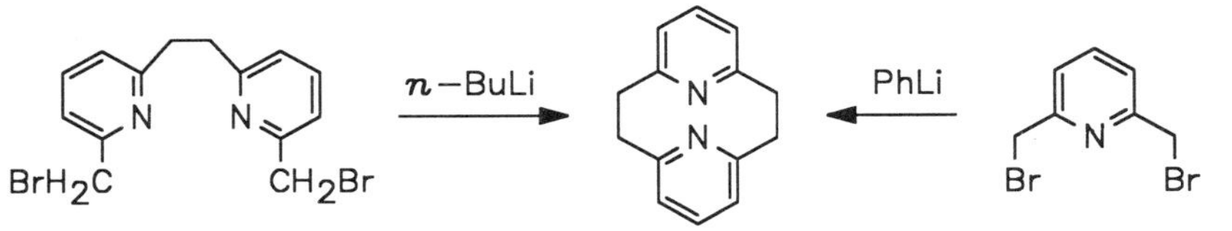

26 **10** **27**

Organosodium and organopotassium compounds lead to very efficient coupling of alkyl and benzyl halides. Even though radicals are assumed to be involved, there is no formation of undesired oligomers, as is the case with the original Wurtz reaction with metallic sodium. This is explained as arising from the formation of solvent cages.[1] Although the Wurtz reaction has often been

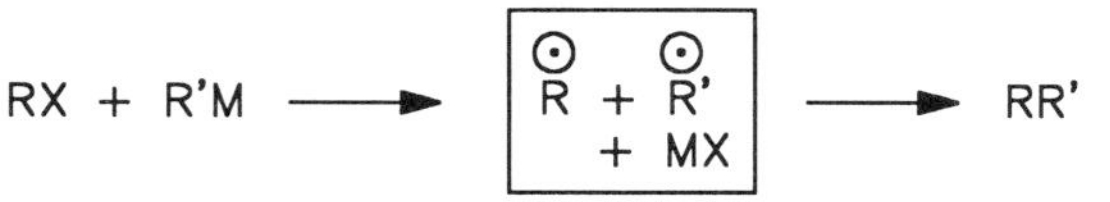

solvent−cage

applied in cyclophane chemistry, it has been of little significance in heterophane synthesis, where the Hofmann elimination and especially the dithia[3.3]phane route are particularly effective. In particular, a series of oligomeric [2.2]- and [2$_n$]heterophanes, prepared by Jenny *et al.* in the 1960s by way of a modified Wurtz coupling, can be mentioned. [2.2](3,5)Pyridinophane **(29)**, for example, was obtained in 2% yield from 3,5-bis(chloromethyl)pyridine (12) with sodium in THF at low temperature.[11]

29

In 1970 Boekelheide *et al.* synthesized (2,6)pyridinophanes such as **31** from the bis(bromomethyl)pyridine compound **30**.[12]

30   **31**

Trimeric and tetrameric macrocycles [2.2.2](3,5)pyridinophane (**32**) and [2.2.2.2](3.5)pyridinophane (**33**) have also been characterized (see Chapter 7, '[m_n]Phanes').[13] Starting from **34** and using the same route, Jenny and

32   **33**

Holzrichter[14] in 1969 prepared [2.2.2.2](2,6)pyridinophane (**35a**) and [2.2.2.2.2](2,6)pyridinophane (**35c**), along with the trimer and pentamer of this series (compare Chapter 7).

The cyclization method of Baker (phenyllithium) was employed to advantage by Jenny for the synthesis of the [2]metacyclo[2]pyridinophane **28** as well.[15]

As shown by Kauffmann in 1970, the C—C coupling could be achieved from methylpyridines by direct lithiation of the corresponding α-methyl-substituted pyridines.[16] The yield of the desired 'dimeric' heterophane **10** in this reaction,

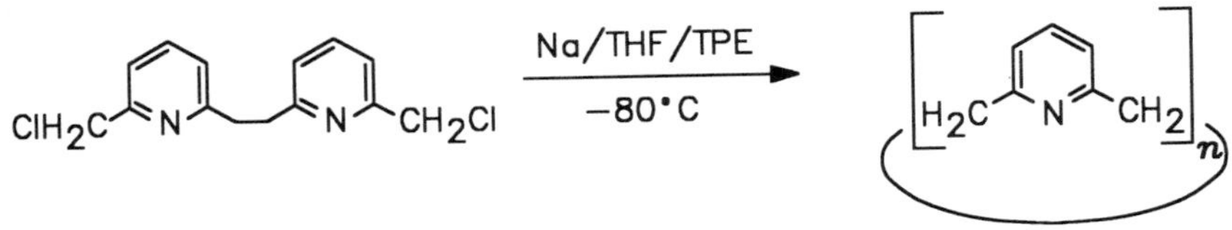

34

Na/THF/TPE
−80°C

35a : n = 4
35b : n = 5
35c : n = 6
35d : n = 7

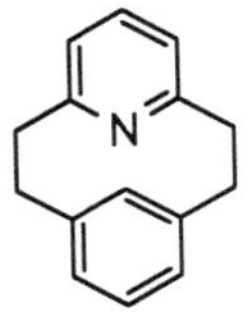

28

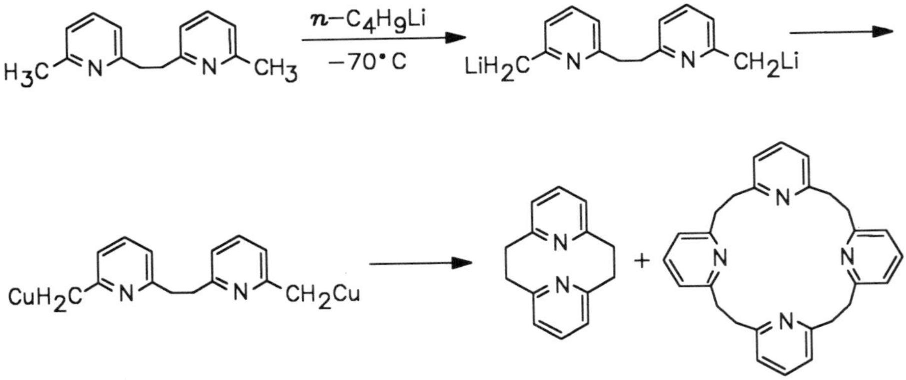

$n-C_4H_9Li$
−70°C

10   35a

however, is only 1%, and the tetramer 35a as major product amounts to only 4%.

2.12.1.4 Hofmann elimination

Jenny *et al.* used the Hofmann elimination for the synthesis of the four possible isomeric [2.2](2,5)pyridinophanes 36–39, which could be separated by

36	**37**	**38**	**39**

chromatography.[1] The [2.2]pyridinophanes **40–44**, with various orientations of the pyridine nitrogens in the phane system, were first described in 1988.[18]

40	**41**	**42**	**43**	**44**

The 1,6-Hofmann elimination was particularly successful in the synthesis of [2.2](2,5)heterophanes. The first known instance was [2.2](2,5)furanophane (**47**; Winberg, 1960), which was formed from the quaternary ammonium hydroxide **45** by way of the presumed intermediate **46** (2,5-dimethylene-2,5-dihydrofuran).[19] This intermediate **46**, stable at −78°C, when heated at reflux in ethanol in the presence of radical inhibitors, is converted in 73% yield to **47**.

45	**46**	**47**

The substituted derivatives **48** and **49** are prepared in an analogous way from the corresponding substituted starting materials. In a similar way Winberg obtained [2.2](2,5)thiophenophane (**50**) in 17% yield by removing the water byproduct through azeotropic distillation. Fletcher and Sutherland first prepared the mixed [2.2](2,5)furanothiophenophane (**51**) in 1969 by a 'cross-breeding'

48a : R = CH$_3$ **49a** : R = CH$_3$ **50**   **51**   **52** : R = H
48b : R = Ph   **49b** : R = Ph   **53** : R = D

reaction in refluxing xylene of equimolar amounts of the corresponding thiophene and furan quaternary ammonium hydroxides.[20] This cross-reaction had already been employed by Cram in 1961 for the synthesis of [2.2](2,5)furanoparacyclophane (**52**);[21] and the deuterium-labelled compound **53** had been synthesized by Whitesides *et al.* in 1968 for purposes of ring inversion studies.[22]

Additional mixed [2.2]phanes of the type **54–56** were later prepared; an example is [2.2]paracyclo(2,5)thiophenophane (**54**), formed in a meagre 1.6% yield by the crossed Hofmann elimination.[1] Multi-layer molecules of this type are known as well (see Chapter 6, 'Multi-layered Cyclophanes').

54   **55**   **56**

Along with [2.2](2,5)furano(1,4)naphthalenophane (**59**) obtained in 11% yield by Wasserman and Keehn in 1969 [2.2]furanophane (**47**) was formed as a byproduct; indeed the cross-reaction typically produces all possible dimers.[23]

The 'crossed' reaction of **57** and **58** in refluxing xylene affords, besides **61**, the isomeric [2.2](1,4)naphthaleno-2,5-thiophenophanes (**60**) in 0.3% yield, with an *anti*:*syn* ratio of 4:1.[24] Similarly, the mixed anthracenophanes **62**, **63**, and **64** were obtained in yields of 0.8, 2.8, and 5.5%, respectively, whereas [2.2](9,10)anthraceno(2,5)furanophane (**64**) could be synthesized in a substantially higher yield of 40%.[25]

CH$_3$
CH$_2$N(CH$_3$)$_3$OH

CH$_3$
CH$_2$N(CH$_3$)$_3$OH

57   **58**   **59**   **60a** : X = S *(anti)*
                                    60b : X = S *(syn)*

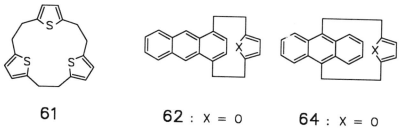

61

62 : X = O

63 : X = S

64 : X = O

65 : X = S

The following cross-reaction affords the triple-layer phanes **66a** and **66b**.[26]

66a : X = O

66b : X = S

In connection with a study of macrocyclic ethers, Cram prepared 1-chloro[2.2]furanophane **67** in 29% yield.[27]

67

178

By opening of the furan ring in the furanocyclophane **52**, Cram was able to prepare paracyclophandione (**68**). This method was extended by Keehn *et al.*, so that one or both furan rings could be transformed to the corresponding pyrrole ring.[1b] The furanophane **47** can thus serve as a synthetic precursor for

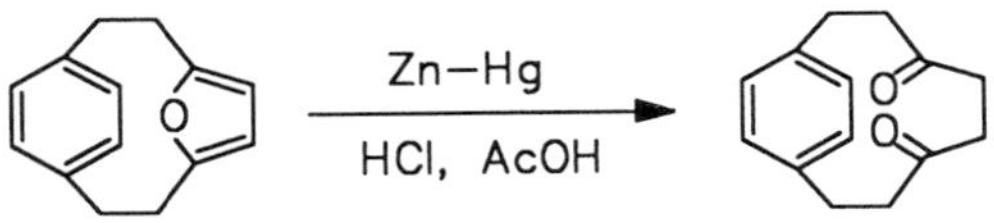

52   **68**

other heterophanes. It was, for example, converted into the tetraketone **71**, which, when subjected to the Paul–Knorr cyclization, leads to **73** as one product.[1b] Condensation of **71** with excess methylamine affords *N,N'*-dimethyl-[2.2](2,5)pyrrolophane (**72**) in 42% yield.[28] The deuteriated derivative **74**,

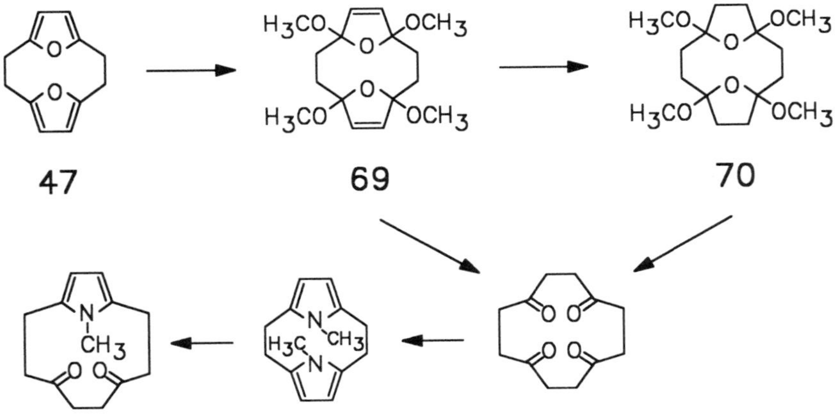

47   **69**   **70**

73   **72**   **71**

useful for conformational studies, could be prepared in an analogous way.[1b] Substituents at the intra-annular nitrogen of the pyrrolophanes can be widely varied[1b] as shown in some of those (**75a–f**) that have been synthesized. A triply bridged pyrrolophane (**76**) has also been prepared.[29]

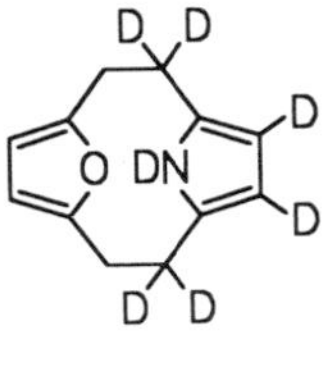

74

75a : R = *p*-PhCH₃; R' = H   **75d** : R = C₂H₅; R' = H   **76**

75b : R = Ph; R' = H   **75e** : R = *p*-BrPh; R' = H

75c : R = C₂H₅; R' = CH₃   **75f** : R = *p*-BrPh; R' = CH₃

The unsubstituted [2.2](2,5)pyrrolophane (**79**) could not be prepared from the tetraketone **71** with ammonia but rather was obtained as the dianion **78**, prepared by reduction of the *N*-benzylpyrrolophane (**75**) with sodium in liquid ammonia (Keehn *et al.*[1b]). Condensation of the tetraketone **71** with *o*-phenylenediamine afforded another triply bridged pyrrolophane **80** (cf. **76**).[1b]

75 : R = CH₂Ph   Na/NH₃   **78**   H₂O   **79**

80

Mention should be made here of the new 'porphycene' redox systems, which can be viewed as [2.2]bipyrrolophandienes such as **82** (E. Vogel). They can be formed by McMurry coupling from the diketone **81**.[30]

2   H₇C₃OC—[pyrrole]—COC₃H₇   TiCl₄/Zn →

81

Chemical structures: H_2O, CH_2Ph

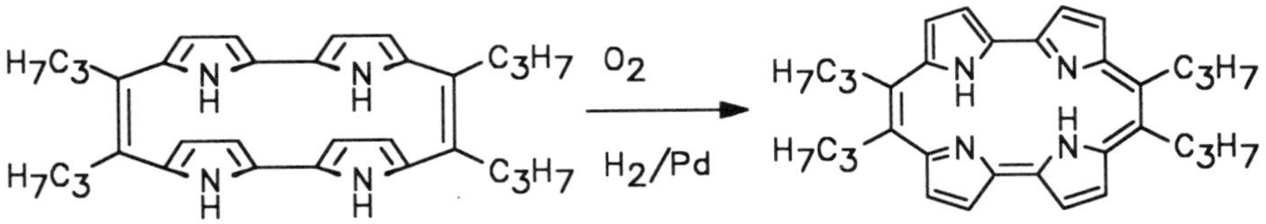

82 → **83**

2.12.2 STEREOCHEMISTRY OF [2.2]HETEROPHANES

If one replaces the benzene rings in [2.2]metacyclophane successively with pyridine rings, the result is [2]metacyclo[2](2,6)pyridinophane (**28**) and [2.2](2,6)pyridinophane [(**10**), see above], with first one and then both intra-annular H-atoms being replaced by the lone pair electrons of the nitrogen. The ^{1}H-NMR spectrum of **28**[31] shows an AX_2 system for the pyridine protons and an AB_2 system for the extra-annular protons of the benzene ring. The intra-annular hydrogen atom H_i at position 8 appears at a resonance shifted considerably to higher field ($\delta = 4.40$). The bridge methylene protons absorb as an ABCD system which is unchanged up to 200°C. From this it is concluded that **28**, like [2.2]metacyclophane, exists as a fixed, *anti*-ladder ring conformer. The free enthalpy of activation for the ring inversion was extrapolated to >113 kJ/mol from the experimental results. On the other hand, a ring inversion

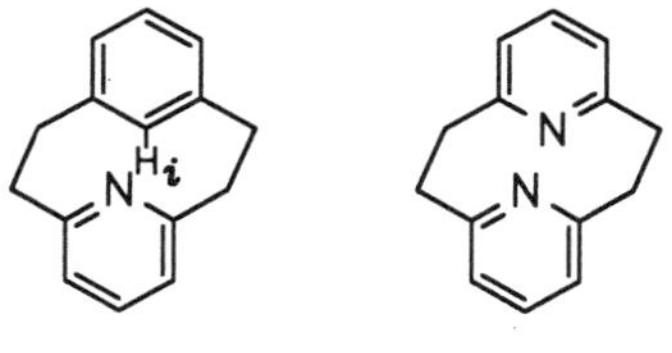

28   **10**

process A ⇌ B is assumed to be taking place for **10** in solution, on the basis of the temperature-dependent ^{1}H-NMR spectrum (Figure 1).[32] The free enthalpy of activation ($\Delta G_c^{\neq}$) for this process was determined experimentally to be 62 kJ/mol, with an Arrhenius energy (E_A) of 64 kJ/mol.

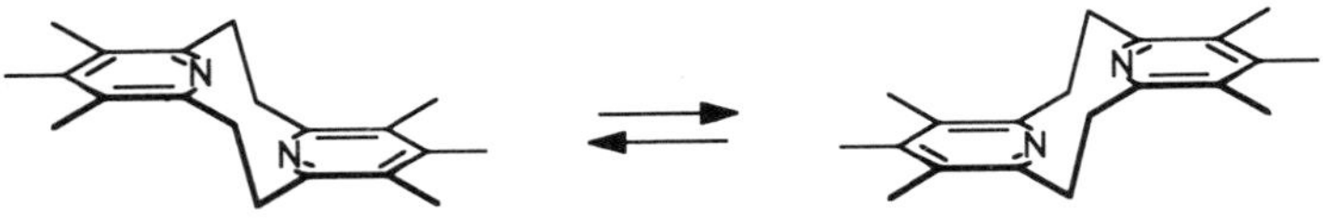

A   B

Figure 1. Schematic representation of the ring inversion process for **10**[33]

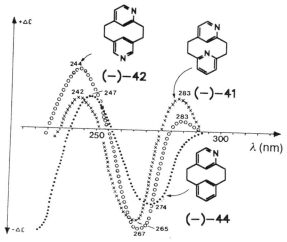

Figure 2. CD Spectra of (−)-**41**(× × ×), (−)-**42**(○○○), and (−)-**44**(●●●) in methanol

On the assumption that the mechanism for ring inversion is analogous in [2.2]metacyclophane (**2**), **28**, and **20**, it would seem that steric hindrance between the intra-annular groups (intra-annular H or pyridine nitrogen) in the transition state is at a minimum in **10**, in which the two lone nitrogen pairs are interacting. From this it follows that the spatial requirement for an electron lone pair on nitrogen in pyridine is smaller than that of an intra-annular hydrogen atom in a benzene ring. Detailed ^{1}H-NMR studies of a series of [2.2]phanes, including also dithia[2.2]phanes, confirm these results. It can be shown that the ring inversion process in [2.2](1,3)- and -(2,6)phanes can be controlled by non-bonded interactions through steric effects of intra-annular groups; that is, the inversion barrier depends on the size of the intra-annular group (see Section 2.13).[33,34] The energy barriers for ring inversion of the pyridinophane isomers **40–44** (see above, Figure 1) were investigated by racemization kinetics. Separation or enrichment of racemates had been accomplished by HPLC on cellulose tris-3,5-dimethylphenylcarbamate and poly(triphenylmethylmethacrylate). The circular dichroism of the structurally isomeric pyridinophanes and their protonated forms was measured, and a comparison made of the effect of the orientation of the chromophore on the CD. A typical example is represented in Figure 2.[18] No effect of orientation between **42** and **41** was observed, the curves being of similar shape. Replacement of a pyridine ring with benzene (**44**) led to a bathochromic shift of 5 nm and disappearance of the first positive Cotton effect, which appears at $\lambda = 283$ nm in **41** and **42** (see Figure 2). The racemization barriers and structure–chiroptical dependence of these [2.2]pyridinophanes were also compared with the corresponding 1-oxa[2.2]pyridinophanes[18] (see Section 2.13, heterocyclic [2.2]phanes).

2.12.2.1 [2.2]Heterophanes with five-membered 'aromatic' rings

The [2.2]heterophanes **47**, **50**, **51**, and **73** can not, of course, be directly compared with [2.2](1,3)benzeno- and pyridinophanes because the geometry of the five-membered rings varies considerably from that in the other series. On the basis of Raman and IR spectra, Winberg *et al.* proposed at an early date the *anti* conformation B with C_{2h} symmetry for [2.2](2,5)furanophane **(47)** and [2.2](2,5)thiophenophane **(50)**, excluding both the *syn* form C and the coplanar conformation A on energetic grounds.[19] The *anti* arrangement for the pyrrole nuclei in **73** was deduced from ¹H-NMR spectra.[28]

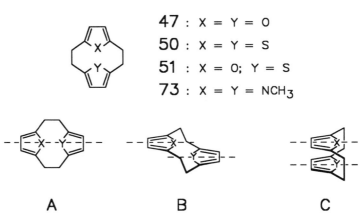

47 : X = Y = O
50 : X = Y = S
51 : X = O; Y = S
73 : X = Y = NCH₃

A   **B**   **C**

From the X-ray crystal structure of the charge-transfer complex between [2.2](2,5)thiophenophane **(50)** and benzotrifuroxan[35] (Figure 3), it was shown that the C—C bond lengths in the phane are essentially identical to those in unbridged thiophene, a situation also true for [2.2]metacyclophane. Nevertheless, the thiophene rings in **50** are significantly distorted, existing in a boat-like form. The bridgehead atoms of the thiophene rings C(2) and C(5) are projected 8–10 pm over the plane of the other ring atoms. These planes of the 'aromatic' rings are situated essentially parallel to each other, on the basis of the least squares method. The distance between two bridgehead C atoms (2,2' or 5,5': thiophene numbering) connected to the same CH_2-CH_2 bridge (280 and 282 pm) is exactly the same as that in [2.2]paracyclophane. The S–S distance of 319 pm is about 100 pm larger than that calculated for an unbent model molecule but about 50 pm less than the van der Waals distance. Likewise the separation between the sulphur atom of one thiophene ring and a bridgehead atom (e.g. C-2) of the second thiophene ring (293–302 pm) is significantly shorter than the sum of the van der Waals radii.

In spite of this detailed knowledge of the skeleton of the [2.2]heterophanes, it is nevertheless not possible to make inferences as to the spatial requirements for sulphur or oxygen because both the bond lengths and bond angles in benzenophanes and pyridinophanes on the one hand and in phanes containing five-membered heterocycles on the other are so different.

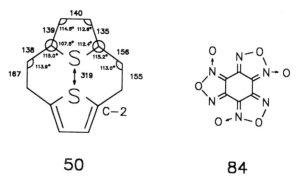

Figure 3. Molecular geometry of [2.2](2,5)thiophenophane (**50**) in a 1:1 adduct with benzotrifuroxan (**84**)[35] (bond lengths in pm)

From ¹H-NMR spectroscopy the thiophenophane **50** proves to be fixed at all temperatures, whereas the furanophane **47** is conformationally flexible. This distinction cannot be attributed solely to the space-filling differences between sulphur and oxygen in a five-membered ring but also to the differing bond lengths and bond angles in [2.2]thiophenophane and [2.2]furanophane. Whereas the ¹H-NMR spectra of compounds **54**, **66**, and **86** (X = S) are temperature-independent because of their conformational rigidity, for the corresponding furanophanes **52**, **66**, and **86** (X = O) one observes coalescence temperatures of −39, −58, and −62°C.[26] This typically more facile ring inversion for furanophanes compared to thienophanes is also manifested in the difference between **87** and **88**. While the coalescence temperature for [3.3](2,5)thiophenophane (**87**) is 105°C, the [3.3](2,5)furanothiophenophane (**88**) is conformationally flexible down to a temperature of −90°C.[36]

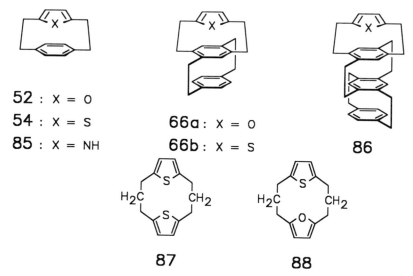

52 : X = O
54 : X = S
85 : X = NH

66a : X = O
66b : X = S

86

87

88

In sharp contrast to this behaviour, the ¹H-NMR spectra of both [2.2](2,5)thiophenofuranophane (**51**) and [2.2](2,5)thiophenophane (**50**) remain unchanged up to 200°C.[32]

The pyrroloparacyclophane **85** shows a coalescence temperature of 105°C.[37] On this basis one can arrange the five-membered heterocycles in the following order, according to 'size': O < NH < S. The larger size of the pyrrole group over the furan oxygen is also consistent with the observation that the furanopyrrolophane **90** exhibits a temperature-independent ¹H-NMR spectrum and is thus conformationally rigid.

85 : R = H
89 : R = CH₃

90

The proton resonance spectrum of **59** suggests that it exists in the *anti* conformation, although it should be noted that in the thiophene analogue **60a,b** there is also a *syn* conformer (see above, Section 2.12.1). It is of interest that the furanopyridinophane **91** is *syn*, according to the X-ray crystal structure; its coalescence temperature is higher than 110°C.[39] Introduction of aromatic methyl substituents, as in **92** and **93**, results in a more flexible furan compound (coalescence temperature −29°C), while the thiophene compound remains conformationally fixed. Both the anthracenofurano- and -thiophenophanes **64** and **65** are conformationally rigid.[24]

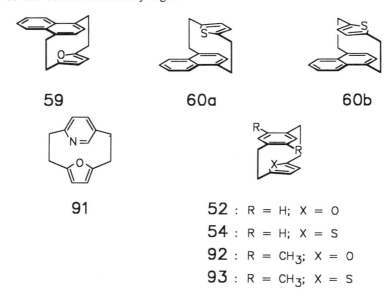

59

60a

60b

91

52 : R = H; X = O
54 : R = H; X = S
92 : R = CH₃; X = O
93 : R = CH₃; X = S

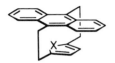

64 : X = O

65 : X = S

The pyrroloparacyclophanes **85** and **89** exist as *anti* conformers. Oxazole and thiazole compounds **94a,b** can be isolated in both *syn-* and *anti-*isomeric forms, their coalescence temperatures being, respectively, 80°C and >150°C.[1b] The *syn* conformation has been assigned to the [2.2]furanopyridazinophane **95**, whose coalescence temperature appears to be >110°C.[40]

85 : R = H

89 : R = CH₃

94a : X = O

94b : X = S

95

The coalescence temperature for the [2.2]parabenzenopyridinophane **17**, as measured in several laboratories, has been found to be −30 to −50°C. The corresponding diene **25a** may be rotationally flexible or conformationally rigid.[1b]

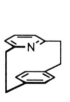

17

25a

R¹

R²   R³

R³   R²

17 : R¹ = R² = R³ = H

24a : R¹ = R³ = H;
R² = CH₃

24b : R¹ = H;
R² = R³ = CH₃

The methyl-substituted pyridinophanes **24a,b** show coalescence temperatures of $-25\,^{\circ}$C. The [2.2](2,5)pyridinophanes **36–39** have been separated into their four isomers, whose ^{1}H-NMR spectra exhibit characteristic differences.

36   **37**   **38**   **39**

UV spectroscopy[1b]

The longest wavelength absorption band in [2.2](2,5)furanophane (**47**; 222 nm) is shifted by 2 nm in the bathochromic direction, as compared to the unbridged parent heterocycle, 2,5-dimethylfuran. Even though this is no indication of any substantial π–π interaction between the furan rings, the ^{1}H-NMR spectrum of this compound does suggest that a transannular electronic interaction involving the lone electron pairs of the furan oxygen atoms is a possibility.

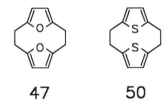

47   **50**

The UV spectrum of [2.2](2,5)thiophenophane (**50**) contains two bands at 245 nm ($\epsilon = 7700$) and at 275 nm ($\epsilon = 5720$), whereas the parent compound 2,5-dimethylthiophene has only one band at 238 nm ($\epsilon = 7250$). From these spectral comparisons it is concluded that [2.2]phanes are typically associated with a red shift and the appearance of a second band. In the case of the thiophenophane system, this pattern is strong evidence for transannular interaction.

A bathochromic shift of 4 nm is found in the furanopyridinophane **91**, as compared to 2,5-dimethylfuran. Likewise one observes a general red shift of all bands in the furano- and thiophenoparacyclophanes **52** and **54** in relationship to their 'half molecules'. Evidence for bathochromic shifts is also present in the spectra of the paracyclopyrrolophanes **85** and **89**. In some cases shoulders appear in the UV curves that are not present in the open-chain analogues. These results can also be interpreted as an indication of enhanced basicity of these [2.2]heterophanes because of the transannular interactions.

The numerous X-ray crystal structural analyses of mixed heterophanes containing five-membered rings indicate little or no distortion of the planar rings.

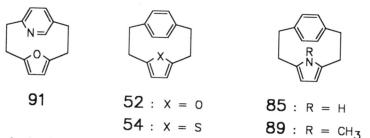

91

52 : X = O
54 : X = S

85 : R = H
89 : R = CH₃

Hence the bathochromic shifts in the UV spectra are more likely attributed to transannular electronic interactions rather than to ring deformations, a conclusion which is supported by the photoelectron spectra.

IR spectroscopy[1b]

There is just one example in the literature in which transannular interactions have been investigated and confirmed by means of infrared spectroscopy.[10] In this study the carbon–deuterium (C—D) frequency shifts in a series of amines were compared to the known pK_a values (in deuteriochloroform), the result being a good correlation. The approach was also applied to the pyridinophanes **17, 24a,b**. It was found that **17** is a stronger base than 2,6-dimethylpyridine

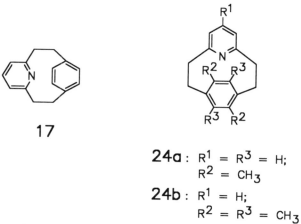

17

24a : R¹ = R³ = H;
      R² = CH₃
24b : R¹ = H;
      R² = R³ = CH₃

by 1.20 pK_a units. Ring substitution of methyl groups, however, lowers basicity. It was concluded that the basicity of these phane compounds was enhanced through electron donation from the benzene ring. Although increasing the number of methyl groups should enhance the electron density of the benzene ring, it has the effect of hindering electron transfer to the pyridine ring.

Photoelectron spectroscopy of [2.2]Heterophanes[1b,42]

Boekelheide, who first applied this method in 1979 to the [2.2]heterophanes, observed nothing abnormal in the spectra of the pyridinophanes **17** and **24**.[10]

On the basis of current literature, it has been concluded that the lowering of ionization potentials in cyclophanes results from transannular electronic interactions. From a theoretical standpoint, this lowering can not be attributed to an increase in substitution on the aromatic rings because the orbitals in question are oriented orthogonally to each other. As an example, the lowering of ionization potential is more pronounced in going from [2.2]paracyclophane to superphane (see below) than it is in going from *p*-xylene to hexamethylbenzene. This appears to be due solely to transannular interactions in this system, and a similar trend is noted in the heterophanes.

2.12.3 CHEMICAL REACTIONS OF [2.2]HETEROPHANES

Irradiation with UV of *trans*-1,3-10b,10c-tetramethyl-2-azadihydropyrene (**96**) converts it to 8,12,14,16-tetramethyl[2]metacyclo[2](3,5)pyridinophan-1,9-diene (**97**). The dark reaction of the hydrochloride of **97** is the fastest known example in the dihydropyrene series.[12] Presumably the presence of a charge in one of the aromatic rings facilitates formation of the new bond to the other ring.

Irradiation of the mono ene **98** sets up a photoequilibrium with the dihydropyrene derivative **99**.

The choice between the two possible isomers [2.2](2,6)pyridinophan-1,9-diene (**22**) and 10,10b-diaza-10b,10c-dihydropyrene (**100**) was made in favour of the diene **22**.[9] No isomerization of **22** to **100** was observed.

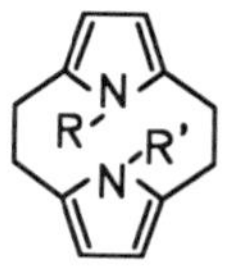

75

a : R = *p*-PhCH₃; R' = H
b : R = Ph; R' = H
c : R = C₂H₅; R' = CH₃
d : R = C₂H₅; R' = H
e : R = *p*-BrPh; R' = H
f : R = *p*-BrPh; R' = CH₃
g : R = CH₃; R' = H

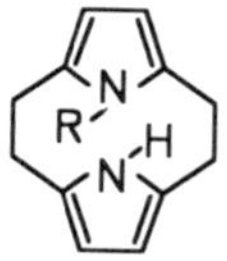

75h

R = CH₂Ph

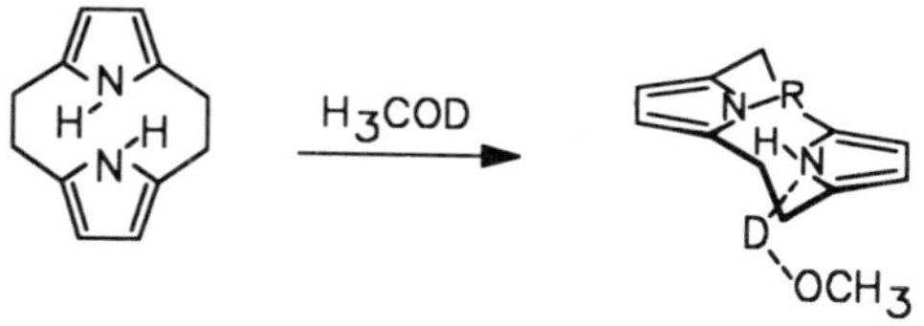

79   →   **101**

The [2.2]pyrrolophanes **75a,d,g** undergo hydrogen–deuterium exchange slowly, over the course of 2 days in CH₃OD. Presumably the nitrogen is quaternized as it is bonded to deuterium, while the NH hydrogen is forced toward the opposing pyrrole ring, causing significant steric interference.

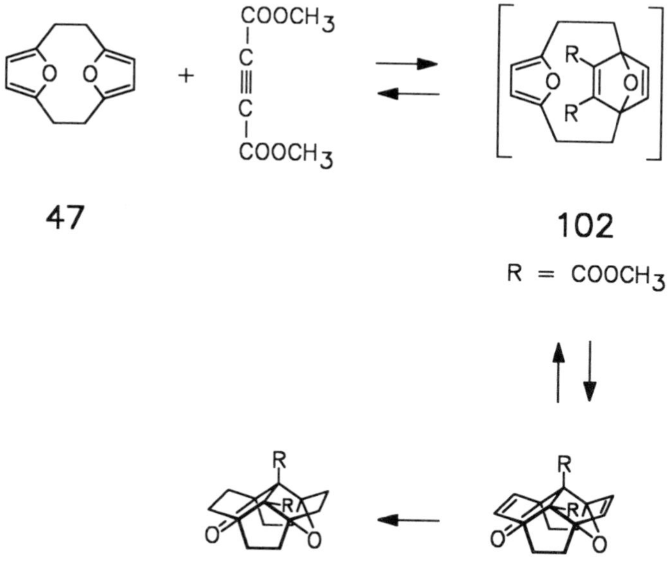

47   +   COOCH₃—C≡C—COOCH₃   ⇌   **102**

R = COOCH₃

104   ←   **103**

190

In some furanophanes interesting transannular reactions have been observed.[1b] Heating an equimolar mixture of [2.2](2,5)furanophane (47) and dimethyl acetylenedicarboxylate at 105°C affords the 1:1 adduct 103, which reverts to the two starting materials at 165°C. Formation of the symmetrical adduct can be explained by a two-step mechanism. The first step is an intermolecular Diels–Alder reaction leading to intermediate 102, which can take part in a second Diels–Alder reaction, this time intramolecular, to afford 103. Hydrogenation of 103 with H_2/Pd gives 104.

In the mixed [2](2,5)furano[2](1,4)naphthalenophane (59) the furan ring, being the more reactive diene component, combines with dimethyl acetylene-dicarboxylate to give the Diels–Alder product 105. This is followed by an intramolecular 1,4-addition at the unsubstituted ring, with formation of 106.[42]

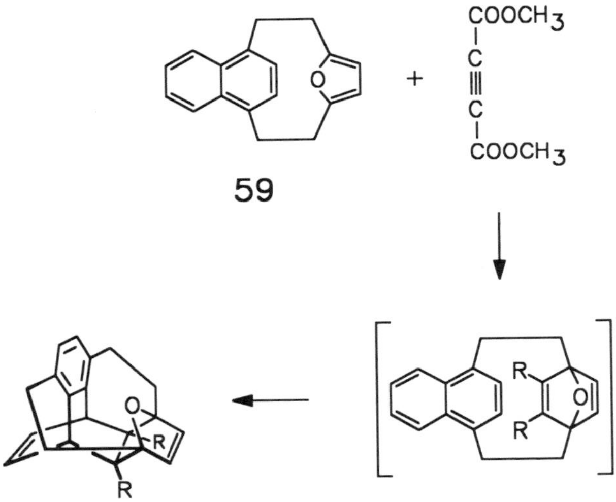

R = COOCH₃

R = $COOCH_3$

2.12.3.1 Intra-annular reactions

The mono- and di-N-oxides 107 and 108 of [2.2](2,6)pyridinophane (10) were obtained with no difficulty,[9] as was also the N-oxide 109 from [2.2](2,5)-

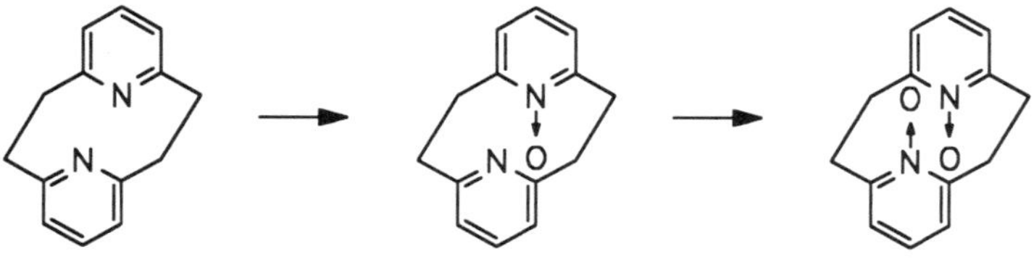

furano(3,6)pyridazinophane (**95**). The ¹H-NMR spectrum of **109** shows that one of the two methylene protons bridged to the *N*-oxidized pyridazine ring experiences a deshielding effect.[40]

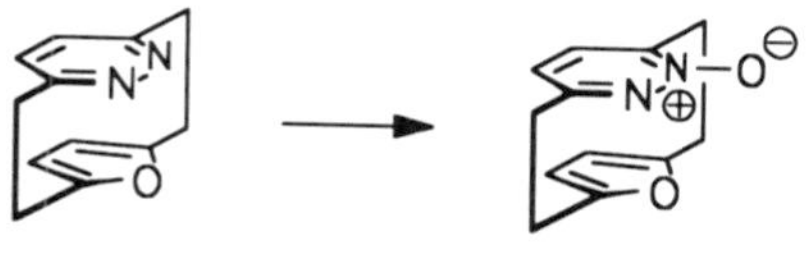

95   **109**

The benzenofuranophane **52** is converted to **110** on treatment with bromine in methanol at low temperature; this, in turn, can be hydrolysed to the *Z*- and *E*-dienones **111** and **112**.[43] Wasserman investigated the reaction of **52** with

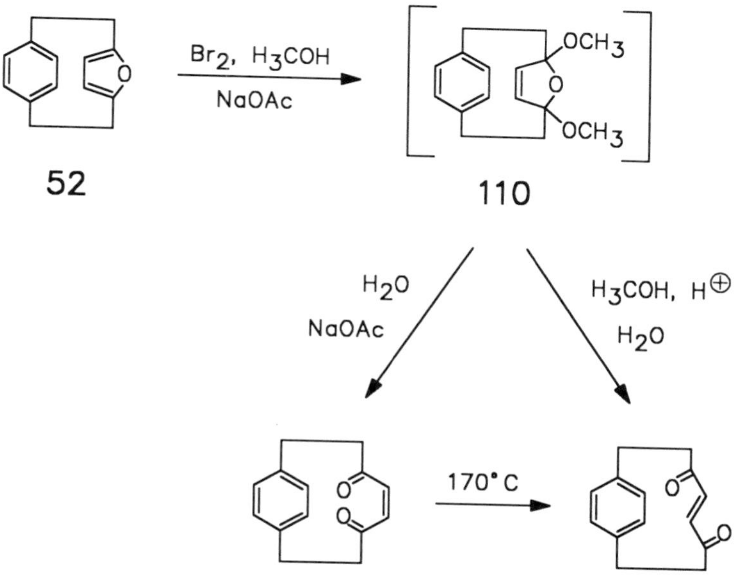

52   **110**

111   **112**

singlet oxygen. From the reaction mixture, obtained after hydrogenation, he isolated three products: **68**, **113**, and **114**.[44] The constitution of the polycyclic compound **114** was established by X-ray crystal analysis. Its formation proceeds through the *endo*-peroxide **115**, which rearranges to **116** and **117**. Isomerization of **117** to the enedione **118**, an intramolecular Diels–Alder reaction, and hydrogenation complete the steps to the final product **114**.[1b]

The diketone **122** is formed by singlet oxygen oxidation in methanol of [2.2](2,5)furanophane (**47**), presumably by way of an intramolecular

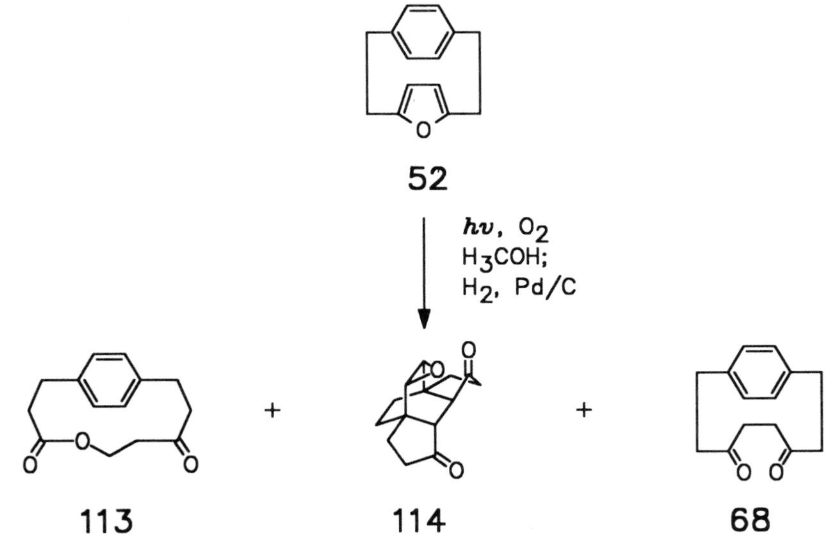
52
hv, O2
H3COH;
H2, Pd/C
113
114
68

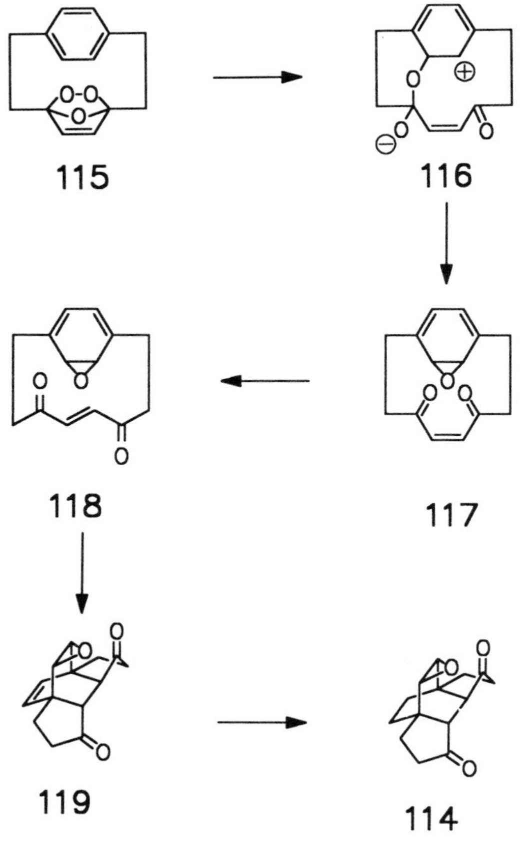
115
116
117
118
119
114

47   120   121

122   123   124

Diels–Alder reaction of the intermediate **121**. **123** was reduced to the known hexahydro-*as*-indacene (**124**).[43] Other authors have also isolated from this reaction a byproduct, presumably with the structure **125**.[43] If the oxidation of

125

47 is carried out in dichloromethane, the product is **127**, which probably results from a rearrangement of the intermediate **126**. X-ray crystal analysis has confirmed the structure of **127**.

47   126   127

Cram discovered a remarkable reaction of the furanoparabenzenophane **52**. Dissolved in cyclohexane and allowed to stand in sunlight, it was converted to [2.2.2]paracyclophane, no trace of residual furan being found.[9]

[2.2]Furanophane (**47**) reacts with benzyne to afford the 1:1 and 2:1 adducts, **129** and **130**.[45] The intramolecular adduct **131** was not found among the

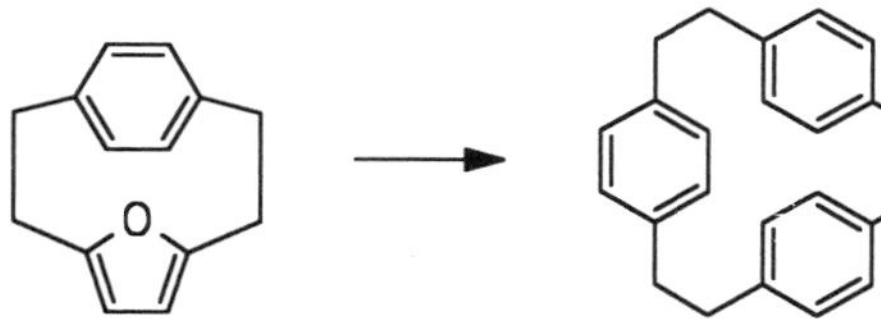

52   128

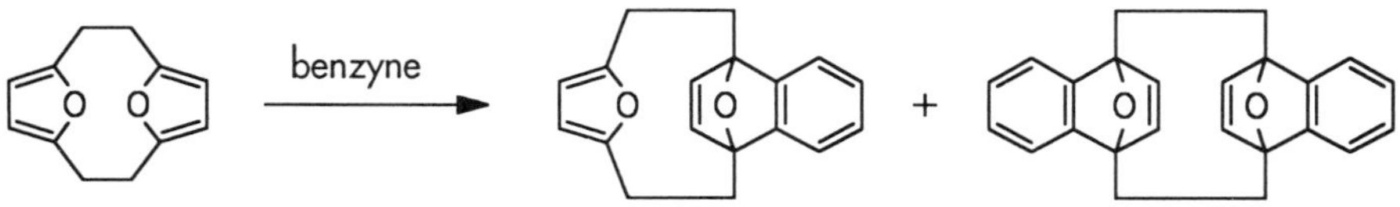

47   129   130

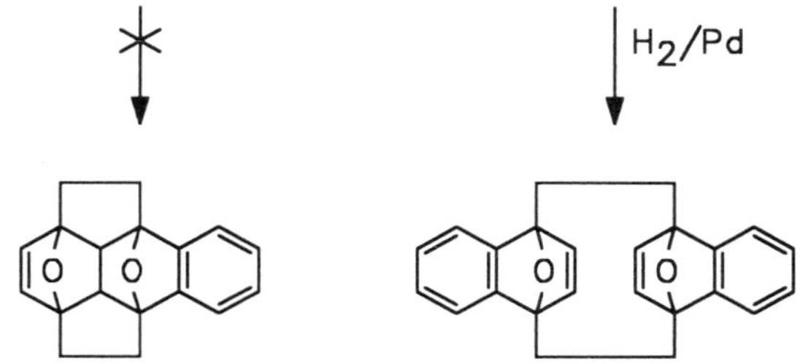

131   132

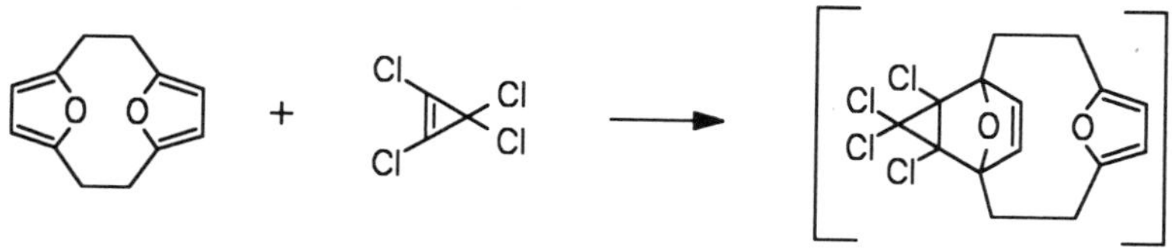

47

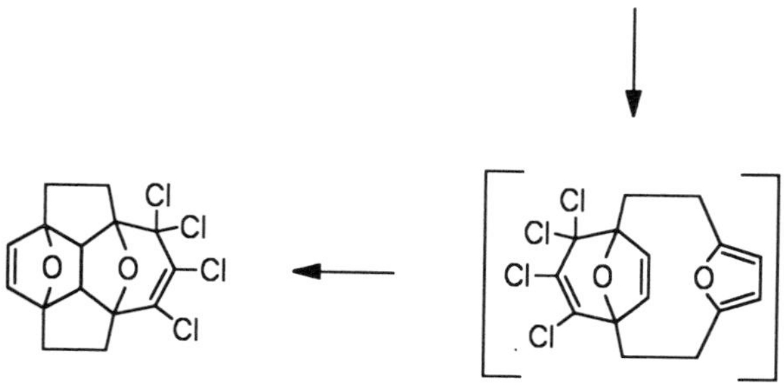

133

products; attempts to convert **130** to [2.2](1,4)naphthalenophane failed. Treatment of the [2.2]furanophane **47** with tetrachlorocyclopropene afforded the product **133**, whose constitution was confirmed by X-ray crystal analysis.[45]

In the cycloaddition of the anthracenofuranophane **64** with dimethyl acetylenedicarboxylate the 2:1 adduct **134** is formed, because the more electron-rich double bond reacts.[42]

Electrophilic substitutions in π-electron-rich multinuclear heterophanes such as **52** and **47** have been described. Products such as **135–137**, for example, are obtained in high yield.[1b]

64 $\longrightarrow$

134

135 : X = O
136 : X = NH

137

References to Section 2.12

1. Reviews:
   (a) F. Vögtle and P. Neumann, *Synthesis*, 85 (1973).
   (b) W. W. Paudler and M. D. Bezoari, in *Cyclophanes* (P. M. Keehn and S. M. Rosenfeld, Eds.), Vol. II, p. 359, Academic Press, New York, 1983; [2.2](2,5)Pyrimidinophanes: F. A. Neugebauer *et al.*, *Chem. Ber.*, **123**, 1885 (1990).
2. W. Baker, K. M. Buggle, J. F. W. McOmie and D. A. M. Watkins, *J. Chem. Soc.*, 3594 (1958).
3. H. J. J.-B. Martel and M. Rasmussen, *Tetrahedron Lett.*, 3843 (1971).
4. F. Vögtle and L. Schunder, *Chem. Ber.*, **102**, 2677 (1969).
5. J. Bruhin, W. Kneubühler and W. Jenny, *Chimia*, **27**, 277 (1973).
6. V. Boekelheide, I. D. Reingold and M. Tuttle, *J. Chem. Soc., Chem. Commun.*, 406 (1973).
7. V. Boekelheide, K. Galuszko and K. S. Szeto, *J. Am. Chem. Soc.*, **96**, 1578 (1974).
8. M. W. Haenel, *Tetrahedron Lett.*, 4007 (1978).
9. V. Boekelheide and J. A. Lawson, *J. Chem. Soc., Chem. Commun.*, 1558 (1970). For [2.2]Pyridonophanes, see F. Vögtle, J. Breitenbach and M. Nieger, *J. Chem. Soc., Chem. Commun.*, 860 (1991).

10. I. D. Reingold, W. Schmidt and V. Boekelheide, *J. Am. Chem. Soc.*, **101**, 2121 (1979).
11. W. Jenny and H. Holzrichter, *Chimia*, **21**, 509 (1967).
12. V. Boekelheide and W. Pepperdine, *J. Am. Chem. Soc.*, **92**, 3684 (1970).
13. W. Jenny and H. Holzrichter, *Chimia*, **22**, 139 (1968).
14. W. Jenny and H. Holzrichter, *Chimia*, **23**, 158 (1969).
15. J. Bruhin and W. Jenny, *Chimia*, **25**, 238 (1971).
16. Th. Kauffmann, G. Beissner, W. Sahm and A. Woltermann, *Angew. Chem.*, **82**, 815 (1970); *Angew. Chem. Int. Ed. Engl.*, **9**, 808 (1970).
17. G. R. Newkome, J. M. Roper and J. M. Robinson, *J. Org. Chem.*, **45**, 4380 (1980).
18. Review: K.-J. Przybilla and F. Vögtle, *Chem. Ber.*, **122**, 347 (1989).
19. H. E. Winberg, F. S. Fawcett, W. E. Mochel and C. W. Theobald, *J. Am. Chem. Soc.*, **82**, 1428 (1960).
20. J. R. Fletcher and I. O. Sutherland, *J. Chem. Soc., Chem. Commun.*, 1504 (1969).
21. D. J. Cram and G. R. Knox, *J. Am. Chem. Soc.*, **83**, 2204 (1961).
22. G. M. Whitesides, B. A. Pawson and A. C. Cope, *J. Am. Chem. Soc.*, **90**, 639 (1968).
23. H. H. Wasserman and P. M. Keehn, *Tetrahedron Lett.*, 3227 (1969).
24. S. Mizogami, N. Osaka, T. Otsubo, Y. Sakata and S. Misumi, *Tetrahedron Lett.*, 799 (1974).
25. H. Wynberg and R. Helder, *Tetrahedron Lett.*, 4317 (1971).
26. S. Mizogami, T. Otsubo, Y. Sakata and S. Misumi, *Tetrahedron Lett.*, 2791 (1971); N. Osaka, S. Mizogami, T. Otsubo, Y. Sakata and S. Misumi, *Chem. Lett.*, 515 (1974).
27. J. M. Timko and D. J. Cram, *J. Am. Chem. Soc.*, **96**, 7159 (1974).
28. H. H. Wasserman and D. T. Bailey, *J. Chem. Soc., Chem. Commun.*, 107 (1970).
29. R. L. Mahaffey, J. L. Atwood, M. B. Humphrey and W. W. Paudler, *J. Org. Chem.*, **41**, 2963 (1976); Pyrrolophanes from aziridinophanes, see F. Müller and J. Mattay, *Angew. Chem.*, **104**, 207 (1992); *Angew. Chem. Int. Ed. Engl.*, **31**, 209 (1992).
30. E. Vogel, I. Grigat, M. Köcher and J. Lex, *Angew. Chem.*, **101**, 1687 (1989); *Angew. Chem. Int. Ed. Engl.*, **28**, 1655 (1989).
31. J. R. Fletcher and I. O. Sutherland, *J. Chem. Soc., Chem. Commun.*, 1504 (1969).
32. I. Gault, B. J. Price and I. O. Sutherland, *J. Chem. Soc., Chem. Commun.*, 540 (1967).
33. Review: F. Vögtle and P. Neumann, *Angew. Chem.*, **84**, 75 (1972); *Angew. Chem. Int. Ed. Engl.*, **11**, 73 (1972).
34. Review: H. Förster and F. Vögtle, *Angew. Chem.*, **89**, 443 (1977); *Angew. Chem. Int. Ed. Engl.*, **16**, 429 (1977); For more recent racemization studies see K.-J. Przybilla and F. Vögtle, *Chem. Ber.*, **122**, 347 (1989); K. Meurer, F. Vögtle, A. Mannschreck, G. Stühler, H. Puff and A. Roloff, *J. Org. Chem.*, **49**, 3484 (1984); F. Vögtle, A. Ostrowicki, P. Knops, P. Fischer, H. Reuter and M. Jansen, *J. Chem. Soc., Chem. Commun.*, 1757 (1989); F. Vögtle, A. Ostrowicki, B. Begemann, M. Jansen, M. Nieger and E. Niecke, *Chem. Ber.*, **123**, 169 (1990).
35. B. Kamenar and C. K. Prout, *J. Chem. Soc.*, 4838 (1965).
36. Y. Miyahara, T. Inazu and T. Yoshino, *Chem. Lett.*, 397 (1980).
37. S. M. Rosenfeld and P. M. Keehn, *Tetrahedron Lett.*, 4021 (1973).
38. J. Bruhin and W. Jenny, *Chimia*, **25**, 238, 308 (1971).
39. C. Wong and W. W. Paudler, *J. Org. Chem.*, **39**, 2570 (1974).
40. M. D. Bezoari and W. W. Paudler, *J. Org. Chem.*, **45**, 4584 (1980).
41. Review: E. Heilbronner and Z. Yang, in *Cyclophanes, I* (F. Vögtle, Ed.), *Top. Curr. Chem.*, **113**, 1 (1983).
42. H. H. Wasserman and R. Kitzing, *Tetrahedron Lett.*, 3343 (1969).
43. D. J. Cram, C. S. Montgomery and G. R. Knox, *J. Am. Chem. Soc.*, **88**, 515 (1966).
44. H. H. Wasserman, A. R. Doumaux and R. E. Davis, *J. Am. Chem. Soc.*, **88**, 4517 (1966); H. H. Wasserman and R. Kitzing, *Tetrahedron Lett.*, 5315 (1969); T. J. Katz, V. Balogh and J. Schulman, *J. Am. Chem. Soc.*, **90**, 734 (1968).

45. L. A. Kapicak and M. A. Battiste, *J. Chem. Soc., Chem. Commun.*, 930 (1973); M. A. Battiste, L. A. Kapicak, M. Mathew and G. J. Palenik, *J. Chem. Soc., Chem. Commun.*, 1536 (1971).

2.13 HETERA[2.2]PHANES

2.13.1 HETERACYCLIC [2.2]METACYCLOPHANES[1]

2.13.1.1 Synthesis

In 1968 Vögtle succeeded in synthesizing 1,10-dithia[2.2]metacyclophane (**6**), at that time the first example of a [2.2]phane containing a hetero atom in the bridge (type I; see Figure 1) and indeed the first medium-ring dithiaphane.[2]

Treatment of 1,3-bis(bromomethyl)benzene with 1,3-dimercaptobenzene under high-dilution (HD) conditions gave, after optimization, the ten-membered ring in 68% yield.[3] Earlier attempts to construct the 1,10-dithia[2.2]metacyclo-phane skeleton **6** had led to the corresponding 20-membered tetrathia-[2.2.2.2]metacyclophane **7**. The formula for **6** is actually presented in B. H. Smith's *Bridged Aromatic Compounds*, the 'cyclophane Bible', but the data for the compound described there belong to the dimer **7**.[4]

With this synthetic methodology it was possible to prepare dithia[2.2](2,6)-pyridinophanes such as **8–10**, which were obtained in 28, 29, and 30% yield, respectively, at that time without benefit of the caesium effect (see below).[5]

Although in the case of the [2.2]metacyclophane hydrocarbon a pronounced ring strain in the ten-membered ring can be demonstrated and the intra-annular hydrogens interfere with each other sterically, it was possible to prepare numerous analogues of the ten-membered **6** containing substituents in the inner cavity; an example is **11**.[6]

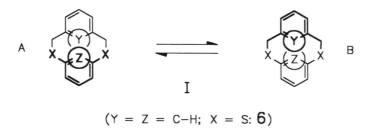

$(Y = Z = C-H; X = S: \mathbf{6})$

Figure 1. General formula for hetero- (Y,Z = hetero atoms) and heteraphanes (X = hetero atom). The ring inversion process represented as A ⇌ B proceeds only if the interfering steric effect between Y and Z is sufficiently small

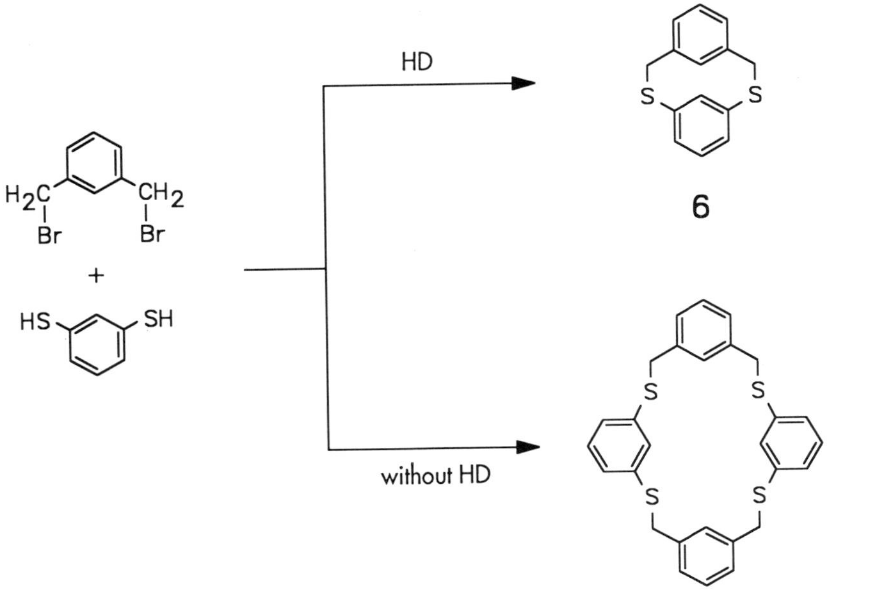

6

7

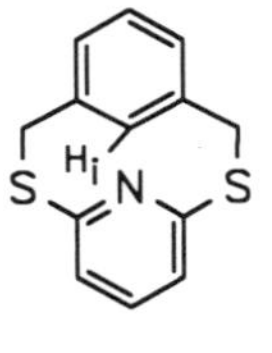

8

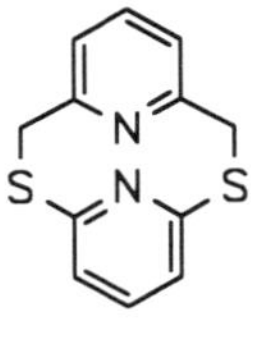

9

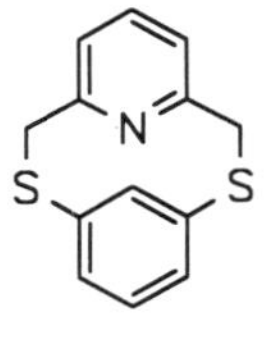

10

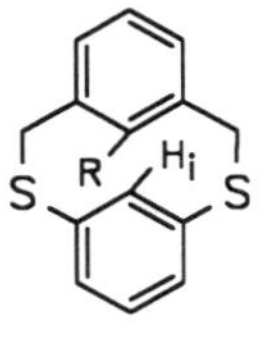

11

a: R = CO₂Me
b: R = NO₂
c: R = CO₂H
d: R = CO₂D
e: R = CO₂-t-Bu
f: R = CO₂Ph

g: R = CH₃
h: R = OCH₃
i: R = CN
k: R = SCH₃
l: R = C₆H₅

Planar chiral dithia[2.2]metacyclophanes such as **12** and **13** were obtained and resolved into their enantiomers.[6]

In most syntheses of such heteraphanes substantial improvements in yield could be achieved by the addition of Cs_2CO_3 (the caesium effect) and by varying the solvent. The use of $CsOH$/ethanol/benzene as reagent with high-dilution

a: R = H
b: R = CO$_2$H
c: R = CO$_2$Me
d: R = CO$_2$Et
e: R = NH$_2$

12

13

conditions (two-component, high-dilution reaction) proved to be the standard method of choice.[7] (DP = dilution principle)

Attempts were made in 1970 and 1973[8] to prepare the [2.2]metacyclophanes **16** and **17**, with still shorter bridges; but the outcome was instead formation of the dimers **18/19** and **20/21**, respectively. Only in the 1980s were Vögtle *et al.* successful in obtaining such strained diheteraphanes as **16** and **17** in small quantities, by optimizing the reaction conditions and adding caesium hydroxide.[9] In this way the oxathiaphane **16** was obtained in 1.6% yield by ring closure between 1,3-bis(bromomethyl)benzene (**14**) and 1-hydroxy-3-thiophenol (**15a**). Treatment of 1-*p*-toluenesulphonylamino-2-thiophenol (**13b**) with **14** under analogous reaction conditions afforded **17** in yields up to 9%. Without 'caesium assistance' **16** and **17** were not obtainable.

A series of investigations into the 'caesium effect' have been undertaken. From these it is concluded that the caesium cation occupies a special place in the periodic table and within the alkali metals. The known differences in cationic radii, in charge density, and polarizability influence the solvation and aggregation properties of the alkali metal salts and thus the nature of ion pair formation in varying solvents. The solvent effect is related to the poor solvation capability of Cs$^+$ in DMF or ethanol/benzene, and there is the possibility of formation of contact ion pairs and 'triple ions'. It has also been proposed that caesium participates in a kind of assistance mechanism in macrocyclic ring formation (Figure 2).[10] In strained medium-size rings such as **16**, the Cs$^+$ ion, acting as a 'homogeneous catalyst', can accelerate the intramolecular cyclization such that it becomes the favoured pathway ('oligomer selectivity').

The influence of the anion should not be overlooked. Indeed, investigations with various anions (fluoride, bromide) or with bases other than OH$^-$ or CO$_3^{2-}$ led to other results. The caesium effect in the formation of heterocyclic [2.2]metacyclophanes, as described here, holds only for strongly basic anions such as OH$^-$ and CO$_3^{2-}$.

Kellogg and Mandolini[11] attributed the uniqueness of the Cs$^+$ ion to its solvation and ion-pair properties, rather than to any stereochemically defined mechanism.

Although of considerable interest from the point of view of stereochemistry and spectroscopy, the monothia- and monooxa[2.2]metacyclophanes **22–25** were unknown until 1986. These structures, like **16** and **17**, constitute a new

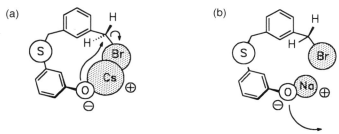

Figure 2. Representation of the (intramolecular) cyclization at the surface of the Cs cation(a) and the Na cation(b) in the case of formation of strained, ten-membered [2.2]metacyclophanes such as **16**[10]

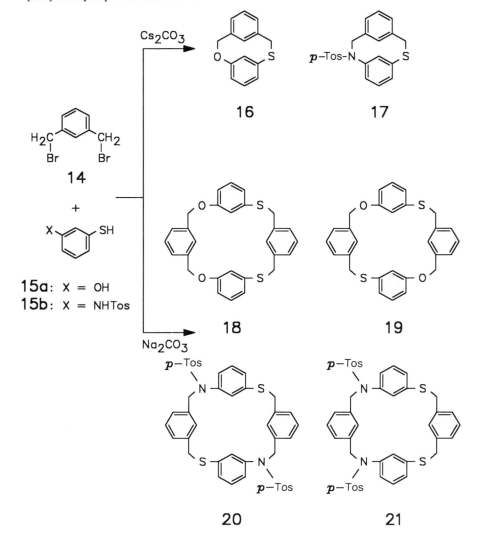

family of low molecular weight helical compounds, with double-wing propeller orientation, in which the ring strain may be dependent on the bond length of the carbon–hetero atom bond in the bridge.

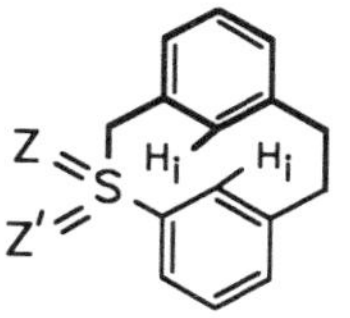

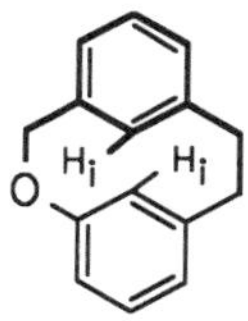

22: $Z = Z' = :$
23: $Z = O, Z' = :$
24: $Z = Z' = O$

25

The heterocyclic [2.2]phanes **22** and **25** (m.p., respectively, 108 and 79–80°C) were obtained by a cyclization reaction of the bis(bromomethyl) compounds **28c** and **28d** with phenyllithium, yields being 65 and <1%, respectively.[12]

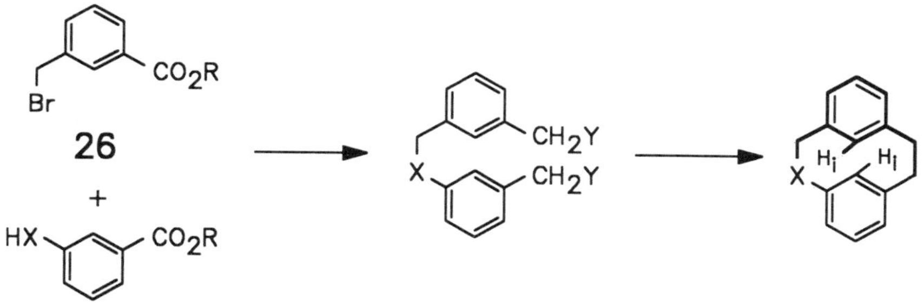

26

+

27a: X = S, R = Et
27b: X = O, R = Et

28a: X = S, Y = OH
28b: X = O, Y = OH
28c: X = S, Y = Br
28d: X = O, Y = Br

22: X = S
25: X = O

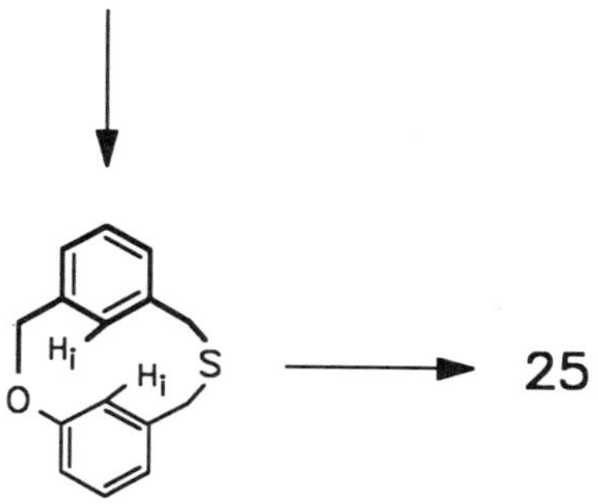

→ **25**

29

The oxa compound **29** was also prepared in 22% yield by photochemical desulphurization of 10-oxa-2-thia[3.2]metacyclophane (**29**).

Oxidation of racemic thiaphane **22** at 70°C with H_2O_2 afforded the four stereoisomeric sulphoxides **30**, with a diastereomeric excess (de) of 76%. By contrast, a higher de of 91% was achieved when the analogous oxidation was carried out on enantiomerically pure sulphide **22** at room temperature.[12]

30a   **30b**

The *N*-tosylaza[2.2]metacyclophane **33** was prepared in the same way as **22** and **25**.[13] In this case the phenyllithium and its solvent play a key role in cyclization of the bis(bromomethyl) compound **32c**. Phenyllithium in benzene/diethyl ether (7:3) affords **33** in 47% yield, whereas yields of only 20% are achievable under the same conditions when the solvent is benzene/*n*-hexane (3:1). It is possible that coordination with the ether oxygen atoms leads to formation of a favourable phenyllithium aggregate.

The helical oxa[2.2]naphthalenophane (**40**), first described by Vögtle in 1988, was prepared by the following pathway involving photolytic desulphurization of **39** with triethyl phosphite.[14]

26a: X = CO_2Et **31**: X = CO_2Et **32a**: X = CO_2Et
26b: Y = CH_2Br **32b**: X = CH_2OH
                                                   32c: Y = CH_2Br

33: R = Tosyl

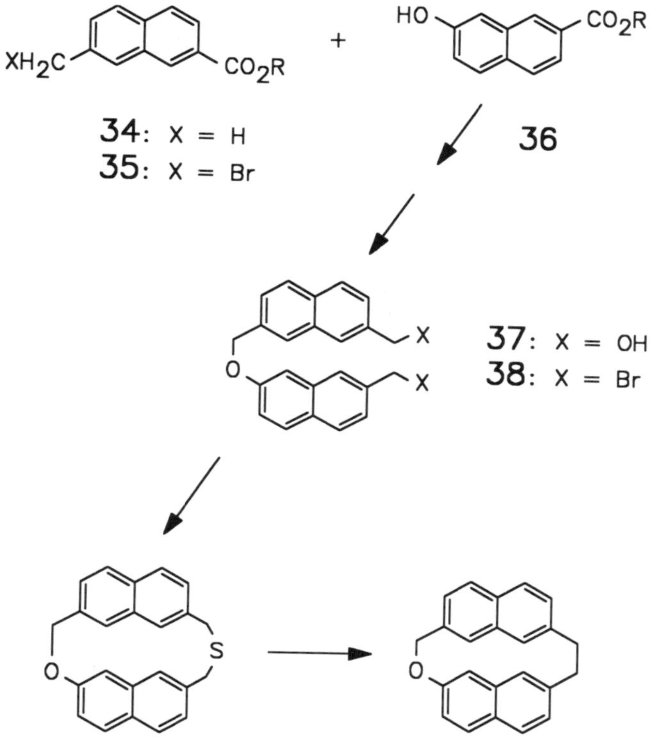
XH2C
CO2R
+
HO
CO2R
34: X = H
35: X = Br
36
X
O
X
37: X = OH
38: X = Br
O
S
O

39 40

The first chiral [2.2](1,3)heter<u>a</u>phanes of the type **42** containing functional groups in the ring cavity were first obtained in 1989, only after cyclization procedures had been improved. This noteworthy achievement was successful through a combination of the high-dilution principle with the 'caesium effect' in the presence of dipolar aprotic solvents such as acetonitrile.[15] Of particular interest are the very large phenyl and *tert*-butyl groups as intra-annular substituents.

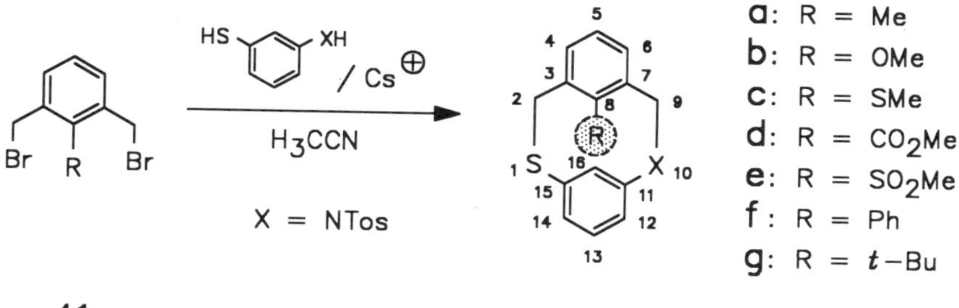
HS
XH
/ Cs⊕
H3CCN
X = NTos
Br
R
Br
S
X
R
a: R = Me
b: R = OMe
c: R = SMe
d: R = CO2Me
e: R = SO2Me
f: R = Ph
g: R = t–Bu

41 42

42f: *anti*   **42g**

By taking advantage of optimized synthetic methodology, it was eventually possible to close the more highly strained condensed aromatics into thiaza-[2.2]metacyclophane units such as **43**.[16a]

43a   **43b**   **43c**

The first *diagonal* di-hetera substituted and chiral [2.2]phanes **44** were all obtained by the following pathway in yields of <1%. The 1,9-oxathia-[2.2]metacyclophane **44a** (m.p. 107–109°C) is not entirely stable at room temperature, in contrast to the sulphone **44b**.

$$Z = \begin{matrix} CO_2Et \\ CH_2OH \\ CH_2Br \end{matrix}$$

44a: X = S
44b: X = SO$_2$

2.13.1.2 Properties

1,10-Dithia[2.2]metacyclophane (**6**) has a melting point of 145–146°C, as compared to a melting point of 132–133°C for the unsubstituted [2.2]metacyclophane (Section 2.2.2). In its stereochemical properties the dithia compound **6** also resembles the hydrocarbon analogue. It exists in a rigid stepladder conformation, which also remains fixed at higher temperatures, as was shown from the coalescence experiments at varying temperatures. The stepladder structure manifests itself in the shift to high field of the intra-annular protons in **6**, such that H(8) absorbs at $\delta = 5.42$ and H(16) at $\delta = 4.4$. By comparison, the value for the corresponding hydrocarbon is $\delta = 4.29$.

The heteracyclic [2.2]pyridinophanes, on the other hand, are conformationally flexible, and this is dependent on the particular structure. While the AB system in the ^{1}H-NMR spectrum of **6** and **45a–d** remains unchanged up to 180°C ($\Delta G_c^{\neq} > 96\,$kJ/mol in diphenyl ether), the CH$_2$ resonances in **10** and **8** are temperature-dependent.[5] The free energy of activation for the ring flipping, which is formulated according to models as the

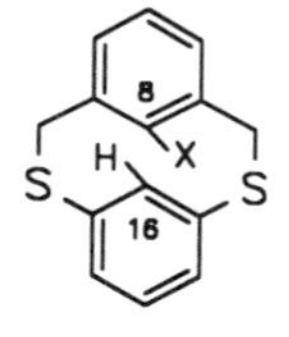

6: X = H	**8**: X = H
45a: X = F	**46a**: X = F
45b: X = Cl	**46b**: X = Cl
45c: X = Br	**46c**: X = Br

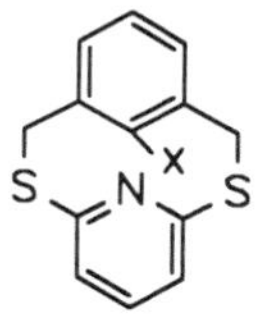

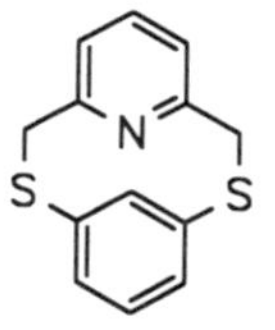

9 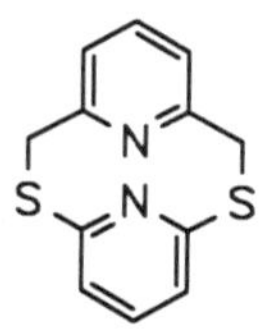   10

process C $\rightleftharpoons$ D, is found to have the value $\Delta G_c^{\neq} = 86\,$kJ/mol for **10** at 140°C (84 kJ/mol at 125°C). In the case of **8** one sees typically a signal broadening but no coalescence; thus $\Delta G_c^{\neq}$ for $T_c > 140$°C has a limiting value of $> 86\,$kJ/mol.

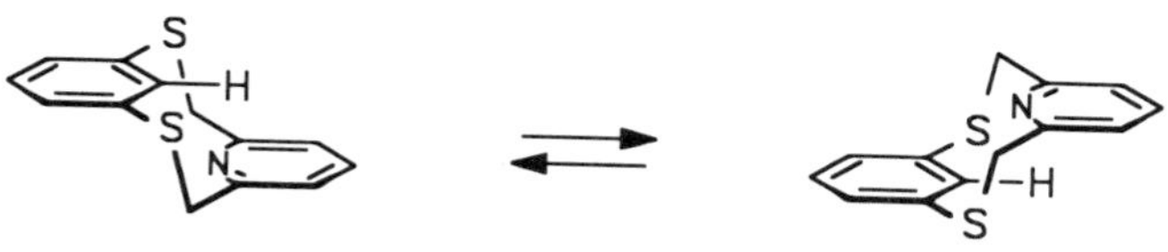

C   9   D

The fluoro compound **46a** is thermally unstable, becoming violet in colour in solution or in the crystalline state at 110°C. Presumably the pyridine nitrogen effects a transannular nucleophilic attack at the nearby carbon of the C—F bond.[5]

The methylene protons in **9** absorb as a singlet in all the common solvents, from which one can estimate $\Delta G_c^{\pm} > 57$ kJ/mol, with $T_c > 0°C$ and $\Delta\nu = 13$ Hz.

The above results lead to the conclusion that substitution of two methylene groups with sulphur atoms in the bridges of [2.2]phanes results in a lowering of the barrier to ring inversion because of greater separation between the interacting (intra-annular) atoms. The same effect is seen by substituting a CH with N in going from **6** to **8** or **9** to **10**. This correlation is in full agreement with the assumption of Gault, Price and Sutherland,[17] that the lone electron pair at pyridine nitrogen occupies less space than an aromatic H atom.[18]

X-ray crystal structural analyses have been reported for several of the heterocyclic [2.2]phanes described here,[19] examples being **6**, **8**, **11b**, **22**, **33**, **42b**, and **42d**. The crystal structure of **11b**, with an intra-annular NO$_2$ group, as well as the solution and solid-state ^{13}C-NMR spectra, show that the NO$_2$ group is bent about 48° away from the phenyl ring. This packing effect leads to a racemic ordering of **11b** in the crystal.[20]

11b

33 : R = Tosyl

In the case of the X-ray structure analysis of *N*-tosylaza[2.2]metacyclophane (**33**), the bond angles in the heterocyclic [2.2]metacyclophane portion are best represented from the perspective of the viewer (Figure 3).[13,15]

Figure 4 illustrates the ladder-like and helical twist skeleton of the phane **33**.[13] The short C—N bond gives rise, not only to a marked steric interference between the two intra-annular H atoms (H$_i$), but to a twisting of the planes of the *m*-phenylene rings and a bending of the benzene rings into a slightly boat shape. The *anti* conformation of the cyclophane framework is clearly seen in the crystal. The relatively short C(8)–C(16) separation of 231.3 pm forces the inner hydrogen atoms H$_i$(8) and H$_i$(16) into the anisotropic region of the overlying benzene rings. The extent of twist δ of the benzene rings from one another is in good agreement with that of the thiaphane **22**. The effect of the neighbouring sp^2-hybridized bridging nitrogen atom on the angle γ in **33** is small, as is the case in **16** and **17**, where $\gamma = 5.6$ and 6.9°, respectively.

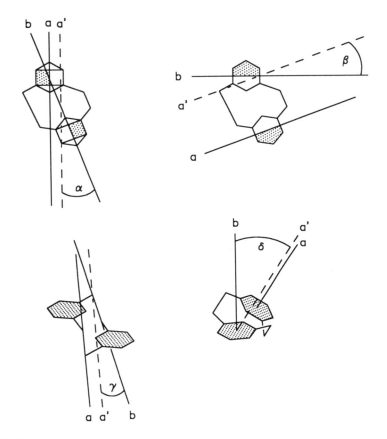

Figure 3. Representation of the bond angles in the heteracyclic [2.2]metacyclophanes on the basis of X-ray crystal structure analysis[9,13,15]

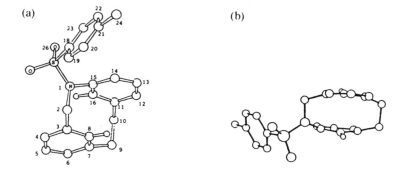

Figure 4. Molecular framework in **33** (X-ray crystal structure): (a) ladder-like arrangement; (b) view along the C(8)–C(16) axis[13]

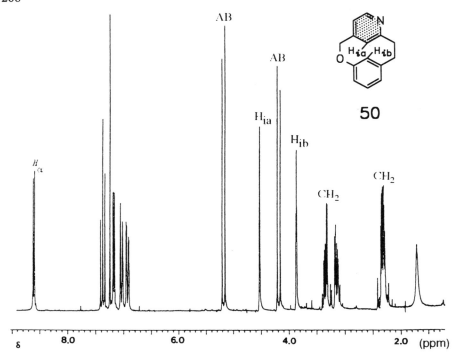

Figure 5. ¹H-NMR spectrum of **50** (200 MHz, CDCl₃)²¹

Static and dynamic stereochemistry

The ¹H-NMR spectra of the oxa[2.2]pyridinophanes **47–50** show the characteristic high-field shifts for the intra-annular protons (Figure 5). A significantly larger high-field shift is associated with the especially close crowding in the oxa[2.2]phanes **47–50**, by comparison with hydrocarbon [2.2]phanes. The shift of $\delta = 3.86$ for **25** is the highest H_i-absorption value recorded to date in a [2.2]phane. Because of the rigid conformation in the [2.2]phane skeleton, the O—CH₂ protons are split as an AB system, the ethane bridge appearing as a complex four-spin system. The ¹H-NMR spectra of all oxa[2.2]phanes **47–50** and **25** are temperature-independent.²¹

Separation of enantiomers

Resolution of oxa[2.2]metacyclophane (**25**) was accomplished on cellulose tris(dimethylphenylcarbamate) as a baseline separation. The heteraphane **47** was obtained in optically pure form by HPLC on polydiphenyl-2-pyridylmethyl-methacrylate.¹² The tosylaza[2.2]metacyclophane **33** was resolved or enriched by flash chromatography on triacetyl cellulose with ethanol as eluent, while attempts with other chiral phases did not lead to enrichment.¹³ A few of the

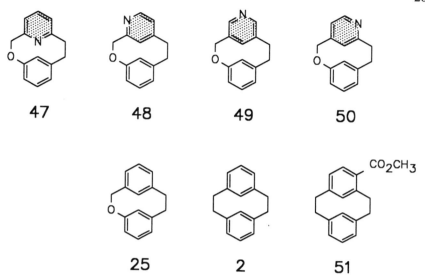

47   **48**   **49**   **50**

25   **2**   **51**

structurally isomeric pyridinophanes could be enantiomerically enriched on poly(triphenylmethylmethacrylate).[21]

Racemization barriers

The energy barrier for [2.2]metacyclophane-4-carboxylic acid methyl ester (**51**) of 131 kJ/mol can be considered as the parent value in [2.2]metacyclophanes. Because of the lower steric demand of the nitrogen free-electron lone pair, the barrier in **47** is 7 kJ/mol lower, at 124 kJ/mol. In the case of oxa[2.2]meta-cyclophane (**25**), where the ring is contracted because of the oxygen atom, the barrier is raised, surprisingly little, to 132 kJ/mol.[21] The value of $\Delta G_{inv}^{\ddagger} = 137$ kJ/mol for the remarkably optically stable tosylazaphane **33** can be compared directly with the value of 128 kJ/mol for the tosylaza compound **17**. The high racemization barriers in **33** and **17**, by comparison to those in the heteraphanes **22** and **16** (123 and 121 kJ/mol, respectively) may be attributed to increased steric interactions but also to electronic effects of the tosyl group in combination with the electron pair on nitrogen. The difference in energy barrier from the monothia compound **22** is about 12 kJ/mol.

	X	Y	Z	R
25	O	CH$_2$	CH$_2$	H
44a	O	CH$_2$	S	H
16	O	S	CH$_2$	H
17	NTos	S	CH$_2$	H
22	S	CH$_2$	CH$_2$	H

46a : X = S
44b : X = SO$_2$

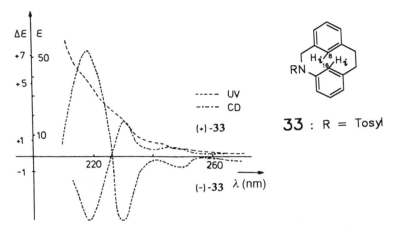

Figure 6. UV and CD curves of (+)- and (−)-**33** in ethanol[12]

The 'diagonal' diheteraphane **44b**, whose crystal structure is known, has been resolved into enantiomers. The barrier to racemization of 127 kJ/mol is similar to that for **16**. The intra-annular hydrogen atoms H_i in **44a** and **44b** absorb, respectively, at $\delta = 5.08$, 3.93 and $\delta = 5.35$, 4.12.

Relationship between structure and chiroptical properties

The helical chiral cyclophanes **16**, **17**, **22**, and **33** show intense Cotton effects in the region between 220 and 260 nm (Figures 6 and 7).[21a]

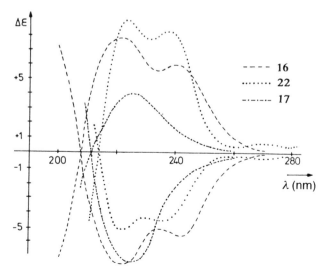

Figure 7. Comparison of the CD curves of both enantiomers of **16**, **17**, and **22** (**16** and **17** in dioxane; **22** in ethanol)[12]

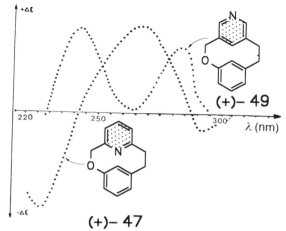

Figure 8. CD spectra of (+)-**47** and (+)-**49** in methanol[12]

In the series of constitutionally isomeric pyridinophanes **47–50**, where there is a rigid helical framework in the absence of chiral centres, it has been possible to observe the exclusive influence of the chromophore orientation. The CD spectra of **47** and **49** are compared in Figure 8.[21] The position of the first Cotton effect for **47** and **49** ($\lambda = 266$ and 284 nm), which suggests similar chromophores in which the pyridine ring is 'turned' 180°, is nevertheless shifted about 1.4 eV. The differences in the CD curves can only be explained by a difference in orientation of the pyridine rings in overall similar skeletal structures. On the contrary, no influence of chromophore orientation is observed in **52** and **53** containing CH_2-CH_2 bridges, the curves showing similar shapes.

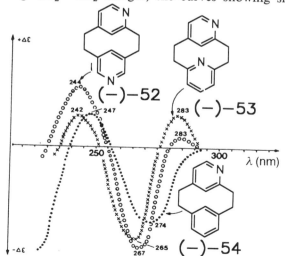

Figure 9. Comparison of the CD spectra of (−)-**53**($\times \times \times$); (−)-**52**(○○○); and (−)-**54**(●●●) in methanol[12]

Replacement of the pyridine ring by benzene (in **54**) causes a bathochromic shift of about 5 nm and disappearance of the first positive Cotton effect, which occurs at $\lambda = 283$ in **53** and **52**. In acid solution the pyridinophanes are protonated to give new CD curves; with **54**, for example, the long wavelength negative maximum disappears.

The CD spectra of **42a–f** resemble that of **33**.[15] Because of the *anti* conformation the H_i intra-annular hydrogen in substituted heterocyclic [2.2]-metacyclophanes appears at noticeably high field. This upfield shift becomes larger as the volume of the other intra-annular substituent increases, the latter forcing the hydrogen into the π cloud of the overhanging aromatic ring. Thus, **42g** exhibits the higher upfield shift in the ^{1}H-NMR for an aromatic proton in this series of [2.2]metacyclophanes. Hindered rotation of the *tert*-butyl group in **42g** is indicated by a characteristic broadening of the singlet of this absorption at $-70°C$ (400 MHz), while the other signals remain sharp. Rotation of the intra-annular phenyl ring in **42f** is impeded, the energy barrier being 65 kJ/mol.

CD spectra of enantiomerically enriched samples of condensed (**43**) and multi-layer [2.2]phanes (see Chapter 6) have been reported.[16]

By means of anomalous X-ray diffraction it was possible to establish the absolute configuration of *N*-toluenesulphonyl-8-thiomethyl-1-thia-10-aza-[2.2]metacyclophane (**42c**).[22] Then the absolute configurations of other similar [2.2]metacyclophanes could be deduced by a comparison of CD curves. For the regioselective syntheses, structures and chiroptical properties of chromium carbonyl complexes of [2.2]metacyclophanes, see reference 16c.

2.13.2 [2.2]METAPARACYCLOPHANES

Only a very few heterocyclic [2.2]phanes with *m,p*-bridging are known, and to date there are no examples of [2.2]paracyclophanes containing hetero atoms in the bridges. 2,9-Dithia[2.2]metaparacyclophane (**55**) was prepared in 9% yield by a high-dilution cyclization of 1,3-dimercaptobenzene and 1,4-bis(bromomethyl)benzene.[23]

55

4 : Y = H
4a : Y = F

The proton resonances of **55**, like those of the [2.2]metaparacyclophane hydrocarbon (**4**), are temperature-dependent (Figure 10).

The *p*-phenylene protons in **55** appear as a very broad absorption at 35°C in CDCl$_3$. Warming the sample in DMSO-d$_6$ causes a disappearance of the original *p*-phenylene signal and eventual generation of a new, sharp singlet at

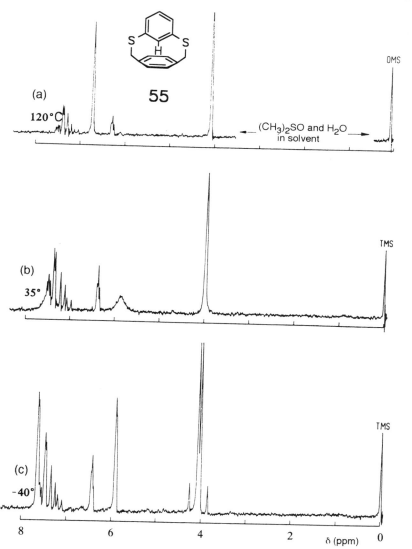

Figure 10. Proton resonances of **55** at various temperatures: (a) DMSO-d$_6$; (b,c) CCl$_4$/CS$_2$

$\delta = 6.70$ at 120°C (coalescence temperature $T_c = 40$°C); the methylene singlet also becomes sharper. Upon cooling of **55** in CCl$_4$/CS$_2$ the p-phenylene protons begin to sharpen, becoming an AA′XX′ system at -40°C. From these data for the p-phenylene signals one can calculate a $\Delta G_c^{\ddagger}$ of 62 kJ/mol. The methylene signal is split upon cooling, with $T_c = -20$°C; at -40°C it is a sharp AB system. This temperature dependence for the methylene absorption leads to a $\Delta G_c^{\ddagger} = 61$ kJ/mol. Inasmuch as a rotation of the p-phenylene ring alone

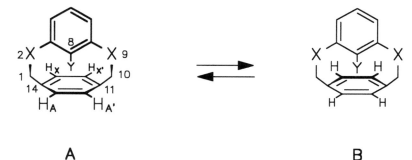

A   **B**

could not give rise to a CH_2 coalescence, it was concluded that the dynamic process A ⇌ B occurs at sufficiently high temperature.[23] This process effects an averaging of the aliphatic as well as p-phenylene protons. One can not conclude from the similar $\Delta G_c^{\ddagger}$ values whether a phenyl ring rotation is simultaneously playing a role. It is reasonable to assume that in fact $\Delta G_c^{\ddagger} > 61$ kJ/mol.

In the case of [2.2]metacyclophane (**4**), a considerably higher free enthalpy of activation of 84.6 kJ/mol was found (Section 2.4). Substitution at H(8) in **4** by a fluorine atom caused a blocking of the ring inversion ($\Delta G_c^{\ddagger} > 95$ kJ/mol), as evident from the temperature-independent proton resonance in **4a**. From this comparison it follows that steric effects in the ring cavity of **55** are considerably less serious than in **4**.[23]

References to Section 2.13

1. (a) Reviews: K. Meurer and F. Vögtle, *Top. Curr. Chem.*, **127**, 1 (1985).
   (b) F. Vögtle and P. Neumann, *Angew. Chem.*, **84**, 75 (1972). *Angew. Chem. Int. Ed. Engl.*, **11**, 78 (1972).
2. F. Vögtle, *Tetrahedron Lett.*, 3623 (1968).
3. F. Vögtle and U. Wolz, *Chem. Exp. Didakt.*, **1**, 15 (1975).
4. B. H. Smith, *Bridged Aromatic Compounds*, Academic Press, New York, 1964.
5. F. Vögtle and A. H. Effler, *Chem. Ber.*, **102**, 3071 (1969).
6. F. Vögtle, J. Grütze, R. Nätscher, W. Wieder, E. Weber and R. Grün, *Chem. Ber.*, **108**, 1694 (1975); K. Böckmann and F. Vögtle, *Chem. Ber.*, **114**, 1048 (1981); F. Vögtle, K. Meurer, A. Mannschreck, G. Stühler, H. Puff, A. Roloff and R. Sievers, *Chem. Ber.*, **116**, 2630 (1983).
7. Review: B. Klieser, L. Rossa and F. Vögtle, *Kontakte (Darmstadt)*, 3 (1984); A. Ostrowicki, E. Koepp and F. Vögtle, *Top. Curr. Chem.*, **161**, 37 (1991).
8. F. Vögtle and P. Neumann, *Tetrahedron Lett.*, **1970**, 115; W. Wieder, Diplomarbeit, Univ. Würzburg, 1973.
9. K. Meurer, F. Vögtle, A. Mannschreck, G. Stühler, H. Puff and A. Roloff, *J. Org. Chem.*, **49**, 3484 (1984).
10. K. Meurer, F. Luppertz and F. Vögtle, *Chem. Ber.*, **118**, 4433 (1985).
11. G. Dijkstra, W. H. Kruizinga and R. M. Kellogg, *J. Org. Chem.*, **52**, 4230 (1987); L. Mandolini and B. Masci, *J. Am. Chem. Soc.*, **106**, 168 (1984).
12. F. Vögtle, J. Struck, H. Puff, P. Woller and H. Reuter, *J. Chem. Soc., Chem. Commun.*, 1248 (1986).
13. F. Vögtle, K.-J. Przybilla, A. Mannschreck, N. Pustet, P. Büllesbach, H. Reuter and H. Puff, *Chem. Ber.*, **121**, 823 (1988).

14. S. Billen and F. Vögtle, *Chem. Ber.*, **122**, 1113 (1989).
15. F. Vögtle, A. Ostrowicki, P. Knops, P. Fischer, H. Reuter and M. Jansen, *J. Chem. Soc., Chem. Commun.*, 1757 (1989); F. Vögtle, P. Knops and A. Ostrowicki, *Chem. Ber.*, **123**, 1859 (1990); New angle definitions: J. Schulz, M. Nieger and F. Vögtle, *Chem. Ber.*, **124**, 2797 (1991); F. Vögtle, J. Schulz and M. Nieger, *Chem. Ber.*, **124**, 1415 (1991); J. Schulz, M. Nieger and F. Vögtle, *Chem. Ber.*, **124**, 2797 (1991); Caesium effect: A Ostrowicki, E. Koepp and F. Vögtle, *Top. Curr. Chem.*, **161**, 37 (1992).
16. (a) F. Vögtle, A. Ostrowicki, B. Begemann, M. Jansen, M. Nieger and E. Niecke, *Chem. Ber.*, **123**, 169; (1990) Review on high dilution reactions: F. Vögtle *et al.*, *Top. Curr. Chem.*, **161**, 1 (1991).
    (b) F. Vögtle, K. Mittelbach, J. Struck and M. Nieger, *J. Chem. Soc., Chem. Commun.*, 65 (1989).
    (c) J. Schulz, M. Nieger and F. Vögtle, *Chem. Ber.*, **124**, 2797 (1991); F. Vögtle, J. Schulz and M. Nieger, *Chem. Ber.*, **124**, 1415 (1991); J. Schulz, M. Nieger and F. Vögtle, *J. Chem. Soc., Perkins. Trans. 2*, 2095 (1992).
17. I. Gault, B. J. Price and I. O. Sutherland, *J. Chem. Soc., Chem. Commun.*, 540 (1967).
18. H. Förster and F. Vögtle, *Angew. Chem.*, **89**, 443 (1977); *Angew. Chem. Int. Ed. Engl.*, **16**, 429 (1977).
19. Review, see References 13, 15.
20. S. Kirgu and W. Nowacki, *Z. Kristallogr.*, **142**, 108 (1975).
21. K.-J. Przybilla and F. Vögtle, *Chem. Ber.*, **122**, 347 (1989).
    (a) S. Grimme, S. Peyerimhoff, S. Bartram, F. Vögtle, A. Breest and I. Hormes, *Chem. Phys. Lett.*, in press.
22. K. Rissanen, A. Ostrowicki and F. Vögtle, *Acta Chem. Scand.*, **44**, 268 (1990).
23. F. Vögtle, *Chem. Ber.*, **102**, 3077 (1969).

2.14 [2.0.0]PHANES

The [2.0.0]phanes described here, like the [2.2](1,3)phanes, contain a fixed ten-membered ring with intra-annular substituents (H, electron pair . . .). Vögtle

56   **57**   **58a** : R = H

58b : R = C_6H_5

59   **60**

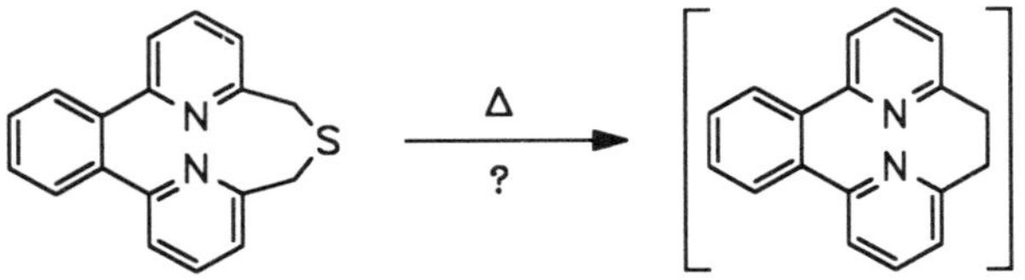

61   **62**

et al. first synthesized 'benzo[2.2]metacyclophane' ('triphenylenicene', [2.0.0]metaorthometacyclophane; **56**) and resolved it into its enantiomers.[1] The X-ray crystal structure shows the ethane bridge and the two benzene rings arranged in a ladder-like skeleton, similar to that for [2.2]metacyclophane (**2**). The three aromatic units, however, are locked in a screw-like twist, a three-winged propeller arrangement. The absolute configuration of **56** was determined by anomalous X-ray diffraction to be: P-(−)-25, M-(+)-25.[2]

The racemization barriers for the heterosubstituted [2.0.0]phanes **57–60** are relatively high ($\Delta G_{int}^{\pm} = 115$–$125$ kJ/mol); that is, the individual enantiomers are stable.[3]

Although the [2.0.0]pyridinophane **62** is not yet known, it is probably easily preparable from the 2-thia[3.0.0]phane **61**, which was obtained by Newkome,[8] albeit in small quantity.

References to Section 2.14

1. (a) M. Wittek, F. Vögtle, G. Stühler, A. Mannschreck, B. M. Lang and H. Irngartinger, *Chem. Ber.*, **116**, 207 (1983); cf. M. Wittek and F. Vögtle, *Chem. Ber.*, **115**, 1363 (1982); E. Hammerschmidt and F. Vögtle, *Chem. Ber.*, **113**, 1125 (1980).
   (b) E. Hammerschmidt and F. Vögtle, *Chem. Ber.*, **113**, 1125 (1980).
2. (a) F. Vögtle, M. Palmer, E. Fritz, U. Lehmann, K. Meurer, A. Mannschreck, F. Kastner, H. Irngartinger, U. Huber-Patz, H. Puff and F. Friedrichs, *Chem. Ber.*, **116**, 3112 (1983).
   (b) A. Aigner, F. Vögtle, S. Franken and H. Puff, *Chem. Ber.*, **118**, 3643 (1985).
3. G. R. Newkome, J. M. Roper and J. M. Robinson, *J. Org. Chem.*, **45**, 4380 (1980).

3 [3.3]Phanes

The most important members of the [3.3]phanes of general formula I are the dithia[3.3]phanes, which have been employed repeatedly as intermediates for the synthesis of the corresponding [3.3]phanes. To be sure a wide variety of interesting [3.3]phanes are known. In the following sections of this chapter a distinction will be made between the 'symmetrical' [3.3]meta-, para-, and orthocyclophanes, the 'unsymmetrical' [3.3]ortho-, metapara-, and ortho-paracyclophanes, and the corresponding [3.3]heterophanes. Examples of all three types have been reported, the best known of which will be selected in the following sections.

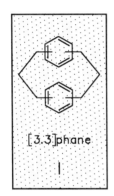

[3.3]phane

I

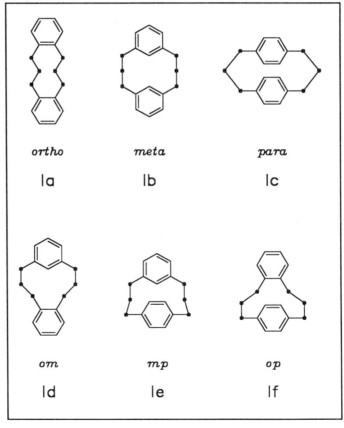

ortho	meta	para
Ia	Ib	Ic

om	mp	op
Id	Ie	If

217

3.1 [3.3]PHANE HYDROCARBONS

[3.3]Paracyclophane (6), described very early by Cram,[1] was prepared by way of an acyloin condensation. It proved to be less strained than [2.2]paracyclophane.[1] The rather limited information on this compound includes an X-ray crystal structure of 6 and of the complex $6\text{-}Cr^+I_3^-$,[2a] as well as more recent NMR-, PES-, and ESR studies of substituted [3.3]paracyclophanes.[2b] Equilibration between A ('chair') and B ('boat') takes place in solution.[3]

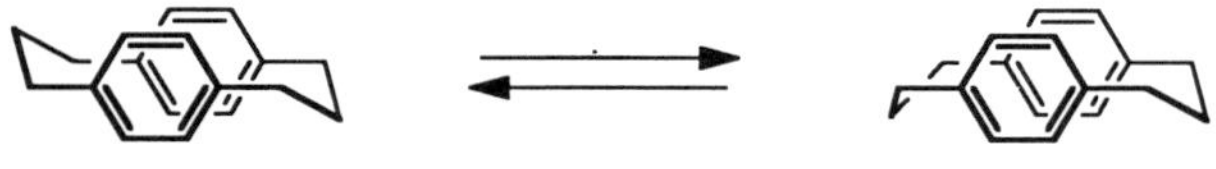

In more recent times [3.3]phane hydrocarbons in the *meta*- and *para*-series[4c] have been prepared by the following methods.

(a) Pyrolysis of 'non-doubly benzylic' sulphones[4a] (and 'photolytic sulphur extrusion)[4b]

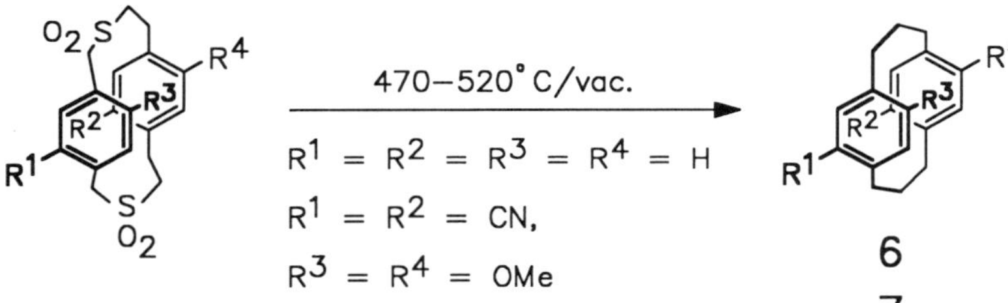

$R^1 = R^2 = R^3 = R^4 = H$

$R^1 = R^2 = CN,$

$R^3 = R^4 = OMe$

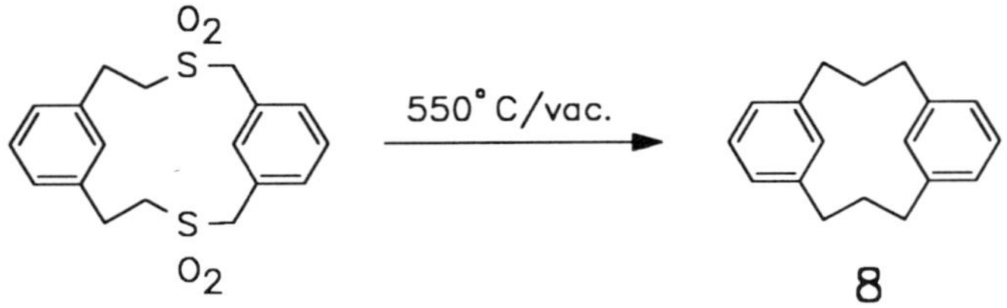

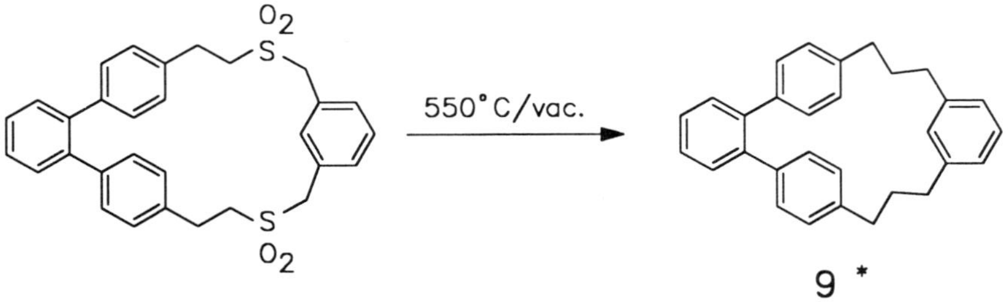

---

*The hydrocarbon 9 can be designated as [3.3.0.0](1,3)(1,4)(1,2)(1,4)benzenophane or, more conveniently, as [3](4,4″)orthoterphenylo[3](1,3)benzenophane.

(b) Cyclization with TosMIC[5]
One-step(Ar = Arene):

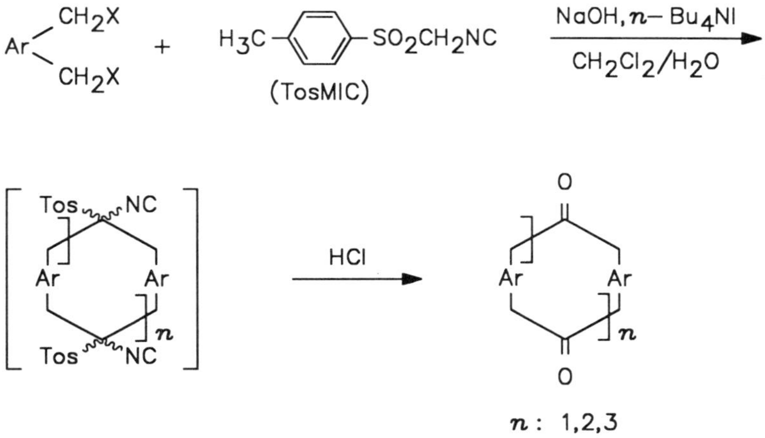

n : 1,2,3

Two-step:

n : 1,2,3

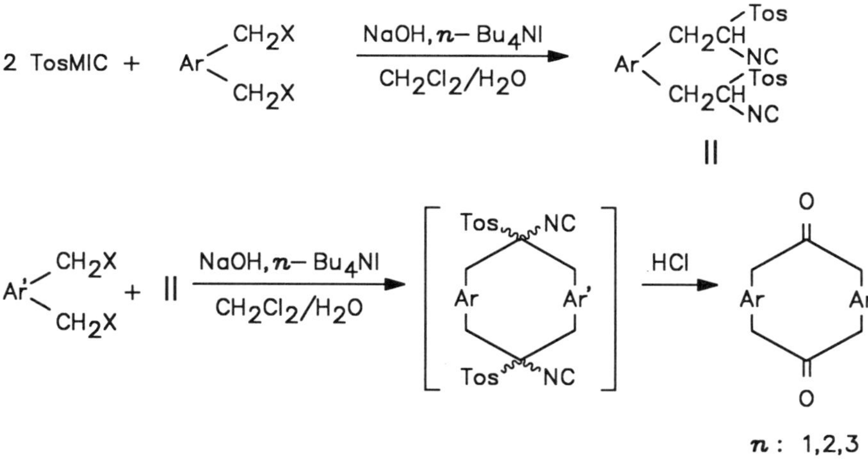

The following are examples of [3.3]phanes {and [3$_n$]phanes} obtained by these routes.

(c) Chromium hexacarbonyl complex method (Semmelhack)[6]

The [3.3]metacyclophane and [3.3]naphthalenophane frameworks have, in addition, been constructed by ring expansion of [2.2]metacyclophan-1,10-dione with diazomethane[5a] and via the malonic ester method.

The structure of [3.3]metacyclophane (**8**) has been confirmed by X-ray crystal analysis.[6] It assumes a *syn* geometry in the crystal, such that the CH$_2$ bridging groups constitute a chair/chair conformation (Figure 1). This same conformation is shown to be an energy minimum by molecular mechanics calculations, from which it also follows that the arene rings are twisted. The calculations indicate

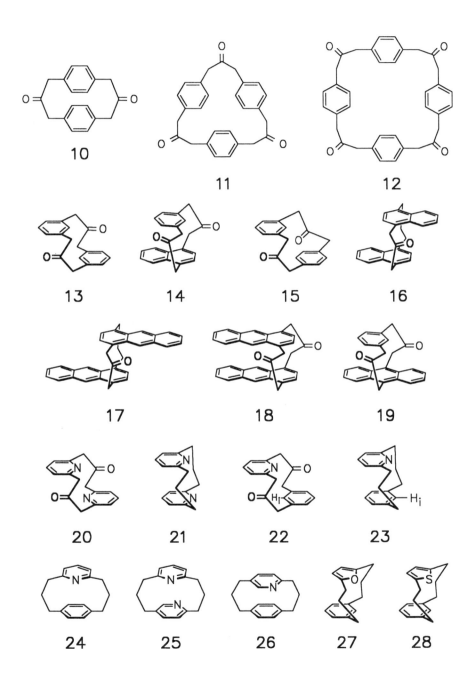

10

11

12

13

14

15

16

17

18

19

20

21

22

23

24

25

26

27

28

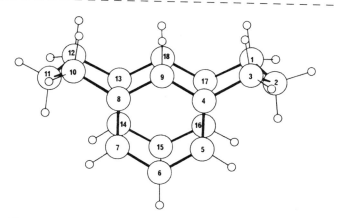

(a) $Cr(CO)_6$; (b) $LiN(iPr)_2$; (c) I_2; (d) $LiN(iPr)_2$

(e) O_2; (f) Na_2SO_3; (g) Na, NH_3

Figure 1. X-ray crystal structure of the [3.3]metacyclophane hydrocarbon (**8**)[6]

that all *anti* conformations lie at about 25 kJ/mol higher energy, as is the case for the 2,11-dithia[3.3]metacyclophanes (see Figure 2 and below).

Temperature-variable ¹H- and ¹³C-NMR spectroscopic studies demonstrate a dynamic process in solution for [3.3]metacyclophane(**8**), that involves

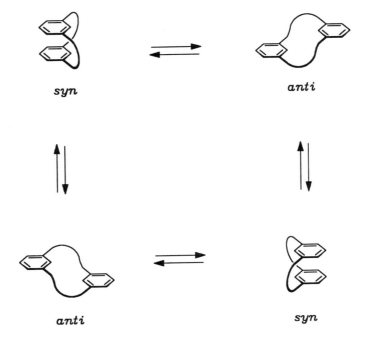

Figure 2. *syn* and *anti* Conformers and schematic *syn/anti*-isomerization of [3.3]metacyclophanes

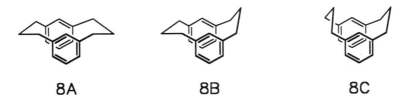

8A   **8B**   **8C**

Figure 3. Conformational isomerization of [3.3]metacyclophane bridges

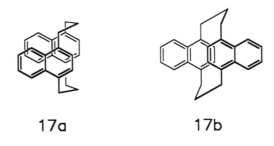

17a   **17b**

the intermediacy of a chair/boat conformation **8B** in the interconversion between *syn* conformers (Figure 3). The isomerization barrier has been estimated to be 42–46 kJ/mol.[6,7]

Description of donor–acceptor type [3.3]phanes can be found in Section 2.11 and in the original literature.[8] For [3.3]phanes with *anti* conformations see reference 8a.

[3.3](1,4)Naphthalenophane exists as both *syn*- and anti-isomers (**17a,b**).[9]

There appear to be only a few known [3.3]orthocyclophanes,[10] which are of limited interest in cyclophane chemistry because they can assume a relatively 'flat' conformation. The following examples, however, exhibit π–π electronic

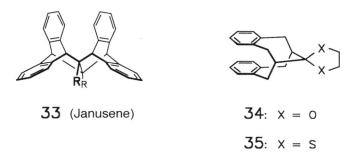

33 (Janusene)   **34**: x = o

35: x = s

interactions between the 'face-to-face' rigidly held benzene rings. 'Janusene' (**33**), considered by Tashiro *et al.* among the [3.3]orthocyclophanes, contains additional bridges.[11] X-ray crystal structure analysis of **34** and **35** indicates the shortest intramolecular distance between two benzene rings to be 303–307 pm, with a dihedral angle between the two phenylene rings of 25°.

[3.3]Azobenzenophane was synthesized by Grützmacher.[12]

References to Section 3.1

1. Review: B. H. Smith, *Bridged Aromatic Compounds*, Academic Press, New York, 1964. Cf. D. J. Cram and R. C. Helgeson, *J. Am. Chem. Soc.*, **88**, 3515 (1966).
2. (a) P. K. Gantzel and K. N. Trueblood, *Acta Crystallogr.*, **18**, 958 (1965); R. Benn, N. E. Blank, M. W. Haenel, J. Klein, A. R. Koray, K. Weidenhammer and M. L. Ziegler, *Angew. Chem.*, **92**, 45 (1980); *Angew. Chem. Int. Ed. Engl.*, **19**, 44 (1980); N. E. Blank, M. W. Haenel, A. R. Koray, K. Weidenhammer and M. L. Ziegler, *Acta Crystallogr.*, **B36**, 2054 (1980).
   (b) M. W. Haenel, A. Flatow, V. Taglieber and H. A. Staab, *Tetrahedron Lett.*, 1733 (1977).
   (c) E. Heilbronner and Z. Yang, in *Cyclophanes, II* (F. Vögtle, Ed.), *Top. Curr. Chem.*, **115**, 1 (1983).
   (d) F. Gerson, in *Cyclophanes, II* (F. Vögtle, Ed.), *Top. Curr. Chem.*, **115**, 57 (1983).
3. F. A. L. Anet and M. A. Brown, *J. Am. Chem. Soc.*, **91**, 2389 (1969).
4. (a) Review: F. Vögtle and L. Rossa, *Angew. Chem.*, **91**, 534 (1979); *Angew. Chem. Int. Ed. Engl.*, **18**, 514 (1979); F. Vögtle and L. Rossa, in *Cyclophanes, I* (F. Vögtle, Ed.), *Top. Curr. Chem.*, **113**, 1 (1983); J. Dohm and F. Vögtle, *Top. Curr. Chem.*, **161**, 69 (1992).
   (b) T. Otsubo, M. Kitasawa and S. Misumi, *Bull. Chem. Soc. Jpn.*, **52**, 1515 (1979).

(c) Orthopara- and orthometa [3.3]phanes: M. Asami, C. Krieger and H. A. Staab, *Tetrahedron Lett.*, **32**, 2117 (1991).

5. K. Kurosawa, M. Suenaga, T. Inazu and T. Yoshino, *Tetrahedron Lett.*, **23**, 5335 (1982); T. Shinmyozu, Y. Hirai and T. Inazu, *J. Org. Chem.*, **51**, 1551 (1986).
(a) D. Krois and H. Lehner, *J. Chem. Soc., Perkin Trans. 1*, 477 (1982).

6. M. F. Semmelhack, J. J. Harrison, D. C. Young, A. Gutiérrez, S. Rafii and J. Clardy, *J. Am. Chem. Soc.*, **107**, 7508 (1985).

7. Cf. Y. Fukazawa *et al.*, *J. Am. Chem. Soc.*, **110**, 7842, 8692 (1988).

8. H. A. Staab, C. P. Hertz and A. Döhling, *Chem. Ber.*, **113**, 233 (1980); H. A. Staab, C. P. Herz and A. Döhling, *Tetrahedron Lett.*, **1979**, 791; H. A. Staab and A. Döhling, *Chem. Ber.*, **1979**, 2019; H. A. Staab, A. Döhling and C. Krieger, *Tetrahedron Lett.*, **32**, 2215 (1991).
(a) J. Breitenbach, R. Hoss, M. Nieger, K. Rissanen and F. Vögtle, *Chem. Ber.*, **125**, 255 (1992).

9. M. Yoshinaga, T. Otsubo, Y. Sakata and S. Misumi, *Bull. Chem. Soc. Jpn.*, **52**, 3759 (1979).

10. For example:
(a) J. H. Golden, *J. Chem. Soc.*, 3741 (1961).
(b) P. J. Collin, D. B. Roberts, G. Sugowdz, D. Wells and W. H. F. Sasse, *Tetrahedron Lett.*, 321 (1972).

11. S. Mataka, K. Takahashi, T. Mimura, T. Hirota, K. Takuma, H. Kobayashi, M. Tashiro, K. Imada and M. Kuniyoshi, *J. Org. Chem.*, **52**, 2653 (1987).

12. J. Schmiegel, H.-F. Grützmacher *et al.*, *Chem. Ber.*, **123**, 1397, 1749 (1990); For further azobenzene switches cf. S. Shinkai *et al.*, *J. Chem. Soc. Perkin Trans. 2*, 1905 (1990); H. L. Ammon and S. Shinkai, *Acta Cryst. Sect. C. Crystal Structure Commun.*, **47**, 623 (1991); Concave dyes: M. Bauer and F. Vögtle, *Chem. Ber.*, **125**, 1675 (1992).

3.2 DITHIA- AND DIAZA[3.3]PHANES

As already mentioned, dithia[3.3]phanes have often been used as intermediates in the synthesis of [2.2]phane hydrocarbons.[2]

2,11-Dithia[3.3]paracyclophane (**7**) was reported in 1970.[1] The analogous ring inversion (A⇌B), formulated for the [3.3]paracyclophane hydrocarbon (**6**), appears to occur at a faster rate in this case.

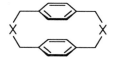

6 : X = CH$_2$

7 : X = S

Of particular interest are two [3.3]phanes: 2,19-dithia[3.3](4,4')-(*E*)-azobenzeno<2>phane* (**8**) and the corresponding quadruply bridged analogue

---

*The symbol <2> denotes that *two* azobenzene units (in the 4,4' positions) are bridged to form a double-layer phane, azobenzene in this case being considered as an 'aromatic unit'. In an alternative designation (more strictly correct but cumbersome), the four benzene rings would be specified as aromatics, such that the compound would be named as a tetraaza[3.2.3.2](1,4)benzeno<4>phane.

To date no decision has been made as to which is the 'right' designation. Ideally a compound should be designated with only one name; and among two alternatives the shorter should be chosen, provided it is unambiguous.

9 (Rau).[3] Both **8** and **9** undergo isomerization in the presence of light ('organic switches').[4] In this process one can demonstrate experimentally the *E/Z* photo-isomerization of azobenzenes through a change of the bond angle at nitrogen. In both **8** and **9** the *E/Z* isomerization by rotation of the N=N double bond is excluded because of the bridging; thus the process must involve an *inversion of the nitrogen atom of the azo group*. Kinetic analysis of irradiations in dilute solution suggests the possibility of formation of a tetra-azacyclobutane.

A large number of intra-annularly substituted 2,11-dithia[3.3]metacyclophanes of type **10** have been prepared because of their conversion, through sulphur extrusion, to [2.2]phane hydrocarbons (see also Section 2.2).[2a] For this reason, this type of phane is described in several places in the text.

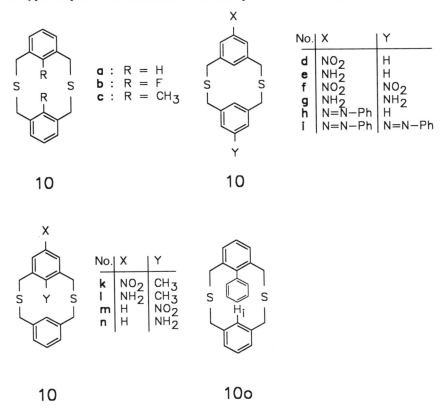

No.	X	Y
d	NO$_2$	H
e	NH$_2$	H
f	NO$_2$	NO$_2$
g	NH$_2$	NH$_2$
h	N=N–Ph	H
i	N=N–Ph	N=N–Ph

No.	X	Y
k	NO$_2$	CH$_3$
l	NH$_2$	CH$_3$
m	H	NO$_2$
n	H	NH$_2$

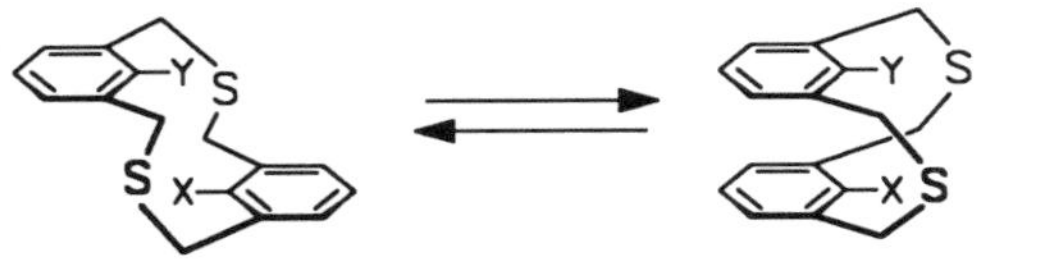

10c : X = Y = CH₃
10p : X = H, Y = CH₃
10m : X = H, Y = NO₂
10n : X = H, Y = NH₂

anti   **10**   *syn*

Vögtle *et al.* first observed the occurrence of *syn* and *anti* conformers of intra-annularly substituted dithia[3.3]metacyclophanes.[5] Conformations of some members of the dithia[3.3]metacyclophane series **10** have been determined by X-ray crystallography.[2b] The unsubstituted **10a** exists exclusively as the *syn* form in the crystal.[7] From dipole moment studies it could be shown that the difluoro compound **10b**[5] exists predominantly in the *syn* conformation in solution. The *syn-* and *anti*-isomers of the dimethyl compound **10c** have been synthesized and separated.[8]

Particularly large substituents have been incorporated into the inner cavity of these [3.3]phanes, an example being **10o**. The phenyl substituent is of particular interest for several reasons. Besides the opportunity to study electronic effects in a novel arrangement, there is the possibility in **10o** to orient benzene rings in layers, without their being multiply bridged, as is the case with other phanes.

Synthesis of **10s** was achieved by way of a benzidine rearrangement **10q** → **10r**, the overall result being the migration of C_6H_5- from 'outside' into the cavity of the 12-membered ring.[9a] Details of dithia[3.3]naphthalenophanes[7] and -pyridinophanes[5] can be found in the specified literature citations.

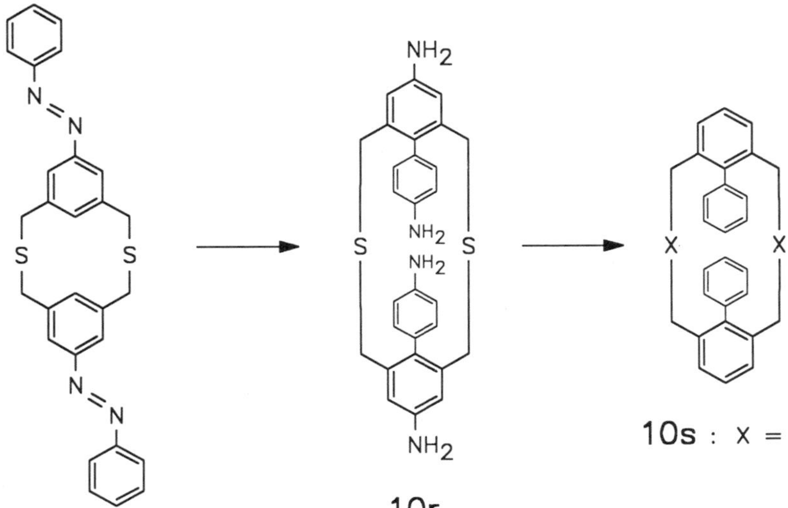

10q   **10r**   **10s** : X = S

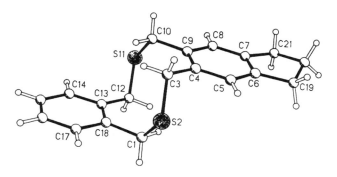

Figure 1. X-ray Crystal Structure of **13**[11]

In a solution of 2,11-dithia[3.3]orthocyclophane **(11)**, first described in 1979,[10] the preferred *anti* conformation appears to be undergoing an *anti/anti* ring inversion of the type A ⇌ B.

| A | **11** | B |

In 1989 the related compounds 2,11-dithia[3.3](5,6)indanocyclophanes **(12, 13)** were reported, along with their X-ray crystal structures (Figure 1; Hopf *et al.*).[11]

| **12** | **13** |

Readers are directed to the literature for details on [3.3]metaparacyclophanes with $(CH_2)_3$-bridges[12] and with CH_2–S–CH_2 bridges.[13] Some tetrathia-[14] and hexathia[3.3]phanes[15] have also been described.

14a : X = S
14b : X = SO_2

15

16

17

228

Both Vögtle et al.[16] and later Takemura et al.[17] synthesized diaza[3.3]-phanes in the *meta-* and *para-*series, with the view to converting them to [2.2]phanes (see Section 2.2). The latter investigators also reported higher cyclooligomers, triaza[3.3.3]-and tetraaza[3.3.3.3]paracyclophanes.[17]

19

a : R = Tos
b : R = NO
c : R = CH₃
d : R = NH₂

18   20

For diaza[3.3](9,10)anthracenophanes and their photocyclization see reference 18.

References to Section 3.2

1. F. Vögtle, *Chemiker-Ztg.*, **94**, 313 (1970).
2. Reviews:
   (a) F. Vögtle and L. Rossa, *Angew. Chem.*, **91**, 534 (1979); *Angew. Chem. Int. Ed. Engl.*, **18**, 514 (1979).
   (b) P. M. Keehn and S. M. Rosenfeld (Eds.), *Cyclophanes, Vols. I, II*, Academic Press, New York, 1983.
   (c) V. Boekelheide, in *Cyclophanes, I* (F. Vögtle, Ed.), *Top. Curr. Chem.*, **113**, 87 (1983).
3. D. Gräf, H. Nitsch, D. Ufermann, G. Sawitzki, H. Patzelt and H. Rau, *Angew. Chem.*, **94**, 385 (1982); Further azobenzenophanes: H.-F. Grützmacher et al., *Chem. Ber.*, **123**, 1397 (1990); N. Tamaoki, K. Koseki and T. Yamaoka, *Tetrahedron Lett.*, **31**, 3309 (1990).
4. Review: F. Vögtle, Supramolecular Chemistry, Wiley, Chichester, 1991; see also S. Shinkai et al., *J. Chem. Soc., Perkin Trans. 2*, 1905 (1990).
5. F. Vögtle and L. Schunder, *Chem. Ber.*, **102**, 2677 (1969); 2,12-Dithia[3.3](1,4)-cycl-[3.2.2]azinophanes: Y. Matsuda et al., *Heterocycles*, **31**, 983 (1990).
6. See T. Sato, M. Wakabayashi, M. Kainosho and K. Hata, *Tetrahedron Lett.*, 4185 (1968), as well as Reference 5.
7. F. Vögtle, R. Schäfer, L. Schunder and P. Neumann, *Liebigs Ann. Chem.*, **734**, 102 (1970); W. Anker, G. M. Bushnell and R. H. Mitchell, *Can. J. Chem.*, **57**, 3080 (1979).
8. R. H. Mitchell and V. Boekelheide, *J. Am. Chem. Soc.*, **96**, 1547 (1974).
9. (a) K. Böckmann and F. Vögtle, *Chem. Ber.*, **114**, 1065 (1981); also contains information on other intra-annular substituents.
   (b) F. Vögtle, J. Grütze, R. Nätscher, R. Grün, W. Wieder and E. Weber, *Chem. Ber.*, **108**, 1694 (1975).
   (c) F. Vögtle, W. Wieder and H. Förster, *Tetrahedron Lett.*, 4361 (1974).
10. M. K. Au, C. W. Mak and T. L. Chan, *J. Chem. Soc., Perkin Trans. 1*, 1475 (1979); S. Mataka, S. T. Lee and M. Tashiro, *J. Chem. Soc., Perkin Trans. 2*, 2017 (1990); Cf. L. R. Hanton and T. Kemmitt, *J. Chem. Soc., Chem. Commun.*, 700 (1991).

11. G. Bodwell, L. Ernst, H. Hopf and P. G. Jones, *Tetrahedron Lett.*, **30**, 6005 (1989); Y. Fukazawa, T. Okajima and S. Usui, *Tetrahedron Lett.*, **31**, 1309 (1990).
12. H. A. Staab, M. Jörns, C. Krieger and M. Rentzea, *Chem. Ber.*, **118**, 796 (1985).
13. F. Vögtle, *Chem. Ber.*, **102**, 3077 (1969).
14. F. Vögtle and R. Lichtenthaler, *Z. Naturforsch*, **26**, 872 (1971).
15. F. Fehér, K. Glinka and F. Malcharek, *Angew. Chem.*, **83**, 439 (1971); *Angew. Chem. Int. Ed. Engl.*, **10**, 413 (1971).
16. F. Vögtle and P. Neumann, *Tetrahedron Lett.*, 115 (1970).
17. H. Takemura, M. Suenaga, K. Sakai, H. Kawachi, T. Shinmyozu, Y. Miyahara and T. Inazu, *J. Incl. Phenom.*, **2**, 207 (1984); P. Finocchiaro *et al.*, *J. Chem. Soc., Perkin Trans. 2*, 449 (1991); Diaza[3.3]anthracenophanes: M. Usui *et al.*, *Chem. Lett.*, 419 (1990).
18. M. Usui *et al.*, *Chem. Lett.*, 419 (1990).

3.3 1.11-DISELENA[3.3]PHANES

The first 1,11-diselena[3.3]metacyclophane (**23**), with no substituents, was prepared from two routes, both low-yield reactions: coupling of dihalides such as **22** with sodium selenide (Mitchell) and deselenization of tetraselena[4.4]cyclophane with P(NEt$_2$)$_3$.[1] In 1987 an improved synthesis of **23** was reported by Misumi.[2] **23** was obtained in relatively higher yield by coupling of bis(selenocyanate)

21 with the dibromide **22** in the presence of NaBH$_4$. In this high-dilution reaction it was advantageous to use a large excess of borohydride, in order to maintain reducing conditions and thereby suppress oxidative polymerization. The conversion of diselena[3.3]metacyclophane **23** to [2.2]metacyclophane (**2**) could be accomplished by (a) a Wittig- or benzyne-induced Stevens rearrangement; (b) photodeselenization with tris(dimethylamino)phosphane; and (c) vacuum flash photolysis.[2]

According to the ^{1}H-NMR spectra of selena[3.3]metacyclophanes, the *syn* conformer predominates as the more stable one;[3] in the crystalline state **23**, like its dithia analogue **10a**, exists exclusively in the *syn* conformation. Through low-temperature ^{77}Se- and ^{1}H NMR spectra it could be shown that 2,11-di-selena[3.3]metacyclophane (**23**) in solution undergoes a *syn/anti* interconversion ($\Delta G_c^{\neq} = 33.5$ kJ/mol). Also observed is a chair ⇌ boat 'bridge wobbling' of the bridging $-CH_2-$ groups, as in the case of [3.3]metacyclophane (**8**, Section 3.1); but the $\Delta G^{\neq}$ value is so low that it is not measurable.

The dimethyl intra-annularly substituted diselena[3.3]metacyclophane is also known.[2]

References to Section 3.3

1. R. H. Mitchell, *Can. J. Chem.*, **58**, 1398 (1980).
2. H. Higuchi, K. Tani, T. Otsubo, Y. Sakata and S. Misumi, *Bull. Chem. Soc. Jpn.*, **60**, 4027 (1987); cf. J. Breitenbach, M. Nieger and F. Vögtle, *Chem. Ber.*, **124**, 2583 (1991).
3. R. H. Mitchell and K. S. Weerawarna, *Tetrahedron Lett.*, **29**, 5587 (1988).

4 [m.n]Phanes (m,n ≥ 2)

4.1 [3.2]PHANES

Griffin and Lehner investigated the ring flipping process in [3.2]metacyclophane hydrocarbons containing bridge substituents by ^{1}H-NMR spectroscopy.[1] They formulated a conformational equilibrium 6A ⇌ 6B, with an energy barrier $\Delta G_{60}^{\ddagger}$ of about 66–70 kJ/mol which is substituent-dependent.

Recently a straightforward synthesis of substituted [3.2]paracylophan-10-enes from bis(4-acylphenyl)propanes via the McMurry reaction was described by Grützmacher.[2]

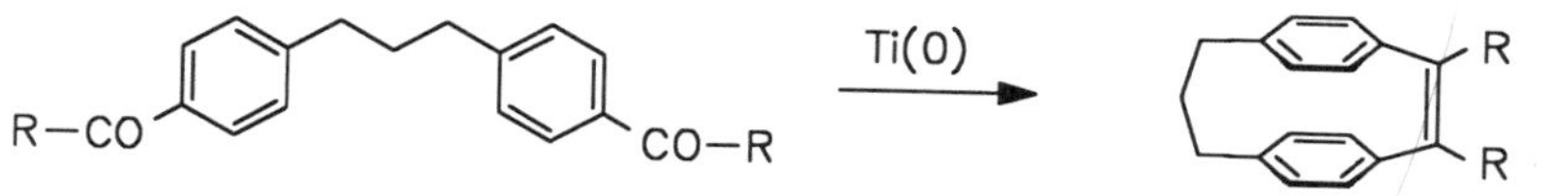

The most common examples of [3.2]phanes are found in the family of 2-thia[3.2]phanes. Sato prepared the first 2-thia[3.2]metacyclophane (8a) by treatment of the bis(halomethyl) compound 7 with Na$_2$S.[3] The ^{1}H-NMR spectra of the sulphide 8a, the sulphoxide 8b, and the sulphone 8c proved to

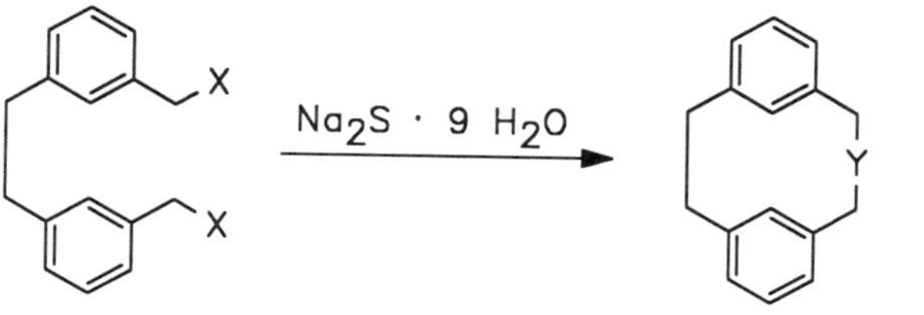

8a : X = S
8b : X = SO
8c : X = SO$_2$

be temperature dependent, with energy barriers for the ring inversion C ⇌ D estimated to be 55, 63, and 68 kJ/mol, respectively. As in the case of [3.3]metacyclophane (see above), the *syn* conformation E appears to be energetically favoured.[3]

C   8   D   E

Vögtle and Przybilla developed a new synthetic route to hetera- and hetero-cyclic oxathia[3.2]phanes by starting with 3-(mercaptomethyl)phenol (**9**).[4] The [3.2]phanes were prepared in good yield under high-dilution conditions in

ethanol solvent with caesium assistance. These tightly bridged oxa[3.2]phanes also proved to be conformationally mobile. Ring inversion barriers were estimated from coalescence of the ^{1}H-NMR signals for the C$\underline{H}_2$O— and sometimes also for the C$\underline{H}_2$S— group.

The 2-thia[3.2]pyridinophanes **11a–c**, prepared analogously, show an approximately 23 kJ/mol higher conformational barrier than the corresponding oxa[3.2]phanes of type **10**.[4] This distinction is attributed to the different geometry of the pyridine ring and also to the shortening of the one bridge in **10** because of the C—O—C bond. The variation in geometry of a pyridine ring from that of benzene also affects the barrier to ring inversion in the [3.2]phane skeleton. Both of the 2,4-bridged pyridinophanes **10b** and **10c** show the same energy value as that for oxa[3.2]metacyclophane **10e**, within the limits of measurement. On the other hand, the barrier in **10c** (95 kJ/mol) is about 7 kJ/mol higher than that in its structurally isomeric **10b** and **10d**. Thus the molecular dynamics in this system serve as a sensitive probe of the molecular fine structure. At coalescence temperatures of 0–3°C in **11a–c** one finds a $\Delta G^{\neq}_{INV}$ value of 78 kJ/mol.

11a   **11b**   **11c**

2-Thia[3.2]naphthalenophane (**13**) was synthesized as already described in Section 2.13.[5] In the high-dilution cyclization of the bis(bromomethyl) compound **12b** with caesium assistance, the choice of solvent is critical. Use of 'standard conditions' [ethanol/benzene(1:1)] leads to a 2.5% yield, while the cyclization in acetonitrile results in an increase in yield to 78%. It appears that a dipolar aprotic solvent enhances the formation of the rings in strained cyclophanes, in particular in the pyridine series. As is the case for all

12a : X = OH
12b : X = Br

13

14

2-thia[3.2]phanes, **13** is an ideal precursor for the synthesis of the corresponding [2.2]phanes (**14**) through desulphurization.

A comparison of the [3.2]phanes **13** and **10e** with the [3.1]phane **14** is revealing.[5] All three compounds show temperature-dependent ^{1}H-NMR spectra, ring inversion barriers being determined from coalescence of the C$\underline{H}_2$O— group. While the barrier for 2-thia[3.2]metacyclophane (**10e**) is 55 kJ/mol, that for the naphthalenophane **13** is surprisingly higher (98 kJ/mol). Its benzo analogue

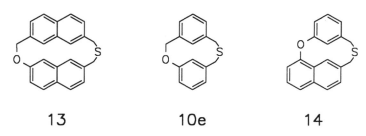

13   **10e**   **14**

14 has a ring inversion barrier about 10 kJ/mol lower (88 kJ/mol). Apparently the steric interference of the four intra-annular hydrogen atoms in **13** is not offset by the conformational flexibility of the two naphthalene rings (as compared to the benzene rings in **10e**).

References to Section 4.1

1. R. W. Griffin Jr and R. A. Coburn, *J. Am. Chem. Soc.*, **89**, 4638 (1967); D. Krois and H. Lehner, *Tetrahedron*, **38**, 2319 (1982).
2. H.-F. Grützmacher, E. Neumann, F. Ebmeyer, K. Albrecht and P. Schelenz, *Chem. Ber.*, **122**, 2291 (1989); Corresponding (*Z*)-2-oxa[3](4,4′)azobenzenophane: N. Tamaoki *et al.*, *Bull. Chem. Soc. Jpn.*, **64**, 2011 (1991).
3. T. Sato, M. Wakabayashi, K. Hata and M. Kainosho, *Tetrahedron*, **27**, 2737 (1971).
4. K.-J. Przybilla and F. Vögtle, *Chem. Ber.*, **122**, 347 (1989).
5. S. Billen and F. Vögtle, *Chem. Ber.*, **122**, 1113 (1989).

4.2 SELECTED HIGHER [*m.n*]PHANES

The series of [*m.n*]paracyclophane hydrocarbons **6** are among the most intensively studied. The consequences of diminished ring strain and π interaction

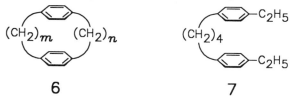

6   **7**

of the 'face-to-face' benzene rings as the bridging is lengthened have been considered in detail. The early methods of UV and NMR[2] were later supplemented by PES[3] and ESR[4] spectroscopy. The electronic spectra of phanes with $m,n > 4$ show absorptions similar to those in the open-chain model compound **7**. In going from [4.4]paracyclophane to the [4.3] bridging, however, one observes a discontinuity in the absorption; this is attributed to the fact that the separation of the benzene rings is less than the normal van der Waals value of about 340 pm. Further shortening of the bridging leads to two new bands of higher intensity: a longer wavelength and a shorter wavelength, the latter appearing about where the longest wavelength band of **7** is found. These results serve as indicators of the increasing deformation and transannular effects between the two 'face-to-face' benzene rings.

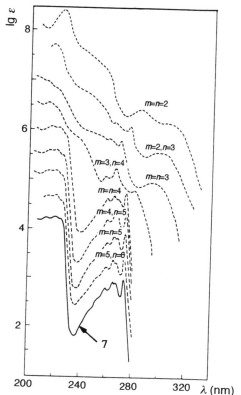

Figure 1. Comparison of the electronic spectra of several [m.n]paracyclophanes in cyclohexane.[1] (All spectra are moved up one ordinate unit in relation to that of the reference compound **7**)

ESR spectroscopic studies have revealed the interesting fact that the frequency of electron exchange between the two benzene rings in the radical anions of [m.n]paracyclophanes diminishes dramatically as m and n increase. In the case where $m = n \geqslant 4$ the exchange frequency has been determined as $< 10^6 \text{s}^{-1}$, it being mainly dependent on the migration rate of the counter ion, usually K^+, between the two equivalent sites of the radical anion.[3]

The conformational analyses of some *syn* and anti-[4.3]- and [4.2]-metacyclophanes were described by Lehner.[4a]

The X-ray crystal structure has been reported for the tetramethoxy[7.7]paracylophandione **8**, whose carbonyl groups were introduced by a double 1,3-dithian reaction.[5] The orange, crystalline [4.4](1,8)anthracenophantetrayne (**9**) was synthesized by the Eglinton oxidative coupling of 1,8-diethynylanthracene.[6]

The ^{1}H-NMR spectra of the heter<u>o</u>-heter<u>a</u>[4.4]phane **10**[7] are temperature-dependent, even though the 14-membered rings contain no blocking intra-annular substituents. Newkome has postulated an 'anti-longitudinal' conformational

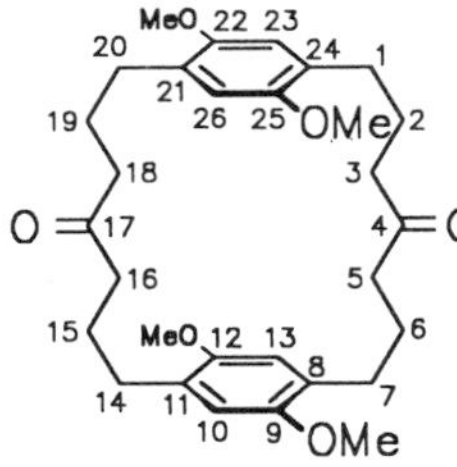

8

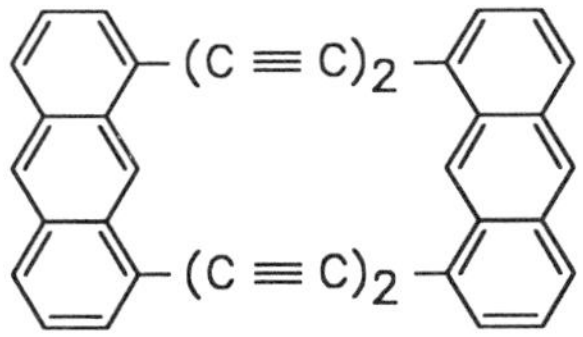

9

process between two essentially planar conformers **10a** and **10b** ($\Delta G_c^{\neq} \approx 70$ kJ/mol).[7] The tetraselena compound **11b** turns out to be conformationally more

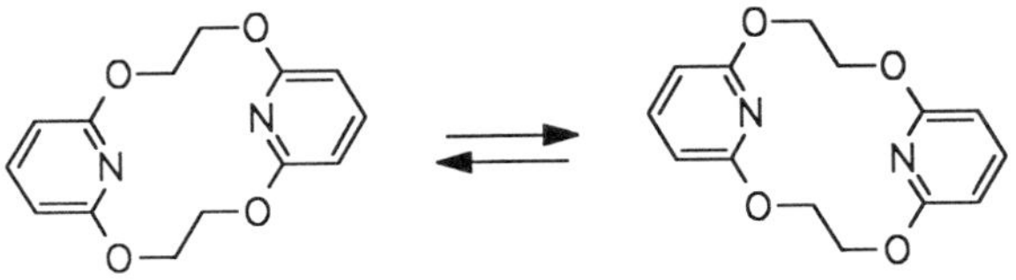

10a
(planar)

10b
(planar)

flexible than the corresponding sulphur analogue **11a** ($\Delta G_{298}^{\neq} = 53$ and 46 kJ/mol, respectively).[8]

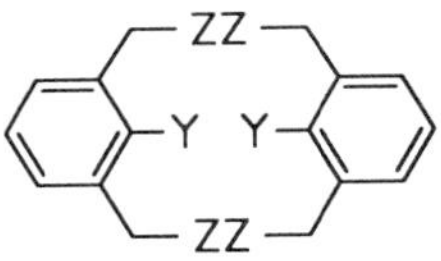

11a : Z = S, Y = H
11b : Z = Se, Y = H
11c : Z = S, Y = CH$_3$
11d : Z = S, Y = C$_6$H$_5$

References to Section 4.2

1. D. J. Cram and H. Steinberg, *J. Am. Chem. Soc.*, **73**, 5691 (1951).
2. Reviews:
   (a) V. Boekelheide, in *Cyclophanes, I* (F. Vögtle, Ed.), *Top. Curr. Chem.*, **113**, 87 (1983).
   (b) F. Vögtle and P. Neumann, *Top. Curr. Chem.*, **48**, 67 (1974).
3. Review: F. Gerson, in *Cyclophanes, Vol. II* (F. Vögtle, Ed.), *Top. Curr. Chem.*, **115**, 1 (1983).
4. Review: E. Heilbronner and Z. Young, in *Cyclophanes, II* (F. Vögtle, Ed.), *Top. Curr. Chem.*, **115**, 57 (1983).
   (a) D. Krois and H. Lehner, *J. Chem. Soc., Perkin Trans, 1*, 2369 (1982).
5. H. A. Staab, G. Matzke and C. Krieger, *Chem. Ber.*, **120**, 89 (1987).
6. S. Akiyama, S. Misumi and M. Nakagawa, *Bull. Chem. Soc. Jpn.*, **33**, 1263 (1960).
7. G. R. Newkome and T. Kawato, *J. Am. Chem. Soc.*, **101**, 7088 (1979).
8. R. H. Mitchell, in *Cyclophanes, II* (P. M. Keehn and S. M. Rosenfeld, Eds.), p. 239, Academic Press, New York, 1983.

5 Multiply Bridged Phanes

INTRODUCTION

The term multiply bridged phanes[1] is understood to connote those with more than two bridges. The first ones to be considered are the $[2_n]$phanes, which contain three to six ethano bridges and two benzene rings as aromatic units. The bridges can be affixed symmetrically, as in (1,3,5)cyclophanes or unsymmetrically as, for example, in a (1,3,5)(1,2,4)cyclophane. Although at an earlier time the [2.2]cyclophane hydrocarbons received the most attention, it is the multiply bridged heterocyclic host compounds with large cavities which are of most interest at the present. This group of compounds, discussed in more detail in Section 12 ('Molecular Recognition with Phanes as Host Molecules'), gained prominence after the successful synthesis of 'superphane (13)', the 'star' among these molecules. The large-cavity compounds typically contain longer bridges, sometimes including hetero atoms. The aromatic nuclei are often larger than benzene (naphthalene, for example) and can also be constructed as 'ring ensembles'.

In Figure 1 are presented the eight possible 'symmetrical' $[2_n]$cyclophane hydrocarbons **6–13** with more than two bridges, in which the substitution pattern of the two bridges is the same for both benzene rings. Figure 2 contains the

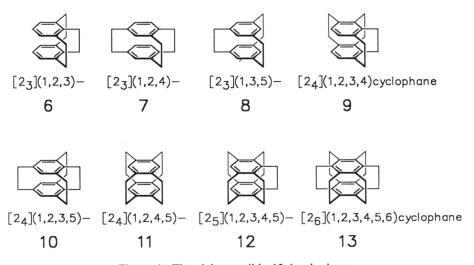

$[2_3](1,2,3)-$ $[2_3](1,2,4)-$ $[2_3](1,3,5)-$ $[2_4](1,2,3,4)$cyclophane

6   **7**   **8**   **9**

$[2_4](1,2,3,5)-$ $[2_4](1,2,4,5)-$ $[2_5](1,2,3,4,5)-$ $[2_6](1,2,3,4,5,6)$cyclophane

10   **11**   **12**   **13**

Figure 1. The eight possible $[2_n]$cyclophanes

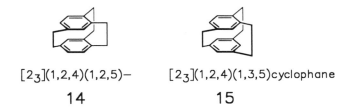

$[2_3](1,2,4)(1,2,5)-$ $[2_3](1,2,4)(1,3,5)$cyclophane

14   **15**

Figure 2. Unsymmetrical $[2_n]$cyclophanes reported to date

known unsymmetrical ('skewed') $[2_n]$cyclophanes, in which the substitution is not the same in both benzene rings.

It should be pointed out that there is a second type of $[m_n]$phane, which is treated in more detail in Chapter 7. The aromatic nuclei in those $[m_n]$ (x,y)phanes are only doubly bridged (in the x,y positions). The difference in the two $[m_n]$phane types is recognized by the notation of the number of bridging positions (the numbers in parentheses). Examples of the two types are $[2.2.2](1,3,5)$phane and $[2.2.2](1,3)$phane or, respectively, $[2_3](1,3,5)$phane and $[2_3](1,3,5)$phane (see Chapter 7).

5.1 [2.2.2]PHANES

5.1.1 SYNTHESIS

The first triply bridged cyclophane hydrocarbon of structure **16** appears to have been synthesized by Hubert and Dale in 1965. A trisubstituted benzene, containing three terminal acetylene functions, was the starting material. The second benzene ring was incorporated into the structure by trimerization of the triple bonds.[2] This method, however, did not allow synthesis of the most interesting, shortest bridged member of the series, **8** (see above).

$(CH_2)_n$ ... $(CH_2)_n$ ... $(CH_2)_n$

16 : $n = 3 - 9$

17a : X = S
17b : X = SO$_2$
17c : X = N—Tos
17d : X = Se

The first triply bridged phane which could be synthesized in one step from an inexpensive starting material was the trithia$[3.3.3]$phane **17a** (Vögtle *et al.*, 1970[3]). Later the aza-[4] and selena-[5] heterocyclic analogues **17c,d** were prepared.* All three serve as convenient starting materials for the $[2.2.2](1,3,5)$ cyclophane hydrocarbon **8** because of the successful elimination of the hetero atom.

---

*Dynamic ^{1}H-NMR studies were carried out with trioxo- and with partially deuterated $[3.3.3](1,3,5)$-cyclophanes. At low temperature ($-70°$C) two conformers exist.[5a]

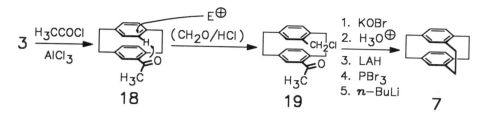

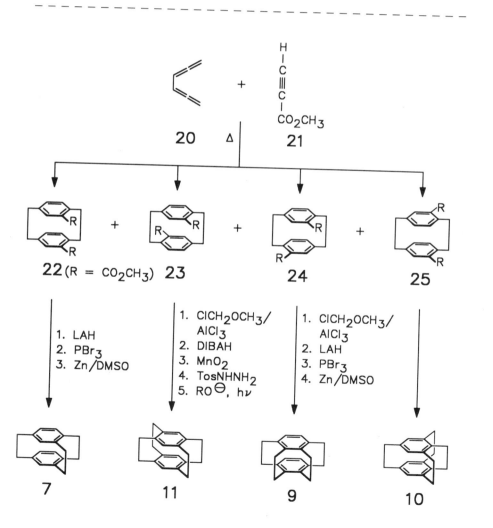

The first triply bridged cyclophane of the [2.2.2]type was [2.2.2](1,2,4)-cyclophane (7), which was obtained by Cram and Truesdale from [2.2] paracyclophane (3) by incorporating an additional ethano bridge into the structure.[1c,d] Of particular interest in the reaction scheme 3→7 is the preparation of the chloromethylated methyl ketone 19, in which the chloromethylation 18→19 proceeds exclusively at the 'pseudo-*geminal*' position. The transannular directing effects responsible for this are well known in cyclophane chemistry. In the present case the directing effect is caused by the basic carbonyl group, which is ideally situated to accept the pseudo-*geminal* proton from the more reactive benzene ring in an intramolecular step, as the electrophile attacks from the outside face. The target molecule 7 was obtained by a series of oxidations, substitutions, and reductions.

The number of steps in the synthesis of multiply bridged cyclophanes could be markedly reduced, once it was possible to obtain pseudo-*geminal* diesters such as 22 by the addition of 1,2,4,5-hexatetraene (20) to methyl propiolate (Hopf).[6]

The Diels–Alder reaction of 20 and 21 afforded, besides 22, the isomeric diesters 23–25, which are ideally suited for conversion into [2₄]cyclophanes (9–11). In the closing of the bridge in all three cases, advantage was taken of the directing influence of a carbonyl group.

Substituted [2₃](1,2,4)cyclophanes can also be obtained through carbene insertion in the following way.[7]

The number of bridges can be extended up to a quadruply bridged compound by repetition of the reaction sequence; the method was not successful, however, for the synthesis of superphane (13; see below).

Sulphone pyrolysis[8] is generally applicable for multiply bridged [2ₙ]-phanes, whereas other synthetic methods are usually multi-step or only convenient for specific [2ₙ]phanes. It does not work, however, for superphane itself. As an example, pyrolysis of the sulphone 31 leads to 14, which is also formed by a rearrangement of 7 under Friedel–Crafts conditions. By contrast, neither reduction of the bis(bromomethyl) compound 32 nor carbene dimerization from 34 leads to [2₃](1,2,4)(1,3,4)cyclophane (33) nor its monoene.[1d,e]

The ketone pyrolysis can be carried out selectively, and the intermediate ketones are isolable:[8a]

In the course of synthesizing $[2_3](1,2,4)$cyclophane (7), Cram and Truesdale also obtained the triply bridged [3.2.2]phane 35.

R = COCH$_3$

The benzcyclobutene cyclization method developed by Boekelheide (see below) was applied to the synthesis of [2₃](1,2,4)cyclophane (7), as well as to the previously unknown [2₃](1,2,3)cyclophane (38).[1d,e] The dimethyl substituted (1,2,3)-bridged ring structure 38a is available by pyrolytic cleavage of an ethano bridge in the quadruply bridged [2₄](1,2,4,5)cyclophane (10).

For the photochemical dimerization of two tris-stilbenes to afford tris-cyclobuta[2.2.2]phanes, see Section 5.5.

5.1.2 PROPERTIES OF [2.2.2]PHANES

The [2.2.2]phane 8 melts at 204–206°C, while the [2.2.2]cyclophantriene 39, which crystallizes in the form of colourless plates, melts at practically the same temperature (203–204°C).

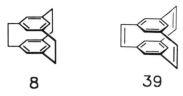

8   39

The UV spectrum of **39** contains maxima at $\lambda = 252$ nm ($\epsilon = 1960$) and 325 nm ($\epsilon = 90$); that of a hexane solution of **8** shows absorptions at $\lambda = 257$ nm ($\epsilon = 1340$) and 312 nm ($\epsilon = 14$).[9]

Ionization energies for **8** lie at 7.70 eV and 8.75 eV.[10] On the other hand, the photoelectron spectrum of **39** shows three bands at 8.06 eV, 9.24 eV, and 9.4 eV, corresponding to the first five ionization energies. Results on ESR spectroscopy have been covered in the review by Gerson.[11]

In the ^{1}H-NMR spectrum of **8** one sees one singlet each for the six aromatic protons ($\delta = 5.73$) and the 12 methylenes ($\delta = 2.75$).[9] The ^{1}H-NMR spectrum of **39**, on the other hand, contains just two singlets of equal intensity at $\delta = 7.37$ and 6.24. The assignment of the two signals could be clarified through synthesis of the deuterated compound **40**. This triene shows only one singlet for

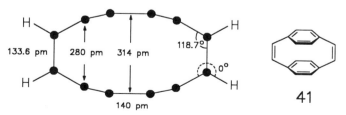

40

the six aromatic protons at $\delta = 6.24$; that is, they absorb at a higher field than do the vinyl protons. This high-field shift of the aromatic protons is attributed to the fact that the two benzene rings are forced into a very close 'face-to-face' orientation. The even more unusual location of the vinyl proton signal is explained by ring strain in the rigid, triply bridged *syn*-cyclophane.

5.1.2.1 X-ray crystal structure analysis

Comparison of the locations of the benzene rings in the series of twofold to sixfold bridged cyclophanes (Figures 3–7) reveals that benzene rings are brought

Figure 3. Geometry of [2.2]paracyclophan-1,9-diene (**41**)[12a]

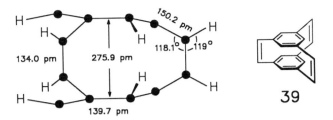

Figure 4. Geometry of [2.2.2](1,3,5)cyclophantriene (**39**)[12b]

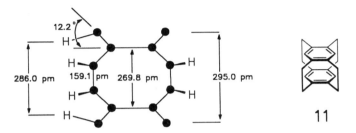

Figure 5. Geometry of [2.2.2.2](1,2,4,5)cyclophane (**11**)[12c]

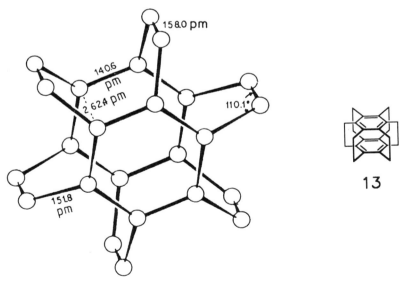

Figure 6. Geometry of superphane (**13**)[12d]

closer and closer together as the number of bridges in the molecule is increased. The distance between benzene rings decreases from 280 to 274 to 268 to 262 pm, respectively, for [2.2]paracyclophan-1,9-diene (**41**), the (1,3,5)cyclophan-1,9,17-triene **39**, the (1,2,4,5)cyclophane **11**, and superphane (**13**).[12] The benzene rings

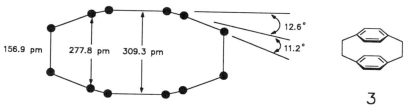

Figure 7. Geometry of [2.2]paracyclophane (3)[12d] (for comparison)

are boat or chair forms or—in the case of superphane **13**—undistorted, depending on the number of bridging groups. The bridging positions in [2.2]paracyclophan-1,9-diene (**41**) are at the bowsprit positions of the boat, whereas in the (1,2,4,5)cyclophane **11** they are attached at the four base carbon atoms. The CH_2—CH_2 single bond distance decreases from 160 pm in [2.2]-paracylophane (**3**) to 158 pm in superphane (**13**). In spite of the shorter vinyl bridges in the cyclophantriene **39**, it is less strained than the cyclophane **8**, with

20 + **21** $\xrightarrow{\Delta}$

22 + **23** + **24** + **25**

(R = CO_2CH_3)

From **23**:
1. $ClCH_2OCH_3$/ $AlCl_3$
2. DIBAH
3. MnO_2
4. TosNHNH$_2$
5. RO$^{\ominus}$, $h\nu$

From **24**:
1. $ClCH_2OCH_3$/ $AlCl_3$
2. LAH
3. PBr$_3$
4. Zn/DMSO

11 **9** **10**

its longer ethano bridges. As is shown from the X-ray crystal structure, the decrease in bond angle from sp^2 carbon atoms (120°) to sp^3 carbon atoms (109°) more than compensates for the effect of the lengthened bond (CH=CH → CH$_2$—CH$_2$).

5.2 [2.2.2.2]CYCLOPHANES

5.2.1 SYNTHESIS

Particularly well suited for this class of compounds is the cycloaddition method of Hopf,[6] by which the isomeric quadruply bridged [2$_4$]cyclophanes 9–11 have been obtained. The isomer 11 was also prepared by a route—historically the

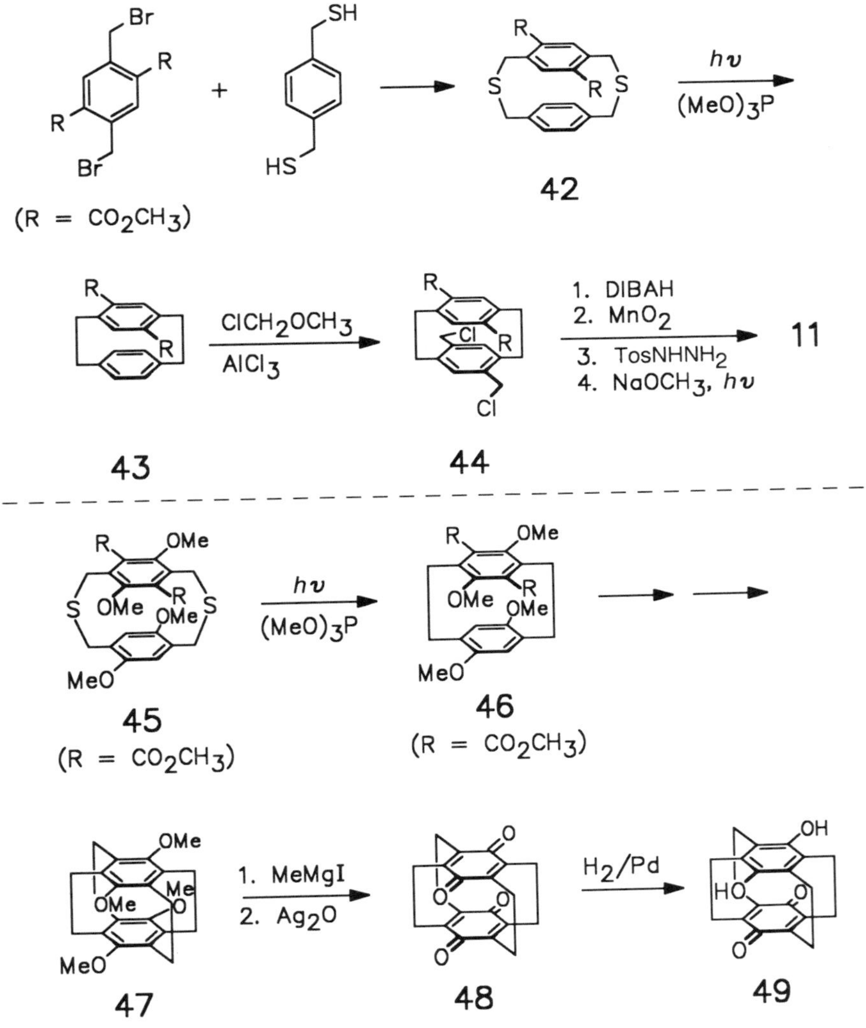

(R = CO$_2$CH$_3$)

42

43   44   11

45
(R = CO$_2$CH$_3$)

46
(R = CO$_2$CH$_3$)

47   48   49

first by Boekelheide *et al.*[13]—in which the transannular directing effect of an ester group was combined with sulphone pyrolysis. The desulphurization of the dithia[3.3]paracyclophane **42** could be achieved photolytically in trimethyl phosphite solvent; chloromethylation again afforded the pseudo-*geminal* substituton product **44**, which was cyclized to **11**. This method was also applied to the synthesis of the quinone **48** and the quinhydrone **49**.[14] The dithiaphane **45** was desulphurized by irradiation in trimethyl phosphite to **46**, which in turn was transformed to the quadruply bridged tetra ether **47**. The latter serves as starting material for the bis(quinone) **48** and for [2₄](1,2,4,5)cyclophanquinhydrone) (**49**).

Boekelheide discovered a novel route[1d,c,15] for constructing the missing members of the [2ₙ]cyclophane family, in particular 'superphane'. Advantage was made of the ease with which *o*-xylylene derivatives undergo [4+4] cycloadditions to [2.2]orthocyclophanes [i.e., benzocyclooctadienes] and of the observation that *o*-chloromethyltoluenes (e.g. **50**, **53**) afford benz-cyclobutenes (i.e. **51**, **55**) on thermal elimination of HCl. Because the *o*-chloromethyltoluenes undergo pyrolytic ring opening to *o*-xylylenes (i.e. **54**), a method is at hand for converting aromatic halogen compounds into these reactive intermediates. If the reaction is directed so as to occur intramolecularly, one can construct two ethano bridges in a single synthetic step. This concept was first made a reality in the synthesis of the tetrasubstituted [2₄]-phane **11** from 2,4-dimethylbenzyl chloride (**50**) as starting material.[15]

Vögtle *et al.* prepared the [2₄]phane **11** in a straightforward reaction sequence by means of sulphone pyrolysis,[16] a method which leads most conveniently to this hydrocarbon. Coupling of the tetrathiol **57** with the tetrabromide **56** by means of caesium carbonate in DMF at moderate temperature affords the isomeric tetrathia[3_4]cyclophanes **58a,b** in 10% yield. The corresponding sulphones **59a,b** on pyrolysis give the [2_4]cyclophane **11**, along with the benzcyclobutene derivative **55**. Thus the whole process involves only three steps. Optimization of conditions for the reaction of **56** and **57** led to an 18.5% yield of the isomeric mixture **58**. Misumi[16] has described separation of the isomers, assignment of their ^{1}H-NMR signals, and an X-ray structural analysis of the pure isomer **58a**, which predominates 3:1 over **58b** in the mixture. He also recommended substitution of the thiouronium salt of **56** for the thiol **57**. When the salt is allowed to react with **56** in the presence of CsOH as base, the yield is increased to 75%. Properties of the [2_4]phanes are described in the literature.[1d,e,10b,11]

5.3 [2.2.2.2.2]PHANES

5.3.1 SYNTHESIS

The elegant synthesis of [2_5](1,2,3,4,5)cyclophane (**12**)—and of superphane (**13**)—illustrates the value of the *o*-xylylene dimerization method for the introduction of multiple bridges. The key intermediate **63** was obtained by a Grignard coupling of **62c**, which can be synthesized from **60** in a series of routine steps. When one realizes that organic chemistry involves to a great extent the joining of carbon atoms, then the last step of this reaction sequence must be considered a synthetic masterpiece: four strained single bonds are created in a single process

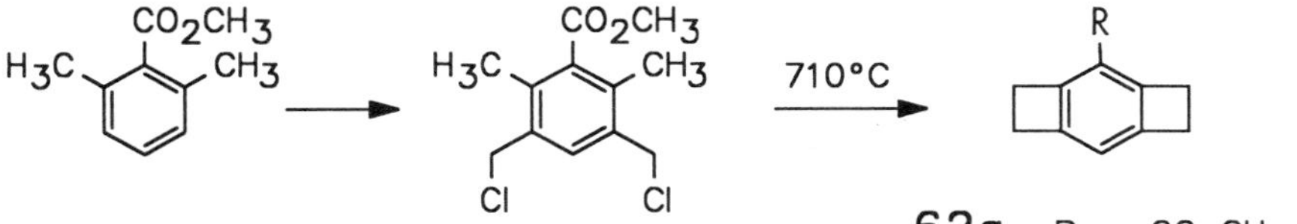

62a : R = CO_2CH_3

62b : R = CH_2OH

62c : R = CH_2Br

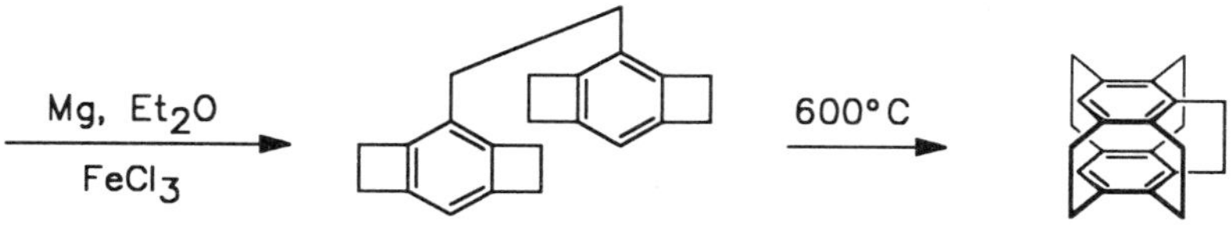

in excellent yield.[1g] An alternative, convenient route to the $[2_5]$cyclophane **11** from **64b** is sulphone pyrolysis, in which the benzcyclobutene **63** can be considered as a plausible intermediate.[16]

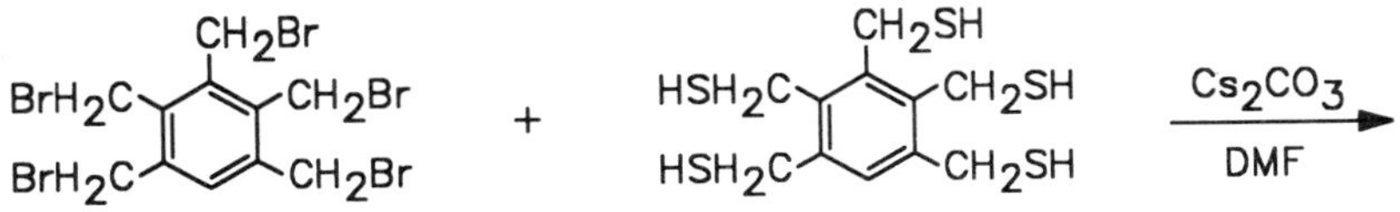

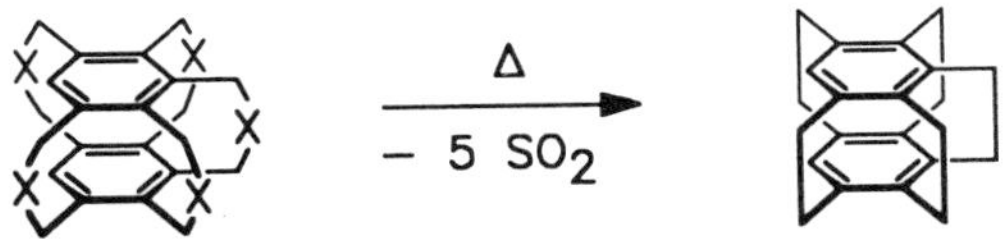

64a : X = S

64b : X = SO_2

12

5.3.2 CHEMICAL REACTIONS OF MULTIPLY BRIDGED PHANES

In the following discussion the chemical reactions of multiply bridged $[2_n]$phanes will be generally discussed according to reaction type.[1d,g] Because of their unique character, the reactions of superphane will be summarized in a separate section (Section 5.4). First the reactions at the bridges of $[2_n]$phanes will be taken up, followed by reaction of the benzene rings.

5.3.2.1 Reactions at the ethano bridges in $[2_n]$phanes

(a) Homolytic cleavage of ethano bridges

As in the case of [2.2]paracyclophane, the bridges in **7** and **10** can be cleaved homolytically, the methylene groups usually being reduced to methyl groups.

Trapping reactions, like those with [2.2]paracyclophane, can also be carried out, such that $[4.2_n]$cyclophanes such as **66** are formed.

By contrast, the $[2_n]$phanes **9** and **12**, as well as superphane (**13**), are thermally much stabler in this series than is **10**. Remaining unchanged up to 350°C, they undergo neither reductive cleavage nor ring enlargement reactions.

(b) Heterolytic cleavage of the ethano bridges

Just as [2.2]paracyclophane can be rearranged to [2.2]metacyclophane with Friedel–Crafts agents (AlCl$_3$/HCl), so it is possible to transform the triply bridged **7** within 30 minutes at $-10°C$ in dichloromethane to the 'twisted' hydrocarbon $[2_3](1,2,4)(1,2,5)$cyclophane **(14)** in 44% yield. This same unsymmetrical cyclophane can be synthesized by sulphone pyrolysis. The tris

alkene **39** can be obtained from the appropriate trithia[3.3.3]cyclophane **17a** by way of a threefold Hofmann elimination of the tris-sulphonium salt **67** with *n*-butyllithium.

5.3.2.2 Reactions of $[2_n]$cyclophanes at the benzene rings

Reactions of multiply bridged phanes are of interest for several reasons, as indeed is the case with doubly bridged cyclophanes, such as [2.2]paracyclophane. In these cases with deformed aromatic rings, the question is raised as to whether reactions in the ring or at the bridges proceed as in open-chain model compounds, or whether strain in the aromatic rings significantly influences reaction behaviour. Cycloaddition reactions at the benzene rings were particularly revealing in this regard.

(a) Diels–Alder additions

Ordinarily benzene undergoes no [4+2] additions in which it might serve as the diene component. Indeed, benzene and other similar aromatics are typically

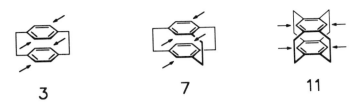

3   **7**   **11**

employed as solvents in such additions because of their low Diels–Alder reactivity. The 'super dienophile' 4-phenyl-1,2,4-trioxazolin-3,5-dione (68) does not add to benzene or oligomethylbenzenes at room temperature, even after a period of several weeks. When aromatic rings are incorporated into a [2$_n$]-phane framework, however, there is noted a dramatic enhancement of the addition rate in some instances. As an example, [2.2]paracyclophane (3) reacts with the dienophile 68 at room temperature in a period of 6 days to give a 1:2 cycloadduct. (The positions of the newly formed C—C bonds in the structures 3, 7, and 11 are marked with arrows.) Incorporation of additional bridges as in 7 and 11 results in a dramatic increase in the addition rate. Upon addition of [2$_4$](1,2,4,5)cyclophane (11), the intense red colour of 68 disappears in a few seconds. That is, this cyclophane can be compared with acyclic polyenes in its tendency for cycloadditions! Similar trends are noted in the additions of 3, 7, and 11 with tetracyanoethene, dicyanoacetylene, maleic anhydride, perfluoro-2-butyne, and dimethyl acetylenedicarboxylate.

Severe strain alone in the substrate molecule does not suffice to induce Diels–Alder addition in multiply bridged phanes, however. This is shown in the behaviour of the hydrocarbons 10, 9, and 13, none of which reacts with the dienophiles named above. One might predict from this behaviour that [2$_5$]-cyclophane 12 would also be a weak diene partner, but this is not the case. It reacts surprisingly with both dicyanoacetylene (60°C, 3 days, 44% yield) and with perfluoro-2-butyne (100°C, 7 days, 100% yield) with formation of 1:1 adducts. Thus it is recognized that formation of [4+2] adducts is controlled by a delicate balance between ring strain in the substrate and in the cyclo adduct. Cycloadditions of superphane induced by Lewis acid catalysts are discussed in Section 5.4.

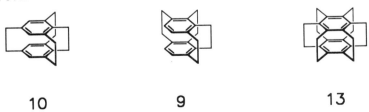

10   **9**   **13**

The [4+2] cycloaddition of 11 and singlet oxygen affords, as the primary product, the *endo* peroxide 68a, which serves as a starting material for a variety of bridged, oxygenated phanes, as is shown in the following sequence.[13]

Irradiation of 11 and dicyanoacetylene in THF at room temperature affords, in 44% yield, the 1:1 adduct 69 containing a cyclooctatetraene ring.

11

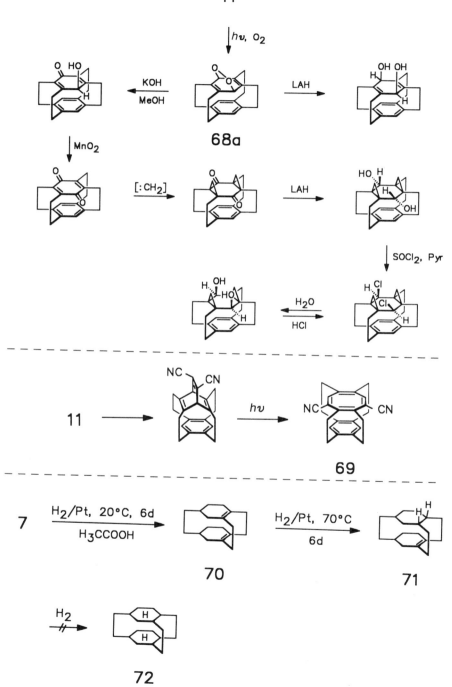

68a

11 → → $h\nu$ → 69

7 → $\dfrac{H_2/Pt,\ 20°C,\ 6d}{H_3CCOOH}$ → 70 → $\dfrac{H_2/Pt,\ 70°C}{6d}$ → 71

72

(b) Hydrogenation of [2ₙ]phanes

The $[2_n]$cyclophanes have been subjected both to catalytic hydrogenation and to Birch reductions, and their reactivity pattern is comparable to that in Diels–Alder additions. In contrast to [2.2]paracyclophane, the catalytic hydrogenation of $[2_3]$(1,2,4)cyclophane (**7**) is complicated, although the diene **70** and monoene **71** can both be isolated. Addition of the fifth equivalent of hydrogen requires drastic conditions (6 days at 70°C), and the monoene **71** is not further reduced. This is reasonable from an examination of molecular models, for the perhydro product **72** is a highly strained hydrocarbon. With regard to hydrogenation reactivity, it can be concluded that neither **10**, **12**, nor superphane (**13**) can be hydrogenated to olefinic products under a variety of conditions with a variety of catalysts. (See below for reduction of superphane with Li/NH_3.)

5.3.2.3 Other addition reactions

Treatment of **11** with ethyl diazoacetate affords initially the product **73**, which, however, spontaneously isomerizes to **74** with a seven-membered ring. For a similar reaction with superphane see below.

$[2_3]$(1,3,5)Cyclophane (**8**) reacts with $AlCl_3/HCl$ at 0°C within 10 minutes to give a halogenated compound, which, on treatment with lithium in *tert*-butyl alcohol, leads to a hydrocarbon $C_{18}H_{24}$ (**78**) with a cage structure.

5.3.2.4 Reactions involving the aromatic ring

(a) Electrophilic aromatic substitution

Bromination of $[2_3](1,2,4)$cyclophane (**7**) leads to the expected monobromo compound **79**, as well as to traces of the dibromide **80**. In this respect **7** is more reactive than either [2.2]paracyclophane (**3**) or the quadruply bridged phane **10**, which does not react with bromine under the same conditions. Friedel–Crafts acylation of **7** affords the analogous acetyl substituted $[2_3]$cyclophane in 79%

yield, whereas the corresponding ketone **81** from $[2_4](1,2,3,5)$cyclophane (**10**) is formed in only 6% yield. The major products are the (1,2,3)cyclophanes **82** and **83**, resulting from *ipso*-substitution.

Rieche formylation of **10** proceeds essentially quantitatively to the aldehyde **84**: nitration with fuming nitric acid and acetic acid at 70°C gives, after just 2 minutes, only traces of the nitro compound **85**.

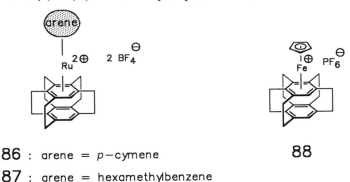

84   **85**

(b) Transition metal complexes

Boekelheide *et al.*[1d] developed a novel method for preparation of ruthenium and iron complexes of $[2_5](1,2,3,4,5)$cyclophane (**12**). It consists of the treatment of arene ruthenium complexes with cyclophanes in the presence of silver salts, by which the ruthenium complexes **86** and **87** were obtained from **12**. The iron complex **88** was isolated in 68% yield by irradiation of η^6-*p*-xylylene(η^5-cyclopentadienyl)-Fe(II) hexafluorophosphate and **12** in dichloromethane.

86 : arene = *p*−cymene   **88**

87 : arene = hexamethylbenzene

References to Sections 5.1–5.3

1. Reviews:
   (a) F. Vögtle, *Chem. Ztg.*, **95**, 668 (1971).
   (b) F. Vögtle and G. Hohner, *Top. Curr. Chem.*, **74**, 1 (1978).
   (c) H. Hopf, *Nachr. Chem. Tech. Lab.*, **28**, 311 (1980).
   (d) V. Boekelheide, in *Cyclophanes, I* (F. Vögtle, Ed.), *Top. Curr. Chem.*, **113**, 87 (1983).
   (e) W. Kiggen and F. Vögtle, in *Synthesis of Macrocycles, The Design of Selective Complexing Agents* (R. M. Izatt and J. J. Christensen, Eds.), Wiley, New York, 1987, Chapter 6, p. 309.
   (f) F. Vögtle, *Reizvolle Moleküle der Organischen Chemie*, Teubner, Stuttgart 1989, pp. 249, 260; *Fascinating Molecules in Organic Chemistry*, Wiley, Chichester, 1992.
   (g) H. Hopf, in *Cyclophanes* (P. M. Keehn and S. M. Rosenfeld, Eds.), Vol. II, p. 521, Academic Press, New York, 1983.
2. A. J. Hubert and J. Dale, *J. Chem. Soc.*, 3160 (1965); see also D. J. Cram and R. A. Reeves, *J. Am. Chem. Soc.*, **80**, 3094 (1958).
3. F. Vögtle, *Liebigs Ann. Chem.*, **735**, 193 (1970).

4. F. Vögtle and P. Neumann, *J. Chem. Soc., Chem. Commun.*, 1464 (1970).
5. H. Higuchi, K. Tani, T. Otsubo, Y. Sakata and S. Misumi, *Bull. Chem. Soc. Jpn.*, 60, 4027 (1987).
   (a) T. Shinmyozu, T. Inazu *et al.*, *Can. J. Chem.*, 68, 440 (1990).
6. A. E. Murad and H. Hopf, *Chem. Ber.*, 113, 2358 (1980); and earlier work cited therein.
7. S. H. Eltamany and H. Hopf, *Tetrahedron Lett.*, 4901 (1980).
8. F. Vögtle and L. Rossa, *Angew. Chem.*, 91, 534 (1979); *Angew. Chem. Int. Ed. Engl.*, 18, 514 (1979); cf. V. Boekelheide and R. A. Hollins, *J. Am. Chem. Soc.*, 95, 3201 (1973); N. Nakazaki, K. Yamamoto and Y. Miura, *J. Chem. Soc., Chem. Commun.*, 1977, 206; J. Dohm and F. Vögtle, *Top. Curr. Chem.*, 161, 69 (1992).
   (a) J. Breitenbach, F. Ott and F. Vögtle, *Angew. Chem.*, 104, 360 (1992); *Angew. Chem. Int. Ed. Engl.*, 31, 307 (1992).
9. V. Boekelheide and R. A. Hollins, *J. Am. Chem. Soc.*, 92, 3512 (1970).
10. (a) B. Kovac *et al.*, *J. Am. Chem. Soc.*, 102, 4314 (1980); V. Boekelheide and W. Schmidt, *Chem. Phys. Lett.*, 17, 410 (1972).
    (b) E. Heilbronner and Z. Yang, in *Cyclophanes, II* (F. Vögtle, Ed.), *Top. Curr. Chem.*, 115, 3 (1983).
11. F. Gerson, in *Cyclophanes, II* (F. Vögtle, Ed.), *Top. Curr. Chem.*, 115, 57 (1983).
12. (a) C. L. Coulter and K. N. Trueblood, *Acta Crystallogr.*, 16, 667 (1963).
    (b) A. W. Hanson and H. Röhrl, *Acta Crystallogr.*, B28, 2287 (1972).
    (c) A. W. Hanson, *Acta Crystallogr.*, B33, 2003 (1977).
    (d) A. W. Hanson and T. S. Cameron, *J. Chem. Res. (S)*, 336 (1980); *(M)*, 4201 (1980); C. J. Brown, *J. Chem. Soc.*, 3265 (1953).
    (e) X-ray structures [2₄](1,2,3,4)- and [2₄](1,2,3,5)cyclophane: H. Irngartinger, J. Hekeler and B. M. Lang, *Chem. Ber.*, 116, 527 (1983).
13. R. Gray and V. Boekelheide, *J. Am. Chem. Soc.*, 101, 2128 (1979).
14. H. A. Staab and V. M. Schwendemann, *Liebigs Ann. Chem.*, 1258 (1979); *Angew. Chem.*, 90, 805 (1978); *Angew. Chem. Int. Ed. Engl.*, 17, 756 (1978).
15. V. Boekelheide and G. Ewing, *Tetrahedron Lett.*, 4245 (1978).
16. B. Klieser and F. Vögtle, *Angew. Chem.*, 94, 922 (1982); *Angew. Chem. Int. Ed. Engl.*, 21, 928 (1982) Suppl. 1392; T. Asoh, K. Tani, H. Higuchi, T. Kaneda, T. Tanaka, M. Sawada and S. Misumi, *Chem. Lett.*, 417 (1988).
17. P. F. T. Schirch and V. Boekelheide, *J. Am. Chem. Soc.*, 101, 3125 (1979).

5.4 SUPERPHANE[1f]

5.4.1 SYNTHESIS

In Boekelheide's synthesis of superphane[1] the strategy of 'o-xylylene dimerization' was repeated three times (see above, Section 5.2), in order to build in the four remaining ethano bridges starting from [2.2]orthocyclophane. Eltamany and Hopf later described a second synthesis of superphane (13) from 4,5,12,13-tetramethyl[2.2](1,4)cyclophane (26), where the remaining bridges were incorporated stepwise.[2,1b-e]

Attempts to synthesize superphane by elimination of sulphur from the corresponding hexathia[3₆]cyclophane (92) were thwarted because this hexathiaphane was not available,[3] even though the corresponding four- and fivefold bridged tetrathia[3₄]- and pentathia[3₅]cyclophanes could be prepared. Apparently closing of the last bridge is difficult, not only because the product is sterically hindered and strained, but because the transition state for the

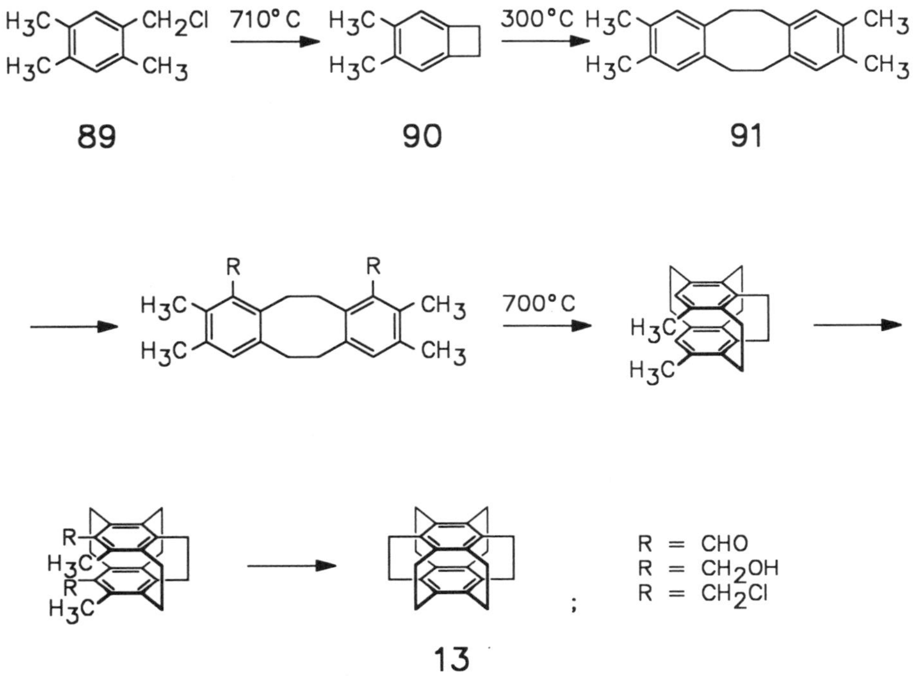
H3C
H3C
CH2Cl
CH3
710° C
300°C
89
90
91
R
700°C
R = CHO
R = CH2OH
R = CH2Cl
13

displacement of Br^- by S^- cannot assume the optimal spatial orientation. Instead a side reaction leads to the tris-hydrothiophene ring compound **93**.

5.4.2 PROPERTIES

Superphane (**13**) is distinct from all the rest of the $[2_n]$phanes in that all arene positions are bridged and thus it possesses a high molecular symmetry with point group D_{6h}. The symmetry is evident in the simple ^1H- and ^{13}C-NMR spectra, which is confirmed unambiguously in the X-ray crystal structure analysis.

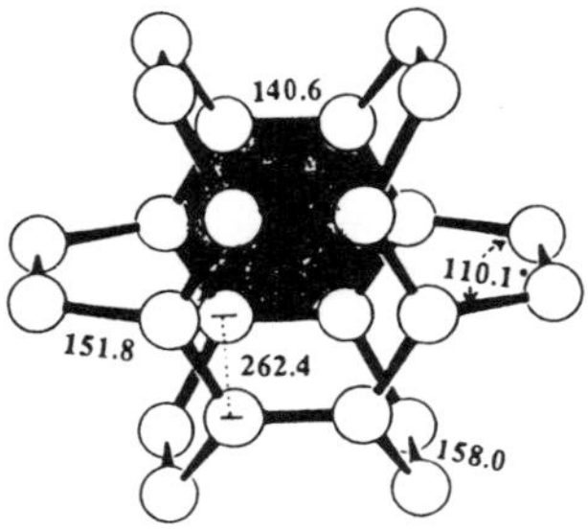
140.6
110.1
151.8
262.4
158.0

Figure 8. Intramolecular distances (pm) in superphane (**13**) (from the X-ray crystal structure)[1a]

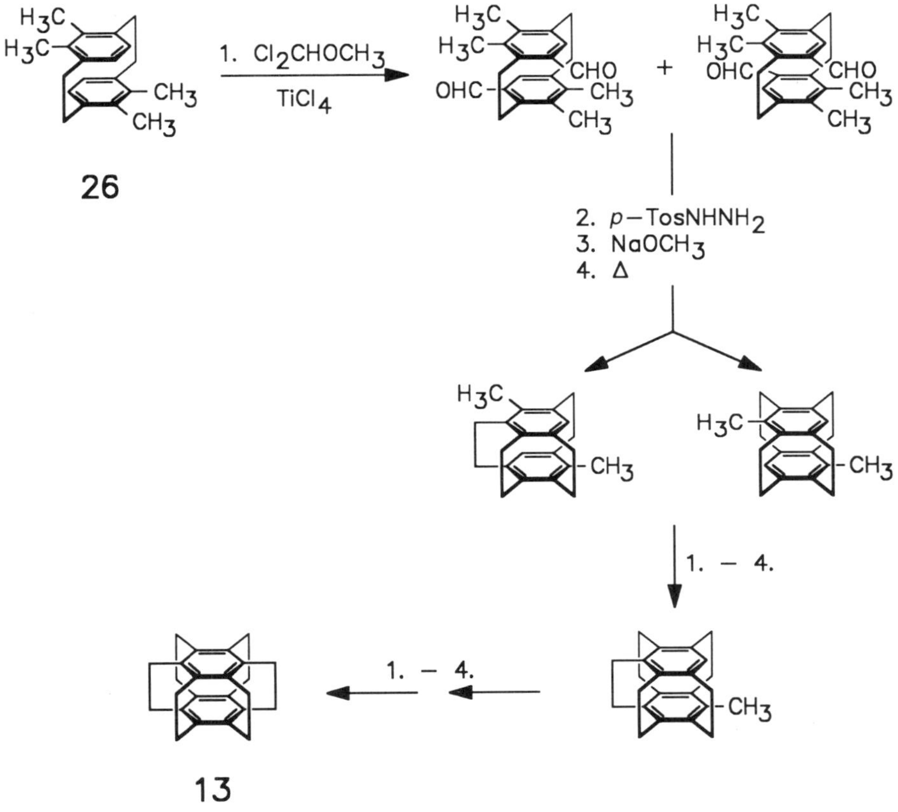
H3C
H3C
CH3
CH3
26
1. Cl2CHOCH3
TiCl4
H3C
H3C
CHO
OHC
CH3
CH3
+
H3C
H3C
OHC
CHO
CH3
CH3
2. p−TosNHNH2
3. NaOCH3
4. Δ
H3C
CH3
H3C
CH3
1. − 4.
CH3
1. − 4.
13

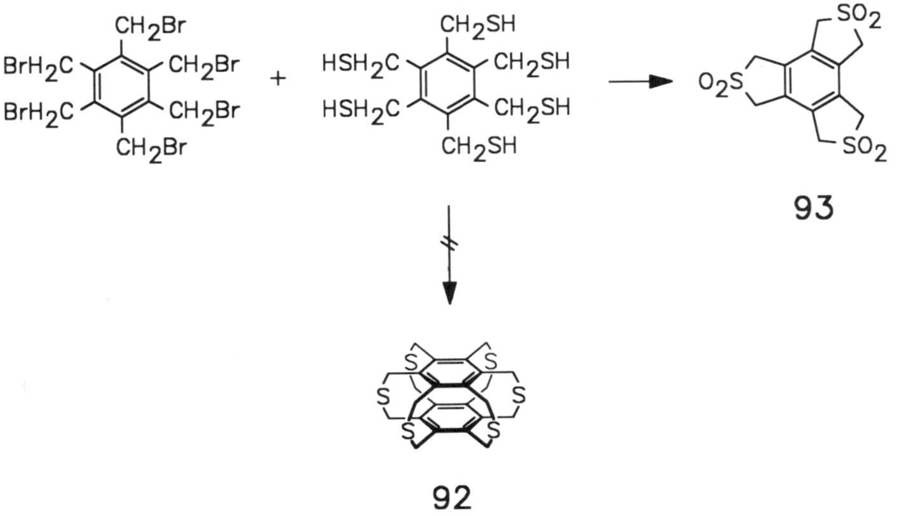
CH2Br
BrH2C
CH2Br
BrH2C
CH2Br
CH2Br
+
CH2SH
HSH2C
CH2SH
HSH2C
CH2SH
CH2SH
SO2
O2S
SO2
93
S
92

Table 1. Comparison of ionization energies I_j [eV] for [2_n]cyclophanes

[2.2]Paracyclophane (3)	[2.2.2](1,3,5)Cyclophane (8)	Superphane (13)
8.1	7.7	7.55
(8.1)	7.7	7.55
8.4	8.75	8.17
9.6	8.75	8.17
10.3	–	9.6

In sharp contrast to the [2_n]phanes prepared before superphane, which are conspicuous for their 'face-to-face' and 'bent and battered' benzene rings, the rings in superphane **13** are planar and undistorted. The significantly shorter intramolecular distance between rings is 262.4 pm (see Figure 8), compared with those in [2.2]paracyclophane of 308.7 and 275.1 pm. The other dimensions determined from the X-ray structure, such as bond lengths and angles, lie within the normal range for phanes.

PES- and ESR-spectra of superphane have been compared with those of other phanes.[1e,4,5] The photoelectron spectrum of superphane (**13**) exhibits three bands at 7.55 eV, 8.17 eV, and about 9.6 eV, corresponding to the first five ionization energies (Table 1, Figure 9). As can be seen, the first five ionization energies decrease by 0.5 eV, 0.2 eV, 1.2 eV, and 0.7 eV, as the number of bridges changes from two to six.[4]

Superphane forms colourless crystals, which are relatively insoluble and melt at 325–327°C without decomposition. In sharp contrast to the bridges in other [2_n]phanes, the six bridges in superphane are thermally stable.

5.4.3 CHEMICAL REACTIONS[1c]

5.4.3.1 Reactions at the bridges

If one attempts to reduce superphane (**13**) with zinc in concentrated sulphuric acid, the remarkable result is formation in 34% yield of the monothia[3.2_5] cyclophane **94** by ring expansion of one of the bridges with sulphur arising from the sulphuric acid. This reaction has been observed only with superphane up to the present time.

The superphane monoene **95** can be prepared from superphane by conventional methods, although the yield is a modest 20%. For this reason it is understandable that synthesis of the fascinating superphanhexaene **96**, containing two orthogonal π-systems, each with 12 electrons, has not yet been achieved.

Anionic cleavage of C_2-bridges in superphane (**13**) proceeds by reduction with lithium/ethylamine in n-propylamine. Formation of the hydrocarbons **97** and **98** in yields of 57% and 7%, respectively, is rationalized according to the following steps.

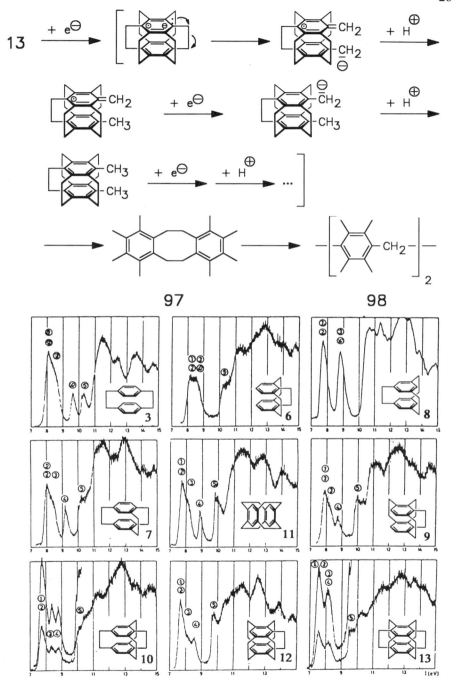

Figure 9. Comparison of the photoelectron spectra of some [2ₙ]phanes (**3–12**) with that of superphane (**13**)[4]

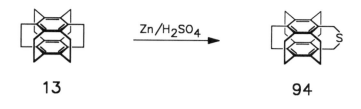

13   Zn/H$_2$SO$_4$ **94**

13   NBS; DBN   **95**   **96**

5.4.3.2 Reactions at the benzene rings

Attempts to facilitate the relatively sluggish [4+2] cycloadditions with superphane by the use of Lewis acid catalysts have led to an interesting series of addition reactions. Treatment of superphane with tetracyanoethene in dichloromethane at room temperature for 3 days in the presence of AlCl$_3$ leads to isolation of the cage compound **100**, whose structure has been confirmed by X-ray crystal structure analysis. It is proposed that **99** is first formed by a

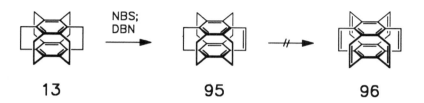

13 (super-phane)   AlCl$_3$ **99**   1. [2+4]   2. HCl   **100**

13   Li/NH$_3$ / MeOH   **101**   O$_2$ **102** $-$ H$_2$O$_2$ **13**

double [2 + 2] cycloaddition and that this 2:1 intermediate is stabilized through an intramolecular [4 + 2] cycloaddition. The resulting tetraene then reacts with HCl to afford the product **100**.

Under Birch reduction conditions superphane (**13**) is converted in a modest 10% yield to the dihydro compound **101**, which, however, is rearomatized rapidly at room temperature to superphane.

Treatment of superphane with diazoacetic ester affords a tropilidene ester **103**. Reduction with LAH gives the alcohol **104**, which isomerizes with BF₃-etherate to the tropyliophane **105**. Recrystallization of the latter with methanol/water or simply exposure to air causes it to revert to superphane (**13**). This might occur by way of the carbinol **106** and its valence isomer **107**, which could lose acetaldehyde to form superphane.

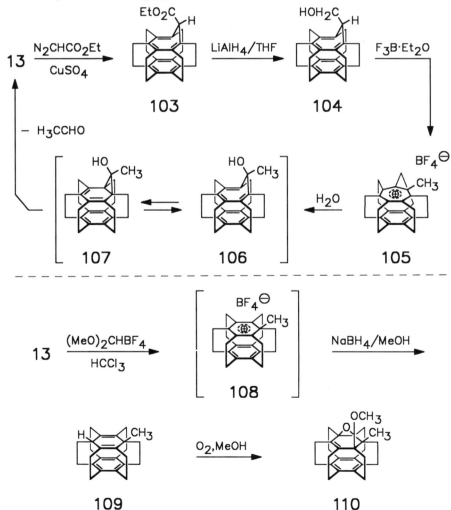

Alkylation of superphane with the Meerwein reagent gives the cation **108**, which is reduced to **109** with NaBH$_4$. Recrystallization of this hydrocarbon affords the oxirane **110**.[1c]

References to Section 5.4

1. (a) Y. Sekine, M. Brown and V. Boekelheide, *J. Am. Chem. Soc.*, **101**, 3126 (1979).
   (b) See review: V. Boekelheide, in *Strategies and Tactics in Organic Synthesis* (Th. Lindberg, Ed.), p. 1, Academic Press, London, 1984.
   (c) V. Boekelheide, in *Cyclophanes, I* (F. Vögtle, Ed.), *Top. Curr. Chem.*, **113**, 87 (1983).
   (d) H. Hopf, in *Cyclophanes*, (P. M. Keehn and S. M. Rosenfeld, Eds.), Vol. I, p. 521, Academic Press, New York, 1983.
   (e) F. Vögtle, *Fascinating Molecules in Organic Chemistry*, Wiley, Chichester, 1992.
   (f) Overview of superphanes: R. Gleiter and D. Kratz, *Acc. Chem. Res.*, **26**, 311 (1993).
2. H. Hopf, *Nachr. Chem. Tech. Lab.*, **28**, 311 (1980); S. El-tamany and H. Hopf, *Chem. Ber.*, **116**, 1682 (1983); W. D. Rohrbach, R. Sheley and V. Boekelheide, *Tetrahedron*, **40**, 4823 (1984).
3. B. Klieser and F. Vögtle, *Angew. Chem.*, **94**, 922 (1982); *Angew. Chem. Int. Ed. Engl.*, **21**, 928 (1982) Suppl. 1392.
4. E. Heilbronner, Z. Yang, in *Cyclophanes, II* (F. Vögtle, Ed.), *Top. Curr. Chem.*, **115**, 1 (1983).
5. F. Gerson, in *Cyclophanes, II* (F. Vögtle, Ed.), *Top. Curr. Chem.*, **115**, 57 (1983).

5.5 OTHER MULTIPLY BRIDGED PHANES

The quadruply bridged [2$_4$](1,4,5,8)(1,2,4,5)biphenylenobenzenophane (**10b**) has been obtained by sulphone pyrolysis as outlined below.[1] Identification of the isomers **8a**, **9a**, and **10b** was accomplished with ^{1}H-NMR spectroscopy. The X-ray analysis of **10b** confirms the first case where a biphenylene is strongly deformed out of planarity.[1]

Irradiation of (*E,E,E*)-1,3,5-tris(styryl)benzene (**11**) in benzene at 317 nm afforded a high-melting dimer **12**, whose constitution has been established by X-ray crystal structure analysis.[2a] Doubly bridged 1,2-ethano[2$_n$]meta- and -paracyclophanes are also available by this route. The second bridge, where $n = 2$, however, can only be introduced in exceptional cases (the arene nucleus = naphthalene, pyridyl, or cyano-substituted ring).[2b]

5.5.1 MULTIPLY BRIDGED PYRIDINOPHANES

Boekelheide's benzcyclobutene/*o*-xylylene route has been utilized to prepare 4,13-diaza[2$_4$](1,2,4,5)cyclophane (**13**) and its 4,16 isomer **14**.[3]

5.5.1.1 [2$_4$]Biphenylophane

Cyclization of the tetrakis(bromomethyl) **15** and the tetrakis(mercaptomethyl) compound **16** was carried out by Vögtle and Weber to give the crossed isomeric

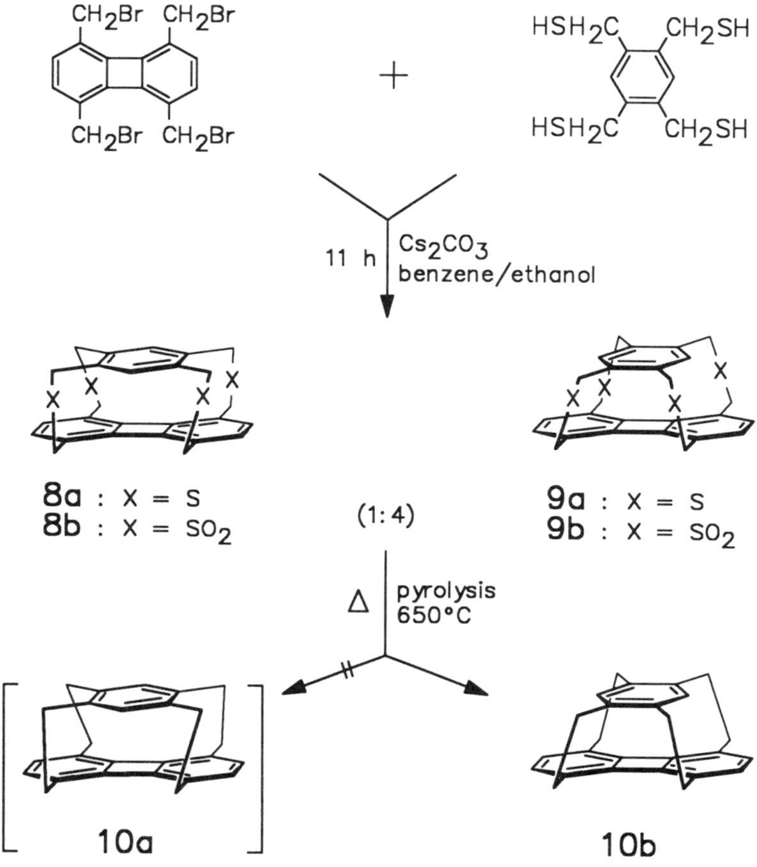

CH2Br CH2Br
CH2Br CH2Br
HSH2C CH2SH
HSH2C CH2SH
11 h
Cs2CO3
benzene/ethanol
8a : X = S
8b : X = SO2
(1:4)
9a : X = S
9b : X = SO2
pyrolysis
650°C
10a
10b

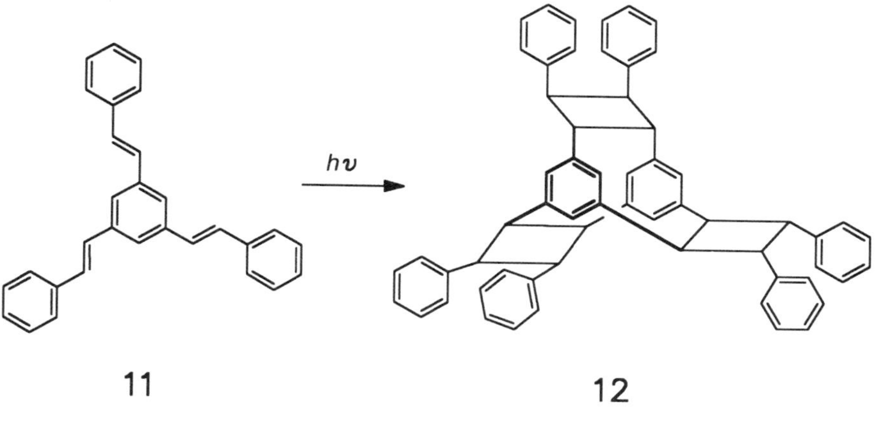

hυ
11
12

13   14

tetrasulphide **18a**,[4] whose structure was confirmed by X-ray crystal structure analysis.[5] Pyrolysis of the corresponding sulphone **18b** gives a hydrocarbon

15   +   16

17a : X = S
17b : X = SO$_2$

18a : X = S
18b : X = SO$_2$

Δ | pyrolysis

19   or   **20**

whose structure has not been clearly defined. While the ^{1}H-NMR spectra are more consistent for the biphenylophane **19** with parallel biphenyl groups, molecular modelling suggests that the crossed isomer **20** would have a higher thermodynamic stability.

The multiply bridged mixed aromatic aliphatic compounds shown below were obtained for the most part through sulphide cyclization followed by sulphur extrusion. Aromatic units include biphenyl, terphenyl, triphenylmethane, and triphenylbenzene nuclei.[6] In the sulphone pyrolysis leading to either isomer **21** or **22**, the former was favoured. Not only are the triply bridged triphenylethane **27**, the triphenylbenzene **28**, and the mixed triphenylethane/triphenylbenzene **29** known, but their corresponding trienes have also been described.[6] The triply

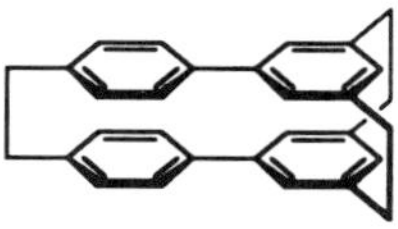

21 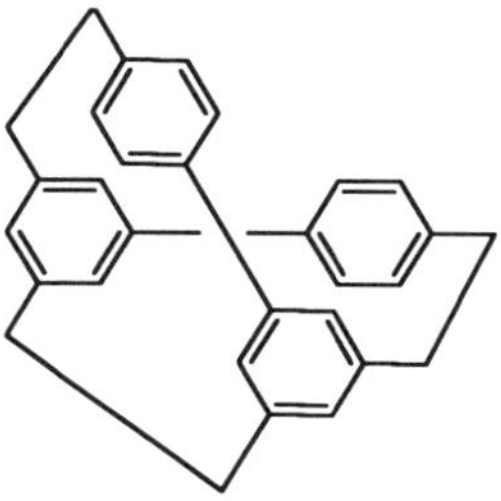　　　**[22]** 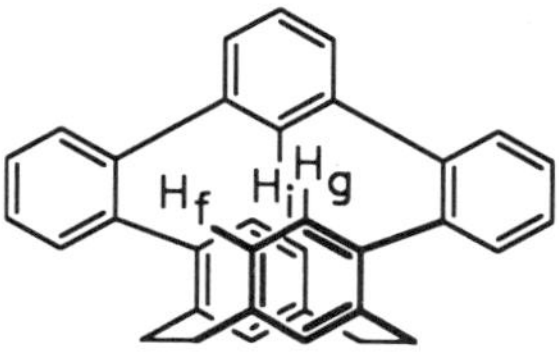　　　**23**

bridged phane **24** derived from triphenylamine can be oxidized to the radical cation **25**.[7] The triply bridged cyclophanes **30**, containing mesitylene and

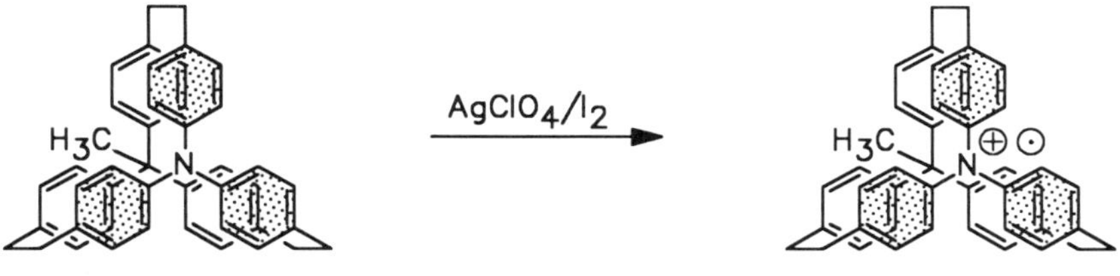

24　　　**25**

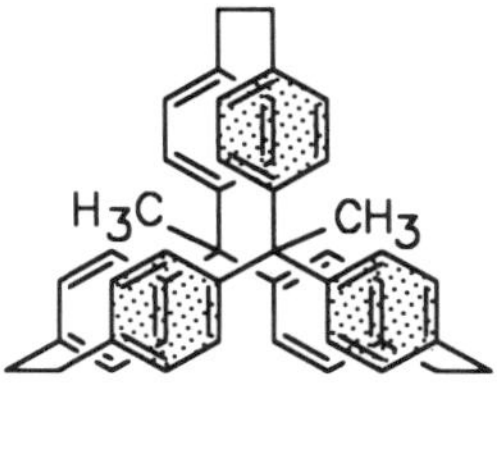

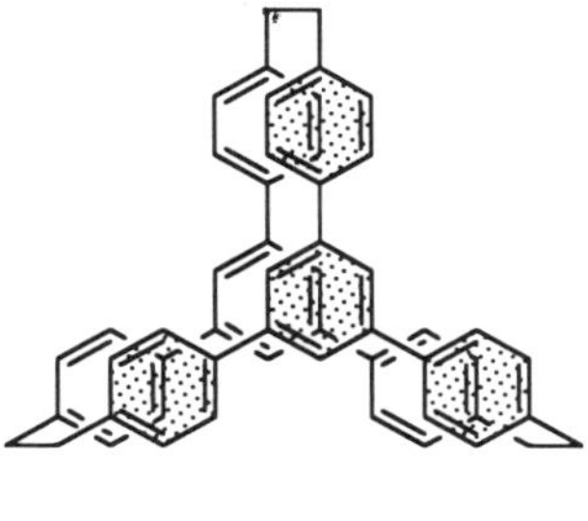

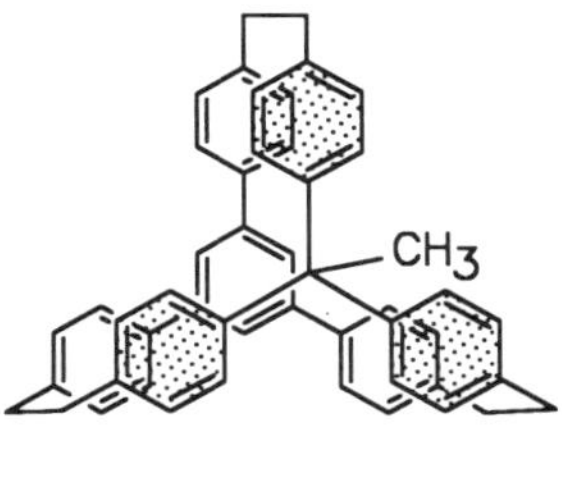

27　　　**28**　　　**29**

meta-substituted triphenylethane units, are formed by sulphone pyrolysis. Although the ether **30** (R = OCH$_3$) can be converted to the bridgehead chloride, it was not possible to form the desired trityl radical.[8] Pascal has described the phane

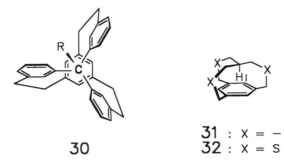

30

31 : X = –
32 : X = S

31 derived from triply substituted benzene, in which the bridgehead consists solely of a methine carbon.[9] It is of great interest that in this type of compound the methine hydrogen can project 'out' as well as 'in'. The intra-annular H_i atom in **31** and **32** experiences a marked high-field shift [$\delta(H_i) = -4.03$ and -1.68, respectively].[9] The distance between H_i (and the corresponding carbon atom C_i) and the opposing benzene ring is accordingly small: 169 and 278 pm, respectively, for **33**, as determined from the X-ray crystal structure. The DNMR

33

measurements for this compound type confirm a ring inversion process, with a barrier $\Delta G_c^{\ddagger}$ of 59 kJ/mol for **33**. Although the quadruply bridged tetraphenylethylene **34** has been made by sulphide cyclization, its conversion to the hydrocarbon has not yet been achieved.[10]

34a

34b

Interesting macro polycycles can be obtained by a Wittig reaction, the details of which were worked out by Wennerström *et al.*[11] The triply bridged cyclophane **37** was made in 1.7% yield by starting with 1,3,5-benzenetricarboxaldehyde

(35) and the ylide derived from the bis(triphenylphosphonium) salt **36** prepared from 1,4-bis(bromomethyl)benzene. The authors proposed the family name for this class of substances, specifying **37** as $[2_6](1,4)_3(1,3,5)_2$bicyclophanhexaene. Catalytic hydrogenation affords the bicyclophane **38** in quantitative yield.

5.5.2 MULTIPLY BRIDGED PHANES CONTAINING HETEROATOMS IN THE BRIDGES

The triazaphane **39** is historically the second example of a triply bridged benzenophane readily obtainable by a cyclization step. It was synthesized in

$R = p-SO_2 \cdot C_6H_4 \cdot Me$

39

A   **39**   B

1970 in 30% yield from 1,3,5-tris(bromomethyl)benzene and the tritosylated derivative of 1,3,5-tris(aminomethyl)benzene.[12] It undergoes a ring inversion

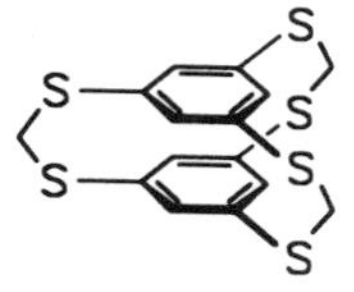

40

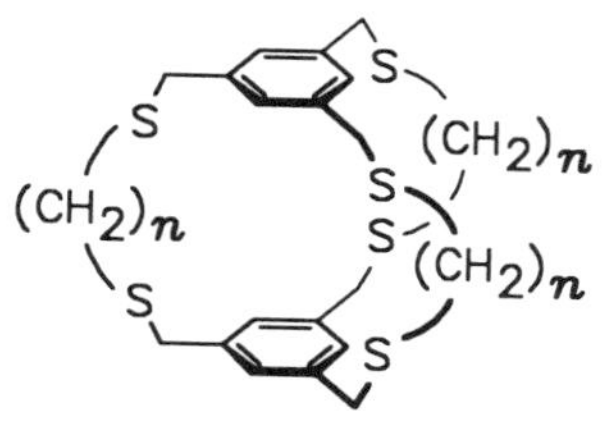

41

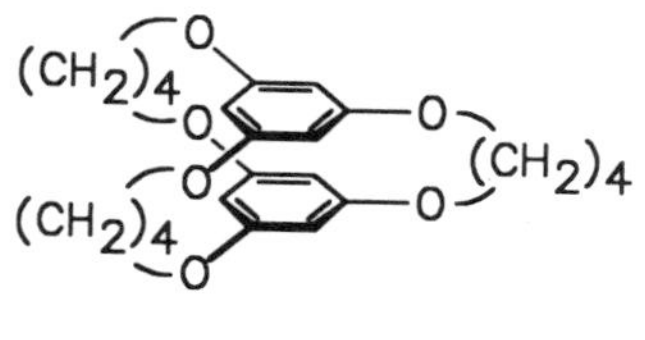

42

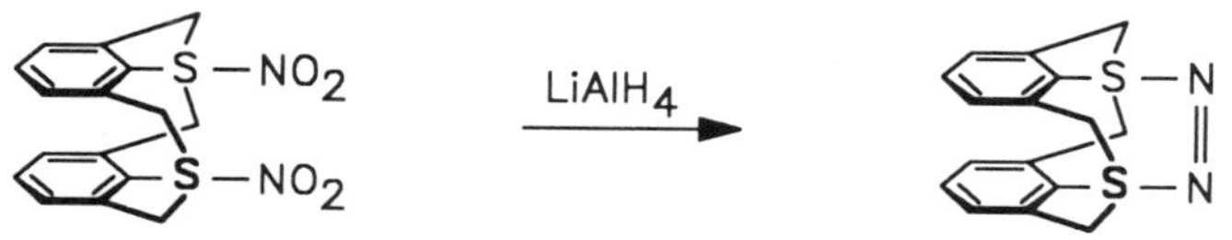

43   **44**

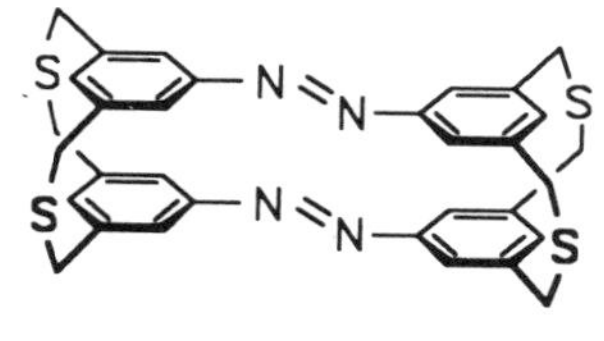

45

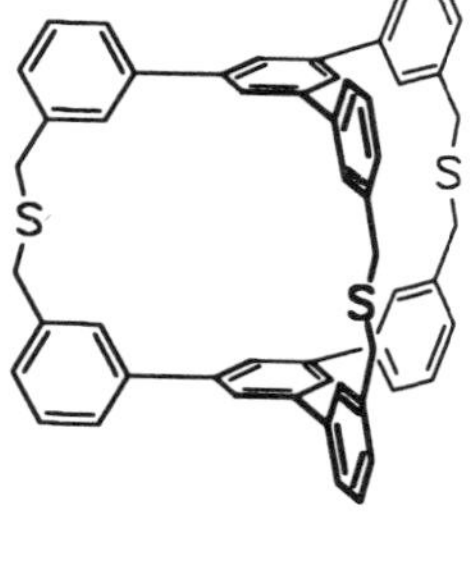

46 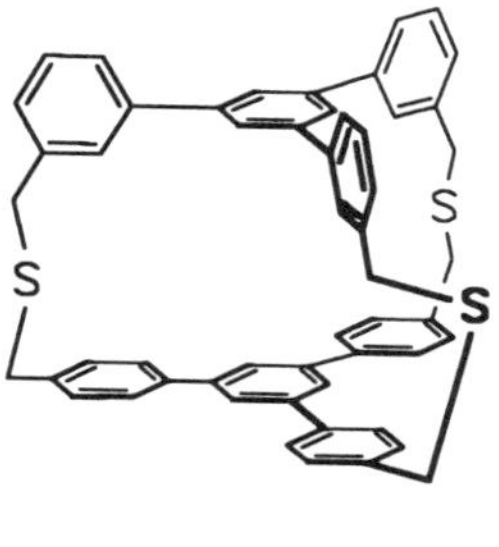   **47**

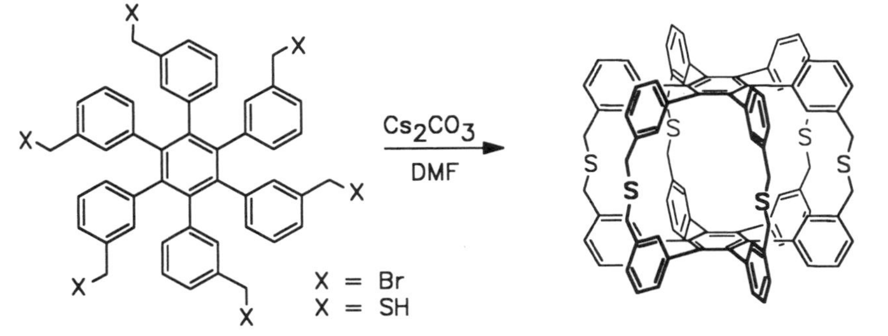

Cs_2CO_3
DMF

X = Br
X = SH

48

+ 2

X = Br
X = SH

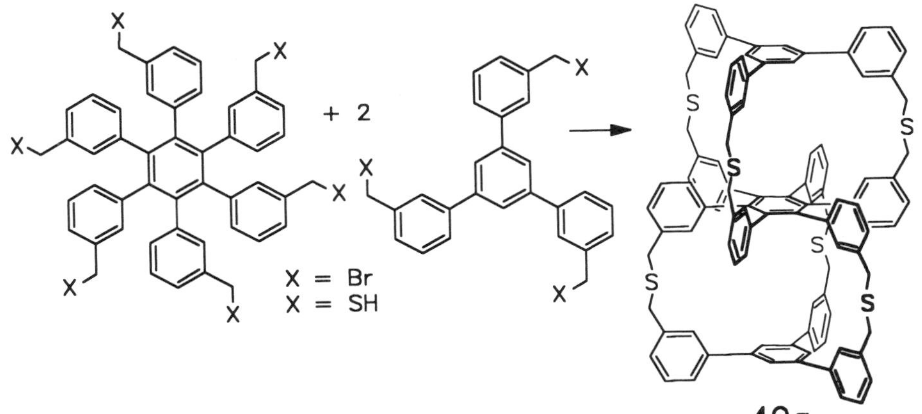

49a

+

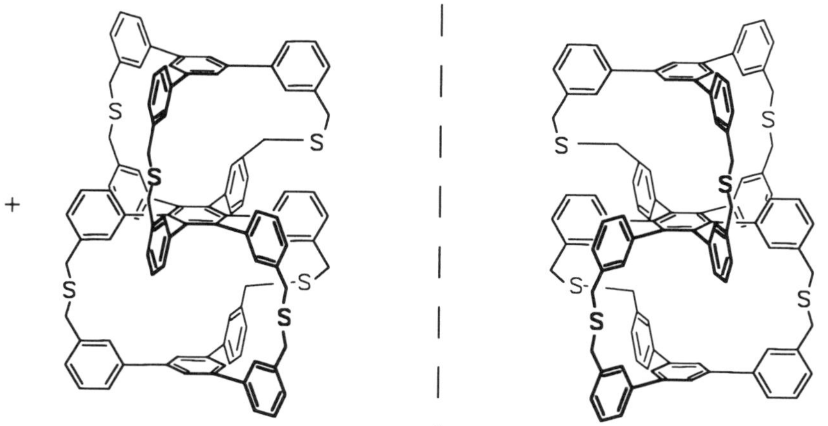

50

50'

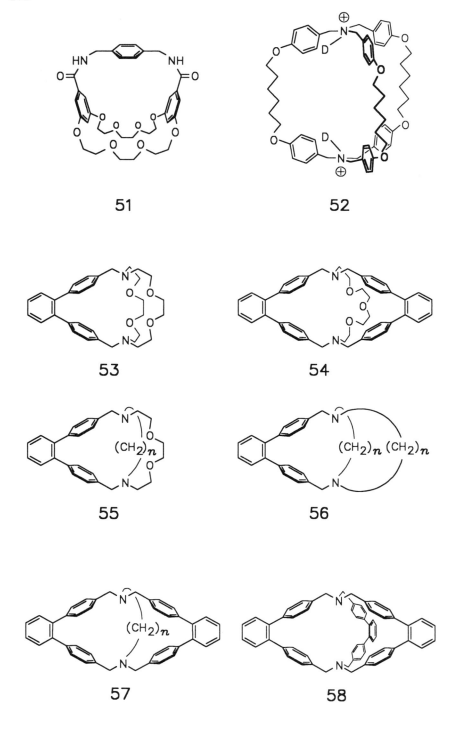

51

52

53

54

55

56

57

58

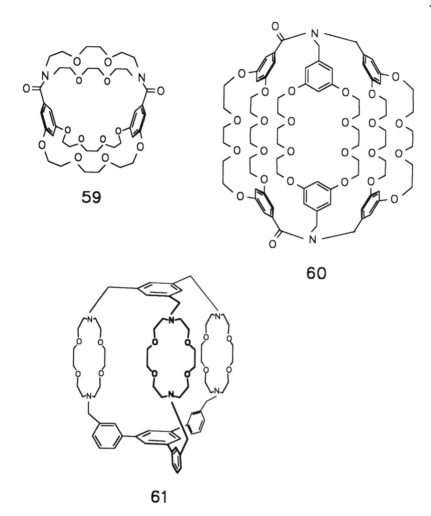

59

60

61

process A ⇌ B whose free enthalpy of activation $\Delta G_c^{\neq} = 57$ kJ/mol. This is a significantly higher barrier than that in the corresponding trisulphide, in which the CH_2 signals in the ^{1}H-NMR spectrum still persist as singlets at $-110°$C; **39** shows up as an AB system at $-57°$C.

The hexathia compound **40** could be synthesized in surprisingly high yield of 25% in 1972, and the hexathia[m_n]phanes of type **41** can be obtained with the aid of high-dilution techniques.[13] Hexaoxaphanes such as **42** can be prepared from phloroglucinol and 1,4-dibromobutane.[14]

The azo[3.2.3]phane **44** was prepared as outlined below by reduction of the two nitro groups in **43**.[15] The quadruply bridged azobenzene **45** has already been mentioned in Section 3.2. A few years ago several large-cavity, multiply bridged thiaphanes were synthesized with the aid of the 'caesium effect'.[16]

These include **46** and **47**, as well as the sixfold bridged molecule **48**, which is reminiscent of a 'benzene analogue' of superphane.

The isomeric three-storey hexathiaphanes **49** and **50** were obtained as outlined above. The isomer **50** proved to be chiral, and enantiomerically enriched samples could be obtained.

The triply bridged, large-cavity phanes **51** and **52**, prepared in connection with host–guest chemistry, are only mentioned here in passing,[17] because large-cavity compounds are described in more detail in Chapter 12. A series of lipophilic, triply bridged ring skeletons (**53–58**), derived from *o*-terphenyl, are illustrated above.[18] Compound **53** proved to be selective for sodium ion.

The structures **59**,[19] **60**,[17] and **61**[20] are shown as examples of multiple crown structures that can be obtained from starting materials with the crown ether skeleton.

Molecules with basket-shaped cavities, such as **62**, are formed by construction of multiple bridges.[21]

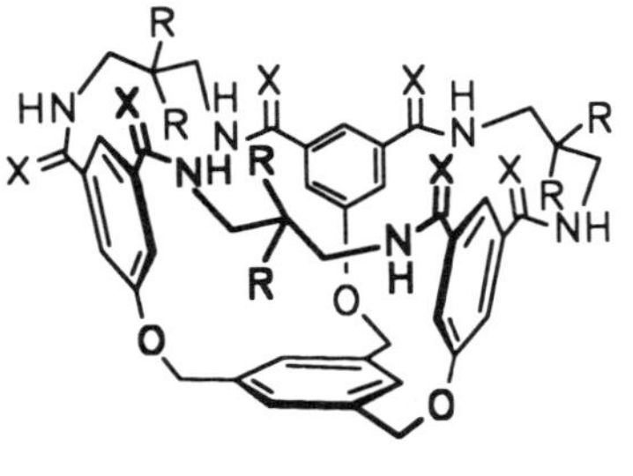

62a 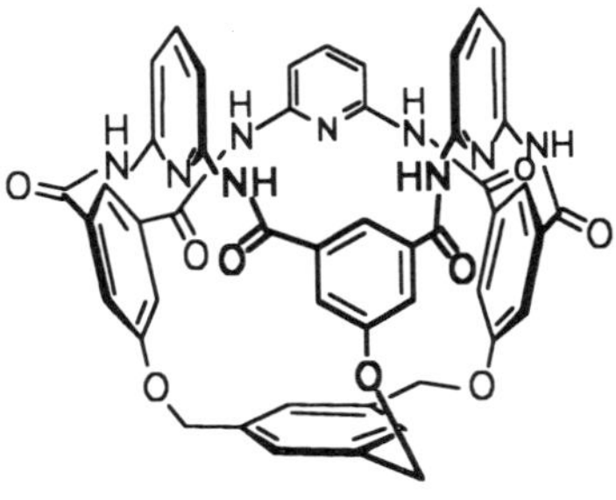   **62b**

Also of interest in connection with the accommodation of guest molecules in large cavities are the in/out isomeric bridged triphenylphosphanes **63** (see Section 12.3).[22]

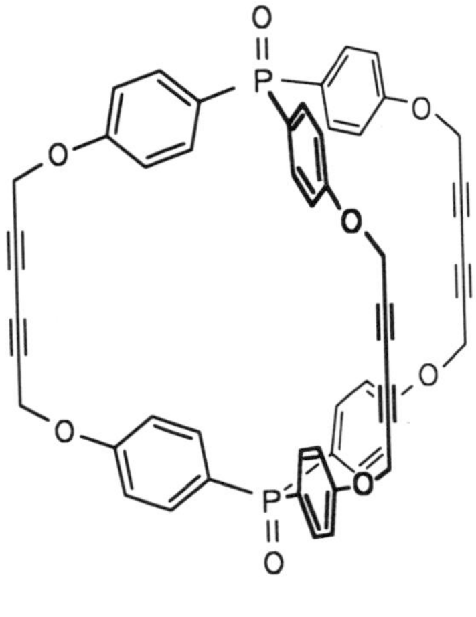

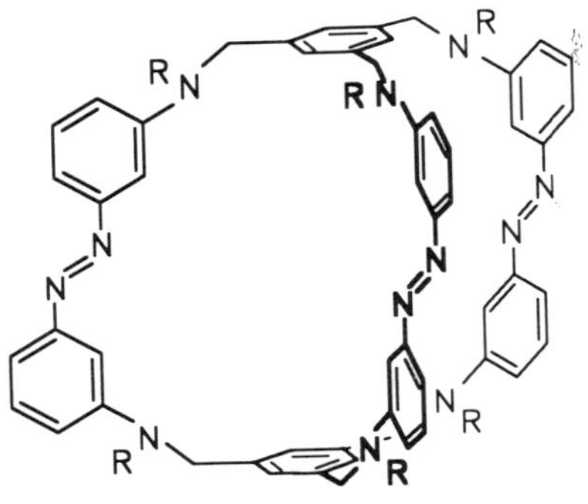

The tris(azobenzene)-bridged cyclophane **64** can undergo E/Z isomerism, thus acting as a 'molecular switch'.[23] Additional examples of multiply bridged phanes are found in Chapters 10 and 12.

References to Section 5.5

1. K. Saitmacher, J. E. Schulz, M. Nieger and F. Vögtle, *J. Chem. Soc., Chem. Commun.*, 175 (1992).
2. (a) J. Juriew, T. Skorochodowa, J. Merkuschew, W. Winter and H. Meier, *Angew. Chem.*, **93**, 285 (1981); *Angew. Chem. Int. Ed. Engl.*, **20**, 269 (1981).
   (b) H. Meier, E. Praß, R. Zertani and H.-L. Eckes, *Chem. Ber.*, **122**, 2139 (1989); M. Hasegawa, Y. Maekawa, S. Kato and K. Saigo, *Chem. Lett.*, 907 (1987); J. Nishimura, A. Ohbayashi, H. Doi, K. Nishimura and A. Oku, *Chem. Ber.*, **121**, 2019, 2025 (1988); J. Nishimura *et al.*, *Tetrahedron Lett.*, **30**, 5439 (1989); K. Nishimura *et al.*, *Chem. Lett.*, 755 (1992).
3. H. C. Kang and V. Boekelheide, *Angew. Chem.*, **93**, 587 (1981); *Angew. Chem. Int. Ed. Engl.*, **20**, 571 (1981).
4. F. Vögtle, G. Hohner and E. Weber, *J. Chem. Soc., Chem. Commun.*, 366 (1973); tube-shaped molecules: F. Vögtle *et al.*, *Chem. Ber.*, **125**, 1881 (1992).
5. K. Matsumoto and W. Nowacki, *Z. Krist.*, **141**, 260 (1975); Cf. [2$_4$]diphenoquinonophane: K. Tani, S. Misumi *et al.*, *Chem. Lett.*, 1461 (1991).
6. Reviews:
   (a) F. Vögtle and G. Hohner, *Top. Curr. Chem.*, **74**, 1 (1978).
   (b) W. Kiggen and F. Vögtle, in *Synthesis of Macrocycles, The Design of Selective Complexing Agents* (R. M. Izatt and J. J. Christensen, Eds.), Chapter 6, p. 309, Wiley, New York, 1987.
   (c) H. Hopf, in *Cyclophanes* (P. M. Keehn and S. M. Rosenfeld, Eds.), Vol. II, p. 521, Academic Press, New York, 1983.
   (d) K. P. Meurer and F. Vögtle, *Top. Curr. Chem.*, **127**, 1 (1985).
   (e) Concave dyes: F. Vögtle *et al.*, *Chem. Ber.*, **125**, 2539 (1992).
7. J. Winkel and F. Vögtle, *Tetrahedron Lett.*, 1561 (1979).
8. M. Nakazaki, K. Yamamoto and T. Toya, *J. Org. Chem.*, **45**, 2553 (1980); *J. Org. Chem.*, **46**, 1611 (1981); Phosphorus analog: R. A. Pascal Jr, *J. Am. Chem. Soc.*, **112**, 6406 (1990); Macrocyclophanes: J. S. Siegel *et al.*, *J. Am. Chem. Soc.*, **113**, 8785 (1991).
9. R. A. Pascal Jr, C. G. Winans and D. Van Engen, *J. Am. Chem. Soc.*, **111**, 3007 (1989); R. A. Pascal Jr *et al.*, *J. Am. Chem. Soc.*, **113**, 2672 (1991); Cf. corresponding triphenylphosphines: R. A. Pascal *et al.*, *J. Am. Chem. Soc.*, **112**, 6407 (1990); cf. R. A. Pascal *et al.*, *J. Org. Chem.*, **58**, 3502 (1993).
10. F. Vögtle and N. Wester, *Liebigs Ann. Chem.*, 545 (1978).
11. H.-E. Högberg, B. Thulin and O. Wennerström, *Tetrahedron Lett.*, 931 (1977); Th. Olsson, D. Tanner, B. Thulin and O. Wennerström, *Tetrahedron*, **37**, 3485, 3491 (1981).
12. F. Vögtle and P. Neumann, *J. Chem. Soc., Chem. Commun.*, 1464 (1970).
13. R. G. Lichtenthaler and F. Vögtle, *Tetrahedron Lett.*, 1905 (1972); *Chem. Ber.*, **106**, 1319 (1973); R. A. Pascal Jr *et al.*, *J. Am. Chem. Soc.*, **113**, 2672 (1991); see corresponding triphenylphosphine, R. A. Pascal Jr *et al.*, *J. Am. Chem. Soc.*, **112** 6407 (1990).
    (a) For new quantitative data and calculations of high-dilution reactions see: G. Ercolani, L. Mandolini and P. Mencarelli, *Macromolecules*, **21**, 1241 (1988); *Gazz. Chim. Ital.*, **119**, 209 (1989); G. Ercolani and P. Mencarelli, *J. Chem. Soc. Perkin*

Trans. 2, 187 (1989); G. Ercolani, L. Mandolini and P. Mencarelli, *J. Chem. Soc. Perkin Trans.* 2, 747 (1990); L. Mandolini *et al.*, *J. Chem. Soc., Chem. Commun.*, 538 (1993).

14. W. D. Curtis, J. F. Stoddart and G. H. Jones, *J. Chem. Soc., Perkin Trans. 1*, 785 (1977).
15. K. Böckmann and F. Vögtle, *Chem. Ber.*, **114**, 1065 (1981).
16. (a) W. Kißener and F. Vögtle, *Angew. Chem.*, **97**, 782 (1985); *Angew. Chem. Int. Ed. Engl.*, **24**, 794 (1985).
    (b) N. Sendhoff, W. Kißener, F. Vögtle, S. Franken and H. Puff, *Chem. Ber.*, **121**, 2179 (1988).
17. N. Wester and F. Vögtle, *Chem. Ber.*, **113**, 1487 (1980); *Chem. Ber.*, **112**, 3723 (1979); *J. Chem. Res. (S)*, 400 (1978); *(M)*, 4856 (1978).
18. L. Rossa and F. Vögtle, *Liebigs Ann. Chem.*, 459 (1981).
19. F. Vögtle, H. Puff, E. Friedrichs and W. M. Müller, *Angew. Chem.*, **94**, 443 (1982); *Angew. Chem. Int. Ed. Engl.*, **21**, 431 (1982) Suppl. 486; F. Vögtle, W. M. Müller, H. Puff and E. Friedrichs, *Chem. Ber.*, **116**, 2344 (1983).
20. F. Vögtle, A. Wallon, W. M. Müller, U. Werner and M. Nieger, *J. Chem. Soc., Chem. Commun.*, 158 (1990); A. Wallon, J. Peter-Katalinic, U. Werner, W. M. Müller and F. Vögtle, *Chem. Ber.*, **123**, 375 (1990).
21. L. Wambach and F. Vögtle, *Tetrahedron Lett.*, **26**, 1483 (1985); B. Dung and F. Vögtle, *J. Incl. Phenom.*, **6**, 429 (1988); J. Breitenbach, K. Rissanen, U. U. Wolf and F. Vögtle, *Chem. Ber.*, **124**, 2323 (1991).
22. B. P. Friederichsen and H. W. Whitlock, *J. Am. Chem. Soc.*, **111**, 9132 (1989); see also R. A. Pascal Jr *et al. J. Am. Chem. Soc.*, **112**, 6406 (1990).
23. H.-W. Losensky, H. Spelthann, A. Ehlen, F. Vögtle and J. Bargon, *Angew. Chem.*, **100**, 1225 (1988); *Angew. Chem. Int. Ed. Engl.*, **27**, 1189 (1988).

6 Multi-layered Cyclophanes

The cyclophanes or phanes discussed thus far have contained at the most two aromatic layers; that is, they are 'double deckers'. In view of the resulting interesting electronic and steric effects, it has been fascinating to consider how to increase the number of layers or 'decks' in such molecules. Of particular interest is the electronic interaction of eclipsed aromatic units in phanes containing more than two layers, where the decks are oriented with their planes 'face-to-face' to one another.[1]

It should be mentioned that there are other types of multi-layered compounds which do not belong to the cyclophanes. These include 1,8,9-triarylanthracene, [14]helicene, triple decker nickelocene, and related triple- and quadruple-deckered transition metal complexes. In the following sections, the presentation will be separated into paracyclophanes, metacyclophanes, metaparacyclophanes, and other multi-layered phanes.

6.1 MULTI-LAYERED PARACYCLOPHANES

6.1.1 MULTILAYERED [2.2]PARACYCLOPHANES

The Hofmann elimination of quaternary ammonium hydroxides has proved to be the best suited and most broadly applicable for the synthesis of multi-layered [2.2]paracyclophanes,[1] one reason being that the necessary intermediates for this method are readily available. Even though we have already considered the Hofmann elimination for synthesis of [2.2]paracyclophanes, we present below for purposes of comparison the straightforward dimerization and the crossed dimerization of *para*-quinodimethanes generated from ammonium hydroxides[1a] (Figure 1).

Because the yields of cyclophanes from Hofmann eliminations, in particular for crossed pyrolysis of two different ammonium bases in 1:1 ratio, were originally minimal,[2] it was necessary to improve the synthetic methodology. Eventually the yields of the crossed phanes **7**, **12**, and **13** were optimized to 15–38% in the following ways: addition of more than a catalytic amount of phenothiazin; use of an excess of the more readily available ammonium base; use of a three-necked round-bottomed flask; and use of the minimum quantity of solvent (Misumi).[1a] In this way the triple- and quadruple-layered [2.2]paracyclophanes in Figure 2 were prepared from the corresponding di- and tetramethyl[2.2]paracyclophanes **7** and **1**.[3] Cross-coupling of **10** with **6** gave the highest yield reaction, with formation of the triply layered

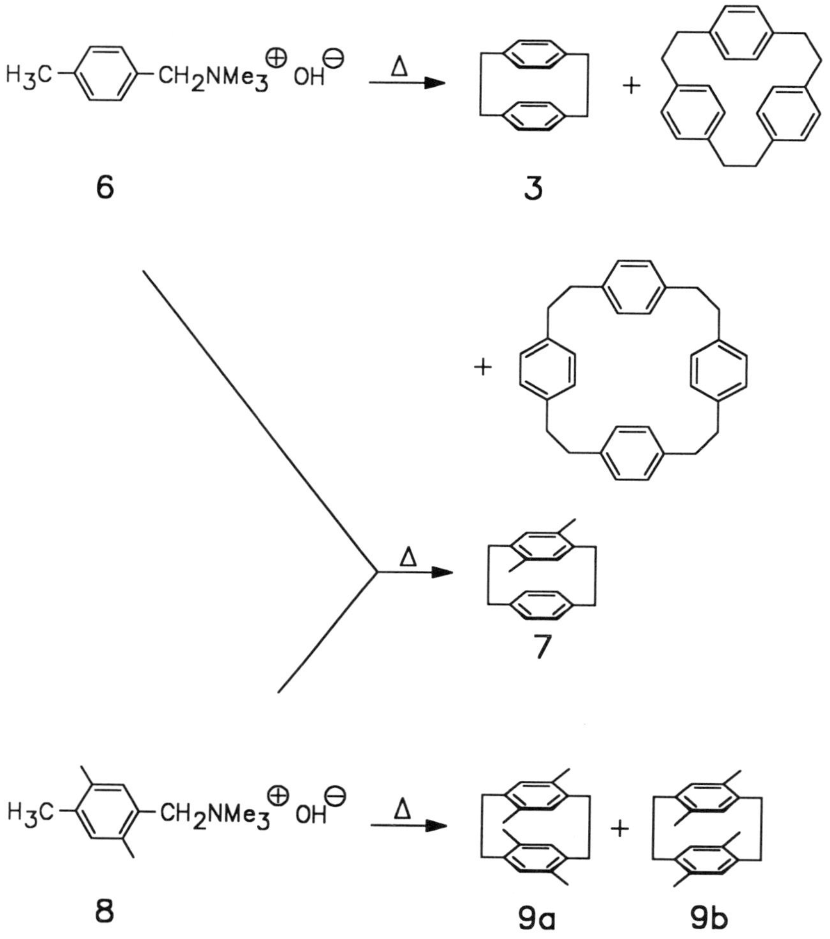

Figure 1. Examples of the preparation of [2ₙ]phanes by the crossed Hofmann elimination

hydrocarbon **12**.[4] The molecules **19** and **20** were the first multi-layered cyclophanes obtained by this method. From Figure 2 it can be seen that elimination of **10** affords the isomeric quadruple-layered molecules **17**(D_2 symmetry) and **18**(C_{2h} symmetry). Pyrolysis of an equimolar mixture of **16** and **10** led to a mixture of the quintuple-layered cyclophanes **21** and **22** (Figure 3). The dimethyl compound **24** with D_2 symmetry geometry could be obtained by pyrolytic cross-coupling of **23** and **6**.[2,5] **16** afforded a 2:1 mixture of sextuple-layered phanes, which could be separated by column chromatography with benzene as eluent. The more soluble, major isomer is **25** with D_2 symmetry, while the less soluble **26**, with C_{2h} symmetry, is formed in smaller amount.[2,6]

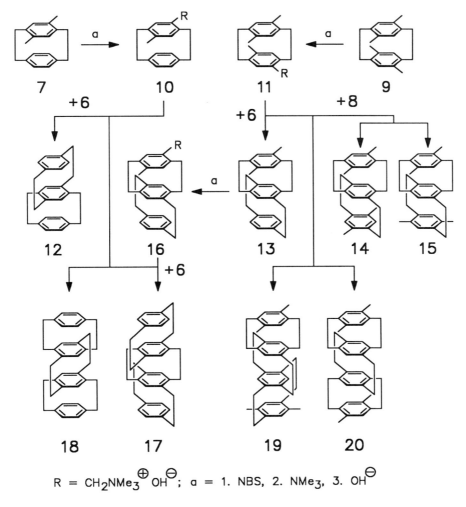

R = CH₂NMe₃⊕ OH⊖; a = 1. NBS, 2. NMe₃, 3. OH⊖

Figure 2. Triple- and quadruple-decker phanes from Hofmann eliminations (yields reported in the literature)[1a]

6.1.2.1. Properties

Structure and strain energy

Figure 4 represents the results of the X-ray crystal structure analysis of the triple-layered bromo[2.2][2.2]paracyclophane **27** (see below). In Figure 5 is provided a representation of the structure of the quadruple-layered tetramethyl[2.2][2.2][2.2]paracyclophane **20**[1a] based upon its X-ray crystal structure analysis, the first carried out on a multi-layered phane. It can be seen

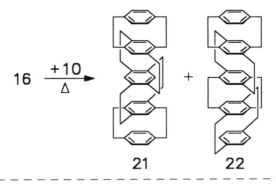

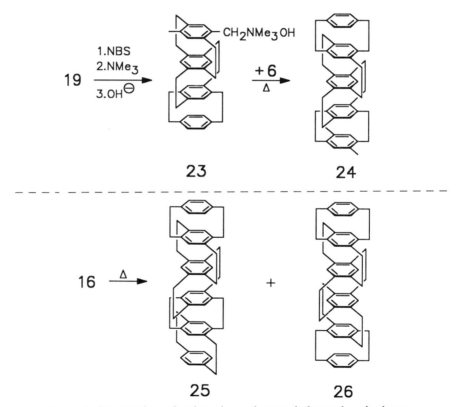

Figure 3. Preparation of quintuple- and sextuple-layered cyclophanes

that the distance between neighbouring benzene rings ranges from 274 to 313 pm; that is, the 'face-to-face' orientation of the benzene rings lies within the van der Waals distance of 340 pm. The X-ray crystal structure of the bromine-substituted triple-layered [2.2]paracyclophanes **27** indicates similar separations of 274–319 pm (Figure 4).

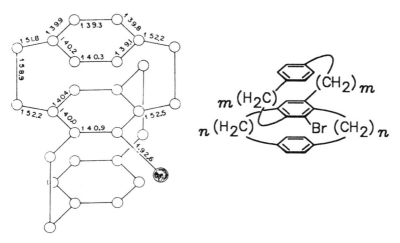

Figure 4. X-ray crystal structure of triple-layered bromo[2.2][2.2]paracyclophane **27**[1] (bond lengths in pm)

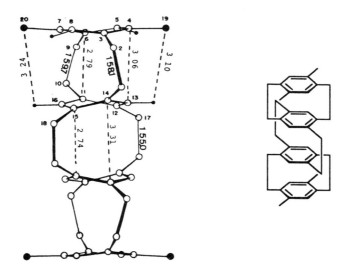

Figure 5. X-ray crystal structure of the quadruple-layered [2.2][2.2][2.2]paracyclophane isomer **20**[7] (bond lengths in pm)

All outside benzene rings in the triple- and quadruple-layered molecules are distorted into boat forms, in which deformation angles are similar to those in the double-layered [2.2]paracyclophane (**3**). In contrast, the inside benzene rings are quite different, being distorted into twist forms, with twist angles γ around 13.5° and angles α and β varying between 10.8 and 12.9 (Figure 6). The bond length for the sp^3–sp^3 single bonds in the ethane bridges connecting the inner

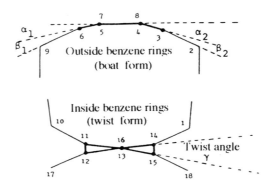

Figure 6. Deformation of the benzene rings in quadruple-layered paracyclophanes[1a]

benzene rings (155 pm) is somewhat shorter than that for the outer benzene bridges or that in the single bridging bonds in **3**. While in the crystal of **3** the symmetrical rotational vibration of both benzene rings (approx. 6.4°) proceeds about a central axis perpendicular to the rings, the rotation of the inner and outer benzene rings in **20** and **27** is limited to one side because of the *o*-methyl group repulsions (angles of 8 and 6.8°, respectively).

Strain energy in the unsubstituted triple-layered [2.2][2.2]paracyclophane **12** of 245 kJ/mol is practically double that in [2.2]paracyclophane (**13**) (138 kJ/mol).[8] Since X-ray crystal analysis indicates that deformation angles in the boat forms of benzene rings in both double- and triple-layered paracyclophanes are similar, it is concluded that the inner, twist form benzene ring in **12** is more highly strained than the outer rings. It follows that the contribution of the inner ring to the total strain energy is double that of the outer, boat form benzene rings.

Spectroscopic properties

The absorption spectra of some multiple-layered [2.2]paracyclophanes are compared with that of [2.2]paracyclophane (**13**) in Figure 7. It can be seen that the UV curves are essentially independent of methyl substitution and configurational stereoisomerism. Particularly noteworthy is the observation that phanes with an increasing number of layers show both bathochromic and hyperchromic shifts. These shifts are especially evident in going from bibenzyl to double-layered and from double-layered to multiple-layered phanes, while the curves become increasingly featureless as well. These spectral characteristics are explained as arising from transannular $\pi-\pi$ electronic interactions or transannular delocalization. This follows from the fact that shifts in absorption bands are only slightly influenced by a second important factor, namely deformation of the benzene rings. Semiempirical calculations of the electronic spectra of double- and multiple-layered [2.2]paracyclophanes, considered from

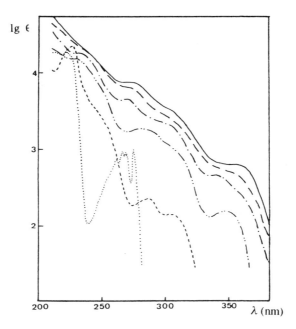

Figure 7. Electronic spectra of multiple-layered [2.2]paracyclophanes (in cyclohexane): (····) 4,4'-Dimethyl-bibenzyl; (-----) **3**; (-·····-) **12**; (-·-·-) **17**; (———) **21**; (———) **25**

the point of view of ground state, excited state, and charge-transfer configurational interactions, are in accord with the above interpretation.

The *emission spectra* provide additional information concerning transannular interactions in multi-layered cyclophanes. Fluorescence and phosphorescence spectra show bathochromic shifts with an increasing number of layers, analogous to the pattern in the electronic spectra.[2] An advantage in the fluorescence spectra is that the maxima can be clearly distinguished, unlike the situation for the long wavelength maxima in the electronic spectra. No additional red shift in the maximum is observed in going from the quintuple- to the sextuple-layered phane, suggesting that no further transannular interaction is observed beyond the quintuple-layered molecule (Misumi). In the phosphorescence spectra the lifetimes become progressively smaller and tend to converge, a pattern similar to that in the electronic and fluorescence spectra. This points to a decrease in the energy gap between singlet and triplet states, such that the probability for a radiationless deactivation increases.[2]

Charge-transfer complexes with π acceptors

A good correlation between charge-transfer (CT) band positions and association constants K of the complexes was noted in 1958 for π-complexes of various oligomethylbenzenes with tetracyanoethene (TCNE).[9] For the quadruple-layered tetramethylparacyclophane (a mixture of **19** and **20**) it was found that the stronger transannular interaction, as compared to that of [2.2]paracyclophane (**3**), is manifested in the π-basicity.[10] As the number of layers is increased, the long wavelength maxima of the TCNE complexes are shifted to longer wavelength, indicating an increasing donor character of the cyclophane. This enhanced π-donor effect is still noticeable in the quintuple-layered complex **28**; only in going from the 1,3,5-trinitrobenzene (TNB) complex of the quintuple-(**28**) to the sextuple-layered phane is there no additional difference seen in the transannular electron-donor capacity.

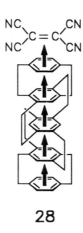

28

Magnetic resonance

It has long been recognized that the aromatic protons in [$m.n$]paracyclophanes absorb at higher field than those in the corresponding alkylbenzenes, and that the extent of high-field shift increases as the number of methylenes in the bridges is reduced.[11] Reasons given for this are the anisotropy effect of one benzene ring exerted on the aromatic protons of the other ring, as well as rehybridization of the benezene carbons because of ring deformation. It was thus to be expected that the aromatic protons in multiple-layered [2.2]paracyclophanes would be influenced by distant benzene rings—as contrasted to the effect of the neighbouring 'face-to-face' benzene rings—and that such compounds would be good models for the study of magnetic anisotropy. Assignment of the many aromatic protons in the spectra was possible because of the observation that protons in the outer rings of multiple-layered [2.2]paracyclophanes are more solvent-dependent than those on the inner rings.[12] A change in solvent from CCl_4 to $CDCl_3$ often results in a chemical shift difference of 0.1–0.15 p.p.m. Identification of some protons has also been accomplished from the nuclear Overhauser effect.[1a]

Magnetic anisotropy

Significant shifts to higher field of all aromatic protons are observed with an increase in the number of layers. Thus, the aromatic protons in [2.2]paracyclophane in CCl$_4$ appear at $\delta = 6.35$, in the triple-layered cyclophane **12** at 5.35, in the quadruple-layered **19** at 5.47, and in the sextuple-layered **25** at 4.8 p.p.m. From this pattern one can in fact predict the position of proton absorption in specific decks of multiple-layered cyclophanes. A second observation is that aryl hydrogens which are pseudo-*geminal* to substituents and aromatically bound methyl groups which are sterically compressed normally experience a downfield shift. This is the case for H$_b$ in **13**, **15**, **19**, and **20**.[11] An average value for the shift of a sterically compressed methyl group, such as that in 4-methyl[2.2]paracyclophane, is $\Delta\delta = -0.36$ p.p.m. It is of interest that the large downfield shift of $\Delta\delta = -0.71$ p.p.m. is observed for H$_b$ in the triple-layered molecule **14**. Thus **14** serves as a novel example of steric compression 'from two sides', and it is indicated that the downfield shift due to compression is additive.[1a]

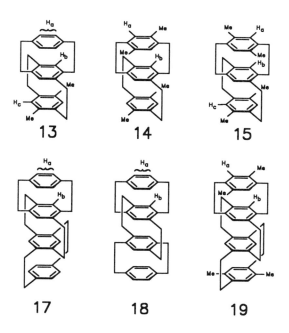

A second effect is observed by the introduction of substituents, found also in [2.2]paracyclophane. One notes a strong chemical shift of the *ortho*-aromatic protons, by comparison with those in a non-cyclic compound.[13] Such enhanced shift values are also observed in methylated multiply layered cyclophanes ($\Delta\delta \approx 0.42$ p.p.m.).

13C-NMR spectra[14]

The $\pi-\pi$ compression exerts a strong effect on the ^{13}C-NMR spectra of multiply layered [2.2]paracyclophanes. Carbons in the outer benzene rings experience practically the same low-field shifts as those in doubly layered cyclophanes. In triply layered compounds such as **12** one notes a doubled compression effect, for the unsubstituted arene carbon atom signals are shifted downfield by $\Delta\delta = 7.95$, as compared to the corresponding carbon resonances in the reference compound 6,9-dimethyl[4.4]paracyclophane. This shift to lower field is about double that of the difference (4.25) between the corresponding carbon atoms of dimethyl[4.4]paracyclophane and 4,7-dimethyl[2.2]paracyclophane (**7**), suggesting an additivity of the contribution of steric compression on the carbon resonances.

ESR spectroscopy

ESR and ENDOR spectra of radical anions of multiply layered [2.2]paracyclophanes[15] show that separation of their spin densities is extensively influenced by interactions with the gegenions. MO calculations suggest that the observed ion pairing effect can be correlated with ion pair models, in which the cation is located above the centre of the outermost benzene ring and that the potassium ion, for example, loosely bound in an ion pair with the triply layered radical anion, migrates back and forth from one side of the molecule to the other (DMF/THF as solvent). A similar situation has been demonstrated for the potassium [2.2]paracyclophane–radical anion ion pair in DMF/THF.

Chiroptical properties

The optically active multiply layered [2.2]paracyclophanes **12**, **19**, and **21**, with known absolute configurations, were chosen to examine the possible correlation

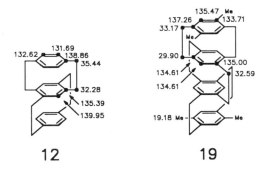

Figure 8. ^{13}C-NMR absorptions of some selected multiply layered [2.2]paracyclophanes[14]

between optical activity and 'stacking interaction'. The synthesis was accomplished by Hofmann elimination from the doubly and triply layered ammonium bases **10** and **16** of known absolute configurations.[16] The optical rotations of compounds with (R) configuration are strongly laevorotatory and become increasingly so with the number of layers. As an example, the rotation for (R)-($-$)-**12** is $[\alpha]_D^{28} = -256$; for (R,R,R)-($-$)-**21** $[\alpha]_D^{22} = -362$ (in CHCl$_3$).

6.1.2.2 Chemical reactions

Chemical reactions in doubly layered [m.n]paracyclophanes, which are attributed to transannular electronic interactions and molecular ring strain, have been extensively investigated, so that they can be compared with those in multi-layered analogues. Triply layered paracyclophanes, like the 'double deckers', exhibit enhanced reactivity in bromination. The low yields from Friedel–Crafts reactions and nitrations are attributed to even higher strain.

Electrophilic substitution; bromination

The triply layered [2.2][2.2][2.2][3.3]- and -[3.3][3.3]paracyclophanes such as **29–31** are converted to the monobromo compounds **27**, **32**, and **33** as single products by uncatalysed bromination with pyridinium hydrobromide perbromide (Figure 9).[17] These brominations are complete in shorter times than required for those with [n]- and doubly layered paracyclophanes. This implies that an accelerated reaction occurs through *transannular electronic stabilization* of the intermediate. Use of excess reagent does not lead to dibrominated product. It is assumed that the first bromo substituent, because of its bulkiness and inductive effect, impedes the second substitution, because the formation of a second δ-complex in the inner (middle) ring is blocked.

It is generally observed that reactions in all triply layered compounds proceed considerably faster than those in doubly layered compounds, including in competition experiments. The greater the number of layers in the molecule, the faster the reaction, no matter what the nature of the reagent. The faster reaction in the triply layered cyclophane as compared to that in the doubly layered analogue is attributed to the difference in stability of the two σ-complexes **42** and **43**, because of *transannular charge delocalization*. The intermediate **43** is stabilized more effectively through the electron-donor properties of the two outer rings, which is not possible for the doubly layered phane **42**.

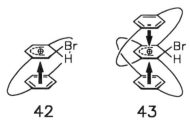

42   **43**

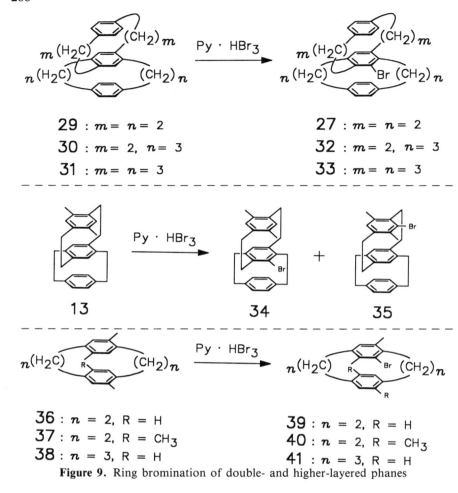

Figure 9. Ring bromination of double- and higher-layered phanes

Skeletal rearrangements

Whereas [2.2]paracylophane rearranges to [2.2]metacyclophane (see Section 2.4), the skeletal rearrangement of tetramethyl[2.2]paracyclophane **9a** with the same catalyst ($AlCl_3$/HCl) leads in 38% yield to tetramethyl[2.2]metapara-cyclophane (**46**), which is formed by an interesting double rearrangement of an ethano bridge at C-3 and of a methyl group at C-4.[18] The triply layered cyclophane **12** afforded polymeric material under the same conditions; but, when the reaction is carried out at lower temperature, the triply layered metaparacyclophane **47** is formed as the main product, along with two isomeric metaparacyclophanes (**48** and **49**). When **12** is treated with weak Friedel–Crafts

47 **12** **48** **49**

catalysts, such as $SnCl_4/HCl$, at room temperature or with $TiCl_4/HCl$, BF_3-etherate/HCl, or I_2, the product in 80% yield is a mixture of **48** and **49**. With the weaker protic acid $SnCl_4/HCl$ **9a**, on the other hand, undergoes no rearrangement: evidence that the triply layered molecule is more reactive than the doubly layered one. The products **48** and **49** come about through a double rearrangement of two ethano bridges at the inner ring of the three benzene rings, by analogy to that described above for the transformation of **9a** to **46**. It is of interest that the ratio of isomers **48** and **49** formed (1.2:1) is independent of reaction conditions—time, temperature, catalyst type.[1a] A similar rearrangement was accomplished with the quadruply layered cyclophane **18** by addition of BF_3/HCl at 0°C. The following mechanism has been

18 **50**

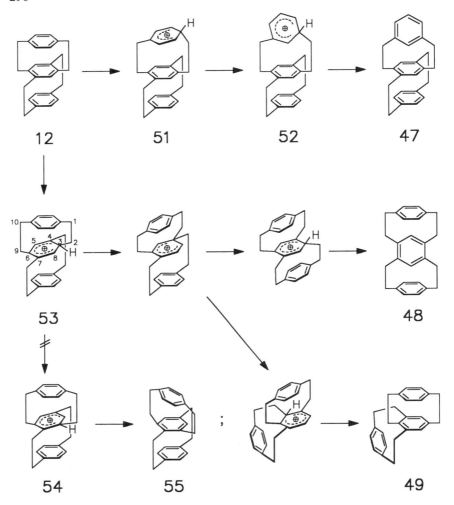

proposed. The structure of the protonated cyclophane **53**, formed by catalytic rearrangement with a weak Friedel–Crafts catalyst, could be substantiated by NMR spectroscopy in $FSO_3H/SO_2ClF/CD_2Cl_2$ at $-100°C$.[18] The preferential protonation at the bridgehead carbon atom of the inner benzene ring is ascribed to the strong π-basicity of this inner ring, as well as the resulting relief of strain and the substantial stabilization of the intermediate σ-complex **53**, as compared to that which would result by protonation of an outer benzene ring. From this investigation it was concluded that the rearrangement with weak protic acids is thermodynamically controlled, whereas kinetic control likely prevails with catalysis by strong acids.[1a]

Carbene Addition

The triply layered phane **12**, like [2.2]paracyclophane (**3**), is attacked by diazomethane/copper chloride. The tropilidenecyclophane **56** so formed, when treated with trityl fluoroborate, loses hydride to give the triply layered tropylioparacyclophane **57**.[19] The 'stacking interaction' between the aromatic rings, in particular between the tropylium ring and the benzene rings, is clearly reflected in the high-field shifts of all aromatic protons in the NMR spectrum, as well as in the bathochromic shifts of the charge-transfer bands ($\lambda = 434$ nm for **57**) in the electronic spectra.

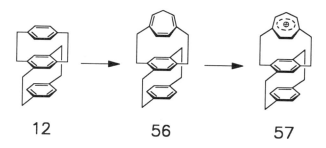

| 12 | 56 | 57 |

Transition metal complexes

In this case attention was drawn to transannular electronic interactions between benzene rings and chromium carbonyl-complexed benzene rings. The appropriate cyclophanes were treated with chromium hexacarbonyl in diglyme at 150°C for 2 hours, after which the complexes **58–63** could be isolated. The yield declines with an increasing number of layers, while the ratio of bis(transition metal):mono(transition metal) increases. This suggests that the initially formed chromium tricarbonyl complex exerts a diminishing transannular electronic effect on the other outer benzene ring as the number of layers increases. From X-ray crystal analysis it can be shown that the interplanar distance between the two benzene rings in [2.2]paracyclophane is shorter than normal after coordination with the chromium tricarbonyl group (e.g. in **58**). The three mono(transition metal carbonyl) complexes **58–60** show significant bathochromic shifts in their charge-transfer bands, by comparison with the chromium tricarbonyl complex of *p*-xylene; it is surprising, however, that there is no difference in the extent of shift with an increase in the number of layers in the cyclophane ligands. Quite the contrary, the charge-transfer bands of the bis complexes **61–63** are shifted to shorter wavelength with an increasing number of layers.

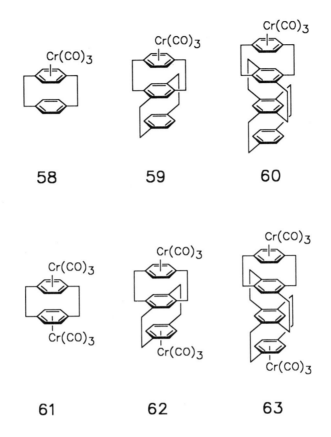

58   59   60

61   62   63

6.1.2 TRIPLY LAYERED [$m.m$][$n.n$]PARACYCLOPHANES

Cyclophanes of this type (m, $n = 2$–4; see **64a–f**), are of interest because one can use them to investigate transannular electronic interactions without having to take into account the ring strain inherent in the more tightly bridged cyclophanes.

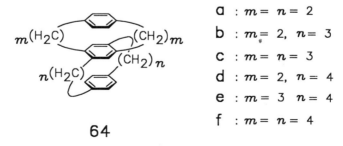

a : $m = n = 2$
b : $m = 2$, $n = 3$
c : $m = n = 3$
d : $m = 2$, $n = 4$
e : $m = 3$ $n = 4$
f : $m = n = 4$

64

The synthetic approach is outlined in the following steps for the triply layered [3.3][3.3]paracyclophane **64c**.[21] Once again, ring contraction from the dithiacyclophane with subsequent sulphone pyrolysis or photodesulphurization was employed.

64c

The UV/VIS spectra of these cyclophanes and their TCNE complexes exhibit the same trend in decreasing interaction as in the case of the doubly layered [n.n]paracyclophanes: [2.2] > [3.3] > [4.4] for the hydrocarbons and [3.3] > [2.2] > [4.4] for the charge-transfer complexes. All five hydrocarbons **64a–e** show characteristic excimer fluorescence bands with broadening and large Stokes shifts; only the [4.4] [4.4] compound **64f** shows a band at 395 nm, which is assigned to an excited state trimer emission.[21]

References to Section 6.1

1. Reviews
    (a) S. Misumi, in *Cyclophanes* (P. M. Keehn and S. M. Rosenfeld, Eds.), Vol. II, p. 573, Academic Press, New York, 1983.
    (b) S. Misumi and Y. Sakata, *J. Synth. Org. Chem. Jpn.*, **29**, 114 (1971); *Hyomen*, **17**, 239 (1979); S. Misumi, *Kagaku no Ryoiki*, **28**, 927 (1974), *Kagaku no Ryoiki*, **32**, 651 (1978); S. Misumi and T. Otsubo, *Acc. Chem. Res.*, **11**, 251 (1978); *Chem. Educ. Tokyo*, **28**, 249 (1980); Y. Sakata, *Kagaku no Ryoiki*, **28**, 947 (1974); *J. Synth. Org. Chem. Jpn.*, **38**, 164 (1980).
2. T. Otsubo, S. Mizogami, I. Otsubo, Z. Tozuka, A. Sakagami, Y. Sakata and S. Misumi, *Bull. Chem. Soc. Jpn.*, **46**, 3519 (1973); T. Otsubo, H. Horita and S. Misumi, *Synth. Commun.*, **6**, 591 (1976).
3. T. Otsubo, S. Mizogami, Y. Sakata and S. Misumi, *Tetrahedron Lett.*, 4803 (1971).
4. H. Higuchi and S. Misumi, *Tetrahedron Lett.*, **23**, 5571 (1982); H. Higuchi, M. Kugimiya, T. Otsubo, Y. Sakata and S. Misumi, *Tetrahedron Lett.*, **24**, 2593 (1983).
5. T. Otsubo, S. Mizogami, Y. Sakata and S. Misumi, *Tetrahedron Lett.*, 2457 (1973).
6. T. Otsubo, Z. Tozuka, S. Mizogami, Y. Sakata and S. Misumi, *Tetrahedron Lett.*, 2927 (1972).

294

7. H. Mizuno, K. Nishiguchi, T. Otsubo, S. Misumi and N. Morimoto, *Tetrahedron Lett.*, 4981 (1972); H. Mizuno, K. Nishiguchi, T. Toyoda, T. Otsubo, S. Misumi and N. Morimoto, *Acta Crystallogr.*, **B33**, 329 (1977).
8. K. Nishiyama, N. Sakiyama, S. Seki, H. Horita, T. Otsubo and S. Misumi *Tetrahedron Lett.*, 3739 (1977); *Bull. Chem. Soc. Jpn.*, **53**, 869 (1980).
9. R. E. Merrifield and W. D. Philip, *J. Am. Chem. Soc.*, **80**, 2778 (1958).
10. D. T. Longone and H. S. Chow, *J. Am. Chem. Soc.*, **86**, 3898 (1964); *J. Am. Chem Soc.*, **92**, 994 (1970).
11. D. J. Cram and R. C. Helgeson, *J. Am. Chem. Soc.*, **88**, 3515 (1966).
12. T. Otsubo, S. Mizogami, Y. Sakata and S. Misumi, *Bull. Chem. Soc. Jpn.*, **46**, 383 (1973).
13. H. J. Reich and D. J. Cram, *J. Am. Chem. Soc.*, **91**, 3534 (1969).
14. T. Kaneda, T. Otsubo, H. Horita and S. Misumi, *Bull. Chem. Soc. Jpn.*, **53**, 101. (1980).
15. Review: F. Gerson, in *Cyclophanes, II* (F. Vögtle, Ed.), *Top. Curr. Chem.*, **115** 57 (1983).
16. M. Nakazaki, K. Yamamoto, S. Tanaka and H. Kametani, *J. Org. Chem.*, **42**, 28" (1977).
17. T. Otsubo, H. Horita, Y. Koizumi and S. Misumi, *Bull. Chem. Soc. Jpn.*, **53**, 167" (1980).
18. H. Horita, Y. Koizumi, T. Otsubo, Y. Sakata and S. Misumi, *Bull. Chem. Soc Jpn.*, **51**, 2668 (1978).
19. H. Horita, T. Otsubo and S. Misumi, *Chem. Lett.*, 1309 (1977).
20. H. Ohno, H. Horita, T. Otsubo, Y. Sakata and S. Misumi, *Tetrahedron Lett.* 265 (1977).
21. T. Otsubo, T. Kohda and S. Misumi, *Tetrahedron Lett.*, 2507 (1978); *Bull. Chem Soc. Jpn.*, **53**, 512 (1980).

6.2 MULTI-LAYERED [2.2]METACYCLOPHANES

The [2.2]metacyclophanes are of particular interest because of their unique molecular framework, unusual electrophilic substitution reactions, transannular reactions, and their potential as starting materials for bridged [14]annulenes, as was mentioned in more detail in Section 2.1. Thus multiply layered molecules of this type should provide further insight into steric and electronic effects.

6.2.1 SYNTHESIS

The synthesis of the first known triply layered [2.2]metacyclophanes is outlined in Figure 1.[1] It can be seen that these molecules can exist in two conformations as **9** and **10** or **14** and **15**, a subject which is discussed in more detail below.

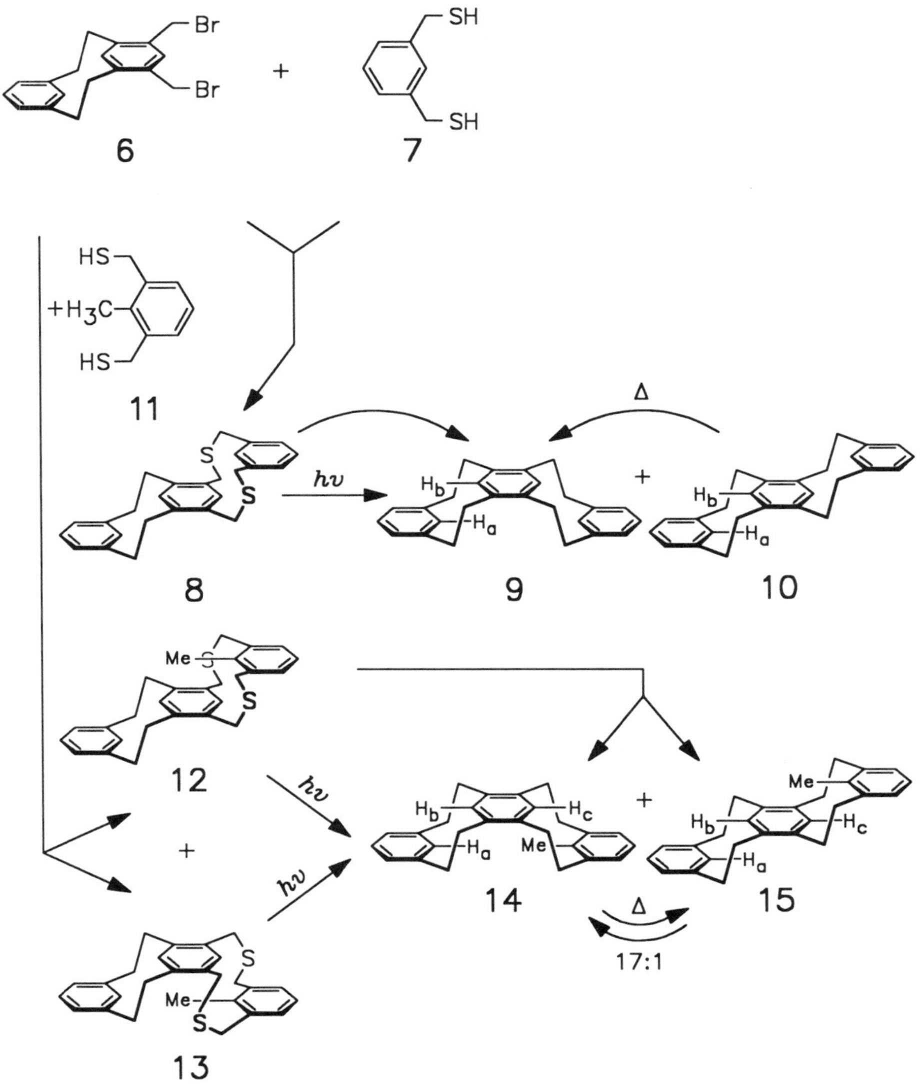

Figure 1. Synthesis of a triply layered [2.2]metacyclophane

In Figure 2 is shown the corresponding synthetic route to quadruply layered [2.2]metacyclophanes, represented in their respective conformers. From both figures it can be seen that the dithia[3.3]phane intermediates and their desulphurization play a major role. The two triply layered cyclophane conformers **9** and **10** were also synthesized by Raney-nickel reduction of the bis(dithian) compound **22**.[2]

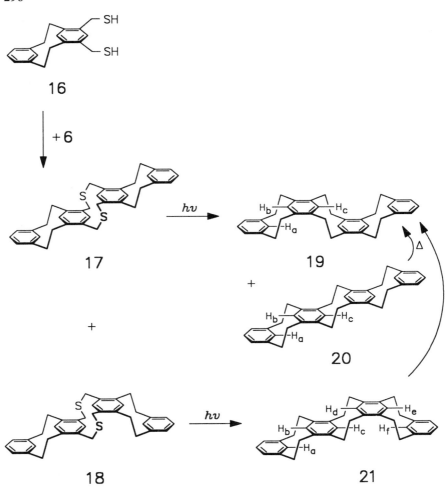

Figure 2. Synthetic route to quadruply layered [2.2]metacyclophanes (**19–21**)

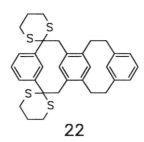

22

6.2.2 PROPERTIES

6.2.2.1 ^{1}H-NMR spectra[1]

One of the best known characteristics of multi-layered metacyclophanes is the chemical shift of the inner aromatic protons, which are located opposite the neighbouring benzene rings and are significantly influenced by the magnetic anisotropy of the rings. While the intra-annular hydrogen atoms (H_i) in [2.2]metacyclophane (2) absorb at $\delta = 4.27$, the inner protons H_a (Figure 3) in 9 ($\delta = 5.03$) are shifted downfield, even by comparison to those in 10 ($\delta = 4.41$). This is explained as a shielding effect of the third, spatially close benzene ring and can be used to assign the two conformers. The downfield shift of H_a in 14 can be attributed both to the shielding effect arising from steric compression between H_a and the nearby methyl group and to the magnetic anisotropy effect exerted by the third benzene ring. The downfield shift of the methyl signals in 14 ($\delta = 1.07$), compared to that in 15 ($\delta = 0.58$) reflects the combination of anisotropy and steric compression effects. Irradiation of the methyl protons in 14 in fact induces a nuclear Overhauser effect (NOE) of 22%, while no effect is observed with 15.

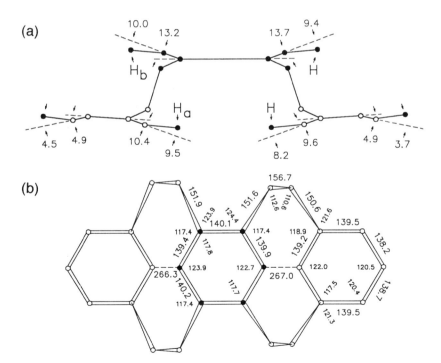

Figure 3. X-ray crystal structure of the *up/down* isomer 9: (a) side view; (b) bond lengths (pm) and angles (degrees)[3]

298

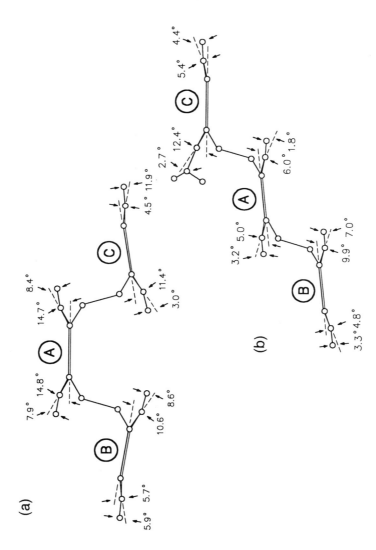

Figure 4. Side view of the triply layered 8-methyl[2.2][2.2]metacyclophane: (a) *up/down* isomer **14** (angle AB = 16.3°, angle AC = 22.5°), (b) *up/up* isomer **15** (angle AB = 3.1°, angle AC = 12.1°)[4]

6.2.2.2 Skeletal structure

The detailed stereochemistry of the *up/down* isomer **9** was determined by X-ray crystallography (Figure 3). It turns out that all three benzene rings are distorted into boat forms, with larger bond angles in the central ring. The central planes of the benzene rings partially stacked over one another are essentially parallel. In contrast the three benzene rings in the methyl-substituted compounds **14** and **15** are deflected from the parallel stacking arrangement because of the repulsive interaction of the methyl group (Figure 4). The pronounced deflection in the *up/down* isomer **14** is attributed to the strong repulsion between the intra-annular hydrogen atom H_a and the methyl group, and this interpretation is supported by NOE experiments.

6.2.3 CHEMICAL REACTIONS

6.2.3.1 Conformational and configurational isomerism

In contrast to the high conformational stability of [2.2]metacyclophane (energy barrier for ring inversion = 138 kJ/mol), the thermal interconversion between conformers of multi-layered metacyclophanes is relatively easy.[1] As indicated in Figure 1, the *up/up* isomer **10** in toluene-d_8 furnishes a 1:1 mixture of **9** and **10**; after being heated at 100°C for 3–4 minutes, it is then converted quantitatively to **9** within 16–18 minutes. Under the same conditions the *up/up* isomeric methylcyclophane **15** affords an equilibrium mixture of **14** and **15** (ratio 17:1), which is obtained by similar treatment of **14**. In an analogous way the quadruply layered isomeric **20** and **21** isomerize thermally under similar conditions to the most stable *up/up/down* isomer **21**.[5]

An interesting skeletal photoisomerization is observed for the triply layered metacyclophane **24**. Irradiation with a medium-pressure mercury lamp leads to a new triply layered metaparacyclophane **27**, which may arise by way of a benzvalene intermediate **26** (Figure 5).[6]

6.2.3.2 Stability of 9 and 10

From NMR investigations of the thermal isomerization of the triply layered metacyclophanes, the *up/down* isomer **9** and the *up/up* isomer **10**, it was concluded that **9** is more stable than **10** by at least 16 kJ/mol. Inasmuch as the two conformers **9** and **10** differ essentially only in the boat or chair deformation of the middle benzene ring, as can be seen in Figures 3 and 4, the difference in their stability is ascribed to the nature of deformation of this central benzene ring. In order to explore this more closely, two types of distortion of the benzene ring into boat and chair forms was tested theoretically with SCF-MO calculations by the MINDO/2 method.[7] The result was that deformation to a boat conformation should proceed more easily with less loss of resonance energy

300

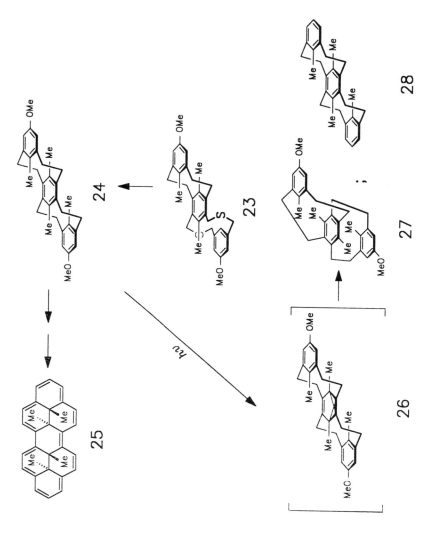

Figure 5. Skeletal photoisomerization of the triply layered phane **24**

than to a chair conformation. From this it was concluded that loss of resonance energy in the middle ring is primarily responsible for the difference in thermal stability of the conformers **9** and **10**.

6.2.3.3 Transannular reactions to pyrenophanes

Transannular formation of tetrahydropyrene (**30**) from [2.2]metacyclophane has been known for some time. It can proceed by elimination of a proton and HX from an intermediate cation **29** or from the corresponding *ipso*-substituted cation **31**; or via a one-electron transfer mechanism (see Section 2.2). Analogous transannular reactions and subsequent dehydrogenations in multi-layered

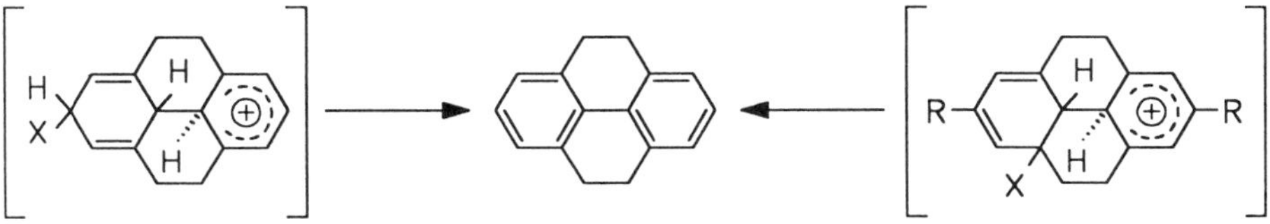

29 **30** **31**

[2.2]metacyclophanes lead to a series of 'condensed' aromatic cyclophanes and polycondensed hydrocarbons, as is represented in Figures 6 and 7.[8]

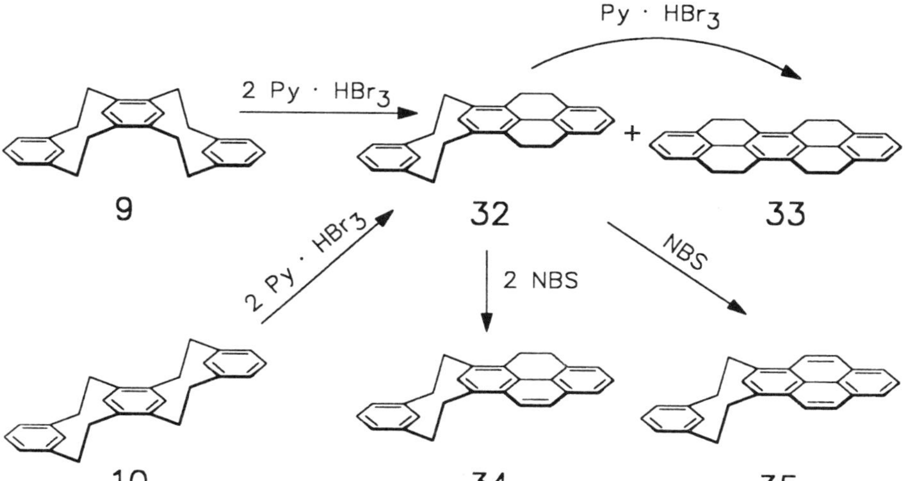

Py · HBr$_3$

2 Py · HBr$_3$

9 **32** + **33**

2 Py · HBr$_3$

2 NBS

NBS

10 **34** **35**

The fact that **40** shows no tendency to isomerize to **42** stands in sharp contrast to the rapid rearrangement of **10** to **9**. This can be interpreted to mean that the distribution of molecular strain, which results from $\pi-\pi$ repulsion, can be

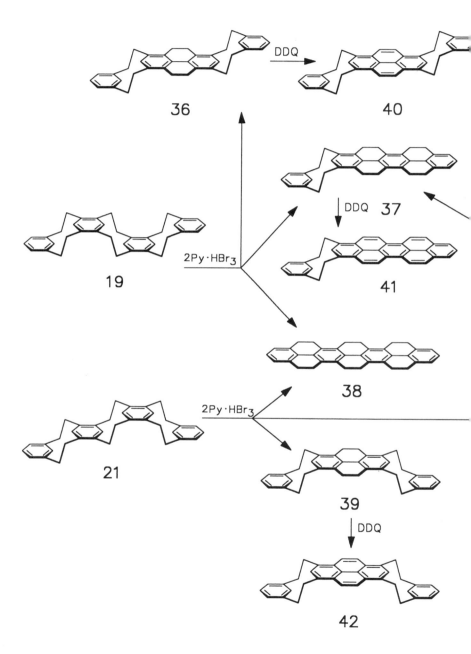

Figure 6. Formation of condensed hydrocarbons and pyrenophanes from triply and quadruply layered [2.2]metacyclophanes[5]

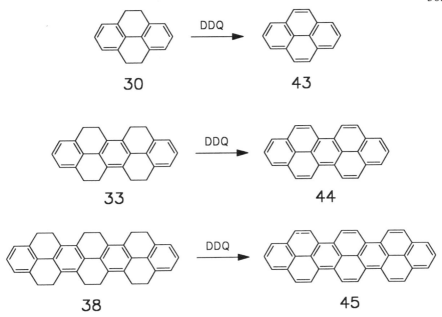

Figure 7. Formation of condensed aromatics by transannular cyclization from doubly to quadruply layered [2.2]metacyclophanes[5]

spread over a large pyrene ring; alternatively, it may be that the loss of resonance energy in the chair- and boat-deformed pyrene rings in **40** and **42** is essentially identical.

As shown in Figure 7, the compounds **30**, **33**, and **38**, formed through transannular reactions, can be dehydrogenated with DDQ, such that pyrene (**43**), peropyrene (**44**), and teropyrene (**45**) are produced in quantitative yield. This provides a convenient synthetic route for pyrene-like condensed hydrocarbons.[9]

The pyrenophanes **46–51**, which differ in their overlapping of the two stacked pyrenes, are of interest because they allow a correlation between structure and fluorescence spectra of pyrene excimers. They have been prepared by a combination of transannular reactions/dehydrogenations and by direct photodesulphurization of the corresponding dithiacyclophanes.[10]

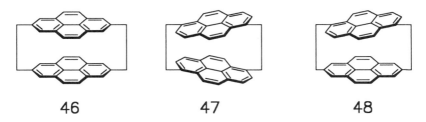

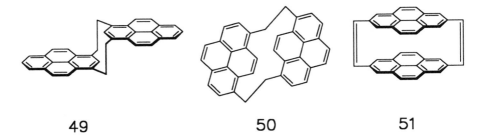

49   50   51

References to Section 6.2

1. T. Umemoto, T. Otsubo and S. Misumi, *Tetrahedron Lett.*, 1573 (1974).
2. H. Lehner, *Monatsh. Chem.*, **107**, 565 (1976).
3. Y. Kai, N. Yasuoka and N. Kasai, *Acta Crystallogr.*, **B33**, 754 (1977).
4. Y. Kai, F. Hama, N. Yasuoka and N. Kasai, *Acta Crystallogr.*, **B34**, 3422 (1978).
5. S. Misumi, in *Cyclophanes* (P. M. Keehn and S. M. Rosenfeld, Eds.) Vol. II, p. 573, Academic Press, New York, 1983.
6. D. Kamp and V. Boekelheide, *J. Org. Chem.*, **43**, 3470 (1978).
7. H. Iwamura, H. Kihara, S. Misumi, Y. Sakata and T. Umemoto, *Tetrahedron*, **34**, 3427 (1978).
8. T. Umemoto, T. Kawashima, Y. Sakata and S. Misumi, *Tetrahedron Lett.*, 463 (1975).
9. T. Umemoto, T. Kawashima, Y. Sakata and S. Misumi, *Tetrahedron Lett.*, 1005 (1975).
10. T. Kawashima, T. Otsubo, Y. Sakata and S. Misumi, *Tetrahedron Lett.*, 5115 (1978).

6.3 MULTI-LAYERED METAPARACYCLOPHANES

Attention was drawn to multi-layered metacyclophanes because of the special characteristics of the simple doubly layered [2.2]metaparacyclophane (**4**); namely, its unique molecular skeleton, its transannular electronic interactions, and the conformational flexibility of the *meta*-phenylene ring. It was expected that in multi-layered molecules of this type certain properties of [2.2]metaparacyclophane would be retained and that others might be more strongly manifested.

6.3.1 SYNTHESIS AND ISOMERIZATION

A considerable number of triply and quadruply layered metaparacyclophanes have been prepared by the usual methods of phane synthesis. Once again, synthesis of dithia[3.3]phanes and their desulphurization-ring contraction

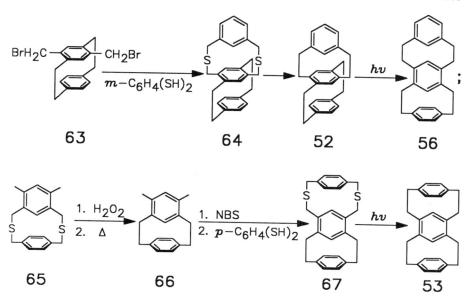

Figure 1. Synthetic route for triply layered metaparacyclophanes

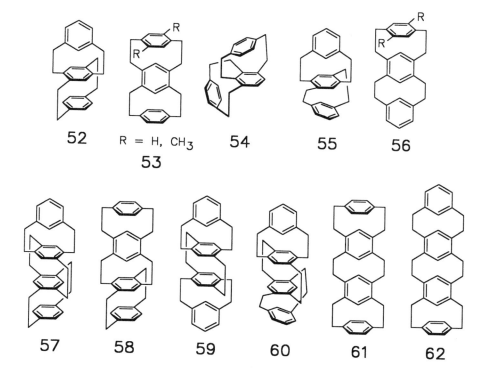

were of major importance. As an example, the synthetic route for some triply layered metaparacyclophanes is depicted in Figure 1.[1]

The unexpected formation of the byproduct **56** during photodesulphurization of **64** to **52** (already mentioned above) may involve a benzvalene intermediate **68** formed from **52** during the irradiation. This assumption is supported by the observation that longer irradiation of **64** results in an increase in **56** (23%) and a concomitant decrease in **52**. Moreover, irradiation of **52** in degassed cyclohexane affords the isomer **56**. A similar skeletal rearrangement is observed in the course of photodesulphurization of the dithiaphanes **69** and **70**,[2] which lead, respectively, to the quadruply layered metaparacyclophanes **61** and **62**; it is also involved, as already described, in the conversion of **24** to **27**.

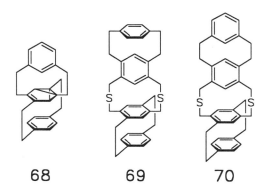

| **68** | **69** | **70** |

A fascinating photoisomerization was discovered during the irradiation of the quadruply layered dithia- and diselenametacyclophanes **71** (X = S and Se, respectively) with a high-pressure mercury lamp in benzene solution. The highly strained cage products **72** are formed, respectively, in 83 and 47% yields.[3] The polycyclenes **72** are stable at room temperature and rearrange back to **71** in THF;

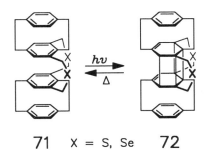

71   X = S, Se   **72**

that is, they exhibit photochromic properties. Where X = S, the yield is quantitative at 90°C, with a half-life of 30 minutes; the yield at 60°C for X = Se is 70%, with a half-life of 25 minutes.[4]

6.3.2 PROPERTIES

6.3.2.1 Skeletal structure and magnetic resonance

The two triply layered metaparacyclophanes **52** and **55** show an ¹H-NMR spectrum pattern similar to that of [2.2]metaparacyclophane itself, with the exception of the high-field shifts of some aromatic protons due to the anisotropy effect of the third benzene ring. The energy barrier for flipping of the *meta* ring has been determined as 75 and 80 kJ/mol for **52** ($T_c = 100°C$) and for **55** ($T_c = 116°C$), respectively; that is, both lie somewhat lower than that for [2.2]metaparacyclophane (88 kJ/mol). The reason for this lowering in energy barrier is understood as arising from the twisting of the central benzene ring because of the bridging attachment of the two *meta*-phenylene rings.

In the case of the cyclophanes **53** and **56** two conformers are possible, the two being depicted for **53** (R = H) as follows.

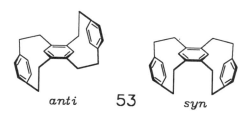

anti   **53**   *syn*

Because **56** (R = H) shows no temperature dependence in the ¹H-NMR spectrum, it is concluded that it consists of a single conformer. In fact, the *syn* orientation, corresponding to the *syn* **53** conformer (except that one of the outer benzene rings is *meta*-bridged), has been confirmed by X-ray crystal structure analysis (Figure 2). The inner benzene ring is bent into a boat form, similar to that in the *up/down* conformer **9** (see above). The corresponding dimethyl compound **56** (R = CH₃) consists of two conformers in a ratio of 1:1, as deduced from NMR spectroscopy. This indicates that the methyl groups render the *syn* and *anti* conformers of equal stability, because they reduce the stability of the *syn* conformer relative to its parent compound **56** (R = H). In contrast to the *syn* structure for **56** (R = H), the *anti* conformer appears likely from NMR spectroscopy. The X-ray crystal structure confirmed this conclusion and also

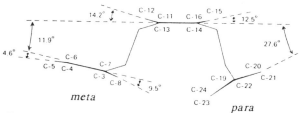

Figure 2. Side view of the triply layered metaparacyclophane *syn*-**56** (R = H)⁴

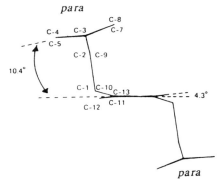

Figure 3. Side view of the triply layered [2.2]metaparacyclophane **53** (R = H)[4]

indicated a partial overlapping of the three benzene rings, as can be seen in Figure 3. A conformational flipping is observed for **53** at higher temperature. Two types of aromatic protons in the two *para*-substituted rings coalesce at 85°C and appear at higher temperature (140°C) as a singlet at the centre of their original position. This indicates a folding of the two *anti* conformations by way of a double flipping of the two *para*-phenylene rings.[5]

Comparison of the NMR spectra for triply and quadruply layered compounds **61** and **62** indicates that the latter two exist in a zigzag conformation, similar to that of the quadruply layered metacyclophane conformer **19**, whereas the hydrocarbon **58** contains as an 'inner unit' the *anti* form **53**.

References to Section 6.3

1. N. Kannen, T. Otsubo, Y. Sakata and S. Misumi, *Bull. Chem. Soc. Jpn.*, **49**, 3307 (1976).
2. N. Kannen, T. Otsubo and S. Misumi, *Bull. Chem. Soc. Jpn.*, **49**, 3208 (1976).
3. H. Higuchi, K. Takatsu, T. Otsubo, Y. Sakata and S. Misumi, *Tetrahedron Lett.*, **23**, 671 (1982).
4. S. Misumi, in *Cyclophanes* (P. M. Keehn and S. M. Rosenfeld, Eds.) Vol. II, p. 573, Academic Press, New York, 1983.
5. N. Kannen, T. Otsubo, Y. Sakata and S. Misumi, *Bull. Chem. Soc. Jpn.*, **49**, 3203 (1976).

6.4 ADDITIONAL MULTI-LAYERED PHANES

6.4.1 HETEROPHANES

In the case of most multi-layered heterophanes one or more of the benzene rings is replaced by furane, thiophene, or pyridine units. These are naturally of interest for comparing transannular electronic effects and steric strain with those in the parent hydrocarbon compounds already discussed above. Heterophanes **8–11**, **13**, and **14** prepared for this purpose were then also compared with the parent doubly layered reference compounds. The synthetic method of choice is once again the Hofmann elimination, in particular crossed reactions of this type from the ammonium hydroxides **12**.[1]

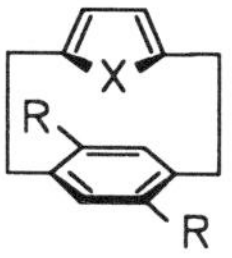

6 : X=O
7 : X=S

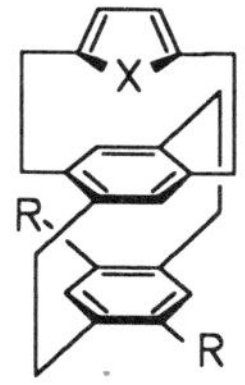

8 : X=O
9 : X=S

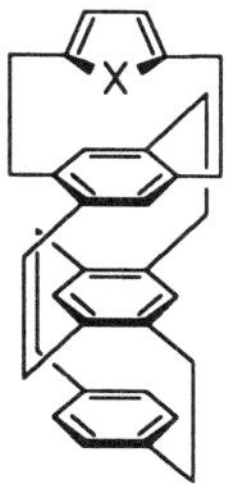

10 : X=O
11 : X=S

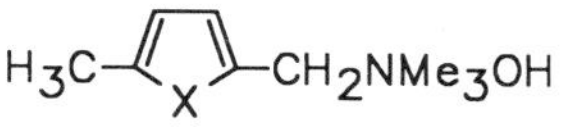

H_3C—CH_2NMe_3OH (X)

12 : X=O
      X=S

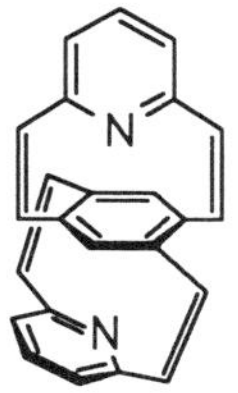

13 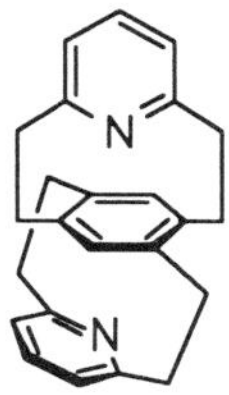   **14**

The characteristic feature in the NMR spectra of multi-layered heterophanes is the shift to higher field of all aromatic protons, although the extent of the shift is relatively small. It is also to be noted that the aromatic protons in the benzene rings situated 'face-to-face' to the heteroaromatic are equivalent in the furane series but not in the thiophene series. This can be attributed to the rigidity of the thiophene ring and the flexibility in the furane series, as confirmed by variable temperature NMR spectra[1] and X-ray crystallography for **8** (R = H) and **9** (R = CH₃) (Figure 1).[2]

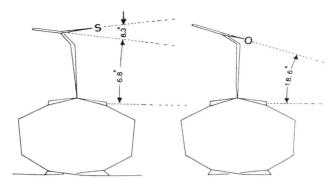

Figure 1. Side view of triply layered paracycloheterophanes **8** (R = H) and **9** (R = CH₃). (Compare the angles between the heteroaromatic and the neighbouring benzene ring)

The electronic spectra for both series of multi-layered paracycloheterophanes show similar characteristics. The typical absorptions of the heteroaromatic rings gradually disappear with an increase in the number of layers. The absorption curves of the thiophene compounds show a distinct similarity to those of the multi-layered paracyclophanes, and the spectra in the furane series are remarkably similar to those of the corresponding metaparacyclophanes. These spectral properties may be the consequence of similar stacking of the aromatic layers in the molecules—their 'stacking mode'. In the case of the pyridinophane **13** experiments were conducted to determine whether a direct interaction between the two pyridine rings, mediated by the π–electron cloud of the inner benzene ring, were possible. The compound shows very low basicity in its conversion to a monoprotonated species. The N–N distance, as determined by X-ray crystallography, is relatively short (504 pm).[3] The NMR spectrum of the triply layered pyridinophane **14** is remarkably similar to that of the simple [2.2](2,6)pyridinoparacyclophane.

6.4.2 DONOR–ACCEPTOR PHANES

Allusion has already been made to this group of compounds in Section 2.11.1.[4–9] A few of the most important representative examples are reproduced here (with the compound numbers as assigned in Chapter 2). Synthesis has already been outlined in Section 2.11.1.1.

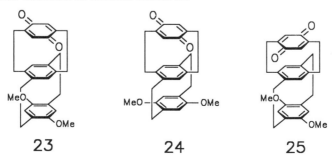

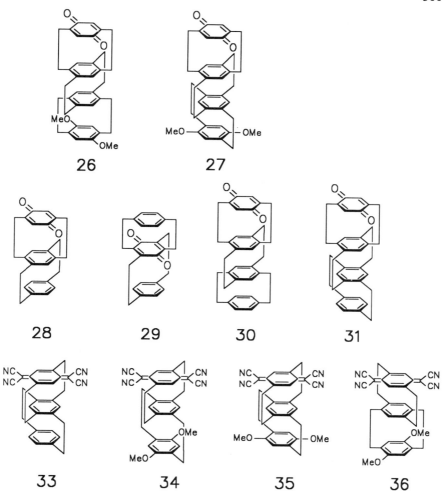

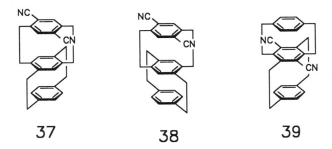

37 38 39

6.4.3 PHANES CONTAINING ANTHRACENE RINGS AND DIACETYLENE GROUPS

The incorporation of anthracene units and triple bonds (diacetylene as chromophore) into phanes has also provided a means of examining electronic interactions. It would be expected, for example, that transannular electronic interaction in the triply layered anthracenophane **6** would be more pronounced than in the isomeric anthracenophane **7**, both, in turn, being compared to the electronic spectra of the corresponding pair of doubly layered anthracenophane compounds. Syntheses of the anthracenophanes **6** and **7** were attempted through crossed pyrolyses of the corresponding quaternary ammonium hydroxides.[10] What resulted instead of **6**, however, was the unexpected product **8** (R = H, CH₃), whose structure has been confirmed not only by UV and ¹H-NMR spectroscopy but also by X-ray crystal structure analysis.[11] Formation of **8** is the

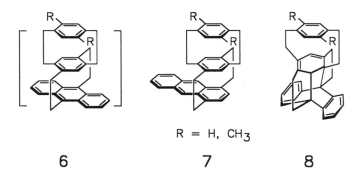

R = H, CH₃

6 7 8

outcome of an intramolecular Diels–Alder reaction of the highly strained inner benzene ring in **6** with the anthracene ring. This unusual cycloaddition appears to be the first example where a benzene ring reacts as dienophile under conventional reaction conditions. Presumably the driving force for the reaction is relief of strain.

The triply layered paracyclophandiyne **11** results from an intramolecular oxidative coupling of the diethynyl compound **10**, which was prepared in an eight-step sequence from dimethyl[2.2]paracyclophane (**9**). As shown by the

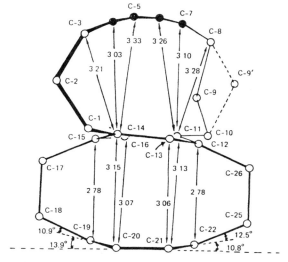

X-ray crystal structure analysis (Figure 2), the diacetylene bonds are bent into a bow shape, and the two benzene rings exist in a boat and twist form. This indicates that strong electronic repulsions are occurring between the three π-systems.

Investigation of the electron spectrum of the **11**-TCNE complex revealed that the originally blue charge-transfer complex rapidly lost its colour at room temperature and was transformed into a colourless solid material. This was

Figure 2. X-Ray crystal structure of paracyclophandiyne **11**[12] (distances in pm). Ethyne carbons in black

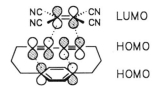

Figure 3. Orbital representation of the interaction between the three π-systems for the 'multi-cycloaddition' of paracyclophandiynes of type **12** with TCNE[14]

characterized as the 1:1 cycloadduct by mass, IR, and NMR spectroscopy.[13] The other paracyclophandiynes **12** were converted to 1:1 cycloadducts in high yield but at higher temperatures. These reactions are interpreted in terms of MO theory as a 'three π-system cycloaddition', that is, the HOMO of a benzene ring, the HOMO of a diacetylene, and the LUMO of TCNE all interact in a concerted process (Figure 3).

6.4.4 MULTI-LAYERED HETERAPHANES

Triply layered phanes of type **18**, with hetero substitutents in the bridges and therefore helical chiral, became available in 1990, once the synthetic methodology

20

19 **a** : X = O, Y=S   **18**
b : X = NTos, Y = S
c : X = NTos, Y = SO$_2$
d : X, Y = CH$_2$

had been optimized.[16] Cyclization of **16b** with **17** affords exclusively the *'up/down'* isomers B of **18a** and **18b** in yields of 5% and 35%, respectively. In the case of the hydrocarbon **18d**, the *up/up* isomer is obtained as well (Misumi, Section 6.2).[17] An X-ray crystal structure for **18b** has been done. Although the circular dichroism curves of **18a** and **18b** strongly resemble those of the doubly layered analogues, they are shifted to longer wavelengths. This may reflect a minimal interaction of π-electrons in the [2.2]metacyclophane hydrocarbon portion of the structure in **18a,b**, which indicates that the third layer exerts little influence on the composite Cotton effect.

A   B

'up − up'   *'up − down'*

316

References to Section 6.4

1. T. Otsubo, S. Mizogami, N. Osaka, Y. Sakata and S. Misumi, *Bull. Chem. Soc. Jpn.*, **50**, 1841 (1977).
   (a) Y. Murakami *et al.*, *Tetrahedron Lett.*, **31**, 1027 (1990).
2. Y. Kai, J. Watanabe, N. Yasuoka and N. Kasai, *Acta Crystallogr.*, **B36**, 2276 (1980).
3. A. W. Hanson, *Acta Crystallogr.*, **B33**, 2657 (1977).
4. H. A. Staab, U. Zapf and A. Gurke, *Angew. Chem.*, **89**, 841 (1977); *Angew. Chem. Int. Ed. Engl.*, **16**, 801 (1977); H. A. Staab and U. Zapf, *Angew. Chem.*, **90**, 807 (1978); *Angew. Chem. Int. Ed. Engl.*, **17**, 757 (1978); H. Machida, H. Tatemitsu, T. Otsubo, Y. Sakata and S. Misumi, *Bull. Chem. Soc. Jpn.*, **53**, 2943 (1980); X-ray structure of triyne: H. Irngartinger, L. Leiserowitz and G. M. J. Schmidt, *Chem. Ber.*, **103**, 1119 (1970).
5. H. A. Staab and U. Zapf, *Angew. Chem.*, **90**, 807 (1978); *Angew. Chem. Int. Ed. Engl.*, **17**, 757 (1978).
6. S. Misumi, in *Cyclophanes* (P. M. Keehn and S. M. Rosenfeld, Eds.) Vol. II, p. 573, Academic Press, New York, 1983.
7. H. Tatemitsu, B. Natsume, M. Yoshida, Y. Sakata and S. Misumi, *Tetrahedron Lett.*, 3459 (1978); M. Yoshida, Y. Tochiaki, H. Tatemitsu, Y. Sakata and S. Misumi, *Chem. Lett.*, 829 (1978).
   (a) Compare intra-annular substituted [2.2.2]phanes: M. Tashiro *et al.*, *J. Chem. Soc., Chem. Commun.*, 1066 (1990).
8. H. A. Staab and H.-E. Henke, *Tetrahedron Lett.*, 1955 (1978).
9. H. Masuhara, N. Mataga, M. Yoshida, H. Tatemitsu, Y. Sakata and S. Misumi, *J. Phys. Chem.*, **81**, 879 (1977).
10. A. Iwama, T. Toyoda, M. Yoshida, T. Otsubo, Y. Sakata and S. Misumi, *Bull. Chem. Soc. Jpn.*, **51**, 2988 (1978).
11. T. Toyoda, A. Iwama, Y. Sakata and S. Misumi, *Tetrahedron Lett.*, 3203 (1975); T. Toyoda, A. Iwama, T. Otsubo and S. Misumi, *Bull. Chem. Soc. Jpn.*, **49**, 3300 (1976).
12. T. Toyoda, Doctoral Dissertation, Univ. Osaka, 1976.
13. T. Kaneda, T. Ogawa and S. Misumi, *Tetrahedron Lett.*, 3373 (1973).
14. S. Inagaki, H. Fujimoto and K. Fukui, *J. Am. Chem. Soc.*, **98**, 4693 (1976).
15. F. Vögtle, A. Ostrowicki, B. Begemann, M. Jansen, M. Nieger and E. Niecke, *Chem. Ber.*, **123**, 169 (1990).
16. A. Ostrowicki and F. Vögtle, *Synthesis*, 1003 (1988).
17. Cf. T. Otsubo, D. Stusche and V. Boekelheide, *J. Org. Chem.*, **43**, 3466 (1978); H. Keller, Ch. Krieger, E. Langer and H. Lehner, *Monatsh. Chem.*, **107**, 1281 (1976).

7 [m_n](m,n)Phanes

7.1 [2_n]PHANES

In order to make a visual distinction between the multiply bridged [2_n]phanes in Chapter 5, the phanes in this chapter are designated with an italicized subscript *n*.*

7.1.1 [2_n]ORTHOCYCLOPHANES

7.1.1.1 Synthesis

As early as 1945 Baker obtained [2.2]orthocyclophane [1,2,5,6-dibenzocyclo-octadiene(**6**)] in 6% yield by cyclization of 1,2-bis(bromomethyl)benzene with an excess of sodium suspended in boiling dioxane (see Section 2.1). He also showed evidence for the presence of the next higher 'oligomer', [2.2.2]ortho-cyclophane (**7**) and open-chain byproducts.[1] Working at relatively high concentrations, Müller and Röscheisen could isolate up to 35% of **7**, along with 40% of **6**.[2] The 'tetramer' [2.2.2.2]orthocyclophane (**8**) was prepared in 40% yield by Bergmann in 1953 by treatment of 1,2-bis[(2-bromomethyl)phenyl]-ethane with phenyl lithium.[3] Among the series of [2.2.2]orthocyclophanes

| 6 | 7 | 8 |

should be mentioned the triyne **9** synthesized by Staab.[4] In 1989 its X-ray crystal structure was reported,[4a] and it was converted to a trigonal planar 'tribenzocyclyne' Ni(0) complex (**10**).[5]

---

*For some triaza [3_2]benzenophanes see, for example, reference 1a.

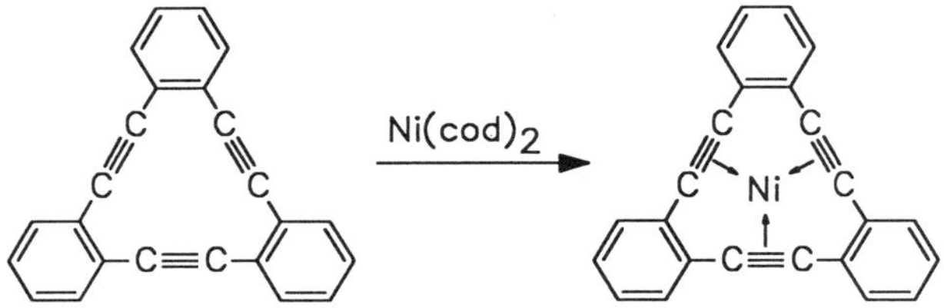

9   **10**

7.1.2 [2_n])METACYCLOPHANES AND ANALOGUES

7.1.2.1 Synthesis

Although Pellegrin had been able to obtain [2.2]metacyclophane (**2**) in 1899
by a Wurtz reaction with 1,3-bis(bromomethyl)benzene (**11**) and excess of sodium
and bromobenzene in anhydrous ether,[6] the higher 'oligomeric' [2_n]metacyclo-
phanes (**12–19**) were only first isolated in 1966 by Jenny *et al.*[7] In this synthesis
1,2-bis(bromomethyl)benzene was cyclized according to Müller–Röscheisen with
sodium tetraphenylethene in THF at − 80°C. From the reaction mixture,
painstakingly chromatographed on Al$_2$O$_3$, could be isolated and characterized
the full series of oligomeric [2_n]metacyclophanes up to the 50-membered
[2_{10}]metacyclophane (**19**). The success of this procedure suggested that
other variations of the Wurtz reaction for [2_n]metacylcophanes were less

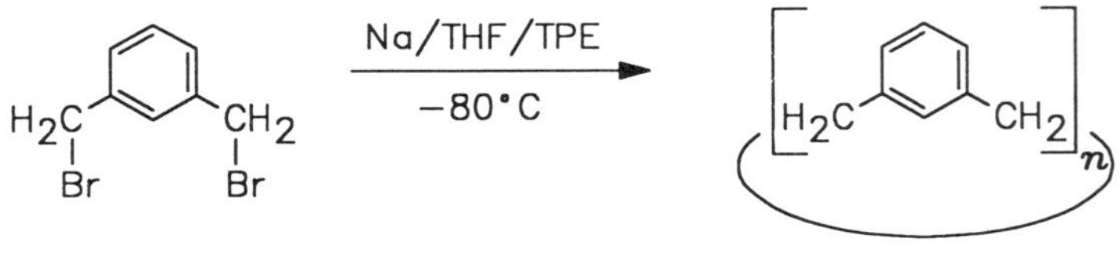

11   **2** : $n = 2$

                                        12–19 : $n = 3$–10

convenient.[7a]. The yields of the various [2_n] metacyclophanes from a typical
reaction mixture are presented in Table 1.[7b,c]

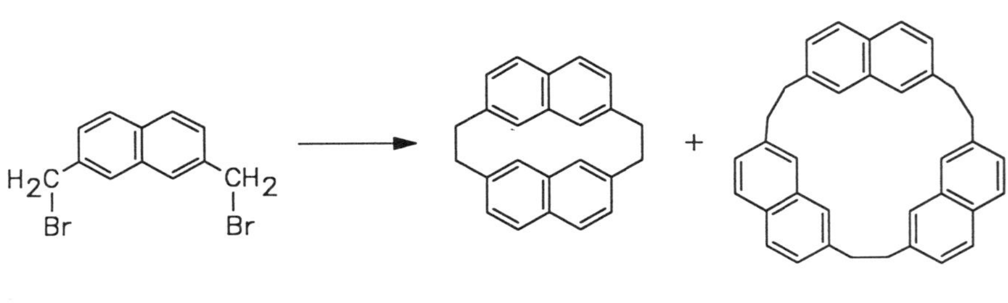

20   **21**   **22**

Table 1. $[2_n]$Metacyclophanes prepared by Jenny *et al.* according to Müller–Röscheisen[7]

n	$[2_n]$Metacyclophane Compound number	% Composition of crude product	Effective yield (%) of purified compound
2	2	44	33
3	12	16	7.5
4	13	13	1.7
5	14	8	5.1
6	15	5.5	3.1
7	16	≈1	0.5
8	17		0.6
9	18		0.5
10	19		0.3

[2.2.2](2,7)Naphthalenophane (**22**) was synthesized in 2% yield by Griffin and Orr by the Wurtz reaction with 2,7-bis(bromomethyl)naphthalene (**20**).[8] Kauffmann *et al.* obtained [2.2.2.2](2,6)pyridinophane (**27**) in 4% yield by way of the bis copper compound **25**.[9]

7.1.2.2 Properties

It is interesting that the higher $[2_n]$metacyclophanes are conveniently drawn on paper in conformations with the benzene rings oriented toward the inside. Even if these conformations may not be the most stable arrangements, the unusual molecular geometry nevertheless suggests the possible construction of hollow cavity or calixarene-type structures. Jenny has proposed conformation A as most likely for [2.2.2.2.2.2.2.2]metacyclophane {**17**; $[2_8]$metacyclophane}. In particular, the fragmentation in the mass spectrum of this octamer **17** can be rationalized on the basis of transannular reactions in the compact, extended conformation (see below).

14   15

17A

The *melting points* of the $[2_n]$metacyclophanes alternate as the ring size increases.

(a) When $n = 2,4,6$, the melting points are shifted only slightly to lower values.
(b) In the odd series, $n = 3,5,7$, the values are significantly lower and are shifted significantly lower as n increases.
(c) The change from $n = 6$ to $n = 8$ is accompanied by a sharp decline in the melting point (n = number of m-xylylene units; see Table 1).

Some examples can be cited. Whereas [2.2]metacyclophane (2) melts at 135°C, the melting points of trimer (12) to octamer (17) are, respectively, 117, 133, 96, 129, and 102°C (heptamer is omitted). From this melting point pattern Jenny concluded, among other things, that the compounds with an odd number of rings are not completely strain free, a conclusion that is also consistent with the yields.[10]

The *electronic spectra* of the oligomeric $[2_n]$metacyclophanes show complete correlation with open-chain model compounds, with none of the peculiarities of the UV spectrum of [2.2]metacyclophane.

NMR spectra[10]

The AB$_2$C spectrum in the aromatic region of [2.2]metacyclophane is also observed for the higher $[2_n]$metacyclophanes, whereas 1,3-diethylbenzene and open-chain reference compounds typically show unresolved multiplets centred around $\delta = 7$ p.p.m. Thus a well resolved AB$_2$C spectrum can serve as evidence for the presence of a ring structure in the $[2_n]$phane series. The aliphatic protons, which appear as an AA'BB' system in [2.2]metacyclophane, become a singlet in [2.2.2]metacyclophane; and it shows no splitting even at $-60°C$. This suggests that the aliphatic protons in the higher oligomers are highly mobile. The high-field shift of the inner protons (H_i) in [2.2]metacyclophane decreases regularly with an increase in ring size. Nevertheless, the shift is still discernible in [2_6]- and [2_8]metacyclophane, because in the AB_2C spectra the A protons absorb at lowest field, the C protons (H_i) at highest field, and the B protons in between. This shielding effect in the large rings is in itself noteworthy. One explanation is the assumption that the benzene rings are oriented out of the plane of the ring or are situated perpendicular to the plane, such that the A protons are directed alternately above and below (see the proposed conformation A for [2_8]metacyclophane above). The spectrum of [2_8]metacyclophane (17) differs from that of [2_6]metacyclophane (15) in that all aromatic proton signals are shifted about 0.1 p.p.m. toward higher field. This likewise supports the stretched conformation A, as do the melting point behaviour and the mass spectrum.

Mass spectrometry

The mass spectra of the [2_n]metacyclophanes are distinctive from the following regular patterns.

(a) The [2_n]metacyclophanes persistently produce an $M^{+\cdot}$ radical ion whose intensity exceeds all other fragmentations. Thus the determination of the molecular mass is unequivocal.

(b) All ion fragments appear in groups of three with masses of $M-(91+104x)$, $M-(105+104x)$, and $M-(119+104x)$, where $x=0, 1, 2 \ldots$ The middle peak in this group of three is always the most intense.

(c) In the lower mass region are found fragments which correspond to a normal decay mechanism for a fragmentation process. These fragments appear at mass numbers $91+104x$, $105+104x$, and $119+104x$. These fragments are less useful, because they only appear for the series up to $x=2$.

The mass spectrum of [2_8]metacyclophane resembles that of the other [2_n]cyclophanes when the inlet temperature exceeds 400°C. Quite surprisingly, however, the spectrum is an exact duplicate of that of [2.2]metacyclophane at a temperature of about 180°C, as noted from the characteristic fragments at 28 (C_2H_4) and $M-1$(H radical). One can conclude from this that the sample was contaminated with [2.2]metacyclophane itself. Conceivably the formation of [2.2]metacyclophane (2) from [2_8]metacyclophane can be traced to the existence of conformations such as 17A, in which transannular ring-closure reactions could take place.[10]

7.1.3 [2_n]PARACYCLOPHANES

Treatment of 1,4-bis(bromomethyl)benzene (28) with sodium/sodium iodide in refluxing dioxane in the absence of bromobenzene affords only

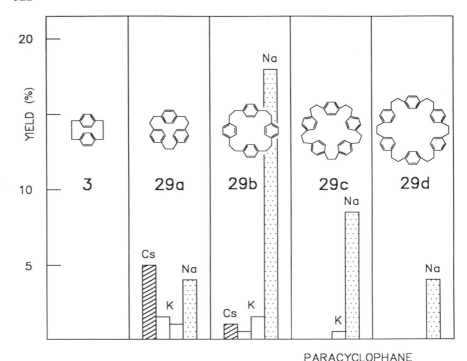

Figure 1. Oligomer selectivity in the Wurtz reaction with varying alkali metals, in particular with Cs. Dependence of yield of the $[2_n]$paracyclophanes **29a–d** on the alkali metal[12]

[2.2.2]paracyclophane (**29a**), along with open-chain products, but no [2.2]paracyclophane (**3**).[11] Vögtle and Kißener noted an oligomer selectivity in the Wurtz reaction with varying alkali metals. Use of Cs metal leads to the highest yield of $[2_3]$paracyclophane (**29a**), as shown in Figure 1.[12]

$[2_4]$Paracyclophantetraene (**30**) was first prepared by Wennerström *et al.* in 1975.[13] The macrocycle is obtained in a simple one-pot Wittig reaction, whereupon four C=C double bonds are created without the need of high-dilution

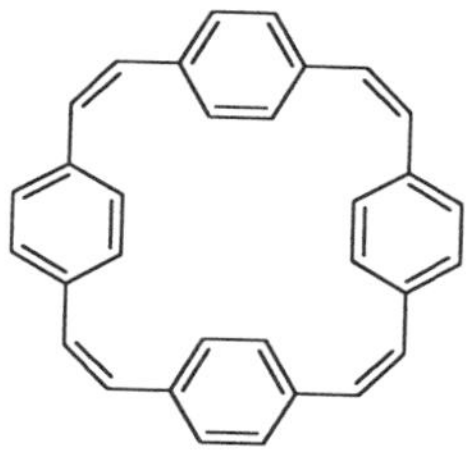

30

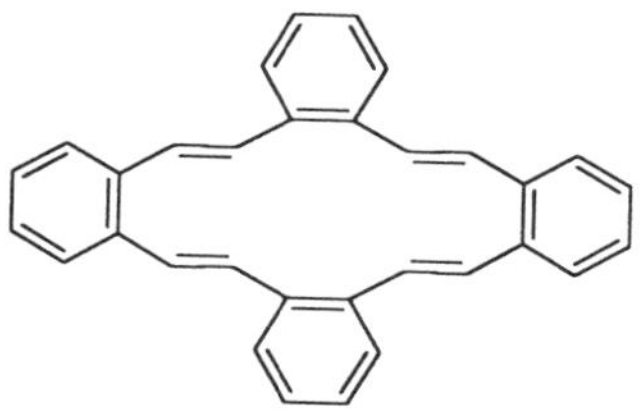

31

techniques. As the formulas **31–39** indicate, this route is generally applicable to the synthesis of o-, m-, and p-bridged macrocyclic polyene hydrocarbons. The methodology is illustrated for the synthesis of the thiopheno macrocycle **36**, which is obtained in 1.6% yield in a one-pot-reaction.[14]

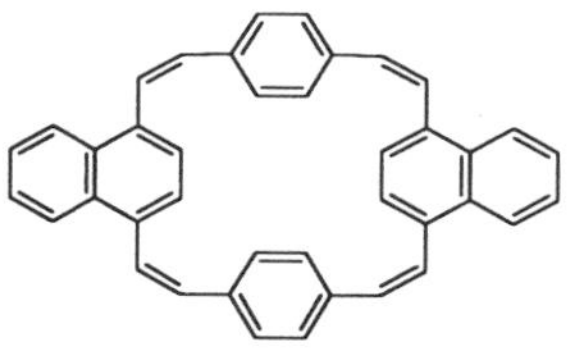

32

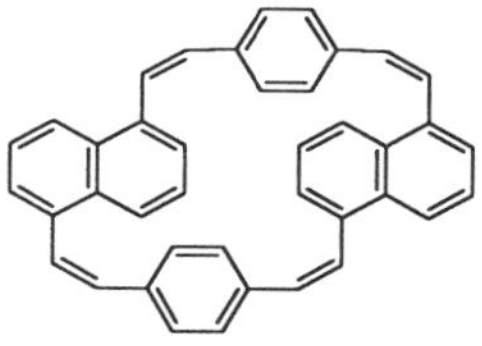

33

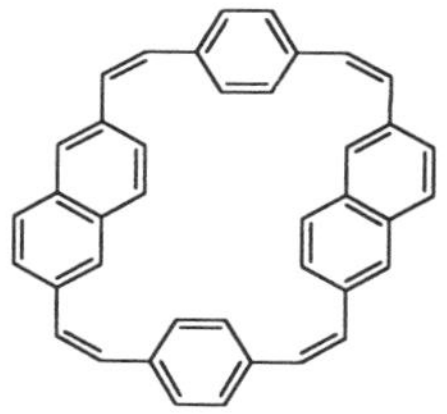

34

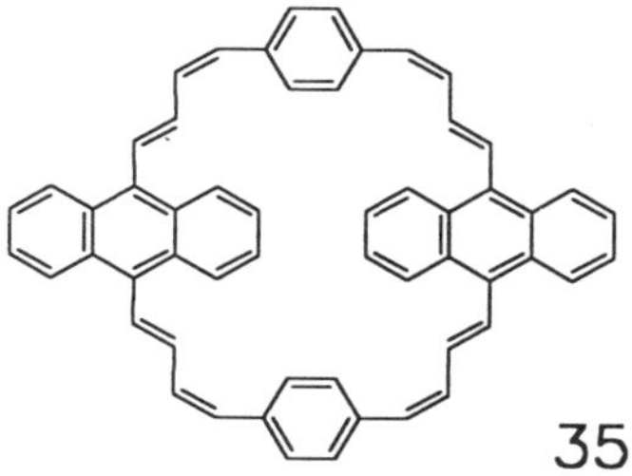

35

Many of the semi-rigid macrocycles with saturated and unsaturated bridges, such as **37–39**, are characterized by their conformational flexibility or stable conformers, which are often determined by temperature-dependent NMR spectroscopy. Some macrocycles containing C=C bonds can be converted through electron transfer reactions with alkali metals into other aromatic compounds or into anionic and dianionic structures; this behaviour has aided in conformational analysis and an understanding of aromaticity in large rings and benzannulenes.[15]

36

A   **37**   B

38   **39**

Recently the first tolanophanes have been prepared:[15a,b]

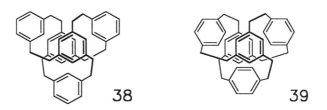

$m = n = 2$
$m = 2, n = 1$

7.1.4 [2$_n$](2,7)PHENANTHRENOPHANES

Hexahydro[2.2.2](2,7)phenanthrenophane (**41a**) and decahydro[2$_5$](2,7)phen-anthrenophane (**41b**) were prepared, along with open-chain products, by the reaction of 2,7-bis(chloromethyl)-9,10-dihydrophenanthrene (**40**) with sodium/tetraphenylethylene in THF. Yields of **41a** and **41b**, as determined by mass spectrometry, were 3.5% and 1.2%, respectively.[16]

7.1.5 TRI-o-THYMOTIDE (TOT)

Tri-o-thymotide (TOT, **43**)[17] was discovered by Spallino and Provençal by the action of concentrated phosphoric acid on o-thymotinic acid (**42**).[18] Baker was the first to show that the product was indeed tri-o-thymotide (**43**) with a melting point of 174°C.[19] Upon recrystallization from n-hexane the clathrate (solvent-inclusion compound) with m.p. 174°C was formed. Only by prolonged heating at 160°C/1 Torr was the hexane removed, the resulting solvent-free tri-o-thymotide melting at 217°C. The crude TOT, usually containing residual di-o-thymotide, was first purified by extraction with acetone in a Soxhlet apparatus, and the resulting acetone clathrate was repeatedly recrystallized from methanol. This resulted in some formation of the clathrate with methanol whose removal was accomplished either by warming *in vacuo* or heating for 24 hours at 130°C. Finally it was recrystallized from isooctane.

42   43

A non-planar, propeller-like conformation has been postulated for TOT (43), a three-blade propeller, in which all three carbonyl oxygen atoms lie on the same side of the 12-membered ring. Aside from a threefold axis, the molecule possesses no element of symmetry. In pure, solvent-free form, TOT forms orthorhombic crystals, which are distinct from the trigonal crystals of the clathrates. Tri-*o*-thymotide appears to form no significant hydrogen bonds, the association of several molecules being affected only by van der Waals forces. This results in cage formation in the crystal lattice of TOT, in which guest molecules can reside, and an overall stabilization of the host–guest lattice. Although pure tri-*o*-thymotide (43) does not form a particularly stable crystal lattice, under certain conditions it provides an expansive host lattice with a cavity for guest molecules. Dimensions of the unit cell of TOT are variable because of the weak van der Waals forces, and so the host lattice can accommodate itself to guest molecules of varying sizes. Formation of the lattice cavity is strongly influenced by the interaction with the guest component.

TOT forms *inclusion compounds* (clathrates) with a variety of guest compounds such as, for example, alkanes, haloalkanes, alcohols, esters, ketones, and ethers.[17] In general the stoichiometric relationship is one guest molecule for two TOT molecules. The length of the guest molecule can apparently not exceed 950 pm; otherwise a channel structure results, such that the composition changes with the length of the guest molecule. Thus the number of guest molecules per host molecule decreases with increasing chain length.

Clathrates of TOT are prepared simply in the following way. One dissolves TOT in the liquid guest molecule, with warming if necessary. When the solution is cooled, the crystalline clathrate precipitates. In some instances the composition of the clathrate depends upon the rate of cooling. Solid guest molecules can be dissolved in 2,2,4-trimethylpentane or 2,3-dimethylpentane, because these hydrocarbons do not form clathrates with TOT. The methanol clathrate is isolated by adding a trace of acetone or some seed crystals of the acetone clathrate.

In its free state, prior to clathrate formation, TOT can exist in either a right- or left-handed screw form. It has not been possible to isolate the antipodes of

TOT by conventional physical chemical methods. The pioneer of clathrate chemistry, Powell,[20] however, succeeded in separating the unstable optical antipodes by means of inclusion compounds. While the methanol clathrate contains equal amounts of right- and left-handed helices, one finds that all six tri-*o*-thymotide molecules in the unit cell of the cyclohexane clathrate possess the same fixed geometry. The clathrate does not form as a simple racemate; rather, each crystal consists exclusively of either the right- or left-handed helix (spontaneous resolution *via* conglomerate formation). For separation of the right- and left-propeller helices, one grows single crystals of the clathrate, which sometimes can be so large that they weigh nearly a gram. When such a homogeneous single crystal is dissolved in chloroform, for example, one can confirm the deflection of plane polarized light either to the right or left.

A second possibility for resolving or enriching TOT consists in forming clathrates of TOT with other racemic guest compounds. In this situation the D-form of the host is generally favoured, and the D-form of the guest component is 'clathrated'. If one adds only the D- or L-form of a guest component, then only the crystals of this form are obtained. In this way tri-*o*-thymotide has been separated into its enantiomers by means of D- or L-*sec*-butyl bromide. With enantiomerically pure TOT one can separate racemic *sec*-butyl bromide into its enantiomers.[20]

Racemization of the 'atropisomers' of tri-*o*-thymotide involves an energy barrier calculated to be 67 kJ/mol. This barrier, somewhat low for atropisomers, is attributed to the steric interaction between the isopropyl and carbonyl groups.

A summary of X-ray crystal structure analyses of TOT clathrates can be found in the literature.[21]

References to Section 7.1

1. W. Baker, R. Banks, D. R. Lyon and F. G. Mann, *J. Chem. Soc.*, 27 (1945).
   (a) Y. Murakami *et al.*, *Tetrahedron Lett.*, **31**, 1027 (1991).
2. E. Müller and G. Röscheisen, *Chem. Ber.*, **90**, 543 (1957).
3. E. D. Bergmann and Z. Pelchowicz, *J. Am. Chem. Soc.*, **75**, 4281 (1953).
4. H. A. Staab and F. Graf, *Chem. Ber.*, **103**, 1107 (1970).
   (a) H. Inrgartinger, L. Leiserowitz and G. M. J. Schmidt, *Chem. Ber.*, **103**, 1119 (1970).
5. J. D. Ferrara, A. A. Tanaka, C. Fierro, C. A. Tessier-Youngs and W. J. Youngs, *Organomet.*, **8**, 2089 (1989); Ag$^+$ complex: R. Dunbar *et al.*, *Organometallics*, **10**, 52 (1991); corresponding trithienocyclotriene: W. J. Youngs *et al.*, *Synlett*, **7**, 427 (1990).
6. M. Pellegrin, *Rec. Trav. Chim. Pays-Bas*, **18**, 457 (1899).
7. K. Burri and W. Jenny, *Helv. Chim. Acta*, **50**, 1978 (1967); *Chimia*, **20**, 403 (1966); W. Jenny and K. Burri, *Chimia*, **21**, 186 (1967); W. Jenny and R. Paioni, *Chimia*, **22**, 142 (1968).
   (a) Compare intra-annularly substituted [2.2.2]phanes: M. Tashiro *et al.*, *J. Chem. Soc., Chem. Commun.*, 1066 (1990).
   (b) [2$_n$]Phanes (homocalixarenes) substituted with OCH$_3$, OH, OCH$_2$CO$_2$H groups: M. Nieger, J. Schmitz and F. Vögtle, *Chem. Ber.*, **125**, 2523 (1992).
   (c) Samarium diiodide-mediated cyclization: S. Takahashi and N. Mori, *J. Chem. Soc., Perkin Trans. 1*, 2029 (1991).

328

8. R. W. Griffin and N. Orr, *Tetrahedron Lett.*, 4567 (1969).
9. Th. Kauffmann, G. Beißner, W. Sahm and A. Woltermann, *Angew. Chem.*, **82**, 815 (1970); *Angew. Chem. Int. Ed. Engl.*, **9**, 808 (1970); donor substituted [2$_n$]pyridinophanes: F. Vögtle, G. Brodesser, M. Nieger and K. Rissanen, *Rec. Trav. Chim. Pays-Bas*, **112**, 325 (1993).
10. K. Burri and W. Jenny, *Helv. Chim. Acta*, **50**, 1978 (1967).
11. W. Baker, J. F. W. McOmie and J. M. Norman, *J. Chem. Soc.*, 1114 (1951).
12. F. Vögtle and W. Kißener, *Chem. Ber.*, **117**, 1538 (1984). For the recently described cyclization method with Cr(0) see H. G. Wey and H. Butenschön, *Chem. Ber.*, **123**, 93 (1990).
   (a) Samarium diiodide-mediated cyclization: S. Takahashi and N. Mori, *J. Chem. Soc., Perkin Trans. 1*, 2029 (1991).
13. B. Thulin, O. Wennerström and H.-E. Högberg, *Acta Chem. Scand.*, **B29**, 138 (1975); see also reference 15.
14. W. Carruthers and M. G. Pellatt, *J. Chem. Soc., Perkin Trans. 1*, 1136 (1973).
15. D. Tanner, B. Thulin and O. Wennerström, *Acta Chem. Scand.*, **B33**, 443 (1979); D. Tanner and O. Wennerström, *Acta Chem. Scand.*, **B34**, 529 (1980); W. Huber, K. Müllen and O. Wennerström, *Angew. Chem.*, **92**, 636 (1980); *Angew. Chem. Int. Ed. Engl.*, **19**, 624 (1980); D. Tanner, Dissertation, Univ. Göteborg, Sweden, 1981.
   (a) H. A. Staab and K. Neunhoeffer, *Synthesis*, 424 (1974).
   (b) M. Bauer, M. Nieger and F. Vögtle, *Chem. Ber.*, **125**, 2533 (1992); [2$_n$]-Azobenzenophanes: N. Tamaoki *et al.*, *Tetrahedron*, **46**, 5931 (1990).
16. F. Vögtle and H. A. Staab, *Chem. Ber.*, **101**, 2709 (1968).
17. (a) Brief review: F. Vögtle, *Supramolecular Chemistry*, Wiley, Chichester, 1993.
   (b) Review: J. L. Atwood, J. E. D. Davies and D. MacNicol (Eds.), *Inclusion Compounds*, Vols. I–III, Wiley, New York, 1984.
18. R. Spallino and R. Provencal, *Gazz. Chim. Ital.*, **39 II**, 325 (1909).
19. W. Baker, B. Gilbert and W. D. Ollis, *J. Chem. Soc.*, 1443 (1952).
20. See, for example, A. C. D. Newman and H. M. Powell, *J. Chem. Soc.*, 3747 (1952); J. E. D. Davies, W. Kemula, H. M. Powell and N. O. Smith, *J. Incl. Phenom.*, **1**, 3 (1983).
21. P. M. Keehn, in *Cyclophanes* (P. M. Keehn and S. M. Rosenfeld, Eds.), Vol. I, p. 69, Academic Press, New York, 1983.

7.2 [1$_n$]PHANES

INTRODUCTION

The main emphasis in this section will be centred on calixarenes, which are becoming increasingly important. Others mentioned only in passing are the dihydropleiadenes (1),[1] cyclic *para*-phenylenesulphides (e.g. 2),[2] the X-ray crystal structure analysis of tetrathia[1.1.1.1]metacyclophane (3),[3] the

1

tetrameric '4,4'-biphenylsulphide' **4**,[2] the [1.1]quaterphenylophane **4a** and the [1.1]diphenyldiacetylenophane **4b**,[3a] the new thiophene analogue of porphyrin **5**[4] and [1.1.1]pyridinophanes.[5]

5a   5b

References to Section 7.2

Introduction

1. Review: P. T. Lansbury, *Acc. Chem. Res.*, **2**, 210; (1969); Cf. T. Kawase, Y. Ohnishi and M. Oda, *J. Chem. Soc., Chem. Commun.*, 702 (1991).
2. J. Franke and F. Vögtle, *Tetrahedron Lett.*, **25**, 3445 (1984). (a) Y. Miyahara, T. Inazu and T. Yoshino, *Tetrahedron Lett.*, **24**, 5277 (1983).
3. I. A. Zamaev, V. E. Shklover, V. I. Nedel'kin and V. A. Sergeyev, *Acta Crystallogr.*, **C45**, 1531 (1989). (a) F. Vögtle and K. Kadei, *Chem. Ber.*, **124**, 903 (1991); *Chem. Ber.*, **124**, 909 (1991).
4. E. Vogel, P. Röhrig, M. Sicken, B. Knipp, A. Herrmann, M. Pohl, H. Schmickler and J. Lex, *Angew. Chem.*, **101**, 1683 (1989); *Angew. Chem. Int. Ed. Engl.*, **28**, 1651 (1989); For the [1.1.1.1]paracyclophane hydrocarbon see Y. Miyahara, T. Inazu and T. Yoshino, *Tetrahedron Lett.*, **24**, 5277 (1983).
5. G. R. Newkome, G. R. Joo, D. W. Evans, F. R. Fronczek and G. R. Baker, *J. Org. Chem.*, **55**, 5714 (1990).

7.2.1 CALIXARENES

The calixarenes,[1] already known for several years, have recently attracted increased interest because they involve many-membered hydrocarbon rings that form a large cavity and also contain functional groups. On the basis of their constitution they belong to the $[1_n]$cyclophanes of type **6**. They were named by Gutsche 'calixarenes' (Gr. calix, chalice; arene: aromatic rings as components of the macrocycle) because of the chalice-like shape of their conformations.[1]

6   7

Increasing interest in calixarenes stems from their basket-like shape being adapted to host–guest or receptor–substrate chemistry and the fact that they are easily synthesized. Some members are in fact commercially available.

7.2.1.1 Nomenclature

The IUPAC nomenclature for even simple calixarenes without the characteristic substituents leads to very complicated names. 7, for example, is: pentacyclo-$[19.3.1.1^{3,7}.1^{9,13}.1^{15,19}]$octacosa-1(25),3,5,7(28),9,11,13(27),15,17,19(26),21,23-dodecaene.

It is understandable that alternative nomenclature was considered very early. Cram and Steinberg named 7 as [1.1.1.1]metacyclophane.[2] According to calixarene nomenclature developed by Gutsche the cyclic tetramer 8, formed from p-tert-butylphenol and methylene units, is named: 5,11,17,23-tetra-tert-butyl-25,26,27,28-tetrahydroxycalix[4]arene. This type of compound is usually abbreviated as p-tert-butylcalix[4]arene.

8

7.2.1.2 Synthesis

Calixarenes are obtained by base- or acid-catalysed condensation of phenols with formaldehyde and other carbonyl compounds. Alkylbenzenes and heterocyclic compounds can also be cyclized with formaldehyde and other aldehydes. Alternatively, calixarenes can be synthesized stepwise by way of open-chain intermediates. They may also be obtained with an ether oxygen in place of the methylene unit. In addition, the hydroxy groups in calixarenes can be etherified and esterified. Of the many preparative methods known today, which are also related to the manufacture of industrial materials (Bakelite, Novo-Lacke, etc.), a few of the characteristic routes will be highlighted.[1]

The story of the calixarenes seems to have begun with Adolf von Baeyer,[3] who heated aqueous formaldehyde with phenol and obtained a solid, resinous, non-crystalline product. Because the methods of that time did not allow characterization of such a material, its structure remained unknown. Three decades later Baekeland developed a process by which this phenol–formaldehyde reaction was utilized for manufacture of a 'phenoplastic' material, which was marketed successfully under the name 'Bakelite'. As a result interest in the phenol–formaldehyde process grew considerably. Zinke, working with various *p*-substituted phenols and aqueous formaldehyde at various temperatures, prepared high-melting, insoluble material, for which at the time the constitution of a cyclic tetramer **8**—that is, a calix[4]arene—was postulated. Later Cornforth tried to show that the material was not homogeneous, but that the higher- and lower-melting products from the condensation of formaldehyde with *p-tert*-butylphenol and other phenols were in fact conformational isomers of calix[4]arenes. This proposal was refuted by the work of Kämmerer (1972)[4] and of Munch (1977),[5] who suggested from temperature-dependent ^{1}H-NMR measurements that conformational reorganization occurs rapidly in calix[4]arenes.

In 1981 Gutsche was finally able to clarify that the mixture consisted of cyclic oligomers of varying ring sizes.[6] In a particularly well studied example, it could be shown that the condensation of *p-tert*-butylphenol and formaldehyde afforded the cyclic tetramer **8** (R = *tert*-butyl throughout), the hexamer **10**, and the octamer **12** as major products, along with small amounts of the pentamer **9** and the heptamer **11**.

9   10

Apparently Raschig had evidence for the existence of cyclic compounds in the Bakelite production from the condensation of *p-tert*-butylphenol and formaldehyde back in 1912, but Baekeland was skeptical. In 1941 Zinke and Ziegler described a product from *p-tert*-butylphenol and formaldehyde, m.p. 340°C, whose acetate had a molecular weight of 1725.[8] Although no structure was proposed for this product, it seems fairly clear that these authors had isolated *p-tert*-butylcalix[8]arene (**12**, R = *tert*-butyl). More details of the history of this development can be found in Gutsche's book.[1]

11

12

In 1975 Gutsche reported the preparation of cyclic tetramers from various *p*-substituted phenols with formaldehyde according to a condensation procedure of the Petrolite Corporation. According to the Petrolite procedure, designed for the synthesis of surface-active compounds, a *p*-substituted phenol was heated at reflux with paraformaldehyde and a trace of 50% NaOH in xylene for several hours. After the reaction mixture had been cooled, an insoluble product separated, which, in the case of *p-tert*-butylphenol, consisted almost exclusively of the cyclic octamer (**12**, R = *tert*-butyl), which thus was readily available in yields of 60–70%. If a stoichiometric amount of base instead of a catalytic amount was used in the same condensation, the product was *p-tert*-butylcalix[6]arene (**10**, R = *tert*-butyl), obtained in pure crystalline form in yields of 70–75%. In contrast, **8**, as the smallest oligomer, was obtained in minimal yield. The procedure is 'capricious', and one must tolerate vacillating yields between 0 and 45%.

Gutsche recommended the following procedure for *tert*-butylcalix[4]arene (**8**). One uses the method of Zinke, modified by Cornforth, and pulverizes the solid, resinous material in the last step. This substance, which cannot be isolated as a base-free solid, is dissolved in an organic solvent and washed with acid. Subsequent evaporation of the solvent affords a base-free resin. Addition of a small amount of base gives **8** in yields of 25–30%.

The calixarenes with odd numbers of aromatic units are more difficult to prepare in moderate amounts than are the even-numbered members of the series. Ninagawa and Matsuda,[9] using a modified Petrolite procedure (6 h at 55°C, finally 6 h at 150°C), obtained a reaction mixture from which they isolated 23% of the tetramer **8**, 5% of the pentamer **9**, and 11% of the octamer **10**. By using dioxane as solvent and heating for 30 hours, Nakamoto and Ishida[10] were able to isolate the hexamer, heptamer, and octamer, with the heptamer (**11**) in 6% yield.

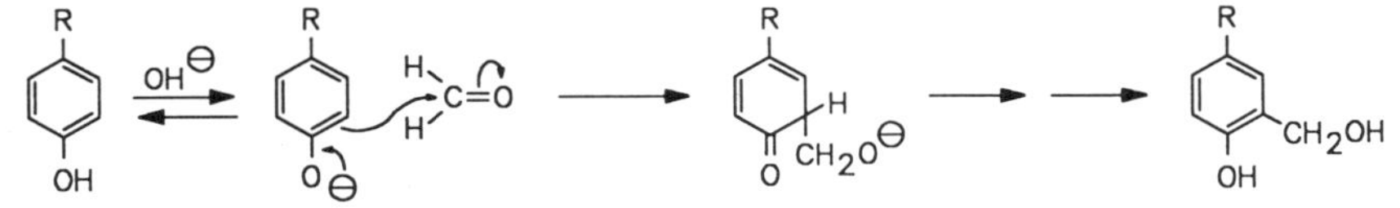

Figure 1. Base-catalysed hydroxymethylation of phenol

Although the mechanism of formation of cyclic oligomers is still not understood, the first steps of the reaction of formaldehyde with phenol have been relatively clearly explained. In a base-catalysed process formaldehyde and phenol produce first 2-hydroxymethyl- and 2,6-bis(hydroxymethyl)phenols (Figure 1).

Subsequent condensation of starting phenol with the hydroxymethylphenols leads to linear dimers, trimers, tetramers, etc., whereby *ortho*-quinonemethide formation can intervene, and these intermediates can react further with phenolate in a Michael-like transformation (Figure 2).

Because the formation of *o*-quinonemethides from *o*-methoxyphenol requires high temperatures (500–600°C), the intermediates proposed above have been disputed. On the other hand, it is known that the oxy Cope rearrangements proceed much faster from anions than from neutral compounds. Thus it seems reasonable that *o*-quinonemethides could be generated from 2-hydroxymethylphenolates.

As outlined in Figure 3, intermolecular dehydration of 2-hydroxymethylphenols can proceed under the Zinke–Cornforth and Petrolite condensation conditions, with formation of dibenzyl ethers. **13a**, for example, can lead to the ether **14**; and heating of **15** affords polycondensed ether. From this it follows that the complex mixture from which the calixarenes arise contains linear oligomers of varying lengths, in which the *o,o'*-bridges consist of CH_2 and CH_2OCH_2 groups.

What occurs in the latter stages of the reaction sequence leading to calixarenes remains a mystery. Although it is known that linear oligomers are transformed into cyclic oligomers in the absence of water and formaldehyde, it is still not

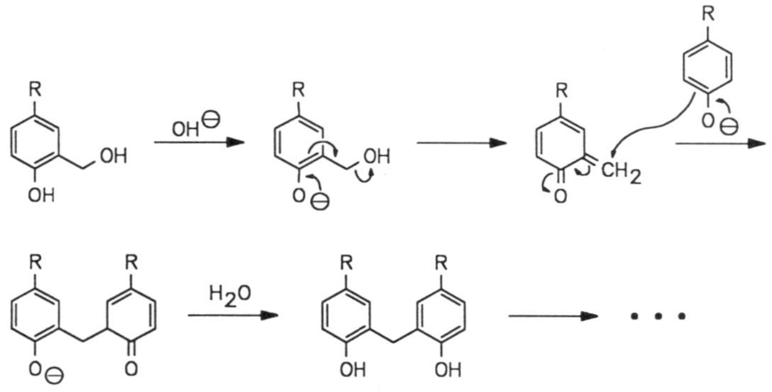

Figure 2. Base-catalysed formation of linear oligomers from phenol and formaldehyde

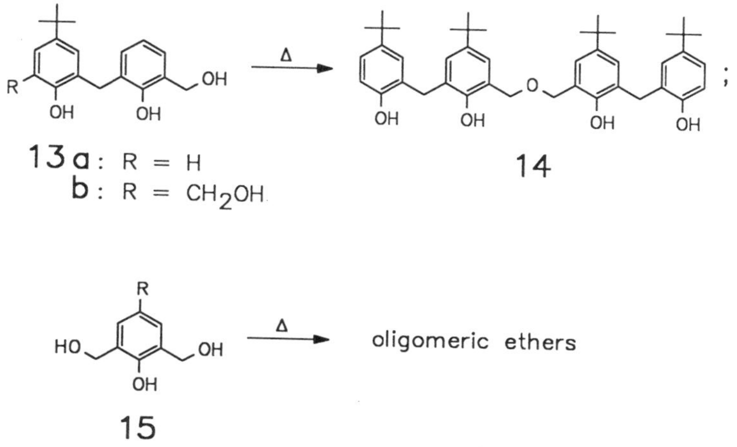

13a: R = H
b: R = CH₂OH

14

15

Figure 3. Formation of dibenzyl ethers from 2-hydroxymethylphenol

clear what the specific precursors of the calixarenes are; but a unifying, common pathway is unlikely.

It is not clear what driving force leads to relatively higher yields of the hexameric and octameric calixarenes. It remains a mystery as to why the higher cyclic oligomers are preferentially formed over the cyclic tetramer, even though the last named should be favoured on entropy grounds. Nevertheless, some arguments of clarification can be offered: intramolecular hydrogen bonding and a cation template effect.

From the concentration independence of the OH stretching vibration in the IR spectrum of **8** ($3160 \, cm^{-1}$), it has been shown that this ring is intra-molecularly hydrogen bonded. It is also indicated from molecular models that the four OH groups in calix[4]arene are in close proximity in the 'cone' conformation (see below). In fact, the more conformationally flexible cyclic hexamer and octamer show similar IR bands (3150 and $3230 \, cm^{-1}$, respectively), suggesting strong intramolecular hydrogen bonding in these ring systems as well. It is very surprising that the linear open-chain tetramer, pentamer, and hexamer exhibit analogous OH stretching vibrations at $3210 \, cm^{-1}$, which are attributed to strong intramolecular hydrogen bonds. Binding through hydrogen bonds in linear oligomers could be either <u>inter</u>molecular ('hemicalixarene'), such that pseudocyclic orientation occurs (Figure 4), or there could be <u>intramolecular</u> hydrogen bonds which likewise would lead to pseudocyclic structures (Figure 4). The latter are designated as 'pseudocalixarenes'.

The template effect mentioned above is appealing as an interpretation for the selectivity in ring and oligomer formation. In support of this explanation is the fact that the yield of calix[6]arene, the major product in the reaction with stoichiometric base, is somewhat higher with RbOH than with CsOH, KOH, or NaOH; LiOH is totally ineffective. In this connection, Izatt has shown that the calixarenes actually exhibit ionophore properties for NaOH, KOH,

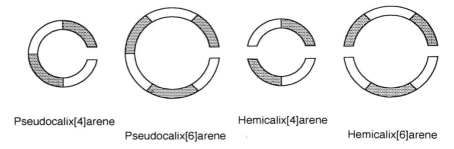

Pseudocalix[4]arene

Pseudocalix[6]arene

Hemicalix[4]arene

Hemicalix[6]arene

Figure 4. Open-chain precursors of calixarenes and their orientation into pseudocyclic hemi- and pseudocalixarenes (according to Gutsche[1]). The darkened and non-darkened parts of the curves represent the arene units

RbOH, and CsOH but not for LiOH. One can conclude from this that the nature of the cation exerts an influence on the ring size, although the details of the role of the cations in the cyclization process remain unclarified.

A recent finding relating to oligomer selectivity warrants mention. Contrary to an earlier report that cyclic oligomers are not interconvertible under the reaction conditions, it has recently been shown that a 20–35% yield of *p-tert*-butylcalix[4]arene is formed when *p-tert*-butylcalix[8]arene and *p-tert*-butylcalix[6]arene are heated in diphenyl ether in the presence of a small amount of potassium *tert*-butoxide. This result demonstrates that a kinetically formed mixture of oligomers from the phenol formaldehyde reaction mixture can be changed to a thermodynamic equilibrium mixture after a prolonged period of time.

Acid-catalysed reactions of resorcinol and aldehydes

A special type of calixarene, with RO groups on the periphery of the macrocycle, is obtained by the reaction of resorcinol with aldehydes (except formaldehyde). In 1883 Michael isolated two crystalline compounds from the reaction of resorcinol with benzaldehyde. The constitution was assigned to a tetrameric 'calix[4]resorcarene' of the type **16** (R' = H), replete with eight extra-annular hydroxyl groups, by Niederl and Vogel in 1940.

The mass spectrum of an octamethyl ether (**16**, R = R' = CH^3) and its X-ray crystal structure confirm this skeletal molecular structure. Högberg[11] has proposed that the acid-catalysed process, by contrast with the base-catalysed oligomerization, involves an aromatic electrophilic substitution initiated by a resonance-stabilized carbocation formed by protonation of the carbonyl oxygen of the aldehyde. The highly reactive formaldehyde apparently leads to substitution at the 2-position (as contrasted to the 4- and 6-positions), the result being a cross-linked macromolecule. The higher aldehydes, with sterically larger alkyl groups, tend not to react at the hindered 2-position between the two OH groups of resorcinol.

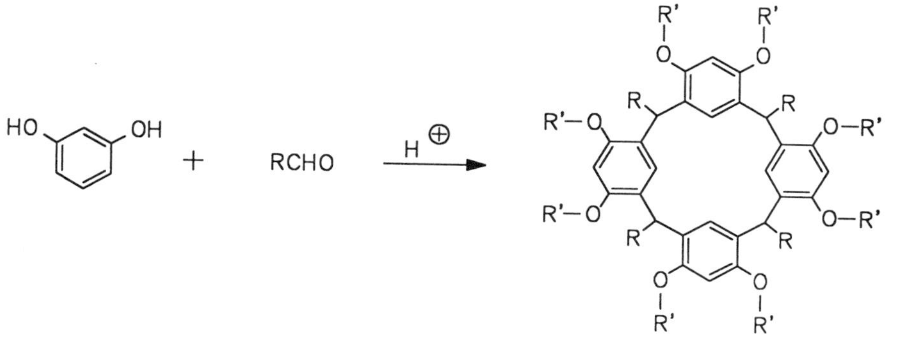

16

The stereochemistry of macrocycles of type **16** is less complicated experimentally than might be thought from a theoretical point of view. Of the four possible diastereomers of **16** predicted from conformational mobility, two predominate: the *cis,cis,cis* and the *cis,trans,cis* products. Högberg could demonstrate that the kinetically favoured *cis,trans,cis* isomer from resorcinol and benzaldehyde can be rearranged *in situ* into the *cis,cis,cis* product because the latter is less soluble. In this way the *cis,cis,cis* isomer can be obtained in greater than 80% yield.

Acid-catalysed condensations of alkylbenzenes and formaldehyde and of heterocyclic arenes with aldehydes

The calix[4]arenes **17a,b** were prepared from mesitylene and 1,2,3,5-tetra-methylbenzene by condensation with formaldehyde in the presence of acetic acid. **17a** can also be prepared by oligomerization via a Friedel–Crafts alkylation of chloromethylmesitylene. Similar macrocycles, in which heterocyclic aromatic

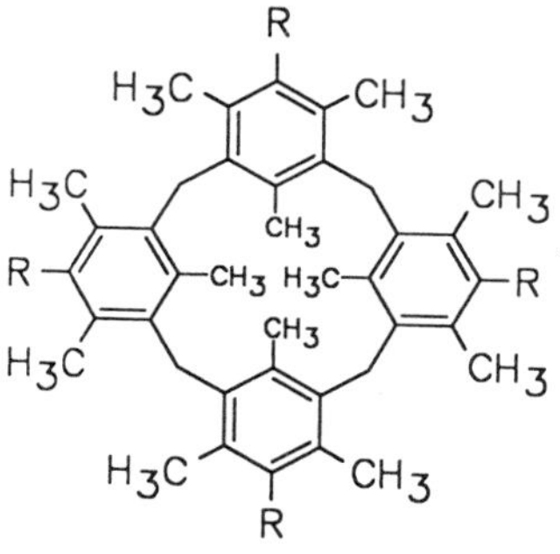

17a: R = H
b: R = CH$_3$

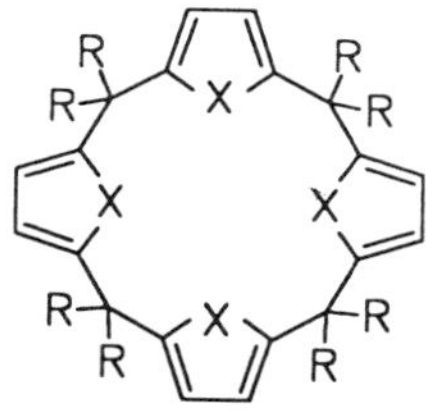

18 a: X = O
b: X = S
c: X = NH

R = CH$_3$, C$_2$H$_5$

units connect one-carbon bridges, are obtained by condensation of furanes, thiophenes, and pyrroles with aldehydes and ketones **(18a–c)**. These compounds are not usually designated as calixarenes, even though they closely resemble them in molecular structure and mode of synthesis. The yield of **18a** from furan and acetone can be increased from about 20% to 40% by the addition of LiClO$_4$. Although this prompted the suggestion that a template effect was responsible, it has more recently been shown that the higher yield is correlated with the acidity of the reaction medium, not with the nature of the metal ion.[12]

Stepwise syntheses of calixarenes

Hayes–Hunter–Kämmerer synthesis These authors developed a 'rational' synthesis in 1956, whereby calixarenes with various substituents and of varying ring size could be obtained by stepwise extension of open-chain precursors and eventual ring closure. One of the *ortho*-positions in the starting *p*-cresol was protected by bromination, after which methylene groups (*via* base-induced hydroxymethylation) and aryl groups (by acid-catalysed arylation) were introduced. The linear bromosubstituted tetramer **19** was debrominated and then cyclized to **20** (Figure 5).

By improving this synthetic method, Kämmerer extended it to the synthesis of pentamers, hexamers, and heptamers of methyl- and *tert*-butylsubstituted calixarenes.[13] The disadvantage of the method lies in the minimal overall yields of end products. The conversion of *p-tert*-butylphenol to *tert*-butylcalix[4]arene is only 11%, that of *p*-phenylphenol to phenyl-calix[4]arene **(26)** only 0.5%. As indicated in Figure 6, the method is complicated by the formation of byproducts in the cyclization step. In the case of the phenyl-substituted calix[4]arene, two isomeric byproducts **27** and **28** are generated besides the main

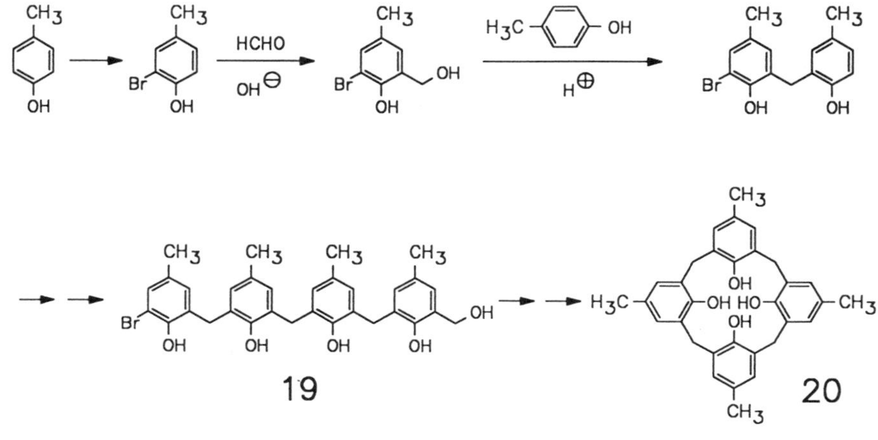

Figure 5. Ten-step synthesis of *p*-methylcalix[4]arene **(20)** according to the Hayes–Hunter method

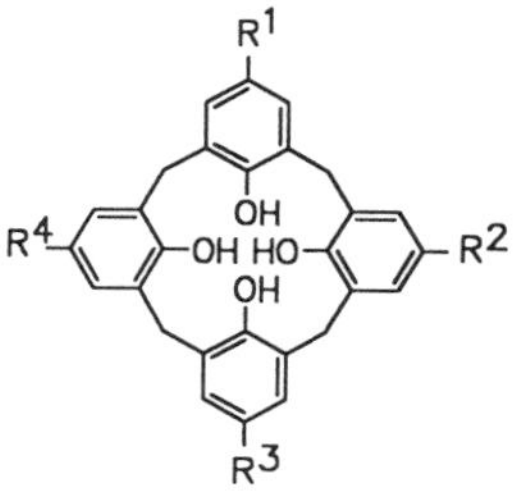

21

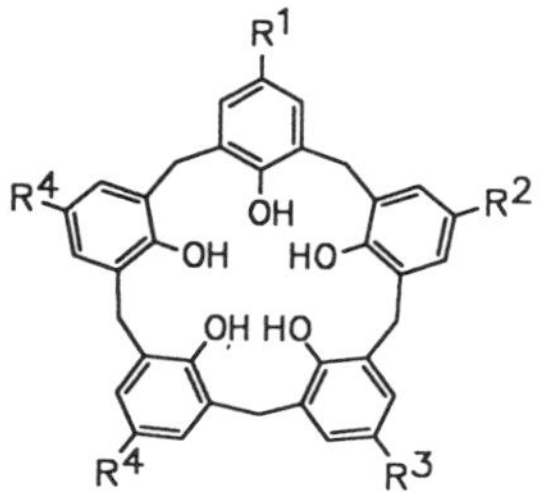

22

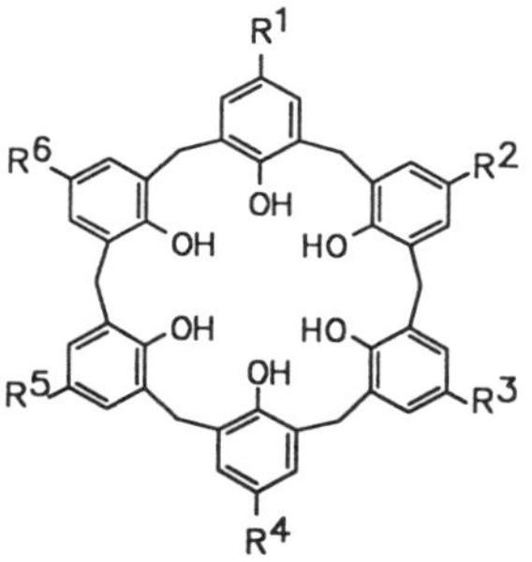

23

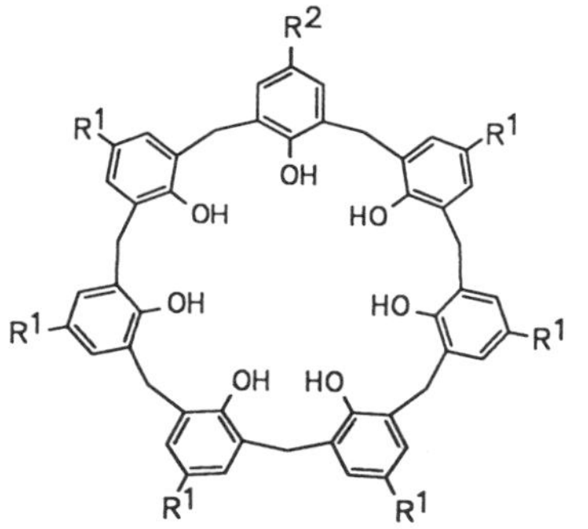

24

a : R¹ = R² = R³ = R⁴ = CH₃

product **26**. For this reason phenylcalix[4]arene (**26**), of considerable interest in host–guest chemistry because of its deep cone-shaped cavity, is not readily available.

Synthesis according to Böhmer, Chhim, and Kämmerer These authors developed a convergent synthetic strategy, with a flexibility lacking in the method just described. First a linear trimer **29** is built up stepwise and then allowed to react with a 2,6-halomethylphenol (**30**) (Figure 7). This strategy is advantageous because of the fewer number of steps but still suffers from the low yields in the final cyclization step. In favourable cases the yield ranges from 10–20%; for synthesis of the interesting examples with Br and NO₂ substituents, the yields are 2–7%.

Synthesis according to Moshfegh and Hakimelahi These authors published the synthesis of a large number of *p*-halocalixarenes in high yield (60%), including the calix[3]arene **37**, whereby side products such as the dioxa compound **35** are formed in 10% yield (Figure 8).[15] Certain points in this procedure are still not clear: for example, the nature of the sterically strained calix[3]arene and comparison of the halocalixarenes reported here and elsewhere.

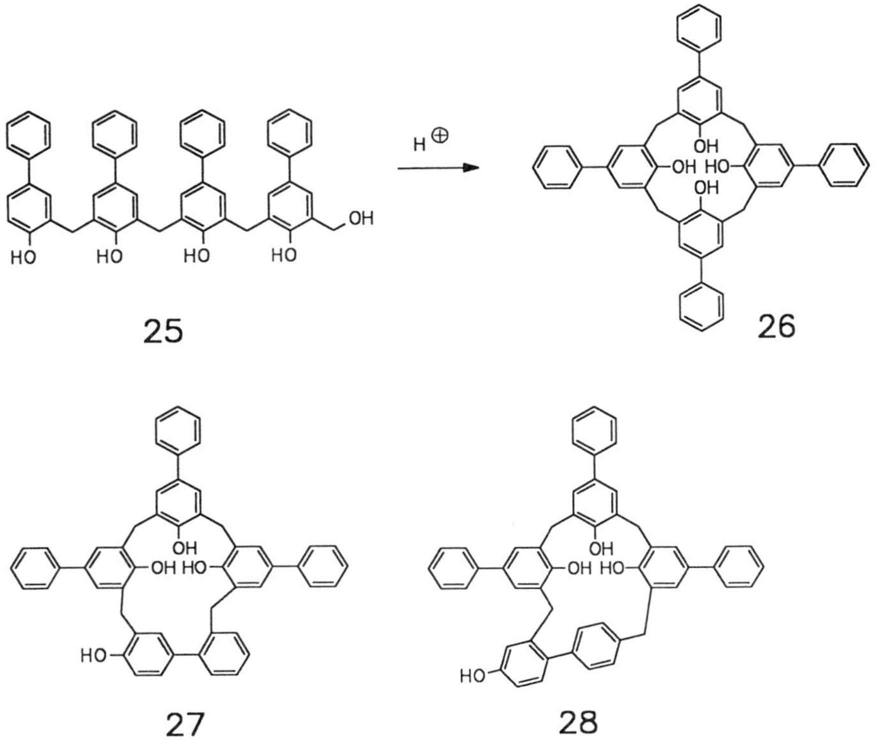

Figure 6. Stepwise formation of phenylcalix[4]arene (**26**) according to the Hayes–Hunter method

Synthesis according to No and Gutsche These authors developed a method leading to calixarenes in relatively high yield from open-chain precursors by way of a convergent synthesis which allowed the introduction of several substituents. Outlined in Figure 9 is a four-step synthesis starting with *p*-phenylphenol and formaldehyde, whose reaction under controlled conditions affords the bis(hydroxymethyl) dimer **38**. The latter, condensed with two equivalents of a phenol (e.g. *p-tert*-butylphenol), affords the linear tetramer **39** (R = *tert*-butyl). Cyclization of the monohydroxymethylated product **40** gives the calix[4]arene **41** in about 10% overall yield. In view of the costly starting materials and the straightforward procedure and workup, this method is very convenient.

Oxacalixarenes Dihomooxacalix[4]arenes, such as the *tert*-butyl compound **47**, are produced from the Petrolite synthetic method (see above), but these macrocycles are difficult to isolate from the reaction mixture. An easier route to **47** is the dehydration of the bis(hydroxymethyl) tetramer **46**, as is outlined

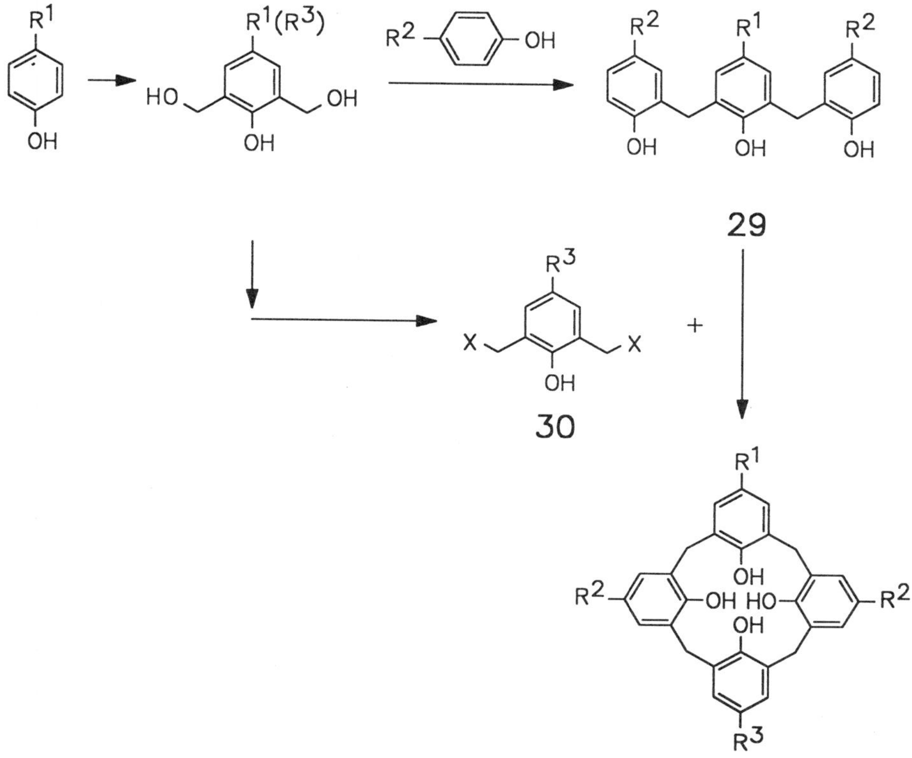

29

30

R^1–R^3 = *t*-Bu, CH_3, Cl, Br

31

Figure 7. Convergent, stepwise synthesis of calix[4]arenes according to Böhmer, Chhim, and Kämmerer[14]

in Figure 10. Other calix[3]- and -[4]arenes such as **43** and **45** are formed by thermally induced dehydration as well.

Calixarene esters and ethers

The phenolic OH groups in calixarenes can readily be transformed into esters and ethers. The corresponding acetates of several calixarenes were prepared, as described by Zinke and Ziegler.[8] The acetates are typically more soluble in organic solvents than the calixarenes with free OH groups. Some calixarenes can be purified indirectly in this way, with the aid of crystallization, column chromatography, or extraction. The ester groups can be cleaved with ethylenediamine in DMF.

Normally the fully acylated calixarene is obtained with an excess of acylating agent, although partially acylated calixarenes have been described. Examples

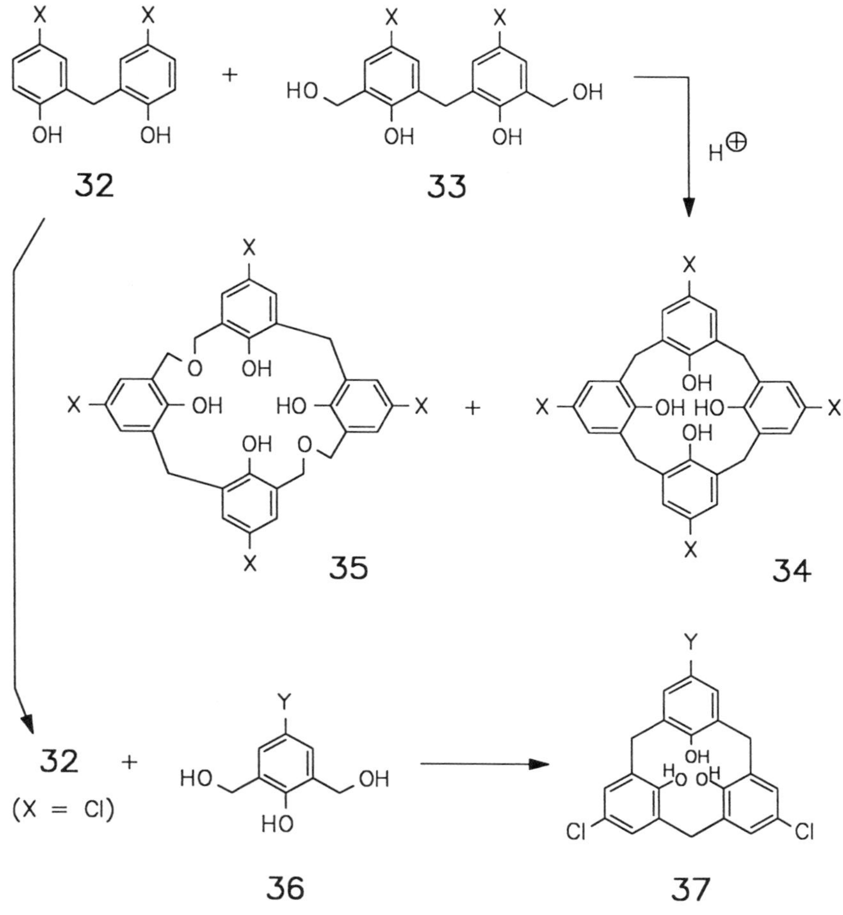

Figure 8. Convergent, stepwise synthesis of calix[3]- and -[4]arenes according to the method of Moshfegh and Hekimelahi[15]

of the latter are two different acetates of *p-tert*-butylcalix[4]arene, which were proved not to be conformational isomers. Their melting points—247–250°C and 383–386°C—are markedly different; the first is attributed to the triacetate, the second to the tetraacetate. Besides acetates the benzoates and *p*-toluenesulphonates are described as easily obtainable, as is also the case for the mono- and dicamphorsulphonate esters of *p-tert*-butylcalix[8]-arene.

Peralkylation of calixarenes to the corresponding ethers is accomplished by treatment in THF/DMF solution with alkyl halides in the presence of NaH. Methyl, ethyl, allyl, and benzyl ethers are formed in very high yields. Other ethers which have been described include oligoethylene glycol

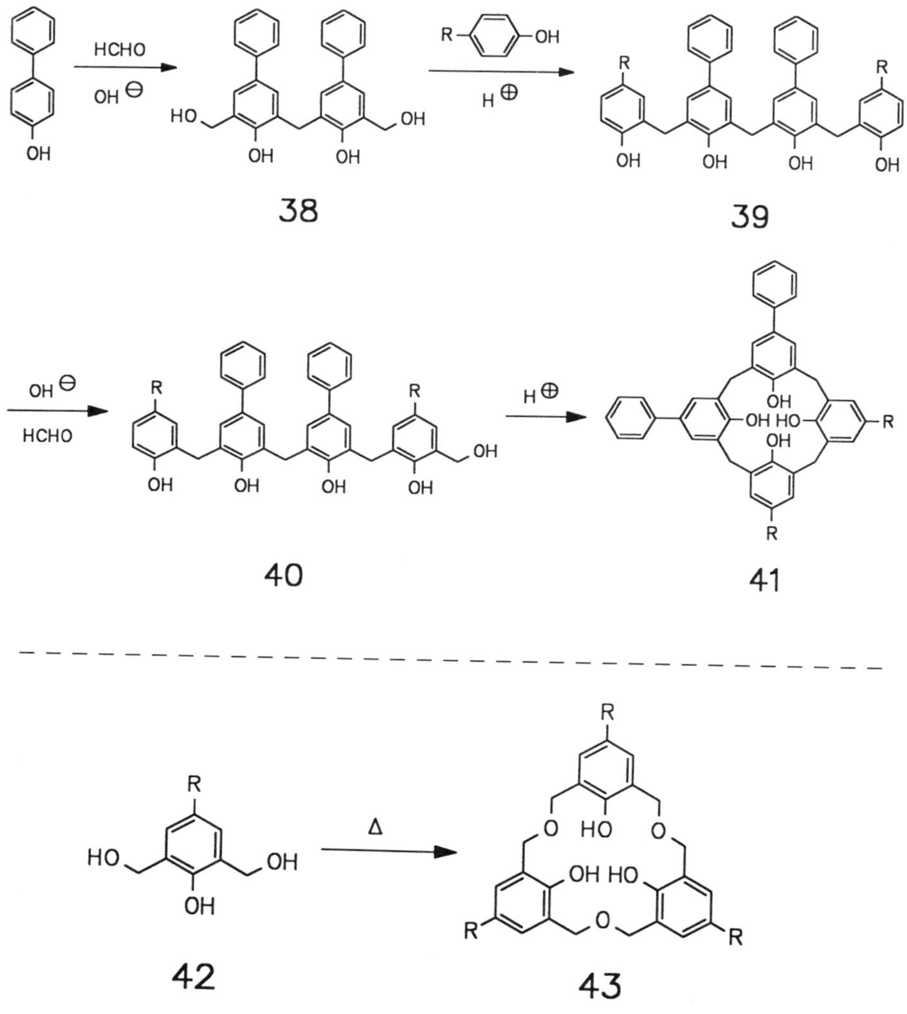

Figure 9. Synthesis of calix[4]arenes

ethers, partially etherified calixarenes, and 2,4-dinitrophenyl ethers (of *p-tert*-butylcalix[8]arene).

7.2.1.3 Physical and spectroscopic properties[1]

Melting points

The calixarenes melt unusually high, by comparison to their open-chain analogues. Practically all calixarenes melt above 250°C, most considerably higher. *p-tert*-Butylcalix[4]arene, for example, melts at 344–346°C; *p-tert*-butylcalix[6]arene

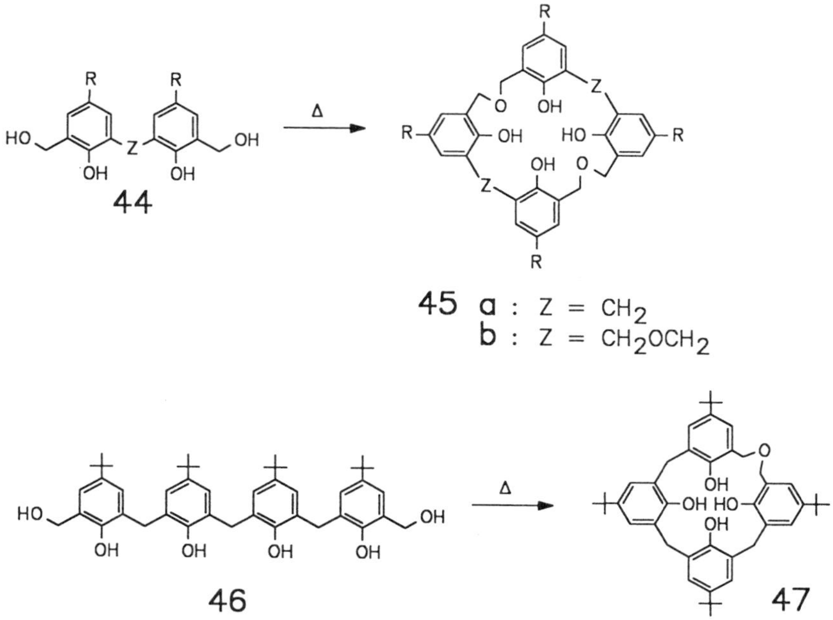

Figure 10. Synthesis of oxacalixarenes

at 380–381°C; and *p-tert*-butylcalix[8]arene at 411–412°C. The melting points of *p*-phenylcalix[4]arene and *p*-phenylcalix[8]arene are, respectively, 407–408°C and 450°C.

Esters and ethers of calixarenes typically melt at lower temperatures. The tetramethyl ether and tetrabenzyl ether of *p-tert*-butylcalix[4]arene melt, respectively, at 226–228°C and 230–231°C. Unusually high melting points have been noted for the tetraacetate of *p-tert*-butylcalix[4]arene (383–386°C) and for the tetrakis(trimethylsilyl) ether of *p-tert*-butylcalix[6]arene (410–412°C).

Solubility

In accord with the high melting points of the calixarenes is their characteristically low solubility in organic solvents. Nevertheless, many calixarenes are sufficiently soluble so that molecular weight determinations (osmometry in CHCl$_3$) and NMR measurements (usually in CDCl$_3$ and pyridine-d$_6$) can be carried out. The nature of the substituent in the position *para-* to the phenolic OH group often has a significant influence on its solubility. The *p*-allylcalixarenes are most easily soluble, whereas the *p*-phenyl- and -adamantylcalixarenes are least soluble. Ethers and esters derived from calixarenes are typically more soluble in nonpolar solvents than the parent calixarenes themselves. In striking contrast to *p*-phenylcalix[8]arene itself, its octamethyl ether is somewhat soluble in CHCl$_3$.

IR spectra

Characteristic of the calixarenes are the solvent-independent OH stretching vibrations in the IR spectra at $3200\,cm^{-1}$, indicative of the strong intramolecular hydrogen bonding. The linear oligomers, however, give rise to stretching vibrations in the same region. The fingerprint regions of the IR spectra of calixarenes are all very similar, although small differences often make it possible to assign a specific ring size to the sample. The cyclic tetramer, for example, is characterized by a relatively strong absorption at $830\,cm^{-1}$, the cyclic hexamer by absorptions at 750 and $800\,cm^{-1}$; the cyclic octamer can be identified from the absence of all three of these absorption bands.

UV spectra

The UV spectra of the calix[4]arenes differ only slightly from those of the linear oligomers, both compound types absorbing at $\lambda_{max} = 280$ and $288\,nm$ with similar extinction coefficients. As the ring size increases, the absorption maxima change very little, although the molar extinction increases (2075–2225 for $n = 4$; 4000 for $n = 8$). A characteristic change with increasing ring size is the difference in relative intensities of the maxima at $\lambda = 280$ and $288\,nm$. For the cyclic tetramer the intensity at $280\,nm$ is higher than that at $288\,nm$, while the reverse is true for the cyclic octamer. In the spectra of the cyclic pentamer, hexamer, and heptamer the bands are of about equal intensity.

NMR spectra

NMR spectroscopy offers the best possibility for identification of the calixarenes. The ^{1}H-NMR spectra of the *p-tert*-butylcalixarenes, for example, show one signal for the aromatic protons, one signal for the *tert*-butyl protons, and temperature-dependent methylene protons. While the ^{13}C-NMR spectra of the linear oligomers become increasingly complicated with increasing size, those of the cyclic oligomers are simple because of the structural symmetry.

Mass spectra

The mass spectra provide the 'parent ion', as is typical of many classes of compounds. This information in the case of calixarenes must be interpreted with caution, however. An example is the mass spectrum of *p-tert*-butylcalix[8]-arene ($m/z = 1872$), which shows a relatively intense peak at $m/z = 648$, which is attributed to the cyclic tetramer. In fact, this easily misinterpreted result was the basis for misassigning the major product in the Petrolite method as the cyclic tetramer.

Kämmerer has noted characteristic differences in the mass spectra of cyclic oligomers by comparison with their analogous linear oligomers.[13] The cyclic oligomers preferentially lose methyl or *tert*-butyl groups and retain their ring

346

Figure 11. *p*-Phenylcalix[4]arene constructed from CPK space-filling molecular models. Left to right: side view; view from above into the cone; view from below (from the side of the OH groups); modified according to Gutsche[1]

structures, whereas the linear oligomers are first cleaved into their smaller phenolic units. In calix[5]arene **22a** the most intense peak appears at $m/z = 480$, corresponding to the loss of one arene unit. The corresponding cyclic octamers show similar behaviour, namely extrusion of one, two, or three arene units.

7.2.1.4 Stereochemistry

Topology of calixarenes

As already mentioned, calixarenes are so named because of the similarity of their molecular shape to that of a Greek vase or chalice. Figure 11 is a representation of the spatial appearance of *p*-phenylcalix[4]arene, as viewed from various perspectives. For comparison the topology and size of the cone and cavity are represented for *p*-phenylcalix[8]arene in Figure 12.

It is recognized that the size of the cavity increases with the number of arene units in the macrocycle, as expected; and intramolecular hydrogen bonding plays a role in the formation of the cone conformation. From the X-ray crystal structure analyses it can be shown that calix[4]arenes and calix[5]arenes exist in this cone conformation in the crystallline state (Figures 11–13). Calix[6]arenes and calix[8]arenes, on the other hand, are held more or less rigidly in an

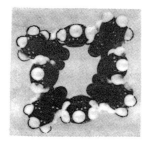

Figure 12. CPK space-filling model of *p*-phenylcalix[8]arene; view from above (left), from below (from the side of the OH groups, right). Same scale as in Figure 11; modified according to Gutsche[1]

'alternating conformation', with OH groups and other substituents oriented alternately up and down (for more details, see below).

Conformationally flexible calixarenes

Cornforth first suggested the possibility of conformational isomerism in calix[4]arenes.[17] He pointed out that four discrete conformations could exist, which were designated 'cone', 'partial cone', '1,2-', and '1,3-alternating' conformations. These are represented in Figure 13.

Kämmerer first demonstrated that the conformers of calix[4]arenes are interconvertible.[4] In the ^{1}H-NMR spectra of p-alkylcalix[4]arenes the Ar-CH_2-Ar protons appear as a sharp singlet above room temperature but as an AB system below room temperature. This finding leads to the conclusion that the molecules are cone conformations, which interconvert rapidly at higher temperatures and more slowly at lower temperatures. One can estimate an interconversion rate of about $100 \, s^{-1}$ on the basis of the coalescence temperature of about 45°C.

It is remarkable that practically the same conformational mobility is noted for p-tert-butylcalix[8]arene as for p-tert-butylcalix[4]arene, as determined by dynamic ^{1}H-NMR spectroscopy in $CDCl_3$ and C_6D_5Br. This is substantiated by the distinctive ^{1}H-NMR spectra with variable temperature in pyridine-d_5 as solvent. While the CH_2 groups in calix[4]arene split into an AB system at

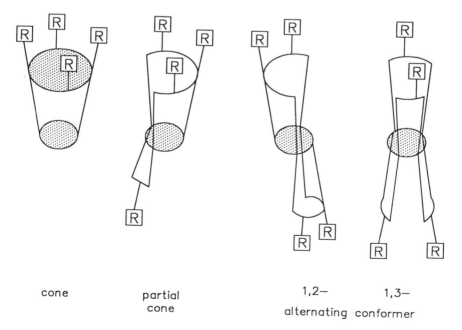

Figure 13. Conformers of calix[4]arene

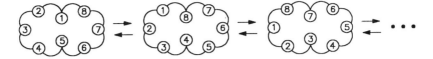

Figure 14. *para*-Substituted calix[8]arene in the compressed ('pinched') conformation. The compressed conformer corresponds formally to a pair of tetramers; thus it can be visualized that the octamer assumes the shape of cyclic tetramers

about 15°C, those in calix[8]arene remain a singlet down to temperatures of − 90°C. The similarity of the dynamic ^{1}H-NMR spectra in non-polar solvents such as CDCl$_3$ is attributed to intramolecular hydrogen bonding in the octamer, although one would have expected conformational flexibility because of the larger ring size. The intramolecular hydrogen bonds nevertheless create a compressed conformation, as represented in Figure 14. The hydrogen bonds in the cyclic octamer have been termed 'circular hydrogen bonds' ('flip-flop'), by analogy to the arrangements in cyclodextrins described by Saenger.[18] Such circular bridges confer a special stability on the structure. Through a 'transannular contraction', involving hydrogen bonds, calix[8]arene can assume a conformation made up of two such circular hydrogen-bonding bridges, each containing four OH groups. This also implies that a pseudorotation of the H-bonding orientation takes place, averaging the CH$_2$ groups at a rate which is fast on the NMR time scale. This averaging of the CH$_2$ groups through pseudorotating, partially transannular hydrogen-bonding bridges can be illustrated as in Figure 14, which shows the eight numbered OH groups of the cyclic octamer between which the hydrogen bonding is occurring. Calix[6]arene likewise assumes a compressed conformation in non-polar solvents, in which case two rings are made up of three intramolecular hydrogen bonds each.

Other calixarenes with varying *p*-substituents and ring sizes have been examined for conformational flexibility by means of dynamic ^{1}H-NMR spectroscopy. The free enthalpy of activation for the ring folding process lies in the range of 38–67 kJ/mol, with coalescence temperatures varying between − 90 and + 53°C.

Conformationally rigid calixarenes

In order to lock the conformations of calixarenes it is necessary to replace the H atoms of the OH groups with larger substituents. This is of interest because conformationally rigid calixarenes with cone-shaped cavities are favourably suited to enclose small molecules as guests. Synthetic compounds possessing a conformationally rigid cavity ('enforced cavity') were named '*cavitands*' by Cram.[19] It can be shown from space-filling molecular models that the OH groups in a calixarene normally offer no substantial resistance to a conformational reorganization. In this reorganization an arene unit rotates about its *meta*-bonding axis, so that the OH groups pass through the centre of the macrocyclic ring. Groups larger than OH, however, completely block this conformational process, as is confirmed from dynamic NMR. Thus one can clearly distinguish in the calix[4]arenes between the conformationally flexible starting material (hydroxy compound) and the tetraacetate **51f**, formed by acetylation of **48**, because **51f** exists as the 1,3-alternating conformation. This result was, in fact, to be expected, for it had already been found that the 'octahydroxycalix[4]arene' **16** is locked in a 1,3-alternating conformation. On the other hand, the 1,3-alternating conformation is seldom seen in derivatives

		Y
	a:	OCH_3
	b:	OC_2H_5
	c:	$OCH_2CH=CH_2$
	d:	$OCH_2C_6H_5$
	e:	$OSi(CH_3)_3$
48	**51** : R = H	f: $OCOCH_3$
49	**52**: R = t-C_4H_9	g: OSO_2⟨⟩CH_3
50	**53**: R = $CH_2CH=CH_2$	

of tetrahydroxycalix[4]arene (e.g. **48–50**), the 'partial cone' or 'cone' conformations being generally favoured. So, for example, the tetraacetate **52** derived from **49** is locked in a partial cone conformation, as substantiated by X-ray crystallography. Methylation and ethylation of **48** and **49** and allylation of **48** afford the corresponding ethers **51a–c** and **52a,b** in partial cone conformations.

It is surprising that the tetramethyl ethers **51a** and **52a** are conformationally more flexible than one would predict from an inspection of the CPK models, an indication that space-filling models tend to overestimate conformational barriers. At room temperature the ^{1}H-NMR spectra of these tetramethyl ethers show only a broad absorption for the methyl groups, analogous to the behaviour of the starting calixarenes (with OH groups in place of the ether functions). These methyl signals become sharper at higher temperature and more complex at lower temperature, the latter being consistent with a partial cone conformation.

Many derivatives are apparently locked in a cone conformation. Examples are the tetraallyl (**52c**), benzyl (**52d**), and trimethylsilyl (**52**) ethers of **49**, the tetrabenzyl ether (**51d**) and tetratosylate (**51e**) of **48**, and the trimethylsilyl ether (**53e**) and tetratosylate (**53g**) of **50**. Only in a few instances are conformational mixtures obtained. Acetylation of **41** (R = *tert*-butyl), for example, results in a 1,3-alternating and two partial cone conformations. By judicious choice of derivatizing agent for calix[4]arenes, one can effect a locking of the conformation into a 'cone' or 'partial cone'. The latter is favoured in many cases by acetylation, while fixation of the cone conformation is achieved by benzylation and trimethylsilylation. In this way one can not only construct cavitands but one has the possibility to model the contour of the cavity by intentional design.

In the case of partially alkylated calixarenes, evidence points to a 'shallow, partial cone conformation' (trimethyl ether), whereas monomethyl and dibenzyl ethers assume a shallow 1,3-alternating conformation. The fact that the trimethyl and monomethyl ethers are conformationally less flexible than the tetramethyl ether is attributed to intramolecular hydrogen-bond bridging involving the free hydroxyl groups.

Although the calix[8]arenes resemble the calix[4]arenes in their conformational mobility (see above), the ester and ether derivatives of the former are not conformationally locked at low temperatures, in contrast to the calix[4]arene ethers and esters. This distinction is attributed to the difference in cavity size in the two series. The small, narrow hole in the calix[4]arenes permits passage of only small groups such as OH and OCH_3, while even large groups like tosyl can penetrate the much larger opening in calix[8]arenes.

Calix[6]arenes are intermediate in their conformational mobility between the calix[4]arenes and the calix[8]arenes. Their derivatives are more flexible than those of the calix[4]arenes; yet they can be frozen out on the NMR time scale at low temperatures. From dynamic ^{1}H-NMR spectroscopic studies of the hexa(2-methoxyethyl) ether of *p-tert*-butylcalix[6]arene, Ungaro concluded the favoured conformations to be either a 1,3,5-alternating (three OH groups up and three down) and/or a 1,4-alternating 'winged' conformation, similar to the case for the free calix[6]arene. The X-ray crystal structure confirms an *up,down,out/down,up,out* conformation in the solid state.[1]

An alternative, recent device for drastically restricting conformational mobility in calixarenes is by bridging two or more regions of the molecule with an appropriate number of connecting atoms. Cram succeeded in converting the

16 $\longrightarrow$

54   55

octahydroxycalix[4]arene **16** into the conformationally rigid **54** (R = H) and **55** by treatment with chlorobromomethane and 2,3-dichloro-1,4-diazanaphthalene, respectively.[19]

Gutsche synthesized the bridged diphenol **56** and then incorporated it into a bis-homooxacalix[4]arene (**58**) in a series of steps. Although the simple bridges in **57** and **58** do not fully prevent conformational mobility, as do the four bridges in **54** and **55**, nevertheless a complete inversion is no longer possible, as is evidenced from the higher coalescence temperature determined by dynamic ¹H-NMR spectroscopy.[1]

56

57: X = CH$_2$

58: X = CH$_2$OCH$_2$

7.2.1.5 Functionalization of calixarenes

The transformation or introduction of new functional groups into calixarenes is of considerable interest because of their potential as enzyme models ('enzyme mimics'). The *p-tert*-butylcalixarenes are particularly well suited for the introduction of functional groups because de-*tert*-butylation of *p-tert*-butylcalix[4]arene (**59**), for example, leads to the ring **60**, in which new *p*-substituents can be incorporated. Bromination of the tetramethyl ether of calix[4]arene **61** proceeds smoothly with formation of the tetrabromo compound **62**, which affords the tetramethyl ether of *p*-carboxycalix[4]arene (**66a**) by lithiation and subsequent carbonation or can be transformed to the tetramethyl ether of *p*-cyanocalix[4]arene (**66b**). Similarly, Friedel–Crafts acylation of **61** leads to *p*-acetylcalix[4]arene, which gives **66a** by haloform oxidation. The bromination and acetylation reactions demonstrate that the calixarenes undergo normal electrophilic substitution reactions without rupture of the macrocyclic ring (Figure 15).

The allyl ethers offer an alternative route to electrophilic substitution for the introduction of functional groups into calixarenes. Heating of the tetra allyl ether **64** in diethylaniline induces a quadruple Claisen rearrangement, with

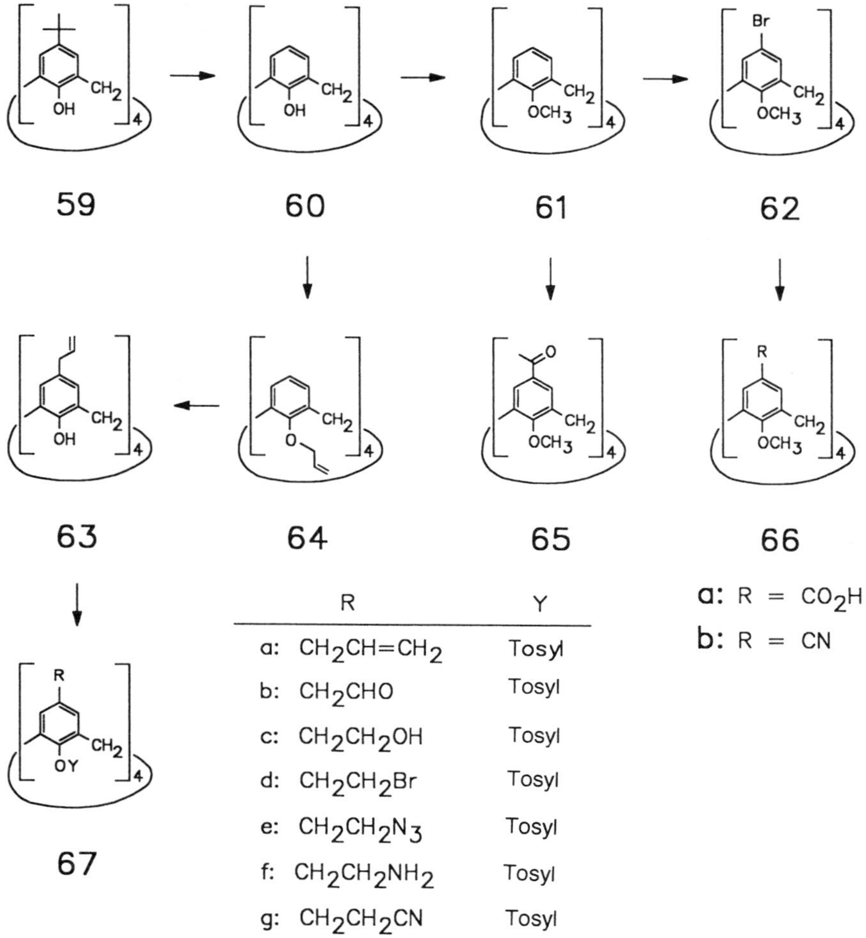

a: R = CO_2H
b: R = CN

	R	Y
a:	$CH_2CH{=}CH_2$	Tosyl
b:	CH_2CHO	Tosyl
c:	CH_2CH_2OH	Tosyl
d:	CH_2CH_2Br	Tosyl
e:	$CH_2CH_2N_3$	Tosyl
f:	$CH_2CH_2NH_2$	Tosyl
g:	CH_2CH_2CN	Tosyl
h:	CH_2CH_2OH	H

Figure 15. Preparation of functionalized calix[4]arenes

formation of *p*-allylcalix[4]arene (**63**) in excellent yield. A large number of functionalized calixarenes are available from the tetratosylate of **63** (**67a**): for example, the aldehyde **67b**, the alcohol **67c**, the bromide **67d**, azide **67e**, amine **67f**, and the nitrile **67g**. Removal of the tosylate group under mild basic conditions affords, for example, *p*-(2-hydroxyethyl)calix[4]arene (**67h**).

The functional interconversions outlined in Figure 15 for the calix[4]arenes should also be applicable to the other larger and smaller calixarenes as well, and some of these have been accomplished. Removal of the *tert*-butyl groups from *p*-*tert*-butylcalix[6]arene and *p*-*tert*-butylcalix[8]arene afford the corresponding calix[6]arene and calix[8]arene, which can be transformed into the corresponding hexaallyl and octaallyl ethers. Claisen rearrangement of the former leads to p-allylcalix[6]arene in modest yield. Analogous reactions in the calix[8]arene series proceed in much lower yield, and the products in many cases have not yet been characterized.

7.2.1.6 Calixarenes as host molecules

Complexation of neutral guest molecules in the solid state

It has been recognized for several years that many calixarenes, upon recrystallization, retain solvent. An example is *p-tert*-butylcalix[4]arene, which forms crystalline adducts (clathrates or complexes) with chloroform, benzene, toluene, xylene, and anisole.[20] *p-tert*-Butylcalix[5]arene forms 'complexes' with isopropyl alcohol[6] and acetone.[20] Complexes* of *p-tert*-butylcalix[6]arene with chloroform and methanol are known, while *o-tert*-butylcalix[8]arene forms adducts with chloroform and *p-tert*-butyldihomooxacalix[4]arene forms a 'complex' with dichloromethane. The tenacity with which guest molecules are held in calixarene hosts varies widely. Whereas the cyclic octameric calixarene loses guest chloroform after a few minutes' standing at room temperature and atmospheric pressure, chloroform can not be removed from the cyclic hexamer even after heating at 275°C in vacuum (1 Torr) for 6 days.

From X-ray crystal structure analyses, particularly those carried out by Andreetti, it can be shown that the guest molecule is to be found in varying locations in the crystal lattice of the calixarenes.[21] In the *p-tert*-butylcalix[4]-arene–toluene 'complex' the toluene is located in the centre of the cavity, with the host assuming the cone conformation.[20] This orientation is characterized as the *endo*-calix conformation (Figure 16).

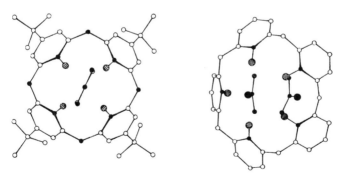

Figure 16. X-ray crystal structure of the 1:1 'complex' of *p-tert*-butylcalix[4]arene with toluene as guest[21] (left) and of the 1:2 complex of calix[5]arene with acetone as guest[20] (right). (Similar complexes are formed with benzene, *p*-xylene, and anisole as guests)

---

*For a definition of the concepts of complex, adduct, clathrate, inclusion compound, see: F. Vögtle, *Supramolecular Chemistry* (Wiley, Chichester, 1991).

exo-Calix complexes (canal-like structures) are formed by the host compound *p*-(1,1,3,3-tetramethylbutyl)calix[4]arene (**8**, R = 1,1,3,3-tetramethylbutyl) with aromatic guest molecules.

Calix[5]arene (**9**, R = H) forms a 1:2 complex with acetone as guest, in which one acetone molecule is located within the cavity and the other outside (Figure 16). The X-ray crystal structure of solvent-free (guest-free) *p*-(1,1,3,3-tetramethylbutyl)calix[4]arene shows that the conformation is a cone and that two of the 1,1,3,3-tetramethylbutyl groups are oriented into the cavity; that is, they are intramolecularly 'complexed' in their own cavity.

Because the unsubstituted calix[4]arene does not form *endo*-calix complexes with aromatic guest molecules, it has been postulated that the *tert*-butyl group plays a specific role in promoting complex formation. The following are considered to be possible reasons:

(a) The *tert*-butyl group is not sufficiently flexible that it can orient itself into the cavity and thus occupy the space.

(b) The *tert*-butyl group is so situated that it can participate in $CH_3-\pi$ interaction involving the methyl groups of the *tert*-butyl and the aromatic ring of the guest molecule.

The *selectivity* of complex or clathrate formation of calixarenes can be estimated by competitive recrystallization of *p-tert*-butylcalix[4]arene from a 1:1 mixture of two guest compounds (benzene *vs.* *p*-xylene). This leads to the generalization that anisole and *p*-xylene are favoured over most other aromatic hydrocarbons. When used as a chromatographic column material, *p-tert*-butylcalix[4]arene effects a decisive selectivity between mixtures of aromatic hydrocarbons.

Complex formation of calixarenes with neutral guests and organic ammonium compounds

The fact that complexes can be isolated and characterized in the crystalline state—most appropriately designated as clathrates or addition compounds usually lacking specific host–guest interactions—does not necessarily imply that similar complexes exist in solution. There is in fact no published evidence for complex formation of calixarenes with dissolved neutral molecules, such as chloroform, benzene, toluene, etc.

Organic amines, on the other hand, do form 'complexes' with calixarene host molecules.[1] The first step in the process, however, is a proton transfer from acidic calixarene to the basic amine, so that the association actually consists of ion pair formation between a calixarene anion and an organic ammonium ion.

Thus, whether one can consider this complex formation between a calixarene and a neutral guest is unclear. In the ^{1}H-NMR spectrum of a mixture of *p*-allylcalix[4]arene and *tert*-butylamine in acetonitrile, the *tert*-butylamine protons are shifted downfield ($\Delta\delta = 0.3$ p.p.m.) and a corresponding downfield shift of the aromatic protons of the calixarene ($\Delta\delta = 0.4$ p.p.m.) is found. Additional evidence for an association is the observation that the rate of conformational inversion of the calixarene is diminished; that the relaxation time (T_1) of the *tert*-butylamine hydrogen atoms is shorter, and that a nuclear Overhauser effect is observed for the allyl protons. Figure 17 represents the initial anion formation (68) of the calixarene host and its association with the resulting ammonium ion in the cavity of the calixarene anion. This is an example of an *endo*-calix complex (70). Additional evidence for the proton transfer between host and guest shown in Figure 17 is the fact that aniline, a weaker base, does

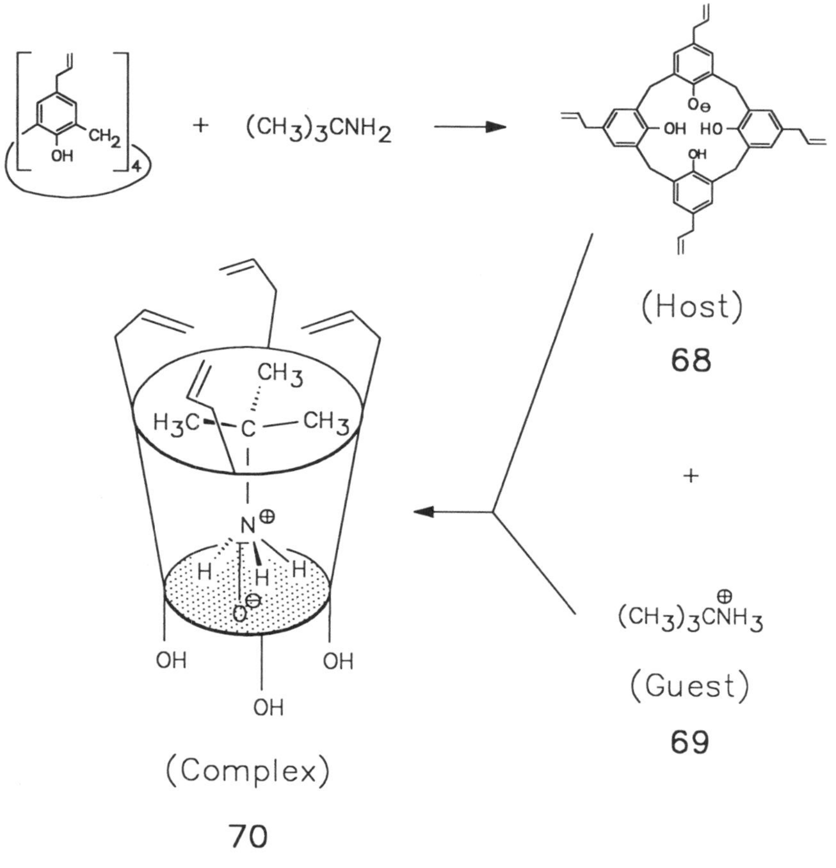

Figure 17. Complex formation between the anion 68 of *p*-allylcalix[4]arene and the *tert*-butylammonium ion (as guest)[1]

not form a complex with calixarenes. Similar ^{1}H-NMR shifts are observed if the calixarene is treated with strong base and the amine with strong acid. Moreover, the fact that the calixarene exhibits a characteristic specificity for amine guests (or ammonium ions) supports the view that the guest is the ammonium ion. Neopentylamine, whose basicity is nearly the same as that of *tert*-butylamine, for example, forms a less stable complex, a fact that can be attributed to the bulky, branched structure of neopentylamine. If one assumes that the three ammonium hydrogen atoms of the amine approach the calixarene oxygen atom as closely as possible, then the *tert*-butyl framework fits comfortably in the cavity. On the other hand, the neopentylamine cation causes repulsive steric interactions with the inside wall of the calixarene.

Complex formation of calixarenes with cations in solution

Izatt and co-workers found it is possible to transport metal ions through hydrophobic liquid membranes with the help of *p-tert*-butylcalixarenes.[22] Interestingly, this complex formation does not occur in neutral media but only in basic solutions, as illustrated by the data in Table 2.

The advantage of calixarene is that it transports ions in *alkaline* solution, whereas [18]crown-6 is much more effective in transporting KNO_3 than KOH. *p-tert*-Butylcalix[4]arene, -[6]arene, and -[8]arene all show a similar capability for ion transport in the order: $Cs^+ > Rb^+ > K^+ > Na^+ > > Li^+$. The cyclic tetramer shows the highest selectivity for Cs^+, while the cyclic octamer functions the best for Cs^+ transport on an absolute scale. From control experiments with *tert*-butylphenol it can be concluded that the macrocycle plays a decisive role, for this open-chain model compound is totally ineffective in alkali metal transport.

Calixarenes are not functioning in a way analogous to the crown ethers in this instance; it is unlikely that they surround the cations with oxygen donors in a planar arrangement. The cavity of calix[4]arene is too small to envelope

Table 2. Comparison of cation transport by calixarenes, [18]crown-6, and *p-tert*-butylphenol in alkaline solutions[1]

Guest ion	Ion flux $[mol/s \cdot m^2 \cdot 10^8]^a$ *p-tert*-Butylcalixarene (as host)		
	8	10	12
Li^+OH^-	—	10^b	2^b
Na^+OH^-	1.5	13	9
K^+OH^-	0.4	22	10
Rb^+OH^-	5.6	71	340
Cs^+OH^-	260	810	1200

^{a}Data rounded off.
^{b}For comparison: [18]crown-6 and *p-tert*-butylphenol with Li^+OH^- as guest ion: $0.9 \ mol/s \cdot m^2 \cdot 10^8$.

the large Cs^+ ion, and that in calix[8]arene is too large to serve as an effective ion cage. The results described earlier for calixarene–amine complex formation rather suggest that the metal complexation involves *endo*-calix complexation; that is, an ion pair interaction is occurring between alkali metal cation and the calixarene anion, such that the cation is imbedded into the lipophilic cone cavity.

Izaat explained the Cs^+ selectivity by noting that Cs ion sheds its hydration sphere more easily than other monovalent cations. If the calix[8]arene assumes a transannular indented conformation (see description earlier of 'pinched' conformation), then it should be able to accommodate two Cs ions per host molecule and thus be twice as effective as calix[4]arene. Experiments indicated that it was in fact about four times as effective.

Izatt notes that calixarenes are useful ion carriers, especially because of their low water solubility and their ability to form neutral complexes with cations (through loss of a proton from an OH group of the calixarene). They also offer the possibility to achieve a coupling of cation transport with reverse proton flow (coupled ion transport).

Water-soluble calixarenes have been described by Shinkai[23] and also Schneider[24] and others.[1a] The original literature should be consulted for information on polymeric calixarenes and calixarenes as catalysts,[1a] as well as for the most recent developments in this field.[25]

7.2.1.7 Physiological properties of calixarenes

It has long been recognized that phenolic compounds exert strong physiological effects. In the case of *p-tert*-butylphenol/formaldehyde resin, particularly the linear tetramer, a contact dermatitis has been noted. *p-tert*-Butylcalix[4]arene and *p-tert*-butylcalix[8]arene have been examined for mutagenesis in the Ames test but without negative results. Halogenated phenols are known to be bacteriostatic agents, and phenol/formaldehyde oligomers have been examined for this property. The linear and cyclic oligomers prepared by Moshfegh, Hakimelahi *et al.* show *in vitro* activity against various pathogenic organisms. An especially high activity was observed for the linear and cyclic tetramers, which appears to be correlated with their chelating ability, in particular towards Fe^{2+}.

Medical and biochemical investigations have been described for a series of phenol/formaldehyde condensation products, including the investigation of tuberculostatic activity, lipophilic/hydrophilic balance, and *in vivo* degradation of calixarene analogs. Carcinogenicity has also been tested.

7.2.1.8 Concluding remarks

Calixarenes offer a simple and economical entry into 'cavity chemistry'. Not only is the simple and efficient cyclization methodology convenient, but also the flexibility in the choice of substituents and the possibility for interconversion of functional groups are advantages. One can choose the size of the cavity and the functional groups attached to the cavity surface. The substances **67c, h** have

358

been shown to function as hydrolysis catalysts; compound **67f** has been utilized as metal chelator and oxygen carrier. Because of their characteristic crystallinity, structures of the calixarenes and of their host–guest complexes can be unequivocally established. The field of calixarenes may well be only in the beginning of its development.[1a]

References to Section 7.2.1

Calixarenes

1. (a) C. D. Gutsche, *Calixarenes, Monographs in Supramolecular Chemistry* (J. F. Stoddart, Ed.), Royal Society of Chemistry, Cambridge, 1989.
   (b) C. D. Gutsche, *Top Curr. Chem.*, **123**, 1 (1984).
   (c) C. D. Gutsche, in *Host Guest Complex Chemistry, Macrocycles* (F. Vögtle and E. Weber, Eds.), p. 375, Springer Verlag, Berlin, 1985.
   (d) C. D. Gutsche, *Acc. Chem. Res.*, **16**, 161 (1983).
   (e) Because hundreds of literature citations are included in review articles (Ref. 1a–c), it does not seem necessary in this section to document each statement or compound with a literature reference.
2. D. J. Cram and H. Steinberg, *J. Am. Chem. Soc.*, **73**, 5691 (1951).
3. A. Baeyer, *Ber. Dtsch. Chem. Ges.*, **5**, 25, 280, 1094 (1872).
4. H. Kämmerer, G. Happel and F. Caesar, *Makromol. Chem.*, **162**, 179 (1972); G. Happel, B. Mathiasch and H. Kämmerer, *Makromol. Chem.*, **176**, 3317 (1975).
5. J. H. Munch, *Makromol. Chem.*, **178**, 69 (1977).
6. C. D. Gutsche, B. Dhawan, K. H. No and R. Muthukrishnan, *J. Am. Chem. Soc.*, **103**, 3782 (1981).
7. F. Raschig, *Z. für Angew. Chem.*, **25**, 1939 (1912).
8. A. Zinke and E. Ziegler, *Ber. Dtsch. Chem. Ges.*, **74**, 1729 (1941).
9. A. Ninagawa and H. Matsuda, *Makromol Chem., Rapid Commun*, **3**, 65 (1982).
10. Y. Nakamoto and S. Ishida, *Makromol. Chem., Rapid Commun.*, **3**, 705 (1982).
11. A. G. S. Högberg, *J. Am. Chem. Soc.*, **102**, 6046 (1980).
12. M. de S. Healy and A. J. Rest, *J. Chem. Soc., Chem. Commun.*, 149 (1981); cf. New O_6 macrocycles: A Gast and E. Breitmaier, *Chem. Ber.*, **124**, 233 (1991).
13. H. Kämmerer, G. Happel and B. Mathiasch, *Makromol. Chem.*, **182**, 1685 (1981).
14. V. Böhmer, P. Chhim and H. Kämmerer, *Makromol. Chem.*, **180**, 2503 (1979).
15. A. A. Moshfegh, E. Beladi, L. Radnia, A. S. Hosseini, S. Tofigh and G. H. Hakimelahi, *Helv. Chim. Acta*, **65**, 1264 (1982).
16. K. H. No and C. D. Gutsche, *J. Org. Chem.*, **47**, 2713 (1982).
17. J. W. Cornforth, P. D'Arcy Hart, G. A. Nicholls, R. J. W. Rees and J. A. Stock, *Br. J. Pharmacol.*, **10**, 73 (1955).
18. W. Saenger, Ch. Betzel, B. Hingerty and G. M. Brown, *Nature*, **296**, 581 (1982); G. A. Jeffrey and W. Saenger, *Hydrogen-bonding in Biological Structures*, Springer, Berlin, 1990.
19. J. R. Moran, S. Karbach and D. J. Cram, *J. Am. Chem. Soc.*, **104**, 5826 (1982).
20. M. Coruzzi, G. D. Andreetti, V. Bocchi, A. Pochini and R. Ungaro, *J. Chem. Soc., Perkin Trans. 2*, 1133 (1982).
21. G. D. Andreetti, R. Ungaro and A. Pochini, *J. Chem. Soc., Chem. Commun.*, 1005 (1979).
22. R. M. Izatt, J. D. Lamb, R. T. Hawkins, P. R. Brown, S. R. Izatt and J. J. Christensen, *J. Am. Chem. Soc.*, **105**, 1782 (1983).
23. S. Shinkai, K. Araki and O. Manabe, *J. Am. Chem. Soc.*, **110**, 7214 (1988); S. Shinkai *et al. J. Am. Chem. Soc.*, **112**, 9053 (1990).

24. H.-J. Schneider, R. Kramer, S. Simova, U. Schneider, *J. Am. Chem. Soc.*, **110**, 6442 (1988).

25. (a) S. Shinkai, K. Araki, J. Shibara and O. Manabe, *J. Chem. Soc., Perkin Trans. 1*, 195 (1989); T. Arimura, T. Nagasaki, S. Shinkai and T. Matsuda, *J. Org. Chem.*, **54**, 3766 (1989).

(b) S. Shinkai, *J. Incl. Phenom. Mol. Recogn.*, **7**, 193 (1989).

(c) S. Shinkai *et al.*, *J. Chem. Soc., Perkin Trans. 2*, 1167 (1989).

(d) S. Shinkai *et al.*, *J. Chem. Soc., Chem. Commun.*, 736 (1989).

(e) D. N. Reinhoudt *et al.*, *Tetrahedron Lett.*, **30**, 2681 (1989).

(f) D. V. Khasnis, M. Lattmann and C. D. Gutsche, *J. Am. Chem. Soc.*, **112**, 9422 (1990).

(g) S. E. Biali *et al.*, *Tetrahedron Lett.*, **32**, 1909 (1991); S. Shinkai, H. Murakami *et al.*, *J. Chem. Soc. Perkin Trans. 1*, 2429 (1991); S. Shinkai *et al.*, *Tetrahedron Lett.*, **33**, 89 (1992); S.-K. Chang, Y. S. Kang *et al.*, *J. Chem. Soc., Chem. Commun.*, 217 (1991); G. R. Newkome, F. R. Fronczek *et al.*, *Tetrahedron Lett.*, **32**, 1133 (1991); F. Grynszpan, O. Aleksiuk and S. E. Biali, *J. Chem. Soc., Chem. Commun.*, 13 (1993).

(h) S. Shinkai *et al.*, *Rec. Trav. Chim. Pays Bas*, **112**, 407 (1993).

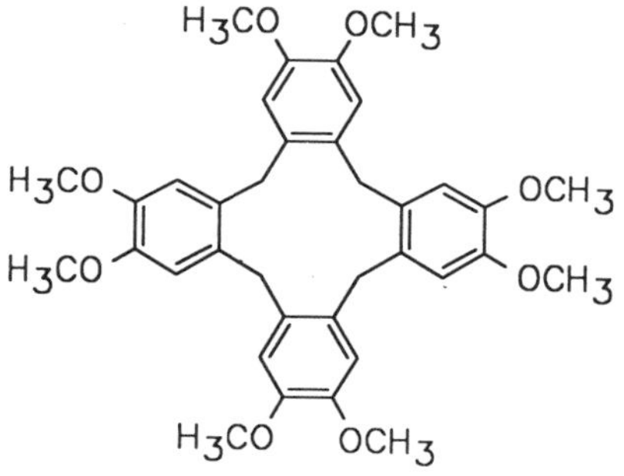

6

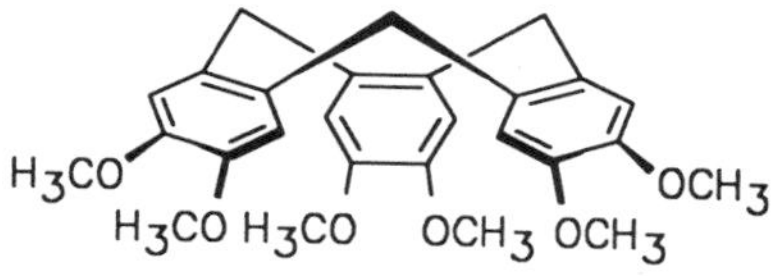

6a

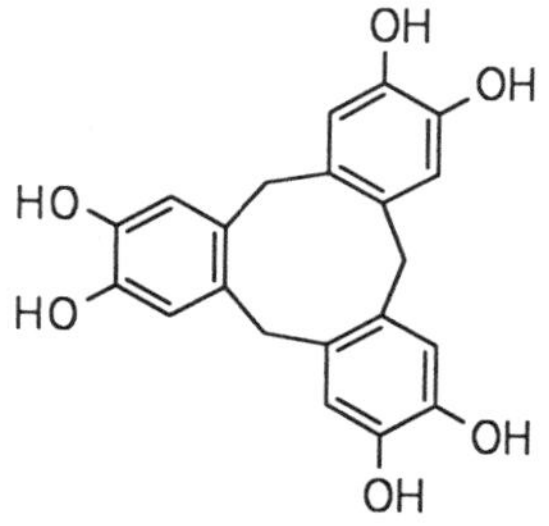

7

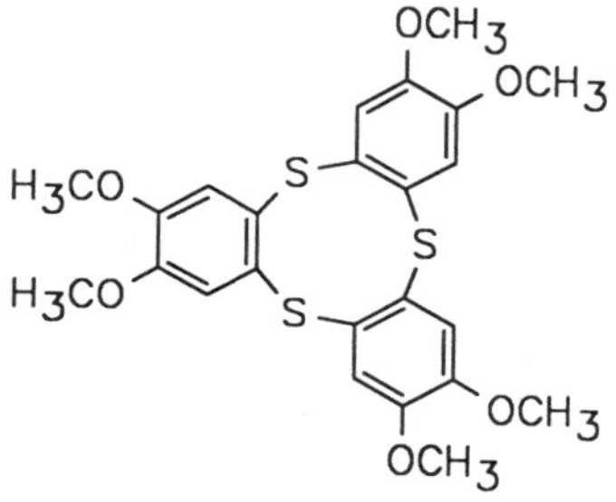

8

7.2.2 CYCLOTRIVERATRYLENE (CTV)

Because cyclotriveratrylene {CTV, '4,5,11,12,18,19-hexamethoxy[1.1.1]ortho-cyclophane'} was described in detail in the textbook *Supramolecular Chemistry* (Wiley, Chichester, 1991), only some mention will be made here for the sake of completeness. CTV (6), first obtained in 1915 by Robinson by the base-catalysed reaction of veratrol with formaldehyde[1] exists in a stable crown conformation (see 6a). The first optically active cyclotriveratrylene derivatives were isolated by Lüttringhaus in 1966.[2] Analogues and derivatives of CTV, like the parent compound itself, are well known for their ability to form inclusion compounds. CTV, for example, forms clathrates with benzene, toluene, chlorobenzene, chloroform, acetone, CS_2, acetic acid, thiophene, decalin, methyl ethyl ketone, ethanol, and other compounds.[3] (See 6 to 8 on page 359.)

Cyclotricatechylene (7) and trithiacyclotriveratrylene (8) also form clathrates with solvent molecules. Crown ether derivatives of CTV such as 9,[4] as well as the speleand 10 and cryptophanes such as 11, prepared by Collet, have been examined for their ability to complex with cations and small neutral molecules. Further information on cryptophanes can be found in the original literature[5] and in the text mentioned above.

For a [1₆]phane oligoketone, see ref. 5e.

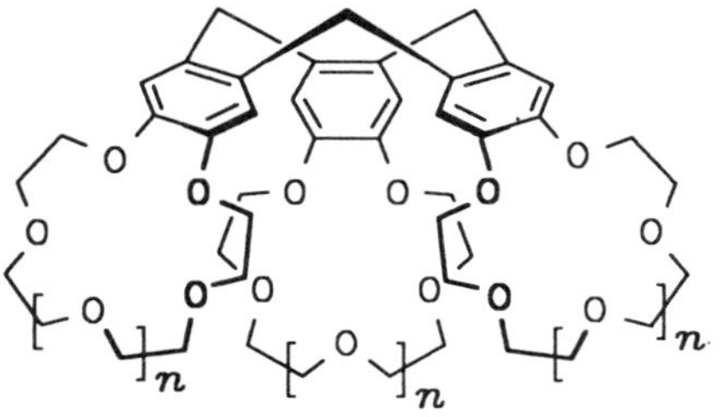

9a : $n = 1$
b : $n = 2$

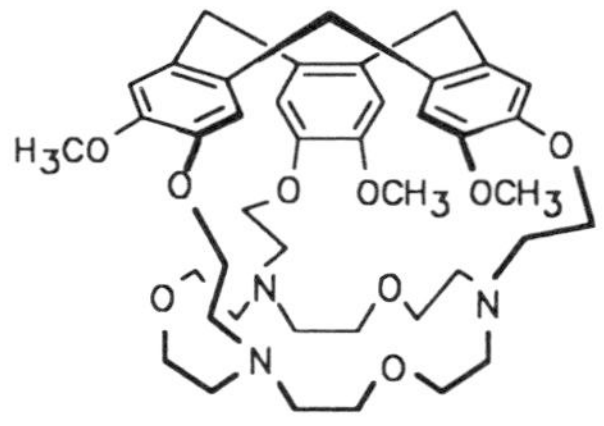

10

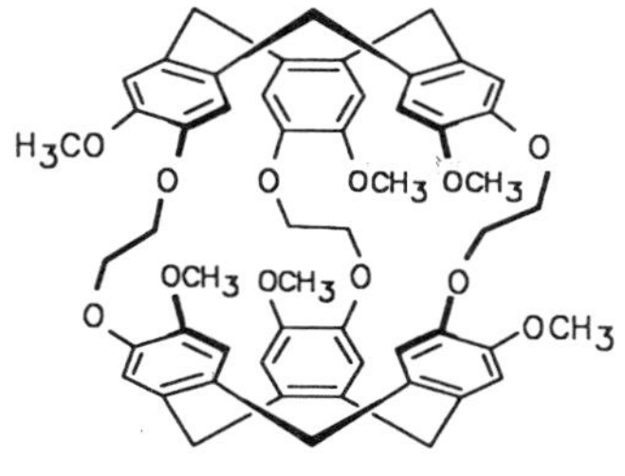

11

7.2.2.1 Cavitands and carcerands

These compounds are mentioned here although they contain bridges in addition to the $[1_n]$phane skeleton. Cavitands, as explained earlier, are understood to be molecules with structurally reinforced cavities which can accommodate ions or molecules. Cram *et al.* carried out the condensation of resorcinol with acetaldehyde followed by bridging of the OH groups in the resulting product to obtain cavitands of the type **12**, some of which crystallize only as 'solvates'.[6] Cavitands of this type can be bridged together 'like watch glasses', whereupon 'carcerands' such as **13** are the result.[7]

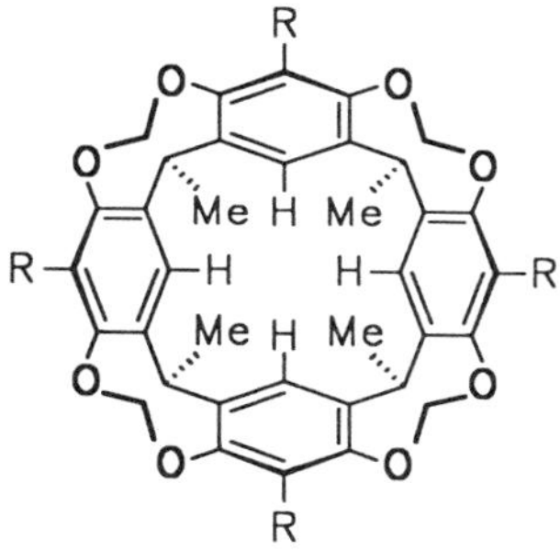

12a : R = H
12b : R = Br
12c : R = CO$_2$Me

As the name implies, carcerands are characterized by such rigid and enclosed cavities, that in the course of the synthesis components from the reaction mixture are trapped in the cavity during the cyclization. Examples of guests captured during such cyclizations are Cs$^+$, argon, and ClCF$_2$CF$_2$Cl. These host compounds are thus cage-like macropolycycles, in which the guest is trapped as in a wire cage. Cram terms the complexes of carcerands 'carceplexes'. Solvent molecules, such as THF and DMF, are also entrapped.

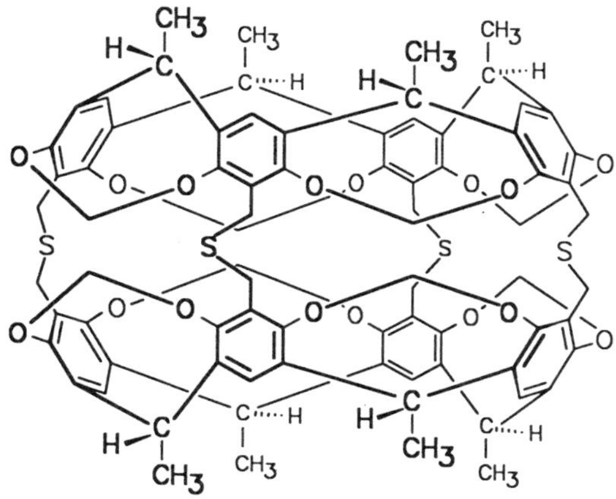

13

362

Characterization of the carceplexes was possible once soluble carceplexes could be obtained in crystalline form. With the aid of NMR spectroscopy Cram could demonstrate that guest molecules have a specific orientation, and that those of considerable length can be placed such that the long axis can rotate. Smaller 'cage inhabitants' move about unhindered in all three directions.

References to Section 7.2.2

1. G. M. Robinson, *J. Chem. Soc.*, **107**, 267 (1915).
2. A. Lüttringhaus and K. C. Peters, *Angew. Chem.*, **78**, 603 (1966); *Angew. Chem. Int. Ed. Engl.*, **5**, 593 (1966).
3. V. K. Bhagwat, D. K. Moore and F. L. Pyman, *J. Chem. Soc.*, 443 (1931).
4. K. Frensch and F. Vögtle, *Liebigs Ann. Chem.*, 2121 (1979).
5. (a) J. Canceill, L. Lacombe and A. Collet, *J. Am. Chem. Soc.*, **107**, 6993 (1985); cf. J. Canceill, A. Collet, J. Gabard, F. Kotzyba-Hibert and J.-M. Lehn, *Helv. Chim. Acta.*, **65**, 1894 (1982).
   (b) A. Collet, in *Inclusion Compounds* (J. L. Atwood, J. E. D. Davies and D. D. MacNicol, Eds.) Vol. II, p. 97, 1984.
   (c) A. Collet, *Tetrahedron*, **43**, 5725 (1987).
   (d) A. Collet *et al.*, *Angew. Chem.*, **101**, 1249 (1989); *Angew. Chem. Int. Ed. Engl.*, **28**, 1246 (1989).
   (e) W. Y. Lee *et al.*, *J. Am. Chem. Soc.*, **115**, 1184 (1993).
6. J. R. Moran, S. Karbach and D. J. Cram, *J. Am. Chem. Soc.*, **104**, 5826 (1982); D. J. Cram, S. Karbach, H.-E. Kim, C. B. Knobler, E. F. Maverick, J. L. Erickson and R. C. Helgeson, *J. Am. Chem. Soc.*, **110**, 2229 (1988); see ref. 1.
7. D. J. Cram, S. Karbach, Y. H. Kim, L. Baczynskyj and G. W. Kalleymeyn, *J. Am. Chem. Soc.*, **107**, 2575 (1985); D. J. Cram, S. Karbach, Y. H. Kim, L. Baczynskyj, K. Marti, R. M. Sampson and G. W. Kalleymeyn, *J. Am. Chem. Soc.*, **110**, 2554 (1988).

7.3 [0ₙ]PHANES

7.3 $[0_n]$PHANES

$[0_n]$ORTHOCYCLOPHANES[1]

For the sake of completeness mention is made here of hexa-*ortho*-phenylene (**9**).[1]

9

7.3.1 OLIGO*META*PHENYLENES

On the basis of their skeletal structure, the spherands described later are examples of oligo-*meta*-phenylenes, which can be designated as $[0_n]$metacyclophanes. The most important member of this series is hexa*meta*phenylene

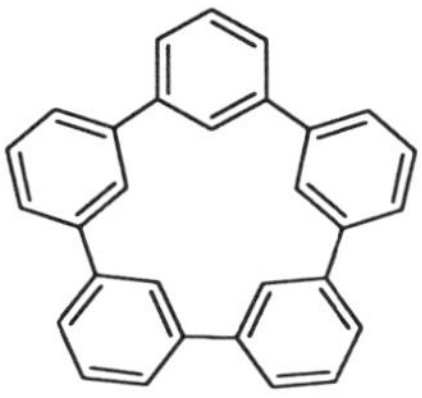

6

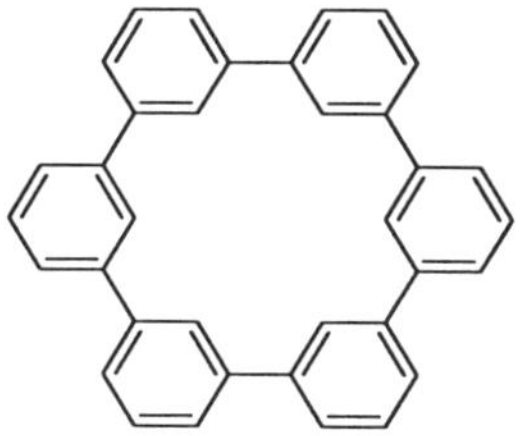

7

{[0$_6$]metacyclophane},[2] first synthesized by Staab and Binnig. It was synthesized by a Kharasch coupling with anhydrous CuCl$_2$ in THF from the di-Grignard reagents derived from 1,3-dibromobenzene, 3,3′-dibromobiphenyl, and 3,3″-dibromo-*m*-terphenyl. The yields fall off dramatically in the conversion from 1,3-dibromobenzene to dibromo-*m*-terphenyl. Following this success

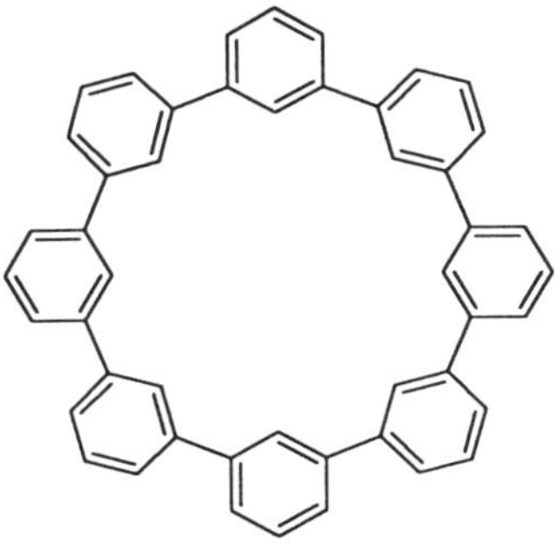

8

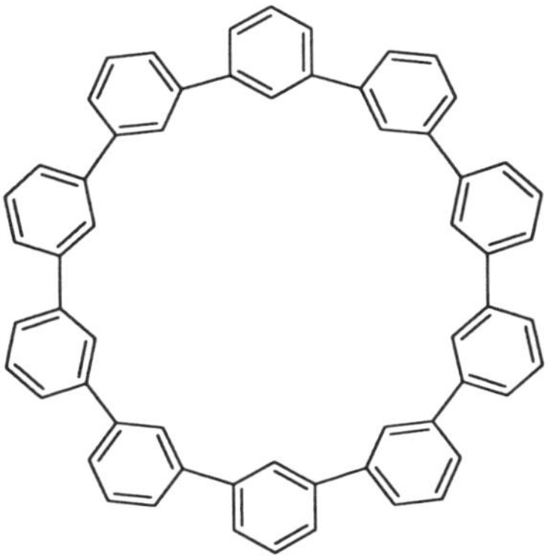

9

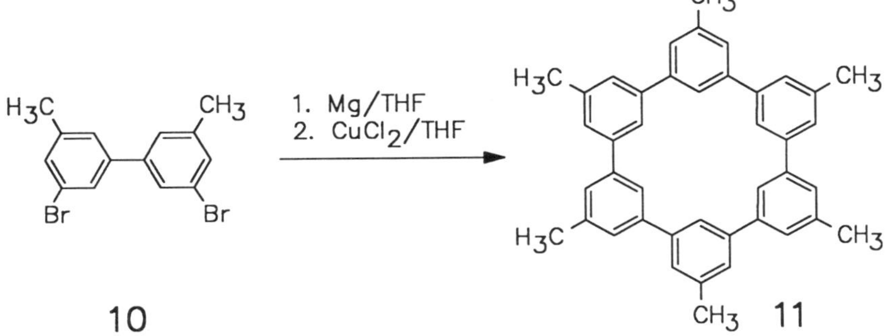

10

1. Mg/THF
2. CuCl$_2$/THF

11

Staab *et al.* were able to synthesize penta- (**6**), octa- (**8**), and deca-*m*-phenylene **9**[3] and also, for example, 5,5′,5″,5‴,5⁗-hexamethylhexa-*m*-phenylene **11**,[4] the last in 14% yield. With CoCl$_2$ a yield of only 1.5% of **11** could be achieved (m.p. 456°C, sublimes at 350°C/10^{-4} Torr, colourless crystals from toluene). None of the substituted hexa-*m*-phenylene **13** could be detected from the Wurtz coupling of the di-Grignard reagent derived from **12**; the exclusive product is the octa-*m*-phenylene **14** (0.47% yield; m.p. 489°C). This is a case of reaction control by steric effects, which are exerted, not at the reaction site itself, but by influencing the conformations of starting materials and intermediates. Similar routes have led to the synthesis of the [0.0.0]dihydrophenanthreno- and -phenanthrenophanes **15** and **16**.[5]

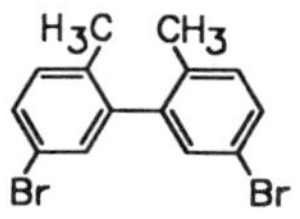

12

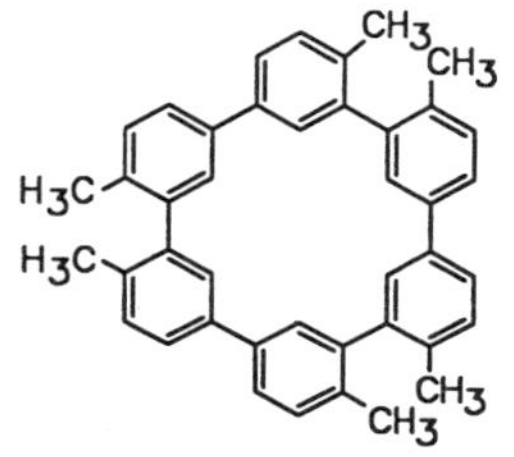

13

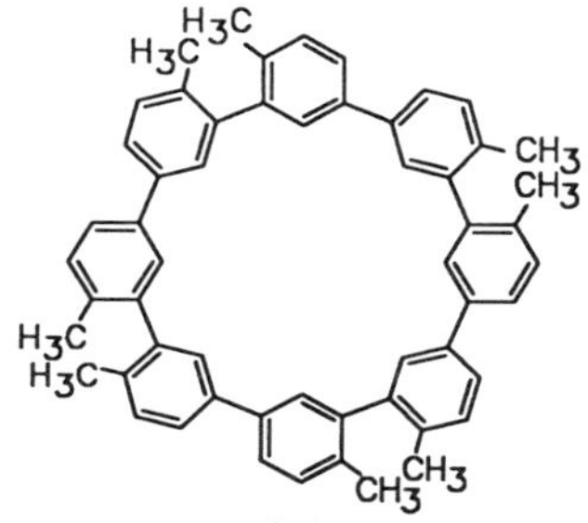

14

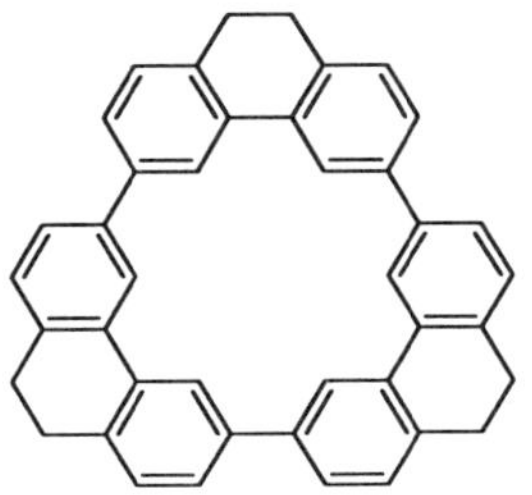

15

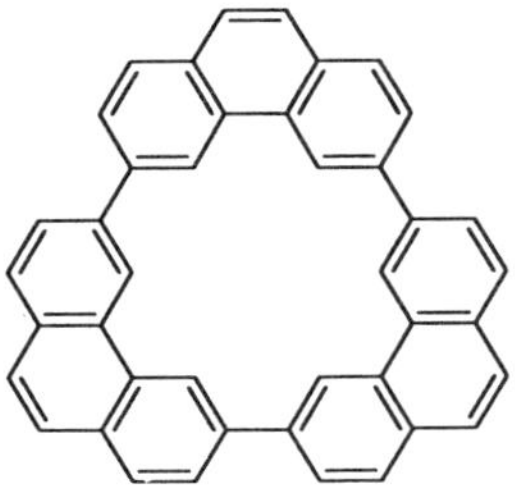

16

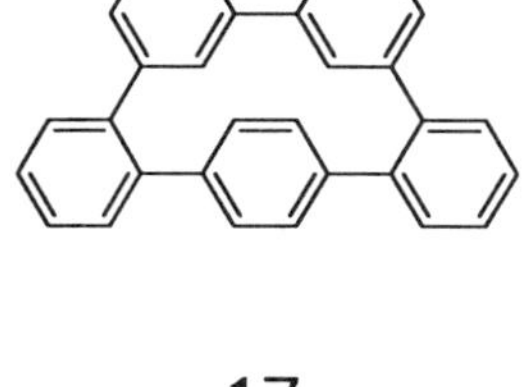

17

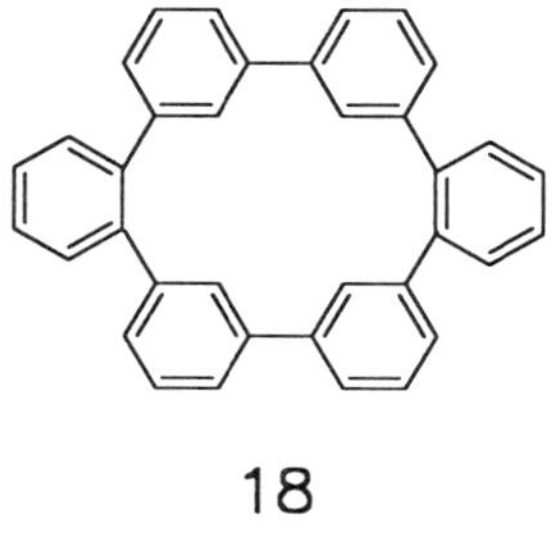

18

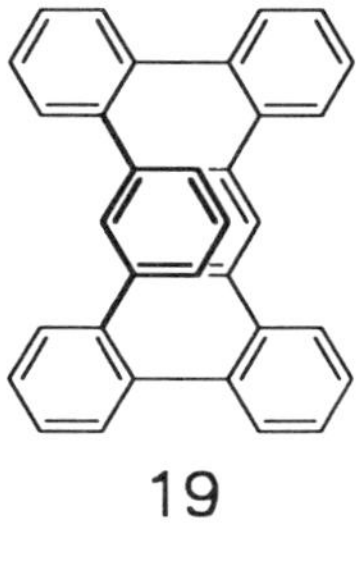

19

An X-ray crystal structure analysis of hexa-*m*-phenylene (**7**) showed that the individual benzenoid rings were twisted away from one another at the exocyclic bonds by an angle of about 30°.

The intra-annular protons in the ^{1}H-NMR spectrum of **15** come at $\delta = 8.63$ p.p.m., a somewhat lower field signal than that in the other hexa-*m*-phenylenes. This may be caused by the higher rigidity of the planar orientation in **15**, with the result that the inner protons are more strongly influenced by

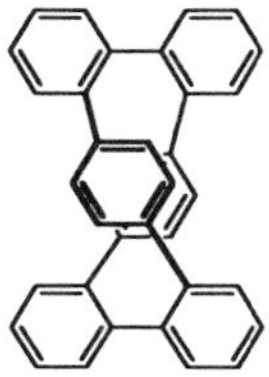

20a

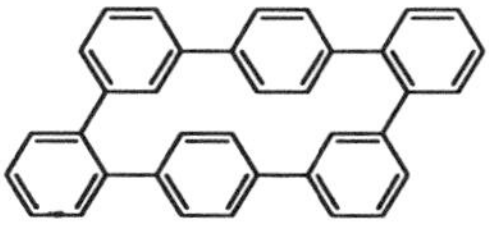

20b

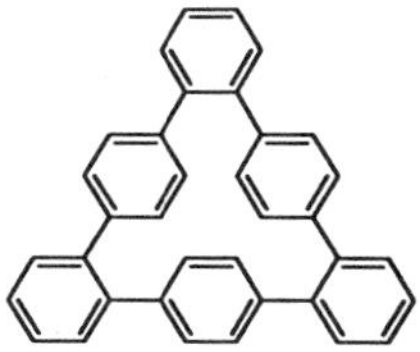

20c

20d

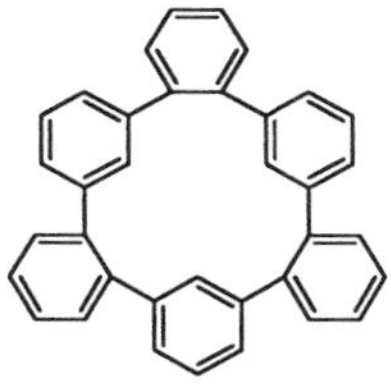

20e

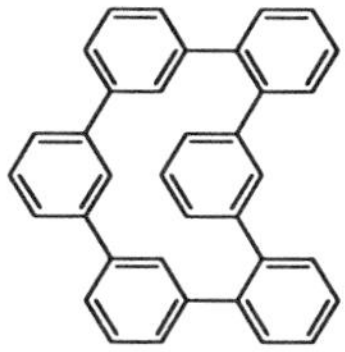

20f

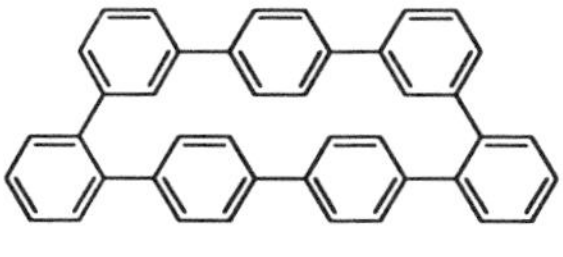

21

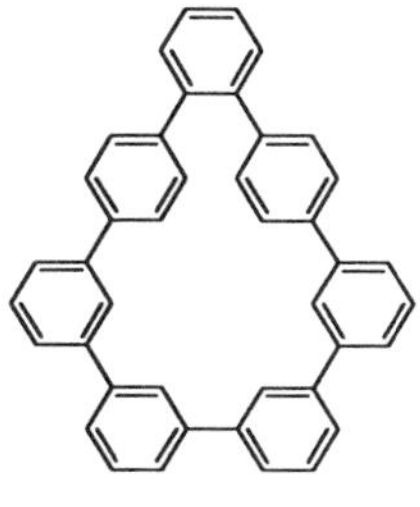

22

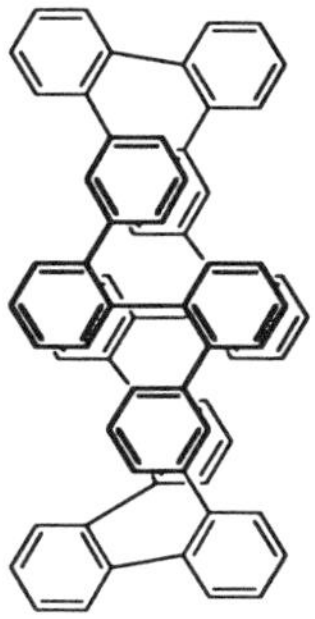

23

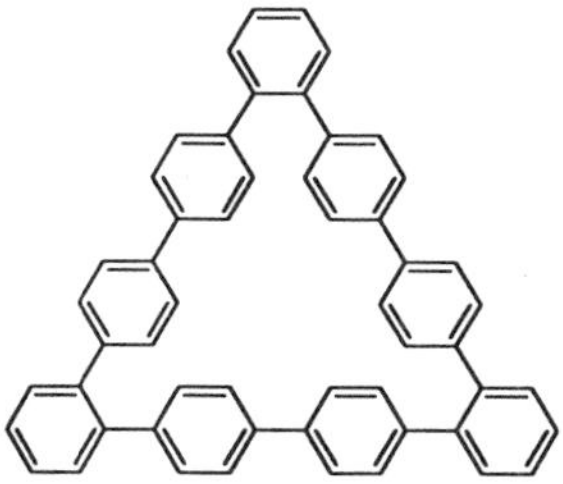

24

the ring current effect from the neighbouring benzenoid rings and they are separated by smaller H–H distances.

The structure of the triphenanthrylene **16** eventually led Staab et al.[6] to a highly successful preparation of kekulene, which actually no longer contains a simple [0]bridged skeletal structure. (Kekulene was described in Section 2.10.2.) More information on the preparation of kekulene and 'small kekulenes' can be found in the textbook *Fascinating Molecules in Organic Chemistry* (Wiley, Chichester, 1992).

To conclude this section, mention is made of several geometrically fascinating macrocyclic ring ensembles belonging to the *ortho-/para-/meta*-types {[0_n]benzenophanes}.[7] In terms of phane nomenclature **20c**,[8] for example, can be called [0_6] (*ortho-para*)$_3$cyclophane.

More difficult to order systematically are the fixed macrocyclic diarylhexatrienes **25** and **26**,[9] whereas **27** again is a genuine [0_n]phane.

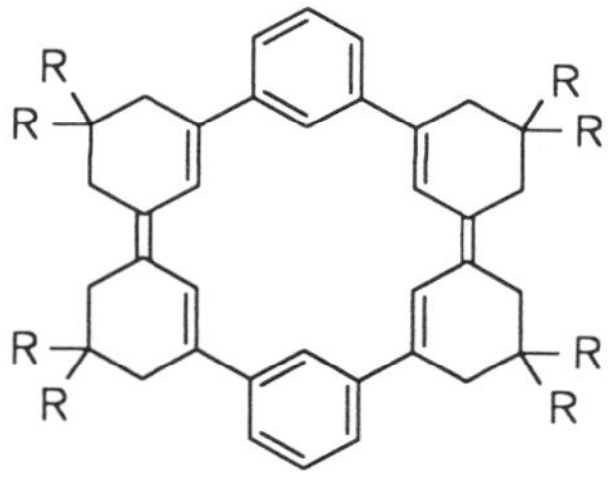

25 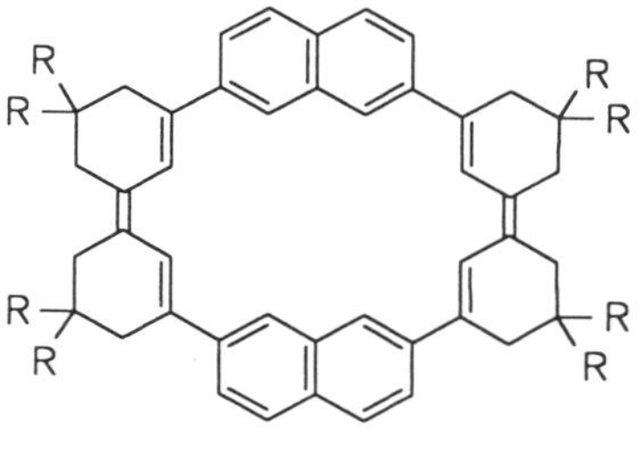   **26**

a : R = H ; **b** : R = CH$_3$

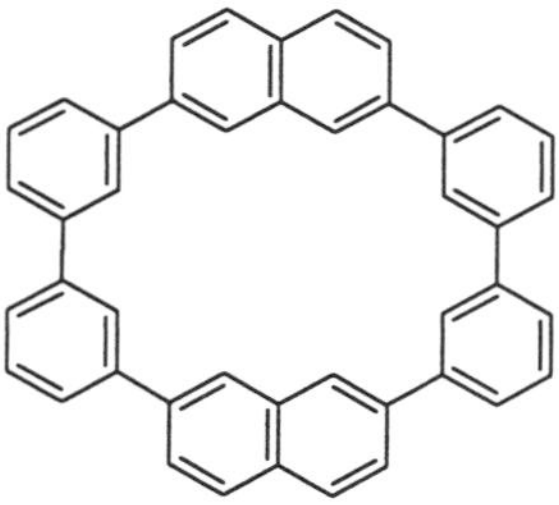

27

References to Section 7.3.1

1. G. Wittig and G. Klar, *Liebigs Ann. Chem.*, **704**, 91 (1967); H. A. Staab and Ch. Wünsche, *Chem. Ber.*, **101**, 887 (1968); X-ray structure of chiral hexa-*o*-phenylenes: H. Irngartinger, *Acta Cryst. B.*, **29**, 894 (1973); centrosymmetric: H. Irngartinger, *Israel J. Chem.*, **10**, 635 (1972).

2. H. A. Staab and F. Binnig, *Tetrahedron Lett.*, 319 (1964).
3. H. A. Staab and F. Binnig, *Chem. Ber.*, **100**, 889 (1967). Cf. Tetraaza[0$_6$]-metacyclophane: G. Sawitzki and H. G. von Schnering, *Chem. Ber.*, **112**, 3104 (1979).
4. F. Binnig, H. Meyer and H. A. Staab, *Liebigs Ann. Chem.*, **724**, 24 (1969); H. Bräunling, F. Binnig and H. A. Staab, *Chem. Ber.*, **100**, 880 (1967).
5. H. A. Staab and H. Bräunling, *Tetrahedron Lett.*, 45 (1965); H. A. Staab, H. Bräunling and K. Schneider, *Chem. Ber.*, **101**, 879 (1968).
6. H. A. Staab and F. Diederich, *Chem. Ber.*, **116**, 3487 (1983); H. A. Staab, F. Diederich, C. Krieger and D. Schweitzer, *Chem. Ber.*, **116**, 3504 (1983); F. Vögtle and H. A. Staab, *Chem. Ber.*, **101**, 2709 (1968); H. A. Staab and M. Sauer, *Liebigs Ann. Chem.*, 742 (1984); for information on superaromaticity and kekulene see J. Aihara, *J. Am. Chem. Soc.*, **114**, 865 (1992).
7. Y. Fujioka, *Bull. Chem. Soc. Jpn.*, **57**, 3494 (1984); *Bull. Chem. Soc. Jpn.*, **58**, 481 (1985).
8. H. Meyer and H. A. Staab, *Liebigs Ann. Chem.*, **724**, 30 (1969).
9. F. Vögtle and C. Thilgen, *Angew. Chem.*, **102**, 1176 (1990); *Angew. Chem. Int. Ed. Engl.*, **29**, 1162 (1990).

7.3.2 SPHERANDS

The *spherands* have been described in detail in the textbook *Supramolecular Chemistry* (Wiley, Chichester, 1991). This family of ligands discovered by Cram[1] has its basis in the rigid structure of oligo-*m*-phenylenes investigated by Staab[2] in combination with the concept of intra-annular functional groups which was developed by Vögtle *et al.*[3]

In a spherand such as **30** the rigid *preorganization* of functional groups (CH$_3$, OH) should produce a nearly perfectly round cavity which is framed by the corresponding binding sites. According to Cram such a cavity should be 'enforced' by a supporting, rigid skeleton of covalent bonds. The conformational flexibility should be so restricted that the cavity of the host compound can not be occupied by rotating components of its skeleton or by substituents. In accord with this concept, the spherand **30**—the first one prepared—contains a rigid hexa-*m*-phenylene backbone with six converging methoxy groups. From CPK models it can be seen that the result is an alternating *up–down* arrangement of the six methoxy groups; that is, the methoxy groups are oriented alternately up and down from one another. Because the conformation is extremely rigid in **30**, not even the methoxy groups are able to rotate through the centre of the cavity. That is, the anisyl building units are *self-organized*, and as a result solvation of the oxygen atoms is hindered by the six phenylene and six methyl groups. The six oxygen atoms assume a nearly perfect octahedral arrangement. The diameter of the cavity varies with the dihedral angle of the six arene groups, the average being about 162 pm. This corresponds roughly to the value of the diameters of Li$^+$ (148 pm) and Na$^+$ (175 pm). Thus the spherand **30** provides a preorganization which is both electronically and sterically complementary to Li$^+$ and Na$^+$ but not adapted for accepting K$^+$ as a guest ion. Indeed, it is found that the free binding enthalpy for the picrate salts of Li$^+$ and Na$^+$ at 25°C in CDCl$_3$ is $-\Delta G° > 96$ and 80 kJ/mol, respectively, while that for other ions is immeasurably small.[1]

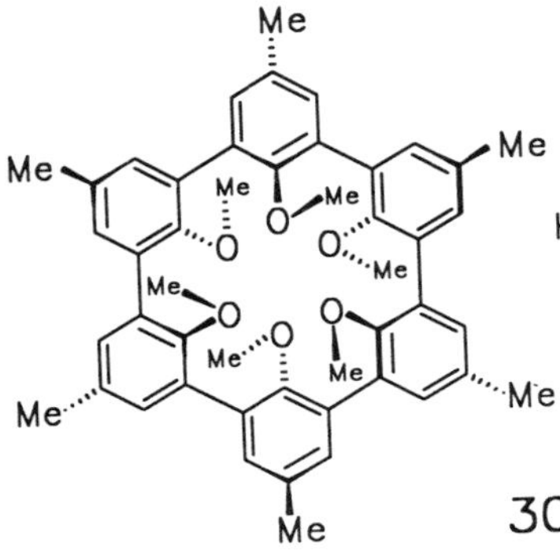

30

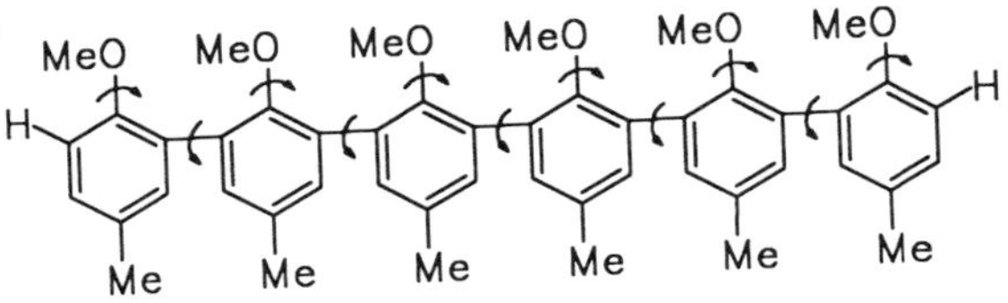

31

The spherand **30** has two advantages over the open-chain model compound, which is not preorganized: sterically rigid, preorganized octahedral arrangement of the six oxygen atoms and resistance to solvation. In fact, the open-chain oligophenol ether **31** binds Li$^+$ and Na$^+$ only with a $-\Delta G° < 25$ kJ/mol. The reason is that **31**, unlike **30**, can exist in more than a thousand conformations resulting from rotations of the arene–arene and the *O*-methyl single bonds, only two of which are ideally suited for cooperative binding. In most of the conformations in the open-chain model compound the donor centres are situated more favourably for solvation than for coordination with a cation.

Figure 1 represents schematically the uptake of an alkali metal cation in the cavity of spherand **30**.[1b]

Since the first spherand was made, several X-ray crystal structure analyses of spherands and their metal complexes have been published.[1] As expected, the free ligands exist in conformations similar to those of the complexes, the difference being only that the free ligand has an empty cavity, which in the case of a complex is filled with a cation.

A large number of additional spherands have been synthesized by Cram, with variations in, for example, the hetero atoms, ring size, binding positions,

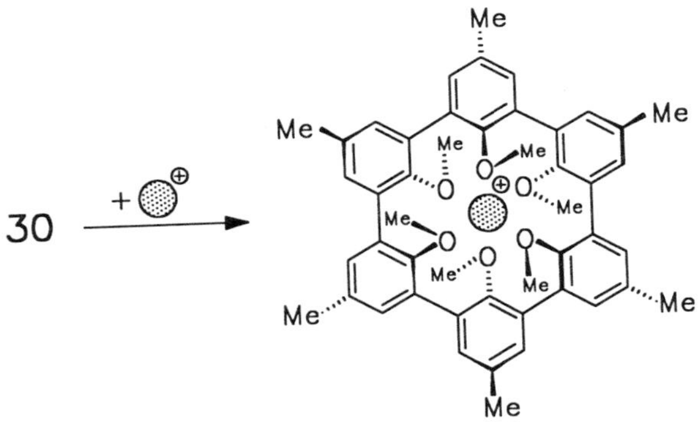

Figure 1. Cation complexation of spherand **30**

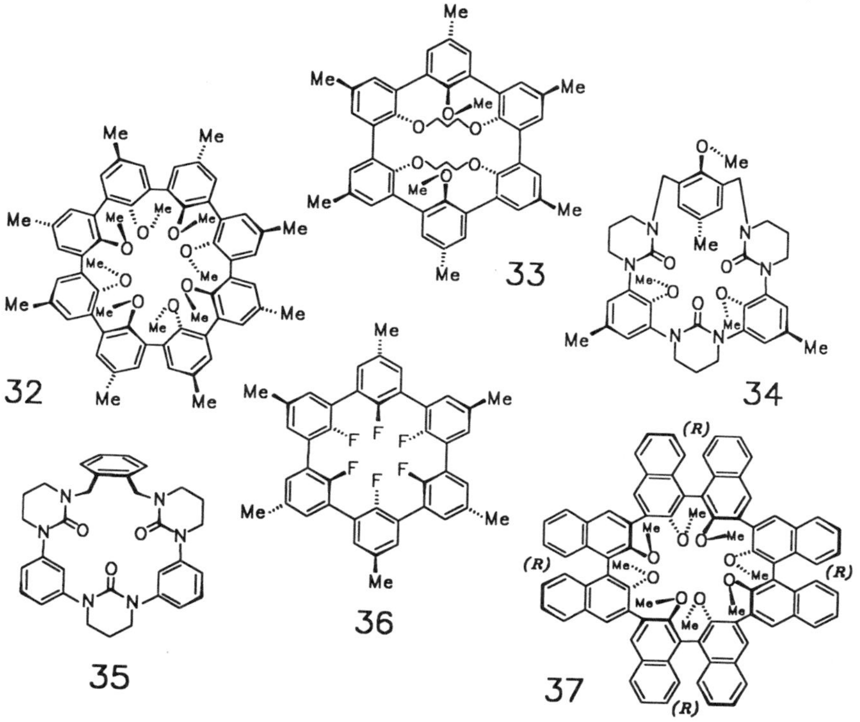

Figure 2. A selection of spherands with varying 'design'

or connectivity. Some examples of these various spherands are given in Figure 2.[1]

In **34** and **35** the 'urea backbone' involves less severe steric hindrance, and the ligands are able to bind to guests through hydrogen bonding. The intra-annular fluoro-substituted spherands **36** prove to be ineffective, their cation binding being very modest. **36** does not form a complex with Li[+]. The chiral binaphthyl spherand **37** was prepared in 1.6% yield by Cram. The reader is referred to reviews and original literature[1] for more details about complexation of various spherand modifications.

In this discussion we shall turn to the 'hemispherands', 'cryptospherands', and 'carcerands',[1] even if they are not genuine [0$_n$]phanes.

Hemispherands (Figure 3) differ from spherands because of their less extensive preorganization. Characteristically made up of a partially flexible rather than fully rigid molecular skeleton, hemispherands are actually a combination of spherands with crowns and cryptands. The special mixing of rigidity and flexibility in hemispherands offers new possibilities for studying structure-binding correlations and for the designed 'breeding' of specific cation selectivity.

Cryptospherands are partially organized cryptands 'mixed' with structural elements from spherands. A selection of characteristic members of these two families of compounds are presented in Figure 3.

From among the extensive investigations of Cram concerning the chemistry of spherands, a few observations involving kinetics and selectivity will be mentioned. Complexation of rigid, preorganized spherands with guest ions was expected to be kinetically slow. In fact, the spherand **30** complexes with Li^+ and Na^+ picrates with a rate constant $k^\rightarrow = 8 \times 10^4 \ 1 \cdot mol^{-1} \cdot s^{-1}$. The decomplexation constant $k^\leftarrow$ for the $30 \cdot Li^+$ complex, on the other hand, is $< 10^{-12}$. These results are reminiscent of the cryptands, which show roughly similar kinetics for ion complexation. In general it is observed that the rates of complexation and decomplexation decrease as the extent of preorganization of the host for complexation is increased.

The highest Li^+/Na^+ selectivity is exhibited by the spherands **30** and **33**. The 'cryptohemispherand' **41b** and the bridged hemispherand **42** show the highest Na^+/Li^+ selectivity. High selectivities for Na^+ over K^+ are provided by the spherands **30** and **31** and the cryptospherand **41a**. The cryptospherand **41c** shows the highest selectivity for K^+/Na^+; the hemispherand **42** for K^+/Rb^+; and the spherand **32** for Rb^+/K^+.

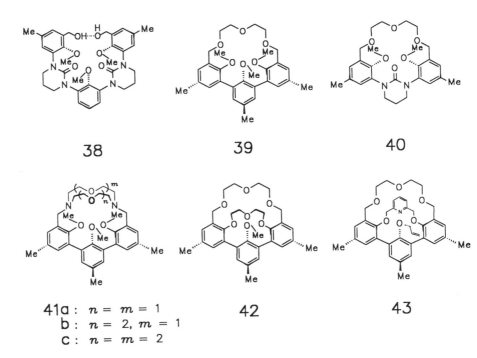

38 **39** **40**

41a : $n = m = 1$
b : $n = 2, \ m = 1$
c : $n = m = 2$

42 **43**

Figure 3. Selected hemispherands (**38–40**) and cryptospherands (**41–43**) with varying extents of preorganization

In summary, it has been found that the spherands **30** and **33**, with their high specificity for binding Li$^+$ over Na$^+$ and for the binding of Na$^+$ over all other ions, are unique. Only the small cryptands '[2.2.1]' or '[2.1.1]' (see *Supramolecular Chemistry*, Wiley, Chichester, 1991) and a few crypto-hemispherands **(41a,b)** show comparably high specificity for Li$^+$ and Na$^+$. It is evident that the principle of preorganization is not only valid for binding power but also controls ion selectivity.

The cyanospherands[4] described in 1989 by Cram offer a fundamentally novel possibility as compared to the spherands and hemispherands. While spherands possess a 'negative hole' for cations, a 'positive' cavity for anions is created in cyanospherands, along with two outer binding regions with high negative charge density for cations. This comes about because of the high dipole moment and orientation of the intra-annular cyano substituents, predicted to afford a new type of cooperative binding for salts—complexation of both cations *and* anions. The cooperative nature implies that coordination at one end of the cyano group enhances the binding capability at the other end. According to Cram, the molecule **44a** exists in a strongly preorganized [*up–down*$_4$] conformation.

44a : $n = 0$
44b : $n = 1$
44c : $n = 2$

Four each of the cyano groups are projected alternately above and below the plane of the ring. The 'positive' carbons of the C≡N groups form a spherical, positively polarized cavity with a diameter of about 220 pm. The negatively polarized nitrogen atoms are located around the opening of the cavity in a crown-like orientation. From the X-ray crystal structure of the di-K$^+$ complex of **44a**, it can be shown that complexation of the cation is actually occurring; but the second step postulated by Cram—namely, anion encapsulation between the two cations—has not yet been substantiated.

The cyanospherands **44b** and **44c**, although conformationally more flexible and thus less well preorganized, nevertheless bind alkali and ammonium ions more strongly than normal crown ethers.

Replacement of anisole units in spherands with pyridines was accomplished by Newkome *et al.*[5] and by Toner *et al.*,[6] with formation of the sodium complex **47** and the so-called cyclosexipyridine **45**. Reference to original papers and reviews[1c] will provide information on the synthetic methodology. Synthesis of bridged pyridinospherands and dodecahydrohexaaza kekulenes **48** have also been reported.[7]

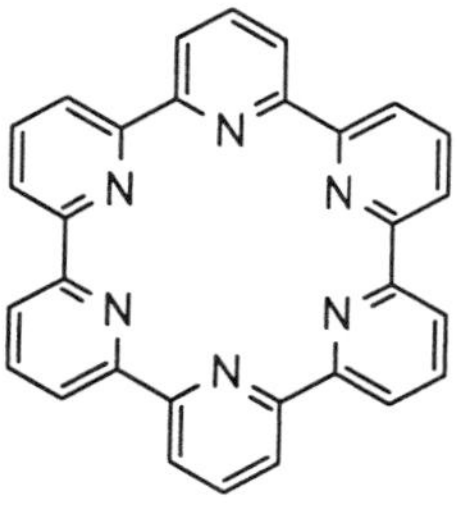

45

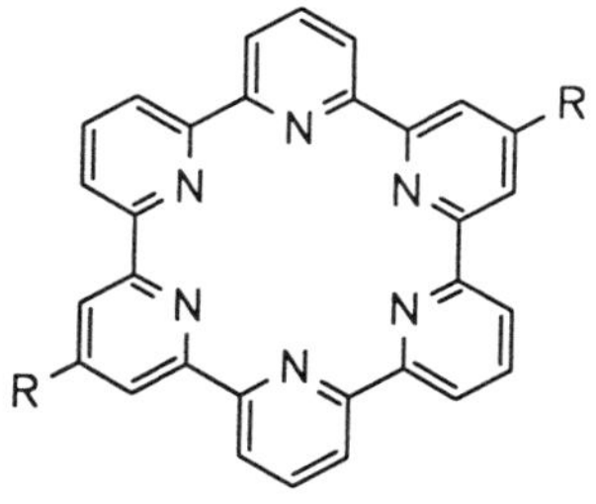

46

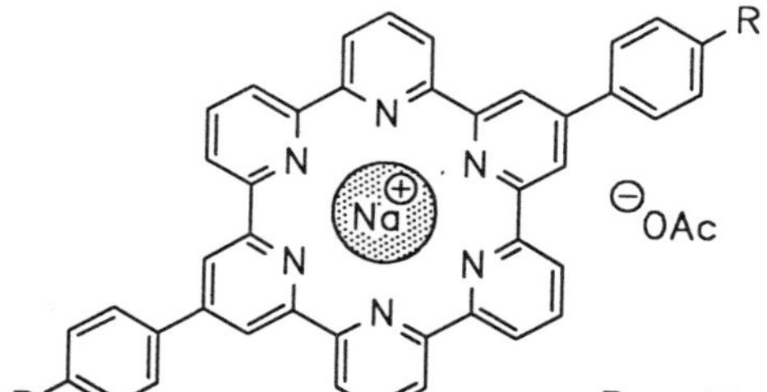

R = Me
R = Et

47

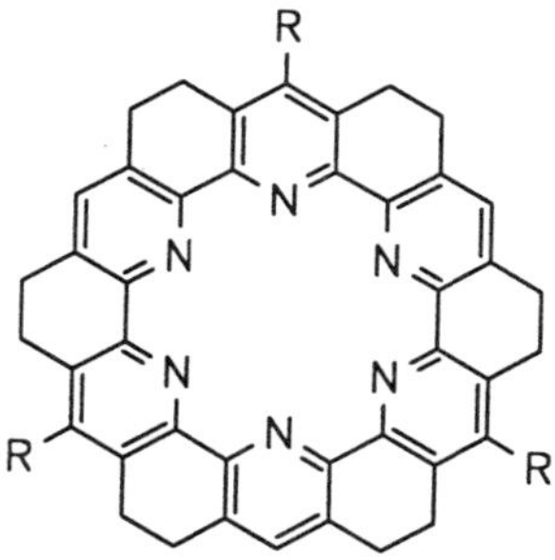

48

R = H
R = Bu

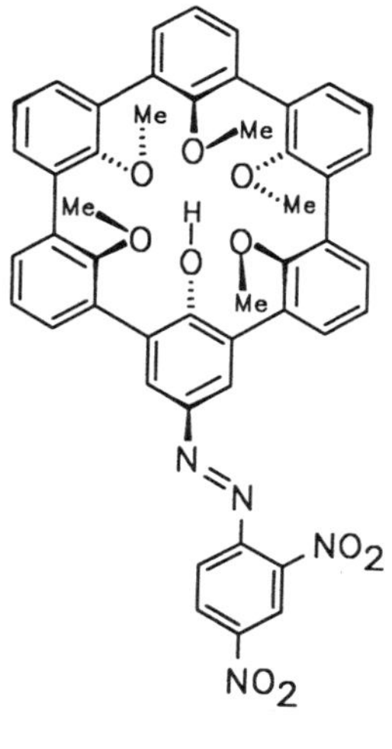

49

An interesting type of compound, based upon the 'chromoionophores' or 'acerands' developed by Vögtle et al.[8] and by Takagi, Ueno, and Misumi et al.[9] is the spherand dye, illustrated by **49**[10] which changes colour in a cation-selective process upon addition of salts.[11,12]
Cavitands and carcerands are discussed in Section 7.2.2.1.

References to Section 7.3.2

1. Reviews:
   (a) D. J. Cram, *Angew. Chem.*, **100**, 1041 (1988); *Angew. Chem. Int. Ed. Engl.*, **27**, 1009 (1988); D. J. Cram and K. N. Trueblood, in *Host Guest Complex Chemistry, Macrocycles* (F. Vögtle and E. Weber, Eds), p. 125, Springer, Berlin, 1985.
   (b) E. Weber, in *Crown Ethers and Analogs* (S. Patai and Z. Rappoport, Eds.), p. 305, Wiley, New York, 1989.
   (c) J. L. Toner, in *Crown Ethers and Analogs* (S. Patai and Z. Rappoport, Eds.), p. 77, Wiley, New York, 1989; see also D. J. Cram *et al.*, *J. Am. Chem. Soc.*, **112** 5837 (1990).
2. H. A. Staab and F. Binnig, *Chem. Ber.*, **100**, 293 (1967).
3. Cf. E. Weber and F. Vögtle, *Chemie in uns. Zeit*, **23**, 210 (1989).
4. K. Paek, C. B. Knobler, E. F. Maverick and D. J. Cram, *J. Am. Chem. Soc.*, **111**, 8662 (1989). See also reference 3.
5. G. R. Newkome and H. W. Lee, *J. Am. Chem. Soc.*, **105**, 5956 (1983).
6. J. L. Toner, *Tetrahedron Lett.*, **24**, 2707 (1983); Cf. reference 1c.
7. J. E. B. Ransohoff and H. A. Staab, *Tetrahedron Lett.*, **26**, 6179 (1985); T. W. Bell and A. Firestone, *J. Am. Chem. Soc.*, **108**, 8109 (1986).
8. Review: H.-G. Löhr and F. Vögtle, *Acc. Chem. Res.*, **18**, 65 (1985).
9. Review: M. Takagi and K. Ueno, in *Host Guest Complex Chemistry, Macrocycles* (F. Vögtle and E. Weber, Eds.), p. 217, Springer, Berlin, 1985.
10. D. J. Cram, R. A. Carmack and R. C. Helgeson, *J. Am. Chem. Soc.*, **110**, 571 (1988).
11. Cf. M. Inouye, K. Kim and T. Kitao, *J. Am. Chem. Soc.*, **114**, 778 (1992).
12. Cf. Ch. Reichardt *et al.*, *Liebigs Ann. Chem.*, 23 (1993); *Chem. Ber.*, **126**, 143 (1993).

8 Porphyrinophanes

INTRODUCTION

The intention in this chapter is not to review the synthesis and chemistry of unsubstituted porphyrins but rather to take up more recent porphyrins that are modified through fence-like substituents or with bridges or caps.[1] Although the unsubstituted porphyrin, with its [1.1.1.1]pyrrolophane skeleton, belongs strictly to the phanes, it is the 'bridged' and 'capped' porphyrins,[1] because of their additional, usually longer bridges, which are more characteristic of the phane series. First, they involve stereochemistry and reaction patterns characteristic of the cyclophanes and phanes as described in this book. Secondly, in addition to the relatively small 'hole' in the centre of the porphyrin skeleton, which typically serves as a cavity for small cations, the additional bridging creates a larger cavity ideal for examining host–guest interactions. In this sense the porphyrin skeleton can serve as a 'spacer plate' for additional bridging, in the same way that a benzene ring occupies the central position in multiple bridging leading to [2.2.2](1,3,5)cyclophane or superphane.

A group of imaginative researchers have developed the approach of using porphyrin not only as the 'anchor group' for multiple bridging but also as a model ligand for the iron complexes of haemoglobin and myoglobins and for cytochromes. This means that bridged, capped, and fenced porphyrins[1] can serve as *functional models* for biological molecules and transformations. The challenge lies in imitating biological processes and perhaps even improving on the modes of action of haemoglobin, myoglobin, the cytochromes, and other natural porphyrins. In this connection the following presentation centres on the basic unit of the [1₄]phane skeletal structure of porphyrin.

8.1 HAEMOGLOBIN- AND MYOGLOBIN MODEL COMPOUNDS

Haemoglobin (Hb), the red component of blood, belongs to the group of metalloproteins. Its major function in vertebrates is the transport of oxygen molecules from the lung to muscle tissue, although it also plays a role in the transport of CO_2 from the muscle to the lungs. *Myoglobin* (Mb), found in skeletal muscle, functions mainly by storing O_2 which has been delivered by haemoglobin.

Some remarks about the structure of Hb and Mb are necessary in order to appreciate the challenge presented in creating synthetic model compounds.[1a] Hb and Mb contain Fe(II)-protoporphyrin (haem, **6**) as the prosthetic group.

Common haemoglobin is a dimer of dimers, consisting of two α- and two β-subunits. Each subunit is made up of a haem group and a polypeptide (globin) chain. Mb consists of one haem group and one globin chain.

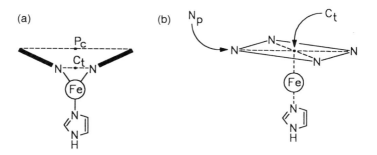

6

Because the X-ray crystal structures are known for both Hb and Mb before and after the uptake of O_2, it is possible to follow the stereochemical consequence of oxygenation. In the deoxygenated form of Hb and Mb the Fe(II) is pentacoordinate. The four planar ligand-donor centres are occupied by the porphyrin nitrogen atoms and the fifth by an axially located imidazole of a histidine ring. Upon oxygenation there is essentially no change in the Fe-imidazole–N distance, but the Fe atom, which experiences a change in spin state from 'high spin' to 'low spin', moves closer to the plane of the porphyrin and brings the proximal imidazole along (Figure 1). According to the mechanical model of Perutz, this migration is an important factor for the cooperativity shown by Hb in its binding with O_2.[2] In the case of Mb this movement is not so pronounced. The fact that the Fe^{2+} remains outside the porphyrin plane can be attributed to the eclipsed orientation of the imidazole plane to the porphyrin N atoms. If this angle is large (see Figure 2), then the interaction between the imidazole and the porphyrin nitrogen atoms becomes smaller. When the Fe-imidazole–N distance remains essentially constant, as is found to be the case, the approach toward the porphyrin plane is facilitated.

Figure 1. (a) Side view; (b) perspective view of the pentacoordinate Fe(II)-porphyrin

(a)   (b)

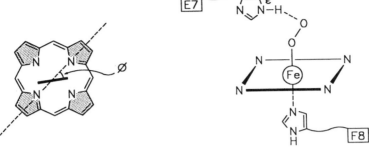

Figure 2. (a) Eclipsing angle (ϕ) between the imidazole plane and the porphyrin axis. (b) Hydrogen bonding between N^{ϵ} of the histidine residue E7 and the bound O_2 molecule

The O_2 molecule is bound in a bent 'end-on' geometry, as was predicted by Pauling (Figure 2b).[3] The hydrogen bonding to histidine E7 was determined by neutron diffraction analysis.

8.2 OXYGEN BINDING IN NATURAL AND SYNTHETIC PORPHYRINS

The most important and fascinating property of haemoglobin is its uptake of O_2.[1a,4] If one plots the saturation of haemoglobin with O_2 against the partial pressure of oxygen, the result is an S-shaped curve (oxygen-binding curve). This means that O_2 is released at low oxygen partial pressure (in the muscle) and that O_2 molecules are bound again at high pressure (in the lungs). By contrast, the oxygen-binding curve for myoglobin is hyperbolic.

Changes in the tertiary and quaternary structures of Hb in the course of O_2 uptake have allowed a distinction between two forms of Hb:

T-state (low O_2 affinity), e.g. deoxy-Hb
R-state (higher O_2 affinity), e.g. oxy-Hb

In the haemoglobin models studied thus far, an attempt has been made to simulate the properties of one or both of these states.

Simple, planar iron(II)-porphyrins (for example, the octaaza macrocycle **7**), react at low temperature rapidly and irreversibly with O_2.[5] Normally the oxidation product is a 'μ-oxo dimer'.

7

The mechanism of this reaction has been well studied, several of the steps postulated in the following equilibria having been characterized by spectroscopic methods. Reaction step (e) in this sequence is irreversible. The μ-peroxo dimer is

$$\text{PFe(II)} + \text{L} \quad \rightleftarrows \quad \text{PFe(II)L} \tag{a}$$

$$\text{PFe(II)L} + \text{L} \quad \rightleftarrows \quad \text{PFe(II)}_2 \tag{b}$$

$$\text{PFe(II)L} + \text{O}_2 \quad \rightleftarrows \quad \text{PFe(II)} \cdot \text{O}_2 \tag{c}$$

$$\text{PFe(II)L} \cdot \text{O}_2 + \text{PFe(II)L} \quad \rightleftarrows \quad \text{LPFe(II)-O-O-Fe(II)PL} \tag{d}$$

$$\text{LPFe(II)-O-O-Fe(II)PL} \quad \rightarrow \quad 2\,[\text{LPFe(IV)} = \text{O}] \tag{e}$$

$$\text{LPFe(IV)} = \text{O} + \text{LPFe(II)} \quad \rightarrow \quad \text{PFe(III)-O-Fe(III)P} + 4\text{L} \tag{f}$$

'decomposed' in step (d), it having been generated in a bimolecular process that involves two metalloporphyrin nuclei [equilibrium (d)]. This reaction step (d) must be avoided, if the decomposition of the Fe(II) porphyrin is to be prevented. Thus a common characteristic goal in the design of all porphyrin models is the prevention of step (d). This can be achieved by steric protection at least on one side of the metalloporphyrin. In haemoglobin itself this bimolecular process is blocked because the protein always holds the two haem units apart, such that the minimum separation is about 2500 pm. The octaaza macrocycle **7** was the first successful porphyrin model in which steric hindrance was utilized in order to prevent or at least slow down the formation of μ-oxo dimer. This iron complex described by Baldwin in 1973 is in fact a reversible oxygen carrier, indeed at relatively low temperature.[5]

8.3 FENCED PORPHYRINS

Since porphyrins can themselves be considered as phanes {[1.1.1.1]pyrrolophane-type structure}, advantage can be taken of 'fenced porphyrins' as an entry into bridged porphyrins, whose phane characteristics are already well known. The first example of a fenced porphyrin is the pivaloyl-substituted **8** published by Collman in 1973.[6] A general scheme for synthesis of fenced porphyrins is

8: M = 2H$^{\oplus}$
9: M = Fe(II)

outlined in Figure 3. The disadvantage in this synthesis is the chromatographic separation and isolation of the four atropisomeric tetrakis(aminophenyl) porphyrins. Although all the other atropisomers can be re-equilibrated, the all *cis*-$\alpha,\alpha,\alpha,\alpha$-tetra-*o*-aminophenylporphyrin (**10**) can be obtained in measurable amounts. In spite of a more recent method, in which all atropisomers can be converted to the $\alpha,\alpha,\alpha,\alpha$-form, the chromatography and workup are nevertheless tedious.

Metallation [i.e. complexation with Fe(II)] can be accomplished in various ways. Normally iron is introduced as Fe^{3+} and reduced to Fe^{2+} before the complexation studies. Because difficulties sometimes arise in the reduction step, methods have been developed for direct introduction of Fe^{2+}.

The Fe complex **9** binds O$_2$ reversibly in the presence of one of many possible axial ligands, including 1-alkylimidazole, 1,2-diimidazole, pyridine, bipyridyl, tetrahydrothiophene, and tetrahydrofuran. In benzene solution at 25°C in an atmosphere of dry oxygen and with three equivalents of 1-methylimidazole, the complex has a half-life of almost 2 months. This indicates that the fenced structure serves to prevent the irreversible formation of the μ-oxo dimer. Since it is known that **9** forms hexacoordinate complexes with additional coordinating ligands, it is likely that this contributes to its remarkable stability. Every hexacoordinate complex of **9** is protected against 'inner-sphere' reactions with O$_2$, which can lead to irreversible oxidation (see Figure 4).

One of the most important contributions in this field was the successful isolation and X-ray crystal structure determination of some hexacoordinate iron fenced-oxy complexes. The first example was **11**[8] which was consequently

380

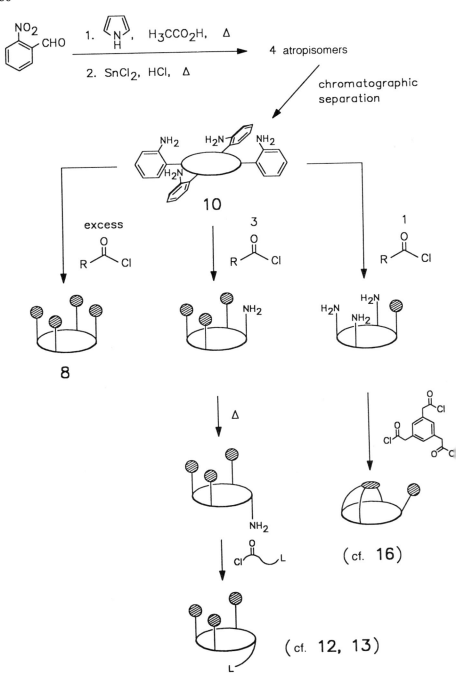

Figure 3. The synthesis of fenced porphyrins

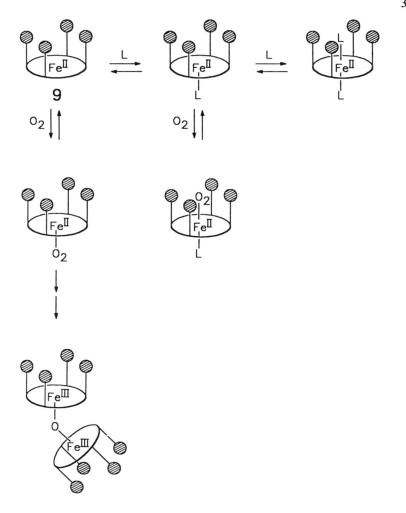

Figure 4. Equilibria in the reaction of **9** with O_2 in the presence of excess ligand (L)

used as a model for the R-state of oxy-Hb and oxy-Mb. Its structure supported Pauling's prediction of a bent, 'end-on' geometry for oxy-Hb.

Some ('tailed') fenced porphyrins equipped with arms, such as **12** and **13**, have proved useful for studies of complexation and oxygen binding.[9] Both Fe(Piv)$_3$(4ClImP)Por (**12**) and Fe(Piv)$_3$(5ClImP)Por (**13**) exist preferentially as pentacoordinate 'high-spin' complexes in dilute solution (approx. 1 mmol) and at room temperature [C = chain length; P = phenyl]. At lower temperatures, however, dimerization takes place; at $-25°C$ there is more than 50% dimer with mixed spin system (S = 2 and S = 1). Oxygen affinity for **12** and **13** is similar

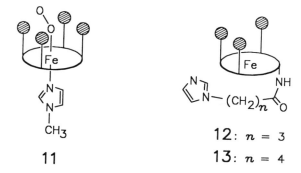

11

12: $n = 3$
13: $n = 4$

to that for the R-state of Hb. With a 'half-life' of about 33 hours, they constitute the most stable mononuclear O_2 carriers described thus far.

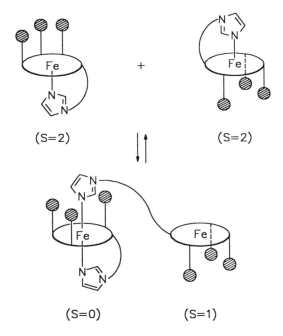

$(S=2)$ + $(S=2)$

$(S=0)$ $(S=1)$

The attachment of an arm to prepare 'tailed' ligands was achieved by coupling the corresponding acid chloride with the $\alpha,\alpha,\alpha,\beta$-atropisomer (Figure 3). Because the porphyrins equipped with an arm are light- and oxygen-sensitive, all reactions must be carried out in an inert atmosphere and in the dark.

Since many fenced porphyrin models show very similar O_2 affinity as do Fe- and Co-Hb and -Mb, it was concluded that the protein portion plays no special role in oxygen binding, other than to hinder the irreversible oxidations. The 'strain' in Hb—that is, the migration of the iron ion toward the porphyrin plane during oxygen binding—ought to be 'recreated' in the model structures. With this goal in mind, models such as **14** and **15** were prepared for the deoxy- and oxy-T-states, in which the axial ligand occupies a substantial space. X-ray crystal structures of these models made it possible for the first time to gain insight

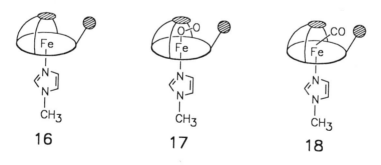

into oxygen binding in pentacoordinate 'high-spin' Fe(II) porphyrins at the molecular level.[10] It could be shown that the O_2 molecule is bound in a bent 'end-on' fashion (see **15**). As anticipated, the Fe ion, which is transformed to the 'low-spin' state, migrates toward the porphyrin plane but does not enter the plane, the axial 3-methylimidazole being carried along.

The binding of CO in these model compounds, by comparison with the CO binding in Hb and Mb, was also of interest. It was shown that the natural systems have somewhat lower affinity. In an attempt to clarify the reason, the 'pocket porphyrin' **16**, among others, was synthesized. The expectation was that this model would support normal oxygen binding, but that binding of CO would be impeded.[11] This proved to be the case. While the O_2 binding of **16** is

comparable to that in the natural systems, the CO affinity of **16** is measurably lower. This was interpreted as indicating that binding of CO in a linear orientation is hindered, such that K_{CO} becomes smaller (see **18**). Because it is known that the O_2 molecule is bound in a bent form (**17**), this mode of binding for CO is less favourable.

Complexes **19** and **20** provide an alternative type of porphyrin, in which both sides carry 'fences'.[12] Pentacoordinate O_2 adducts such as **21** and **22** are formed at low temperature in the presence of oxygen. Upon warming of the solutions of **21** and **22**, the complexes dissociate to their deoxy forms or are oxidized irreversibly to Fe(III) compounds; but μ-peroxo- or μ-oxo dimers are not formed, however, in sharp contrast to the case with other pentacoordinate O_2 adducts (see above). Thus the type of compound represented by **19** and **20** is of interest for future design of new O_2 carriers.

19: R = CH_3
20: R = C_2H_5

O_2, < −70°C

21: R = CH_3
22: R = C_2H_5

8.4 CAPPED PORPHYRINS

The rationale for designing porphyrins with 'caps' is to create a binding position for O_2 which is fully protected against μ-oxo dimer formation. The idea is that a relatively rigidly attached, impenetrable steric barrier on the side of the cap would serve to shield against dimerization. The synthesis of capped porphyrins follow the Rothemund strategy. All four aldehyde groups, which are to become the *meso* positions in the porphyrin system, originate from the pyromellitic acid derivative **23**, which is available in two steps from salicylaldehyde.[13] Capped porphyrins containing naphthalene rings have also been prepared (see **25**).

Strategies were subsequently developed to attach arms on the non-capped side of the porphyrin ring, in which case it would be necessary to functionalize the capped porphyrin at the periphery. In order to synthesize the mononitro

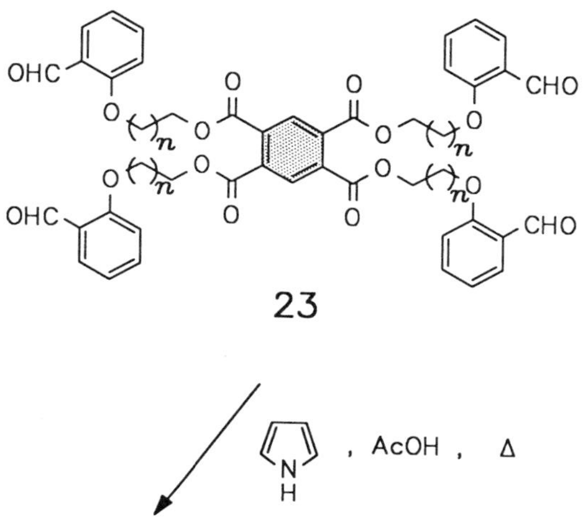

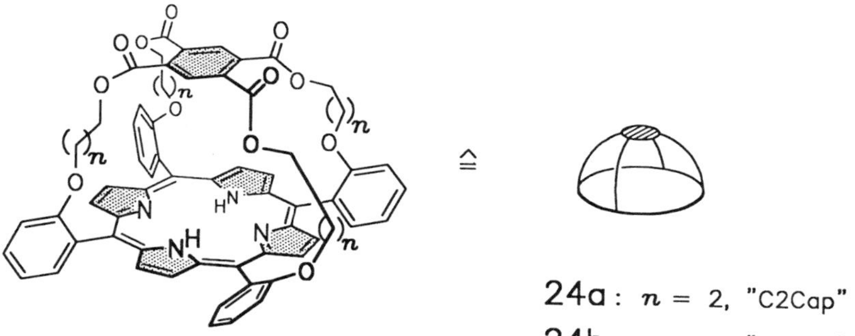

23

$\nearrow$, AcOH , Δ

24a : $n = 2$, "C2Cap"
24b : $n = 3$, "C3Cap"

compounds 26 and 27, the capped porphyrin was subjected to nucleophilic nitration; subsequent reduction led to the monoamino capped porphyrin.

The 'FeC2Cap' 24a is converted with excess 1-methylimidazole in benzene solution to the pentacoordinate complex, which is a reversible O_2 carrier at room temperature. The affinity of 'FeC2Cap-1MeIM' for O_2 is significantly smaller than that of comparable open-chain systems. Its inefficiency to form hexacoordinate complexes is reflected in a relatively short half-life of 5 hours.[14]

By contrast, the 'FeC3Cap' (24b), designed with a larger 'cap', can form hexacoordinate complexes which in solution bind O_2 reversibly, whereby heptacoordinate oxy complexes result. Because the binding of the second ligand is weak and therefore not consistent with axial coordination, it is presumed that the second ligand does not occupy an axial position.[1a]

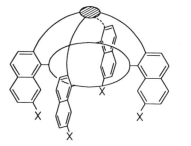

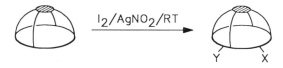

25 : X = H

X = OBz

X = OH

26 : X = H, Y = NO$_2$

27 : X = H, Y = NH$_2$

28 : X = NO$_2$, Y = H

29 : X = NH$_2$, Y = H

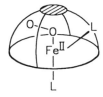

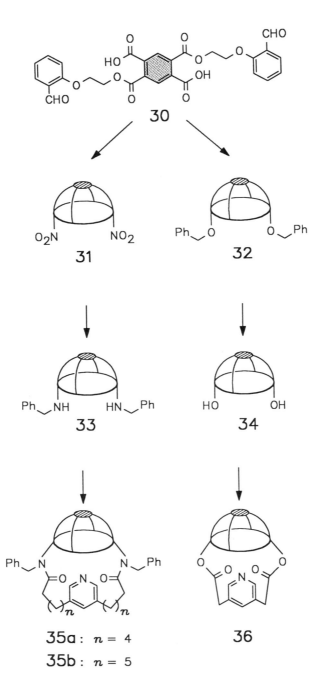

Figure 5. Synthesis of 'capped strapped' porphyrins. For more details, consult the original literature[1a]

It has been shown that the CO affinity of both capped porphyrins **24** is comparable to that of open-chain systems, whereas the O_2 affinity is significantly lower. X-ray crystal structures and NMR studies suggest a substantial conformational mobility in the capped structures. It could be the case that the 'cap' assumes a conformation in solution which impedes O_2 binding while the CO binding is less hindered.

It has been possible recently to introduce yet another bridge into capped porphyrins on the other side from the cap, the result being so-called 'capped strapped porphyrins' (see the bottom of p. 386).[1a] A synthetic scheme is outlined in Figure 5. Key intermediates are the α,γ-difunctionalized, capped porphyrins **33** and **34**. Formation of the second cap is accomplished by cyclization of the pyridine dicarboxylic acid under high-dilution conditions and by ester activation with 2-chloromethylpyridinium iodide. The porphyrins obtained, of type **35** and **36**, are stable, reversible O_2 carriers at room temperature in solution.

8.5 BRIDGED PORPHYRINS

8.5.1 SIMPLE BRIDGED PORPHYRINS

Since 1971 several dozen assorted porphyrins with simple bridges have been synthesized.[1] Besides oligomethylene bridges, some have included bulky units such as biphenyl, anthracene, naphthalene, etc. in the bridging structures. The synthesis can be approached from one of two directions. Either a bridge can be attached to the preformed porphyrin or, alternatively, the bridge can be synthesized and then the porphyrin constructed afterwards. The former method—construction of a bridge from a prefunctionalized porphyrin—has been

37

the almost exclusive method of choice in the last few years, often by starting with the readily available diesters of *meso*-porphyrin II (**37**, $R' = H$). A successive bridging has been described by Battersby, as outlined in Figure 6.[15]

Figure 6. Synthesis of bridged porphyrins

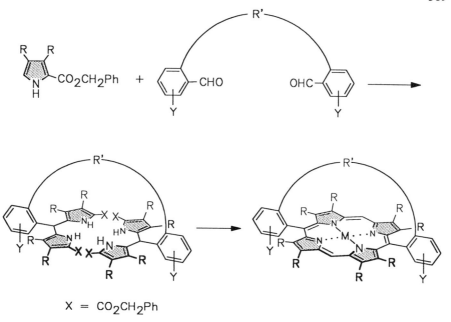

X = CO$_2$CH$_2$Ph

Figure 7. Synthesis of bridged porphyrins by construction of the porphyrin system in the second step

An example of prior formation of the bridge and subsequent construction of the porphyrin system is the synthesis formulated by Baldwin of α,γ-diphenylporphyrin derivatives.[16]

In the case of several simple bridged Fe(II) porphyrins the binding of a second axial ligand is significantly impaired. This suggests that the bridge is shielding the binding position which it encloses. If a bulky ligand such as 1-tritylimidazole is introduced, it is possible to obtain pentacoordinate Fe(II) porphyrins containing a simple bridge. The O$_2$ adduct of the 'crowned porphyrin' **38**

38

prepared by Chang has a half-life in solution at 1 atm O_2 of >1 hour at room temperature.[17]

Most of the bridged porphyrins have been prepared in order to estimate the extent of steric shielding, which can result in a differentiation between CO and O_2 binding, as is the case with haemoproteins. Traylor *et al.* made two observations concerning the 'cyclophane porphyrins' **39** and **40**:

(a) the smaller cavity in **39** reduced both the CO and O_2 affinity, as compared to that in **40** and in other R-state Hb models;
(b) the impairment in affinities due to reduction in the cavity size is similar for both gases (CO and O_2) and is reflected in the association constants.[18]

From these two models it appears that remote steric effects do not differentiate between O_2 and CO.

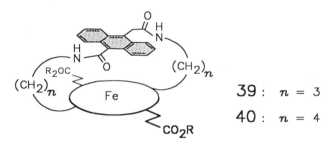

39 : $n = 3$

40 : $n = 4$

8.5.2 DOUBLY BRIDGED PORPHYRINS

The two doubly bridged porphyrin systems **41** and **42** were obtained according to the strategy devised by Battersby.[19] The doubly *meso*-bridged 'basket handle porphyrins' were obtained by two different pathways, as outlined in Figure 8.[20] Additional doubly bridged porphyrins (**47** and **48**) have been described by Momenteau.[21]

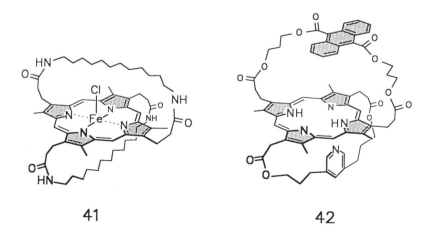

41   42

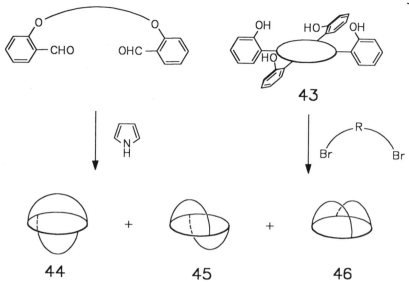

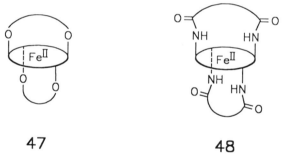

Figure 8. 'Basket handle porphyrins'

Those bridged porphyrins containing ligand donor centres in the bridging units constitute important progress in the design of model oxygen carriers, particularly of mononuclear O_2 carriers. With the ligand affixed in the axial position, the problems accompanying the open-chain donor arms are excluded.

The half-life of **42** in the presence of O_2 is only 15 minutes.[19] On changing the solvent from dimethylformamide to dichloromethane the $t_{1/2}$ increased to >2 hours, probably a consequence of the ability of the solvent to compete with O_2 and the sixth coordination site. It is noteworthy that CO in the **42**–CO complex could be replaced by O_2, in contrast to all other model systems, in which the affinity for CO is somewhat higher than that for O_2. This constitutes an important analogy to the behaviour of myoglobin.

It was found through ^{1}H-NMR and X-ray analysis that the model compounds **49** and **50**, which represent 'half' of **42**, are conformationally flexible. The bridge is capable of oscillating from one side of the porphyrin plane to the other. This flexibility in solution explains the autoxidation which **42** undergoes.

49   50

With a half-life of 25 minutes in the presence of 1-methylimidazole and an atmosphere of O_2, the *trans/trans*-isomer **44** proves to be the most stable of this group of porphyrins. The isomer **46**, which is only shielded on one side of the porphyrin ring, rapidly forms the μ-oxo dimer, an observation that supports the inhibition of μ-oxo dimerization through steric protection of both sides of the porphyrin ring. The kinetics of CO and O_2 binding for both 'basket handle' compounds **47** and **48** have been determined, with the observation that both gases are bound more strongly than by unhindered Fe(II) complexes. Of the two, **48** shows a stronger affinity for O_2 than does **47** by an order of magnitude. This is attributed to the difference in O_2 dissociation rates, which reflects the more polar surroundings provided by the two amide bonds in each bridge in **48**.

8.6 CYTOCHROME MODELS

8.6.1 CYTOCHROME P450 MODEL COMPOUNDS

Cytochromes P450 constitute a group of monooxygenases which are widespread in animal tissue, in plants, and in microorganisms.[1a] The number 450 refers to the UV/VIS absorption of the carbonyl adduct at $\lambda = 450$ nm. The designation 'cytochrome' is somewhat misleading, because the function of this enzyme is totally different from that of cytochromes in the electron transport chain. The cytochromes P450 catalyse hydroxylation of many substrates, including aliphatic and aromatic hydrocarbons. They are responsible for the degradation of drugs and for the hydroxylation of steroids.

The prosthetic group in cytochrome P450 is protoporphyrin IX. Spectroscopic results suggest that a thiolate group (probably cysteinyl) occupies an axial coordination site. The additional, sixth coordination site could be imidazole, tyrosine, serine, or water. Because cytochrome P450$_{cam}$ can be obtained in a high state of purity from *Pseudomonas putida* bacteria, several of the steps in the complicated catalytic cycle of this enzyme are known. In the first step a hydrocarbon (RH) reacts with the 'low-spin' hexacoordinate Fe(III) complex to give a 'high-spin' pentacoordinate Fe(III)–RH adduct. This is reduced in the second step, such that iron is transformed to Fe(II) and remains a 'high-spin' pentacoordinate. Then follows oxidation with molecular oxygen, with formation of a 'low-spin' hexacoordinate Fe(II) (O$_2$)–RH. After combination with two protons, this is decomposed by way of an active oxy intermediate to ROH + H$_2$O, whereupon the original Fe(III) complex is reformed.

The major goal in the synthesis of Fe-porphyrin–oxygen models is directed toward the last two steps; that is, the active oxygen intermediate capable of oxidizing C—H bonds in substrates.

8.6.2 FENCED PORPHYRINS

Groves synthesized an Fe(IV)–oxygen complex with the structure **51** as a model for the active oxy intermediate E.[23] Treatment of a solution of **51** with norbornene affords norbornene oxide in 78% yield. If the solution of **51** is pretreated with H$_2$[18]O before the reaction with norbornene is carried out, one finds that norbornene oxide incorporates O[18] to the extent of 99%. This points to a facile exchange of oxygen with water.

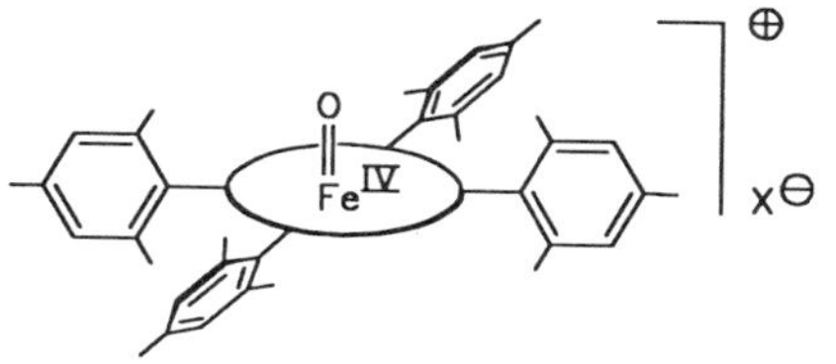

51

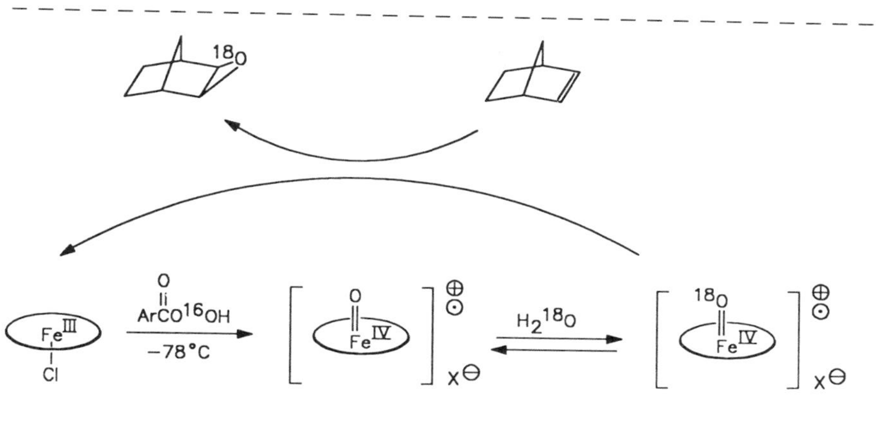

51

^{1}H-NMR, electronic, Mössbauer, and EPR results all support the involvement of an Fe(IV)–porphyrin π-cation radical. Because of the stability of the system resulting from the mesityl fence, such an intermediate could be detected for the first time. **51** can thus be considered as a good model not only for cytochrome P450 but also for peroxidase compounds.

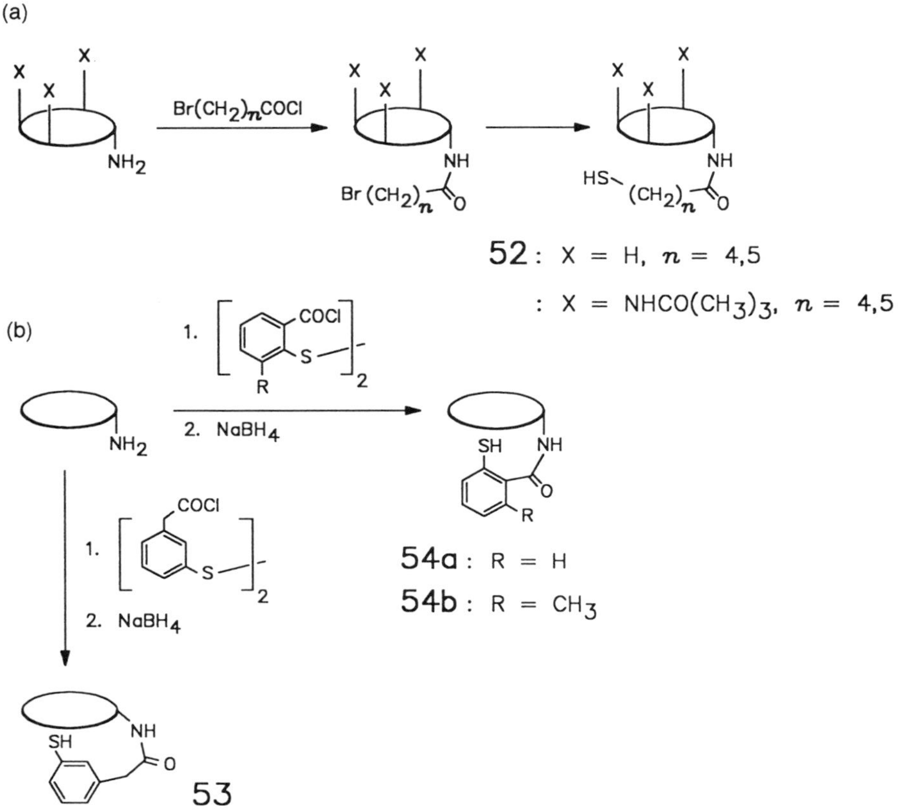
(a)
Br(CH2)nCOCl
NH2
NH
Br (CH2)n
HS
(CH2)n
52 : X = H, n = 4,5
: X = NHCO(CH3)3, n = 4,5
(b)
1.
COCl
S
R
2
2. NaBH4
SH
NH
54a : R = H
54b : R = CH3
53

Figure 9. Synthesis of cytochrome P450 porphyrin models containing a thiol group

Collman has described cytochrome P450 models based upon the TPivPP and TPP types, which possess a side arm carrying a thiolate group (Figure 9).[24] In order to minimize the competing thioester formation in the course of synthesis, the thiol groups were introduced as *S*-trityl or *S*-acetyl groups and as disulphide derivatives. It was found that the *S*-trityl group is a more convenient protecting group than the *S*-acetyl, because it can be removed later in higher yield.

All porphyrins functionalized with a thiol arm prove to be sensitive to air and light. This is traced to the sequence of reactions induced by the porphyrin-sensitized generation of singlet oxygen. The undesirable decomposition can be counteracted, however, by working in an inert atmosphere and only with red light. **53** forms the pentacoordinate complex **55**. In the presence of CO hexacoordinate 'low-spin' Fe(II) complexes are formed from both porphyrins **52**, containing an alkylthiol side chain, and the aryl compound **53**. The other arylthiols **54** exist as an equilibrium mixture with the pentacoordinate 'tail-off' complex at 25°C, whereas at 0°C, they are fully hexacoordinate. Treatment of **53** with the base K^+ $CH_3(CO)N^-(C_6H_5)$ in the presence of CO induces

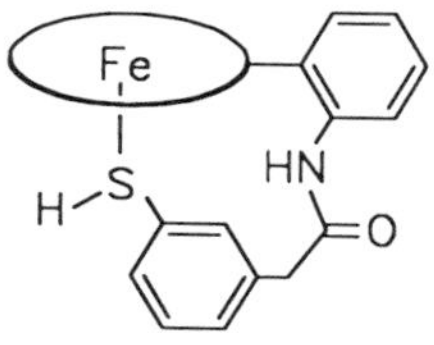

55

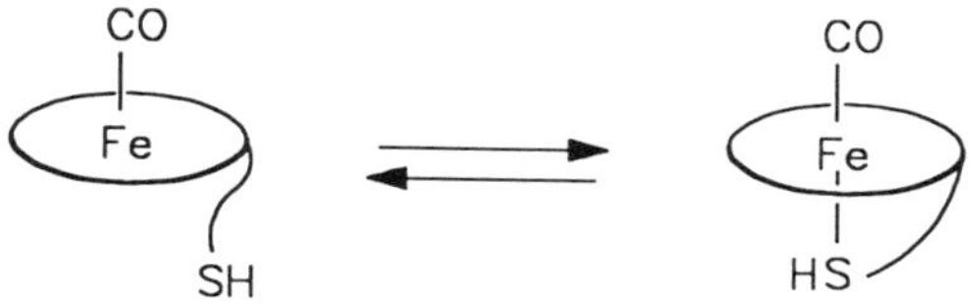

'tail–off'   *'tail–on'*

complete conversion into the thiolate complex **56**. Its UV/VIS and MCD (magneto circular dichroism) spectra prove to be similar to those of reduced CO-complexed cytochrome P450. Such hexacoordinate Fe(II)–thiolate–CO adducts result in 'hyperporphyrin' spectra with split Soret bands at about 450 and 380 nm.

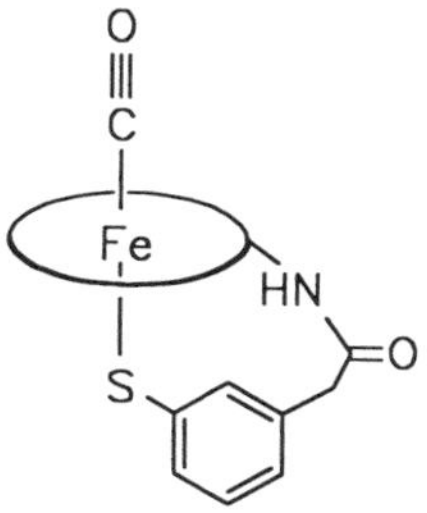

56

8.6.3 BRIDGED PORPHYRINS

The synthesis of a bridged Fe(II) porphyrin containing a thiolate arm has been described by Battersby.[25] The thiolate group, affixed in the middle of the bridge, can coordinate with Fe(II). During the synthesis the SH group was protected in the form of its acetyl derivative (Figure 10). The UV/VIS spectrum of the pentacoordinate Fe(II) derivative **57** resembles that of the reduced form of cytochrome P450$_{cam}$. In the presence of CO **57** was converted to the hexacoordinate **58**, which showed the characteristic hyperporphyrin spectrum of P450. The ^{13}CO complex of **57** displays a resonance at $\delta_c = 196.8$, a signal shifted to significantly higher field, by comparison to that in the corresponding haem compound. The methylene protons in the bridge are strongly shielded

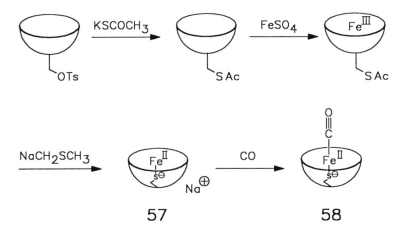

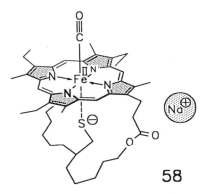

Figure 10. Synthesis of the cytochrome P450 model compound **57** by Battersby

58

$[\delta(CH_2) = 0.5$ to $-3.9]$, by comparison to the chemical shifts in **57**. This reflects the closer proximity of the bridge to the porphyrin as a result of the strong binding of the thiolate with the metal.

8.7 CYTOCHROME *c* OXIDASE MODELS

Cytochrome *c*, which occurs in animals, plants, yeasts, and bacteria, is an enzyme responsible for the catalysis of four-electron reduction of O_2 to H_2O, the last reaction in the mitochondrial electron transport chain. This occurs by the acceptance of four reducing equivalents of cytochrome *c* and their transfer to the bound O_2 molecule, with formation of water.[1a] This redox process has

$$4 \; \text{Cyt C}^{2\oplus} + O_2 + 4 \; H^{\oplus} \rightarrow 4 \; \text{Cyt C}^{3\oplus} + 2 \; H_2O$$

physiological significance, in that free energy is produced which can be stored through the conversion of two equivalents of ADP to ATP.

The intensely coloured cytochrome c oxidase has a molecular weight of approximately 140 000. It contains two haem units, two Cu ions, and seven peptide subunits. While both haem units have the same structure (haem a), the two cytochromes are different (a and a_3). Model studies have been designed to provide the following characteristic properties of cytochrome c oxidase:

(a) the catalytic four-electron reduction of O_2 to H_2O;
(b) the electron-transfer and spectroscopic properties of cytochrome a;
(c) the properties of the Fe(III)/Cu(II) complex, which shows strong antiferromagnetic coupling.[1a]

8.7.1 FENCED PORPHYRINS AS MODELS FOR CYTOCHROME c OXIDASE

Mixed complexes as models for cytochrome a_3 models were described by Gunter, an example being the fenced porphyrin complex **59**. Although it could be demonstrated that incorporation of Cu(II) into the complex significantly perturbs the electronic state of the Fe(III), there appears to be essentially no spin coupling between the metal ions.

59

8.7.2 BRIDGED PORPHYRINS AS MODELS FOR CYTOCHROME c OXIDASE

Gunter et al.[27] also prepared bridged porphyrins of the type **61** as models for cytochrome a_3. The synthesis, beginning with o-nitrobenzaldehyde and tetramethyldipyrromethane, involved as an intermediate the α,γ-diarylporphyrin **60**. The bridging step, carried out under high dilution, proceeded in high yield (70%). No spin coupling was found in **61**, however. A cytochrome a_3 model (**62**), which did show significant spin coupling (132 cm^{-1}), was described by Chang.[28] This provides one of the few examples of a non-symmetrically bridged porphyrin, where the bridging atoms are attached to adjacent rather than opposing pyrrole units. The formation of a square planar Cu(II) complex was intentionally avoided in this complex.

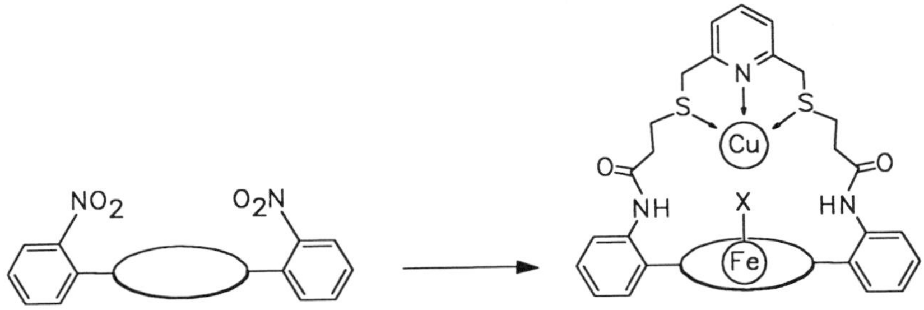

60 → 61

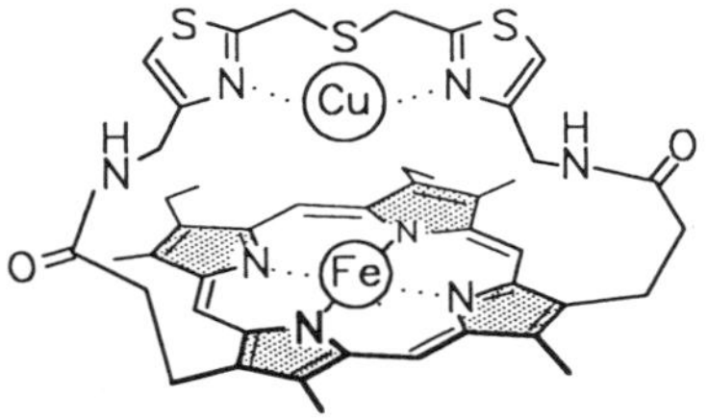

62

8.7.3 DIMERIC PORPHYRINS AS MODELS FOR CYTOCHROME *c* OXIDASE

In the Fe and Mg imidazolate dimers **63** and **64,** prepared by Reed *et al.*, the antiferromagnetic coupling proved to be minimal.[29] In order to study models

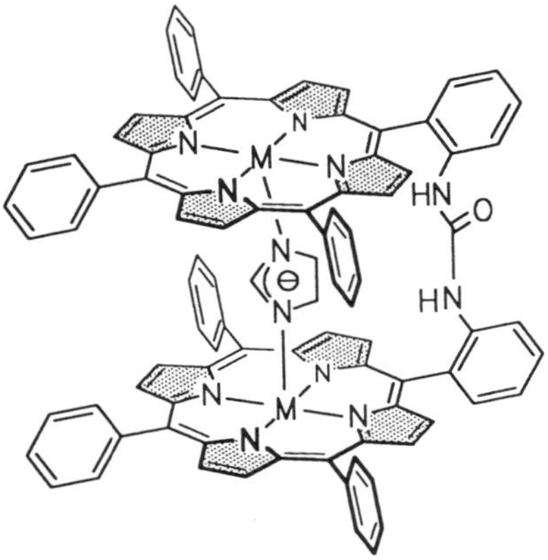

63 : M = Fe

64 : M = Mn

for the four-electron reduction of O_2 to H_2O, Collman synthesized 'face-to-face' dimeric porphyrins of the type **65**. Not only could the nature of the metal ion be varied, but also the distance between the two porphyrin planes.[1c] Compounds of this type prove to be effective catalysts for the O_2 reduction. In terms of their rates of reduction, they surpass platinum, the best known catalyst up to that time.

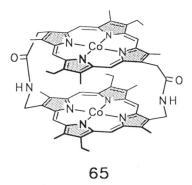

65

8.8 MISCELLANEOUS PORPHYRINS

Several fenced, bridged, 'capped', and dimeric (doubly and triply layered) porphyrins have attracted considerable attention because of their potential donor–acceptor interactions, their use in studying primary reactions in bacterial photosynthesis, and perhaps also because of their fascinating symmetrical structures.[1] In the concluding section of this overview of porphyrinophanes, some of these will be presented.

Exotic examples of fenced porphyrins are the ferrocenyl-(**66**) and tetra-carboranyl-(**67**) substituted porphyrins.[30,31] Kagan has synthesized the 'strati-bis-porphyrin' **68** by fourfold bridging of two tetra-*meso*-substituted porphyrins.[32]

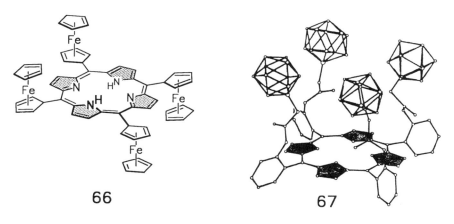

66   **67**

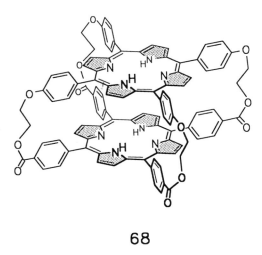

68

The quinone-bridged (**69**) and -capped (**70**) porphyrins, of interest in connection with charge-transfer interactions, have already been described as donor–acceptor phanes in Section 2.11.[33] The synthesis of **70** provides an example where yields can be surprisingly high in macrocyclizations involving rigid subunits and reversible reaction types.

69   **70**

Staab described the first three-layer donor–acceptor porphyrinophanes **71** and **72**.[34] The 'aromatic planes', separated by 342 pm, are parallel to one another, as determined by X-ray crystallography. An electron transfer could be detected by fluorescence quenching in the quinone **71** but not in **72**.[34a]

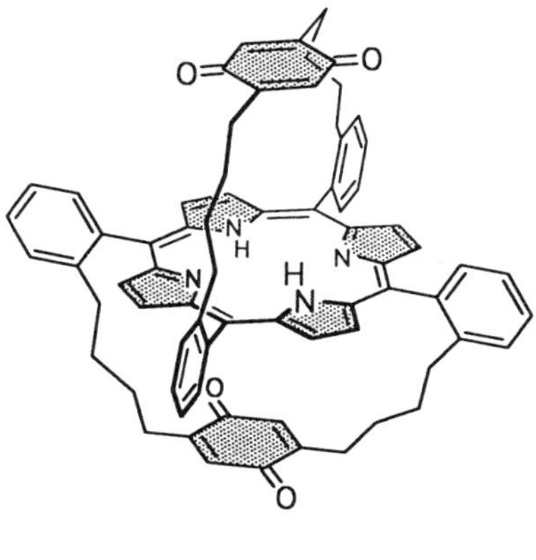

71 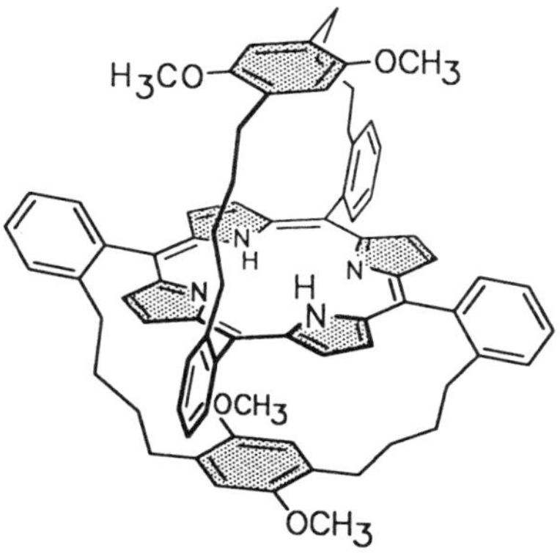   72

Various isomers of quadruply bridged porphyrinophanes containing azobenzene units as optical switches have been synthesized by Vögtle *et al.* Among these are the isomers **73a** and **73b**.[35]

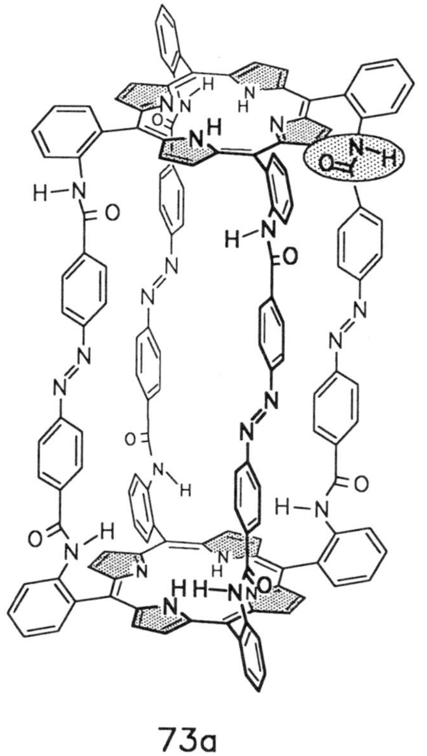

73a 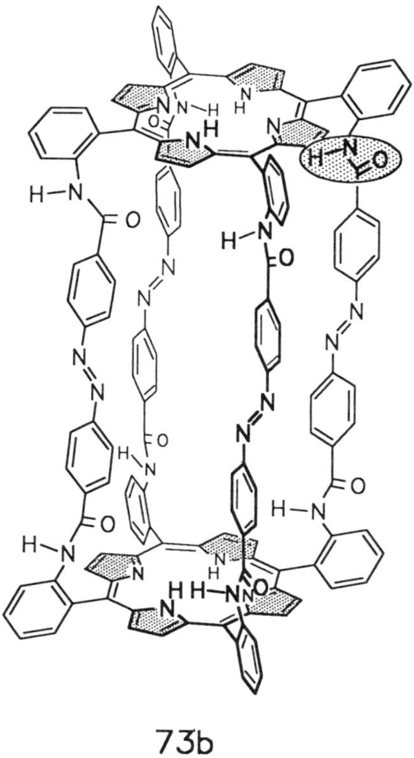   73b

Diederich recently reported a porphyrin-bridged cyclophane **74** as a model for cyctochrome P450 enzymes.[36] Additional novel porphyrinophanes can be found in the original literature.[37]

74

References

1. (a) Extensive review: J. E. Baldwin and P. Perlmutter, *Top. Curr. Chem.*, **121**, 181 (1984); extensive additional information.
   (b) H. Ogoshi and Y. Kuroda, *Yuki Gosei Kagaku Kyokaishi*, **47**, 514 (1989).
   (c) Metal complexes with tetrapyrrole ligands: J. W. Buchler (Ed.), *Struct. Bonding*, **64**, Springer, Berlin, 1987.
   (d) J. P. Collman, F. C. Anson, S. Bencosme, A. Chong, T. Collins, P. Denisevich, E. Evitt, T. Geiger, J. A. Ibers, G. Jameson, Y. Konai, C. Koval, K. Meier, P. Oakley, R. Pettman, E. Schmittun and J. Sessler, *Organic Synthesis Today and Tomorrow* (B. M. Trost and C. R. Hutchinson, Eds.), Pergamon Press, Oxford, 1981.
   (e) D. Dolphin, J. Hiom and J. B. Paine, III, *Heterocycles*, **16**, 417 (1981).
2. M. F. Perutz, *Sci. Amer.*, **239**, 68 (1978).
3. L. Pauling, *Nature (London)*, **203**, 61 (1964).
4. (a) J. P. Collman, T. R. Halbert and K. S. Suslick, *Metal Ion Activation of Dioxygen* (T. G. Spiro, Ed.), Wiley, New York, 1980.
   (b) R. D. Jones, D. A. Summerville and F. Basolo, *Chem. Rev.*, **79**, 139 (1979).
   (c) T. G. Traylor and P. S. Traylor, *Ann. Rev. Biophys. Bioeng.*, **11**, 105 (1982).
5. J. E. Baldwin and J. Huff, *J. Am. Chem. Soc.*, **95**, 5757 (1973).
6. J. P. Collman, R. R. Gagne, T. R. Halbert, J.-C. Marchon and Ch. A. Reed, *J. Am. Chem. Soc.*, **95**, 7868 (1973).
7. J. Lindsey, *J. Org. Chem.*, **45**, 5215 (1980).
8. G. B. Jameson, G. A. Rodley, W. T. Robinson, R. R. Gagne, Ch. A. Reed and J. P. Collman, *Inorg. Chem.*, **17**, 850 (1978).
9. J. P. Collman, J. I. Brauman, K. M. Doxsee, T. R. Halbert, E. Bunnenberg, R. E. Linder, G. N. LaMar, J. Del Gaudio, G. Lang and K. Spartalian, *J. Am. Chem. Soc.*, **102**, 4182 (1980).
10. G. B. Jameson, F. S. Molinaro, J. A. Ibers, J. P. Collman, J. I. Brauman, E. Rose and K. S. Suslick, *J. Am. Chem. Soc.*, **102**, 3224 (1980).

11. J. P. Collman, J. I. Brauman, T. J. Collins, B. Iverson and J. L. Sessler, *J. Am. Chem. Soc.*, **103**, 2450 (1981).
12. A. R. Amundsen and L. Vaska, *Inorg. Chim.*, **14**, L49 (1975).
13. J. Almog, J. E. Baldwin, M. J. Crossley, J. F. DeBernardis, R. L. Dyer, J. R. Huff and M. K. Peters, *Tetrahedron*, **37**, 3589 (1981).
14. J. Almog, J. E. Baldwin and J. Huff, *J. Am. Chem. Soc.*, **97**, 227 (1975).
15. A. R. Battersby, D. G. Buckley, S. G. Hartley and M. D. Turnbull, *J. Chem. Soc., Chem. Commun.*, 879 (1976).
16. J. E. Baldwin, M. J. Crossley, T. Klose, A. O'Rear, III and M. K. Peters, *Tetrahedron*, **38**, 27 (1982).
17. C. K. Chang, *J. Am. Chem. Soc.*, **99**, 2819 (1977).
18. T. G. Traylor, D. Campbell, S. Tsuchiya, M. Mitchell and D. V. Stynes, *J. Am. Chem. Soc.*, **102**, 5939 (1980).
19. A. R. Battersby and A. D. Hamilton, *J. Chem. Soc., Chem. Commun.*, 117 (1980).
20. M. Momenteau, J. Mispelter, B. Loock and E. Bisagni, *J. Chem. Soc., Perkin Trans. 1*, 189 (1983).
21. M. Momenteau and D. Lavalette, *J. Chem. Soc., Chem. Commun.*, 341 (1982).
22. W. B. Cruse, O. Kennard, G. M. Sheldrick, A. D. Hamilton, S. G. Hartley and A. R. Battersby, *J. Chem. Soc., Chem. Commun.*, 700 (1980).
23. J. T. Groves, R. C. Haushalter, M. Nakamura, T. E. Nemo and B. J. Evans, *J. Am. Chem. Soc.*, **103**, 2884 (1981).
24. J. P. Collman and S. E. Groh, *J. Am. Chem. Soc.*, **104**, 1391 (1982).
25. A. R. Battersby, W. Howson and A. D. Hamilton, *J. Chem. Soc., Chem. Commun.*, 1266 (1982).
26. D. A. Buckingham, M. J. Gunter and L. N. Mander, *J. Am. Chem. Soc.*, **100**, 2899 (1978); M. J. Gunter, L. N. Mander, G. M. McLaughlin, K. S. Murray, K. J. Berry, P. E. Clark and D. A. Buckingham, *J. Am. Chem. Soc.*, **102**, 1470 (1980).
27. M. J. Gunter and L. N. Mander, *J. Org. Chem.*, **46**, 4792 (1981).
28. C. K. Chang, M. S. Koo and B. Ward, *J. Chem. Soc., Chem. Commun.*, 716 (1982).
29. J. T. Landrum, C. A. Reed, K. Hatano and W. R. Scheidt, *J. Am. Chem. Soc.*, **100**, 3232 (1978).
30. R. G. Wollmann and D. N. Hendrickson, *Inorg. Chem.*, **16**, 3079 (1977).
31. R. C. Haushalter, W. M. Butler and R. W. Rudolph, *J. Am. Chem. Soc.*, **103**, 2620 (1981).
32. N. E. Kagan, D. Mauzerall and R. B. Merrifield, *J. Am. Chem. Soc.*, **99**, 5484 (1977).
33. K. N. Ganesh and J. K. M. Sanders, *J. Chem. Soc., Chem. Commun.*, 1129 (1980).
34. C. Krieger, J. Weiser and H. A. Staab, *Tetrahedron Lett.*, **26**, 6059 (1985); D. Mauzerall, J. Weiser and H. A. Staab, *Tetrahedron*, **45**, 4807 (1989).
    (a) H. A. Staab *et al.*, *Chem. Ber.*, **126**, 811 (1993).
35. K.-H. Neumann and F. Vögtle, *J. Chem. Soc., Chem. Commun.*, 520 (1988).
36. D. R. Benson, R. Valentekovich and F. Diederich, *Angew. Chem.*, **102**, 213 (1990); *Angew. Chem. Int. Ed. Engl.*, **29**, 191 (1990).
37. See, for example, J. P. Collman, P. S. Wagenknecht, R. T. Hembre and N. S. Lewis, *J. Am. Chem. Soc.*, **112**, 1294 (1990); J. T. Groves and P. Viski, *J. Am. Chem. Soc.*, **111**, 8537 (1989); C. A. Quintana, R. A. Assink and J. A. Shelnutt, *Inorg. Chem.*, **28**, 3421 (1989); D. Mandon, R. Weiss, M. Franke, E. Bill and A. X. Trautwein, *Angew. Chem.*, **101**, 1747 (1989); *Angew. Chem. Int. Ed. Engl.*, **28**, 1709 (1989); D. Gust, T. A. Moore, A. L. Moore, G. Seely, P. Liddell, D. Barrett, L. O. Harding, X. C. Ma, S.-J. Lee and F. Gao, *Tetrahedron*, **45**, 4867 (1989); A. Osuka, K. Maruyama and S. Hirayama, *Tetrahedron*, **45**, 4815 (1989); J. S. Lindsey, P. A. Brown and D. A. Siesel, *Tetrahedron*, **45**, 4845 (1989); M. Momenteau, B. Loock, P. Seta, E. Bienvenue and B. d'Epenoux, *Tetrahedron*, **45**, 4893 (1989); S. O'Malley and T. Kodadek, *J. Am. Chem. Soc.*, **111**, 9116 (1989); P. Maillard, C. Schaefer, C. Tétreau, D. Lavalette, J.-M. Lhoste and M. Momenteau,

J. Chem. Soc., Perkin Trans. 2, 1437 (1989); J. P. Collman *et al.*, *J. Am. Chem. Soc.*, **112**, 8206 (1990); H. L. Anderson, J. K. M. Sanders *et al.*, *Angew. Chem.*, **102**, 1478 (1990); *Angew. Chem. Int. Ed. Engl.*, **29**, 1400 (1990); M. Kreysel and F. Vögtle, *Synthesis*, **8**, 733 (1992); Z. Asfari, J. Vicens and J. Weiss, *Tetrahedron Lett.*, **34**, 627 (1993).

9 'Protophanes' and 'Aliphanes'

This chapter is concerned with certain borderline cases of molecules, in particular the 'open-chain phanes' ('protophanes') and aliphatic analogues of phanes ('aliphanes').

9.1 PROTOPHANES

The designation 'open-chain (cyclo)phane'[1a] conveys, of course, a contradiction which is intentional. What is implied is an orientation of structural units or functional groups sterically analogous to those in cyclophanes, for example in [2.2]paracyclophane. The difference is that the preorganization of the groups is not brought about through ring closure. It is a fact that the rigid, 'face-to-face' orientation of aromatic nuclei, such as benzene rings, characteristic of cyclophanes, is also possible through other arrangements, including open-chain structures. As already mentioned in Chapter 6, layered benzene rings are found not only in helicenes (and, for example, metallocenes), but also in certain orthocyclophanes (Chapter 13) and in 1,8-diarylnaphthalenes or 1,8-diarylanthracenes. For this reason the term 'open-chain cyclophane', although misleading, is logical.

The orientation of functional groups, such as is found in pseudo-*geminal* [2.2]paracyclophanedicarboxylic acid, is also fully realized in an open-chain arrangement, for example in the 1,8-diarylnaphthalene skeleton. The analogy has also been extended, as illustrated by the 'stereology concept' introduced in 1981.[2] The characteristic properties of stereologic molecules depend upon the cooperative interaction between two or more functions attached to an 'anchor

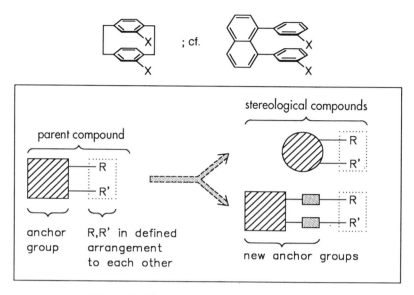

Figure 1. The stereology concept (schematic)

group'. A 'stereological molecule' is characterized as one which can assume a similar steric orientation of the functional groups to that in the parent compound (Figure 1).

The familiar relationship in homologues, vinylogues, 'phenylogues', 'arenologues', and 'heteroanalogues' is the basis for grouping molecules together that exhibit similar properties simply because of common functional groups, even though they may differ sterically and electronically. The stereology concept involves a more complex and subtle relationship between functional groups. For example, the functional groups are separated from one another by a similar distance when attached to various 'anchor groups'; or they are fixed ('preorganized') in a linear or convergent fashion to one another. This can be illustrated in a concrete way with the EDTA stereological complexones.[2b] In order for there to be effective complexation with cations, it is imperative that the two aminodiacetic acid units, like those in the parent compound EDTA (6), be able to assume a favourable orientation to one another for the formation of the chelate ring. The metal cation is then complexed in a complementary and coordinative fashion by the six donor sites in an octahedral arrangement. In principle it should be possible to attach the ligands to other anchor groups besides the ethano chain such that they bind cations in a cooperative fashion.

In fact, the EDTA stereological ligand **7** shows a high selectivity for calcium, as does EDTA itself.

A stereological orientation of functional groups or even portions of a molecule can be achieved in other ways, as is indicated in Figure 1. The anchor group of the reference substance (parent compound) is retained; then advantage is taken of classical approaches for varying the structure by inserting an extension between the anchor group and the functional groups. An example is the insertion of phenylene units to produce a 'double phenylogue' (see **8** and **9**). The stereological concept thus offers the possibility intentionally to design new

8   **9**

compounds with closely predicted and widely varying properties. The stereological oxaacetic acid **11**, in which the two functional groups are somewhat farther apart than in the parent compound **10**, is consequently not calcium-selective like **10**, but barium-selective because the barium ion fits better in the somewhat larger cavity of **11** than in the smaller cavity in **10**. Another possibility for designing novel EDTA compounds is illustrated by the EDTA stereological compound **12**, containing the quaterphenyl anchor group.

10

11   **12**

The stereology concept reached its peak with the introduction by Whitlock in 1978 of 'molecular tweezers'. In the first such host compounds of type **13** (schematic) described by Whitlock, the donor functions were too flexibly arranged to justify calling them effective 'tweezers'.[3] Nevertheless, heterocyclic guests such as caffeine were enclosed between two upper and lower tweezer plates (cross-hatched) in a sandwich-like fashion. The disadvantages in these early host molecules were the free rotation and consequent lack of preorganization of the two acceptor plates. The 4,4'-functionalized *peri*-diarylnaphthalenes **7** mentioned

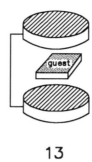

13

earlier, with their rigid, fork-like preorganization, appear more promising than the tweezers. Indeed, it has been demonstrated that these functional groups are cooperative and ion selective.[2]

In 1985 Rebek succeeded in arranging two functional groups in a practically ideal colinear, convergent fashion to give what can justifiably be termed highly efficient *molecular tweezers*[4] (Figure 2). A concrete example of Rebek's molecular tweezers is represented by structure **15**. As can be seen, the two colinear and convergent carboxyl groups selectively bind cations, in the case of small spacers such as naphthalene units; but they can also bind difunctional guests with complementary donor groups (carboxyls) in the 'molecular cleft' (see also Chapter 12).

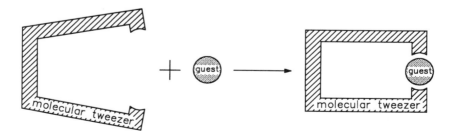

Figure 2. Scheme for 'molecular tweezers' and their molecular recognition and binding of a guest

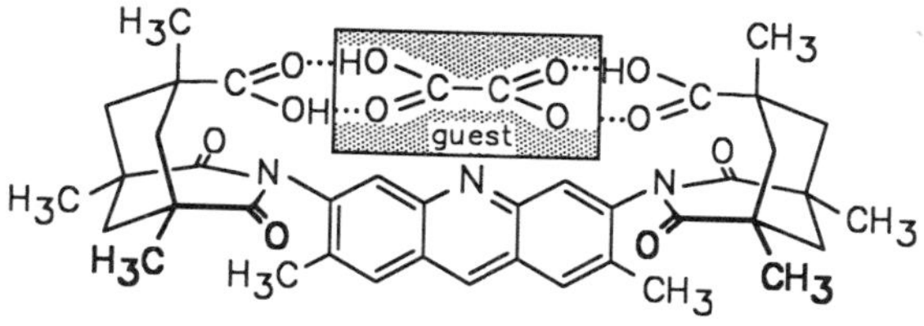

15

In the case of molecular π-tweezers **16**,[2d] the guest is accommodated cooperatively by the two triple bonds in the 'claws' of the tweezers. The hydrocarbons **16**, however, prove to be 'one-way' tweezers. They react with Fe(CO)$_6$ to afford an iron pentacarbonyl complex, and a C–C single bond is formed between the two triple bonds which can not be reversibly removed.[2d] In **16a** the triple bonds are likewise rigidly preorganized but separated by a greater distance from one another.[2e]

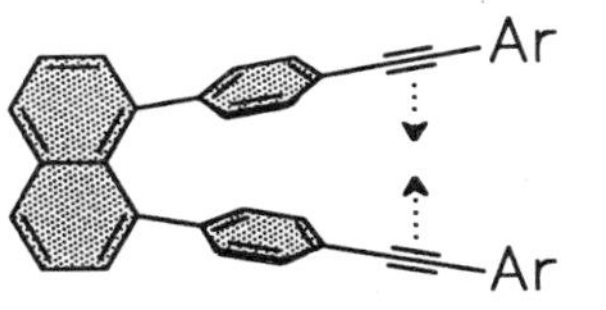

16

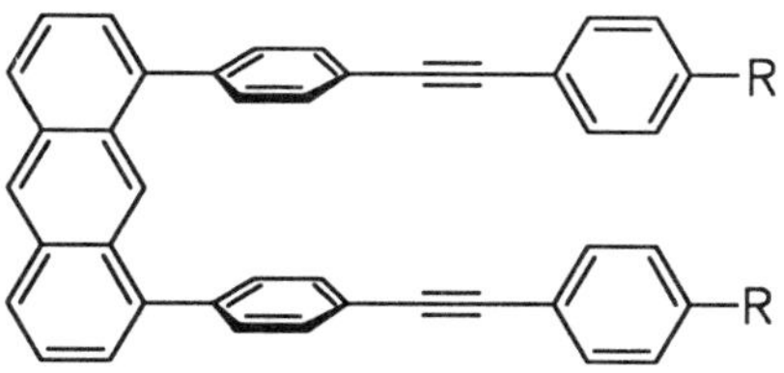

$$R=C(CH)_3$$

The known preorganization and 'π-stacking' also play a role in the molecular tweezers of type **17**, as described by Zimmerman *et al.*[6] The aromatic rings create a rigidly preformed cleft, in which planar, π-deficient guest molecules (e.g. 2,4,5,7-tetranitrofluorenone, **18**) can 'nest'. Investigation of such compounds showed that preorganization and rigidity are important factors in designing 'receptor substances' for uncharged guests, just as they are in cation binding in cyanospherands (see Section 7.3.2).

In 1971 Kauffmann designated as protophanes[1b] the open-chain precursors which, on cyclization, lead to phanes. '*Protophanes* are defined as open-chain arenes bridged with aliphatic groups, each containing at least three arene units (single rings or condensed arene systems). Their relationship to phanes (cyclic, bridged arenes) is analogous to that of alkanes to cycloalkanes.

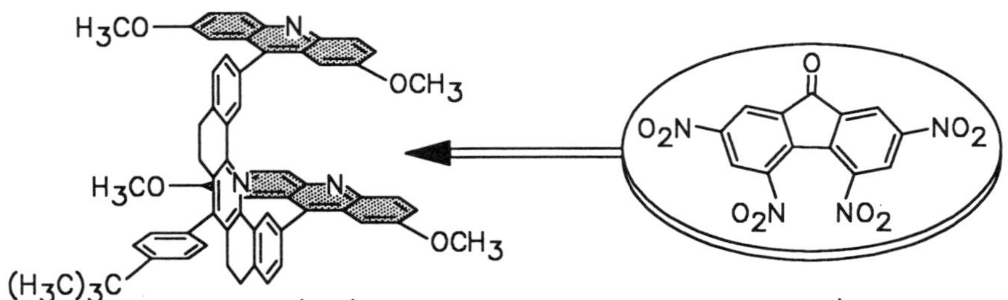

host
17

guest
18

'A nomenclature is proposed ("a[arene]-nomenclature") for protophanes, phanes, polyarenes, and cyclopolyarenes, in which the substitution of an arene nucleus for a C atom in hydrocarbons is expressed by the "arene"-notation such as "benzena", "pyridina", or "thiophena".'* This is analogous to the a-nomenclature of R. Stelzner, where substitution of a C atom in hydrocarbons by hetero atoms is designated by 'aza', 'oxa', or 'thia'. This new nomenclature, with its clarity and close similarity to familiar nomenclature systems, should be easily translated into formulas and have extensive application. It will be especially useful for the above named classes of compounds where IUPAC nomenclature or the 'phane nomenclature' is inapplicable or too cumbersome. It can be combined with Stelzner's a-nomenclature (designation: a[atom] nomenclature) for heteraprotophanes and heteraphanes to an even more all encompassing 'a[atom/arene]nomenclature'.

Examples[1b]

Parent alkane: octane
Name: 1-(2)furana-4,5-di(2,6)pyridina-7-(7,2)quinolinaoctane

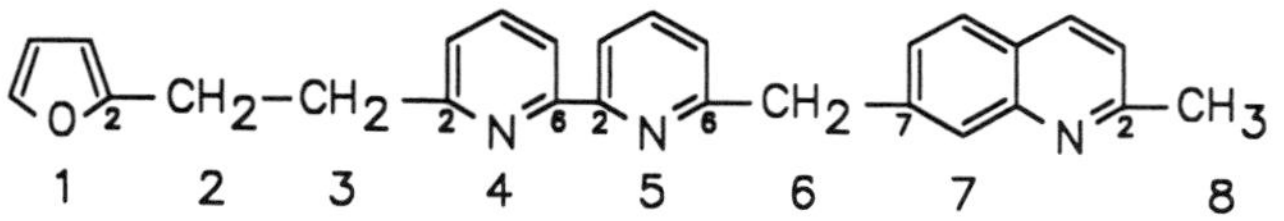

Parent alkane: octane
Name: 1,8-di(2)furana-2,4,6-tri(2,6)pyridina-3,5,7-tri(4,6)pyrimidinaoctane

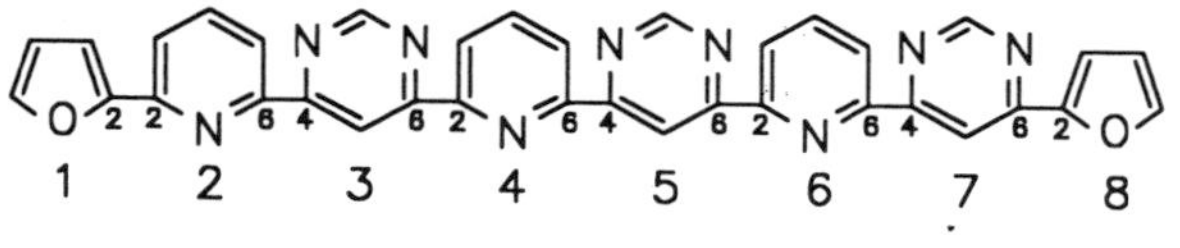

---

*According to Kauffman, modified.

H_3C—⟨2 S 5⟩—CH_2—CH_2—CH_2—CH_2—⟨2 O 5⟩⟨2 O 5⟩—CH_2—CH_3

1 2 3 4 5 6 7 8 9 10

2-(2,5)thiophena-7,8-di(2,5)furanadecane
[not: 3,4-di(2,5)furana-9-(2,5)thiophenadecane]

A broad view of the concept 'aromatic' extends the applicability of the a[arene]nomenclature, without, however, introducing any serious disadvantages.[1b] This is also the case for the 'aromatic part' of phanes, as can be shown below.

9.2 ALIPHANES

In the course of describing the perhydro[2.2]metacyclophane (**19**) obtained from [2.2]metacyclophane (**2**), the question arises whether one could not use aliphatic structural units such as the cyclohexane ring, analogous to arene rings, for a nomenclature of purely aliphatic 'aliphanes'. According to this, the molecule **19** would be called '[2.2](1,3)cyclohexanophane'. Because these names are easily formulated and are useful to designate a family, a few examples to illustrate this recent suggestion are presented.

2   **19**

In this way particularly characteristic aliphatic units, such as bridged molecules containing adamantane or cubane, are conveniently designated and classified as *adamantanophanes*[7] and *cubanophanes*.

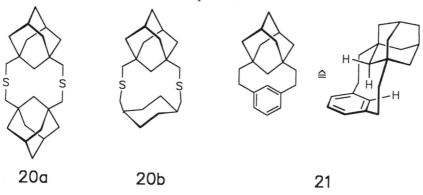

20a   **20b**   **21**

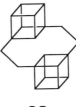

22

In the case of phanes with mixed aliphatic/aromatic units such as **23a** and **23b**, the term 'araliphanes' can be used;[8] the name 'arophanes' can be given to those phanes with arene units linked with aromatic bridges. To this group would belong hexametaphenylene and kekulene.

This naming scheme, like the phane nomenclature itself and other nomenclature proposals, is useful for a limited number of selected compounds and is not claimed to be a universal nomenclature. For certain limited examples, however, like the adamantano- and cubanophanes described above, it offers an ideal, manageable, abbreviated designation for a family of compounds. Such a classification according to family is also sensible from the point of view of its chemistry. When one considers the cyclohexanophane **23a**, which is chemically derived by catalytic hydrogenation from **2** (see Section 2.2), then the relationship between the names is consistent.

The bridged steroid **24** might be included in the group of araliphanes as a 'steroido-benzenophane'.[9] The following known paddlanes **25–27** are

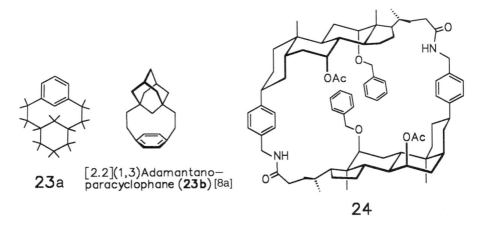

23a [2.2](1,3)Adamantano–
paracyclophane **(23b)** [8a]

24

'aliphanes'.[20] Both the dithiatriquinacenophane **(28)**[11] and curcurbituril **29**[12] can be considered to be 'aliphanes'.

Not only **28** but also the recently described [2.2](1,5)cyclooctatetraenophane **(30)**[13] have been named in the literature as phanes; that is, the aliphatic anchor groups, in keeping with the aliphane concept, were incorporated into their description. This illustrates anew the value and wide application of phane

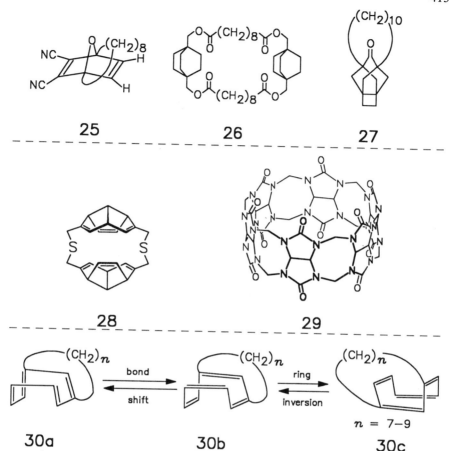

nomenclature. The 'troponophanes' and 'cycloheptatrienophanes' described in Section 2.7 are also examples of how bridged aliphatic compounds are widely counted among (ali-)phanes.

References

1. (a) W. Bieber and F. Vögtle, *Angew. Chem.*, **89**, 199 (1977); *Angew. Chem. Int. Ed. Engl.*, **16**, 175 (1977).
   (b) Th. Kauffmann, *Tetrahedron*, **28**, 5183 (1972); cf. K. Hirayama, *Tetrahedron Lett.*, 2109 (1982). The 'nodal nomenclature' evolved later from this proposed nomenclature. N. Lozac'h and A. L. Goodson, *Angew. Chem.*, **96**, 1 (1984); *Angew. Chem. Int. Ed. Engl.*, **23**, 33 (1984).
2. (a) R. Leppkes and F. Vögtle, *Angew. Chem.*, **93**, 404 (1981); *Angew. Chem. Int. Ed. Engl.*, **20**, 396 (1981).
   (b) R. Leppkes, F. Vögtle and F. Luppertz, *Chem. Ber.*, **115**, 926 (1982).
   (c) R. Leppkes and F. Vögtle, *Chem. Ber.*, **116**, 215 (1983).
   (d) F. Vögtle, Th. Papkalla, H. Koch and M. Nieger, *Chem. Ber.*, **123**, 1097 (1990).
   (e) F. Vögtle, H. Koch and K. Rissanen, *Chem. Ber.*, **125**, 2129 (1992).
   (f) Cf. K. L. Cole, M. A. Farran and K. Deshayes, *Tetrahedron Lett.*, **33**, 599 (1992).

3. C.-W. Chen and H. W. Whitlock Jr, *J. Am. Chem. Soc.*, **100**, 4921 (1978).
4. J. Rebek Jr, B. Askew, N. Islam, M. Killoran, D. Nemeth and R. Wolak, *J. Am. Chem. Soc.*, **107**, 6736 (1985); D. P. Curran, K.-S. Jeong, T. A. Heffner and J. Rebek Jr, *J. Am. Chem. Soc.*, **111**, 9238 (1989); J. Rebek Jr, *Acc. Chem. Res.*, **23**, 399 (1990); J. Rebek *et al.*, *Rec. Trav. Chim. Pays Bas*, **112**, 330 (1993); cf. F. Vögtle *et al.*, *Angew. Chem.*, **105**, 647 (1993); *Angew. Chem. Int. Ed. Engl.*, **32**, 601 (1993).
5. E. Weber and F. Vögtle, *Chemie in uns. Zeit.*, **23**, 210 (1989).
6. S. C. Zimmerman, M. Mrksich and M. Baloga, *J. Am. Chem. Soc.*, **111**, 8528 (1989); S. C. Zimmerman and W. Wu, *J. Am. Chem. Soc.*, **111**, 8054 (1989); J. P. Collman *et al.*, *J. Am. Chem. Soc.*, **112**, 8206 (1990).
7. F. Vögtle, J. Dohm and K. Rissanen, *Angew. Chem.*, **102**, 943 (1990); *Angew. Chem. Int. Ed. Engl.*, **29**, 902 (1990); J. Dohm, M. Nieger, K. Rissanen and F. Vögtle, *Chem. Ber.*, **124**, 915 (1991).
8. Compare, for example, a cyclohexanoparacyclophane: A. W. Cordes, S.-T. Lin and L.-H. Lin, *Acta Crystallogr.*, **C46**, 170 (1990).
   (a) R. Lemmerz, M. Nieger and F. Vögtle, *J. Chem. Soc., Chem. Commun.*, in press.
9. R. P. Bonar-Law, A. P. Davis and M. G. Orchard, Workshop 'Supra-molecular Organic Chemistry and Photochemistry', Saarbrücken, 27.8–1.9, 1989; R. P. Bonar-Law, A. P. Davies and B. A. Murray, *Angew. Chem.*, **102**, 1479 (1990); *Angew. Chem. Int. Ed. Engl.*, **29**, 1407 (1990); 'Cholaphanes': A. P. Davis, M. G. Orchard, A. M. Z. Slawlin and D. J. Williams, *J. Chem. Soc., Chem. Commun.*, 612 (1991); R. P. Bonar-Law and J. K. M. Sanders, *Tetrahedron Lett.*, **33**, 2071 (1992); For chiral cyclophanes derived from α,α'-trehalose see S. Penadés and J. M. Coterón, *J. Chem. Soc., Chem. Commun.*, 683 (1992).
10. Cf. F. Vögtle and P. Koo Tze Mew, *Angew. Chem.*, **90**, 58 (1978); *Angew. Chem. Int. Ed. Engl.*, **17**, 60 (1978).
11. W. P. Roberts and G. Shoham, *Tetrahedron Lett.*, **22**, 4895 (1981).
12. W. A. Freeman, *Acta Crystallogr.*, **B40**, 382 (1984); cf. J. F. Stoddart, *Angew. Chem.*, **104**, 1550 (1992); *Angew. Chem. Int. Ed. Engl.*, **31**, 1475 (1992).
13. L. A. Paquette and M. A. Kesselmayer, *J. Am. Chem. Soc.*, **112**, 1258 (1990); [2.2]Cyclooctatetraenophane: L. A. Paquette *et al.*, *J. Am. Chem. Soc.*, **114**, 2644 (1992).

10 Exotic Phanes

Some more or less exotic appearing cyclophanes and phanes are compiled in this chapter; their selection is subjective. Some of the molecules already described in earlier sections perhaps also belong here: superphane, double-decker porphyrins, etc. 'Exotic' can include an unexpected or complex connection of atoms or bridges; sometimes the name refers to a peculiarity of its formation, structure, or properties. Some phanes are included here which did not find a place in earlier sections or are difficult to classify elsewhere.

The 'texaphyrins' developed by Sessler carry a special name.[1] The name stems from the state of Texas which, because of its size, is intended to characterize this new type of compound type with an 'expanded porphyrin' ring. Of particular interest is the use of a 'proton sponge' base for deprotonation of the pyrrole NH hydrogen in formation of the gadolinium complex. Such

6 : R = H, CH$_3$

7 : M = Gd, Eu, Nd, Sm, Nd, Cd

gadolinium complexes are of interest for diagnostic imaging with magnetic resonance, being appropriate ligands for photodynamic therapy.[1a]

Among the class of superphanes should be mentioned the as yet unknown cyclobutadieno-superphane hydrocarbon **8**, for which the limiting dipolar structure **8b** can be formulated. This might be stabilized by an intra-annular electron transfer, with both subunits becoming aromatic.[2] Attempts to prepare **8** by vacuum flash photolysis of **9**, led instead to the [4]radialene **10**, which is also the starting material for **9**. For reference to a transition metal-complexed 'cyclobutadienophane', see below.[8] So far it has likewise not been possible to bridge two cyclooctatetraene units at all eight corners.

415

'Cryptanal' (**11**: Newkome) is particularly fascinating because of the ease of its formation.[3] The basket-shaped cavity compounds **12–14** illustrate to what

extent one can control bridging of all sorts at the present time. These molecular 'baskets' can be varied in the size of the base of the basket as well as the width of the sides, as can be seen from the structures.[3a]

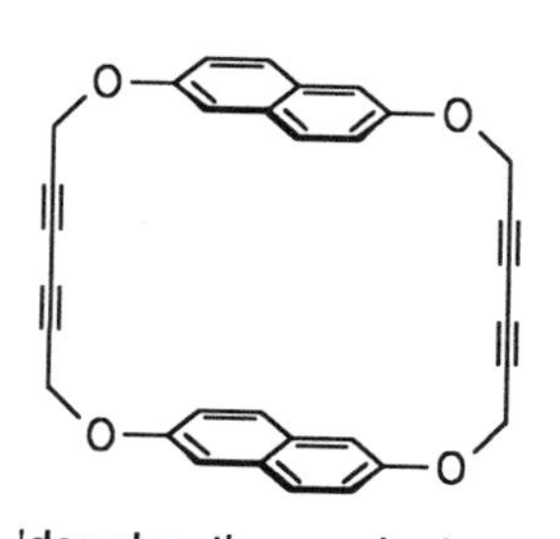

'doughnut' — cyclophane

15

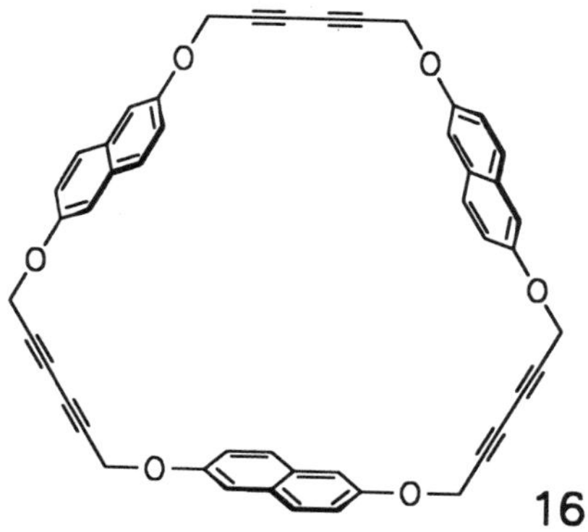

16

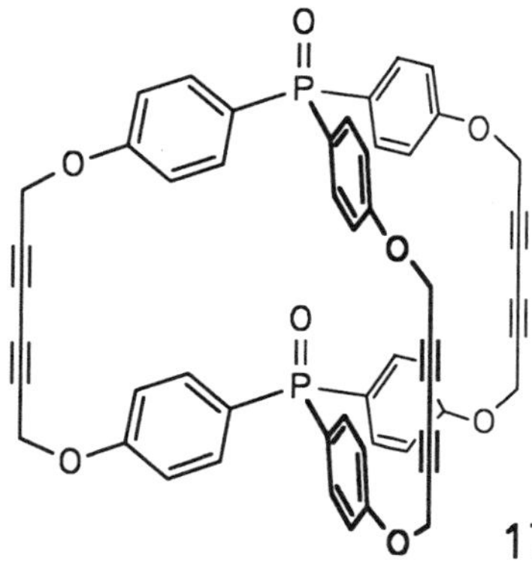

17

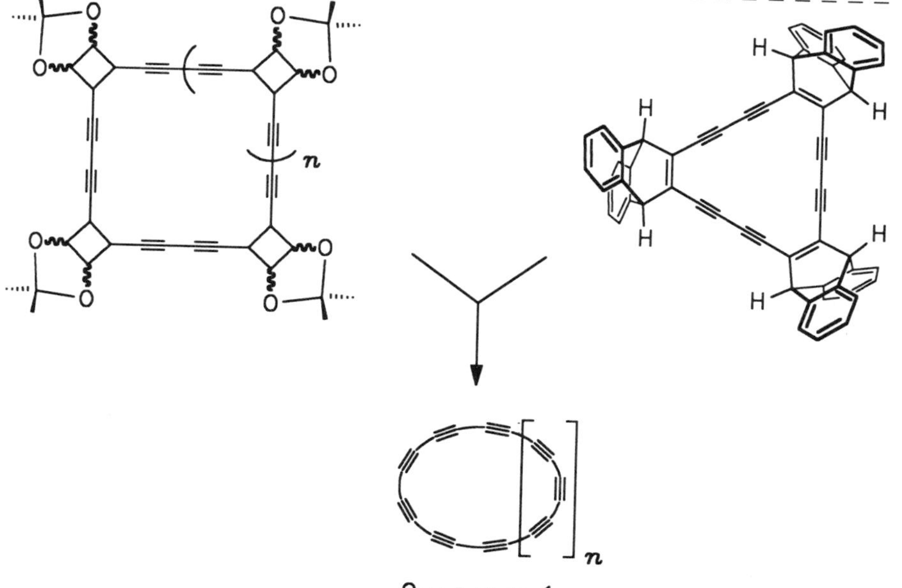

$C_{18} : n = 1$

A favoured, efficient cyclization method for macrocycles is the Glaser/Eglinton coupling.[4,5] The above polyacetylenes[6] are precursors for 'Hückel aromatic' cyclo[n] hydrocarbon molecules, consisting only of carbon, lacking hydrogen or other substituents.

Molecules of the superferrocenophane type (**19, 20**) have recently been synthesized (see *Fascinating Molecules in Organic Chemistry*, Wiley, Chichester, 1991). The synthesis of **19** by way of the partially bridged [4.4.4.3]-ferrocenophane (**18**) is outlined below.[7] Because the iron atom is fully enclosed

$$\xrightarrow[\text{BF}_3\cdot\text{Et}_2\text{O}]{\text{H}_2\text{CN}_2} \qquad \xrightarrow[\text{AlCl}_3]{\text{LiAlH}_4}$$

1. formylation
2. BrCH$_2$CO$_2$Et/ Zn–I$_2$
3. H$_2$/Pd–C

$$\xrightarrow{\hspace{2cm}}$$

1. NaOH
2. ClCO$_2$Et/Et$_3$N
3. AlCl$_3$

$$\xrightarrow{\hspace{2cm}}$$

R = CH$_2$CH$_2$CO$_2$Et

$$\xrightarrow[\text{AlCl}_3]{\text{LiAlH}_4}$$

18 ≙ **18**

19

as in a cage, the superferrocenophane **19** was designated as a 'package compound'. Gleiter has described the transition metal-complexed 'super-cyclobutadienophane' **21** and the supercubanophane **21a**.[8]

Fascinating geometry because of their symmetry is provided by the tris-(chromocarbonyl) complex of [2.2.2]paracyclophane and the chromium complex of hexasila [2.2.2](2,3,5)cyclophane.[7b]

The trimeric [1.1.1]ferrocenophane **23** could be isolated and characterized, along with the di-, tetra-, and pentamers.[9]

A Cope rearrangement leading to a large ring has been described by Vögtle *et al.*[10] In this case a strained cyclophane ring, rather than a cyclopropane or cyclobutane ring, provides the driving force for the [3.3]sigmatropic rearrangement **25 → 26**.

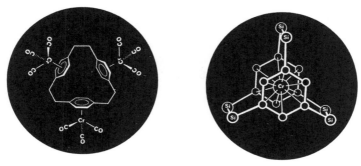

Figure 1. Illustrations (phane–transition metal complexes and metallocenophanes) from various editions of the textbook by Elschenbroich/Salzer[7b]

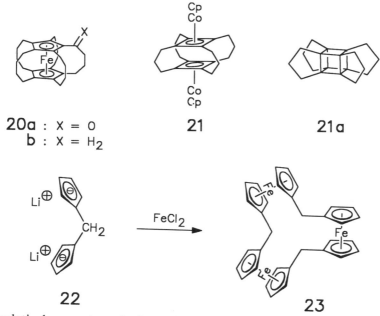

20a : X = O
   b : X = H$_2$

21

21a

22

23

A relatively recent synthetic method for cyclophanes is the intramolecular [2 + 2] photocycloaddition, which affords geometrically attractive 'cyclobutano-*meta*- or *para*-cyclophanes' such as **27**. This, in turn, can be converted to the corresponding [4.4]metacyclophane **28** by ring opening.[11]

The following anthracenophanes are aesthetically pleasing structures,[12] some of which have already been described earlier. Mention can also be made of the dihydro-*sym*-indacenophanes **29**, as well as the methano- and otherwise bridged [10]annulenes, only a few examples of which have been selected (see pages 420, 421).[14,15]

Some structures representing helicenophanes, alias 'bastards' between helices and cyclophanes include propellicene [**30**, 2,13-bis(pentahelicene)];

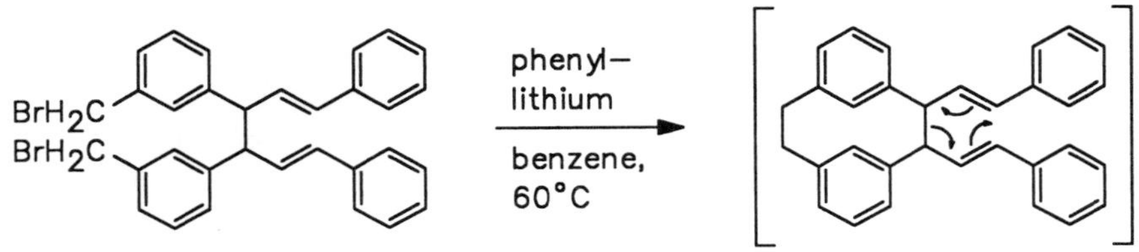

BrH₂C

BrH₂C

phenyl–
lithium

benzene,
60°C

24

25

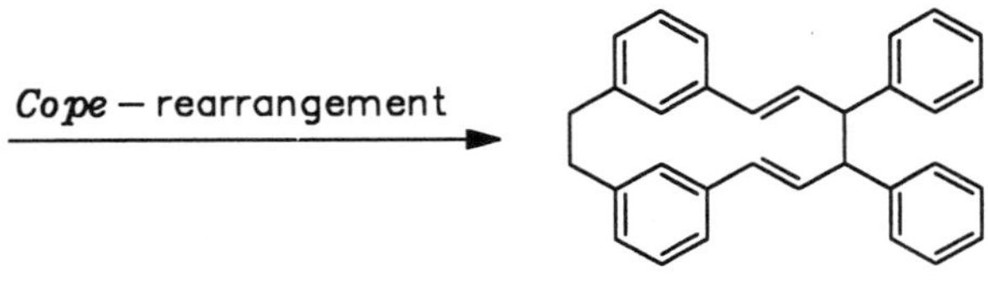

Cope – rearrangement

26

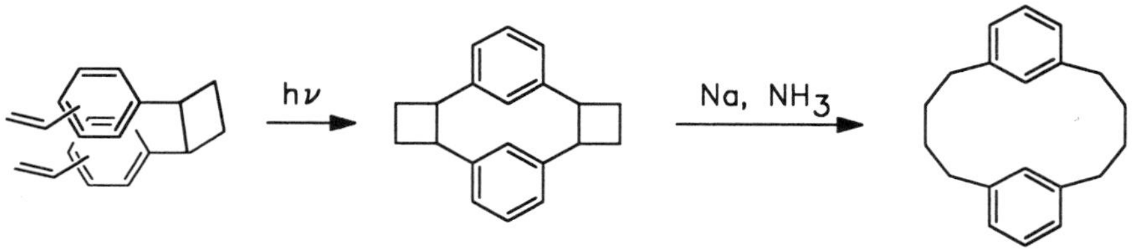

hν

Na, NH₃

27

28

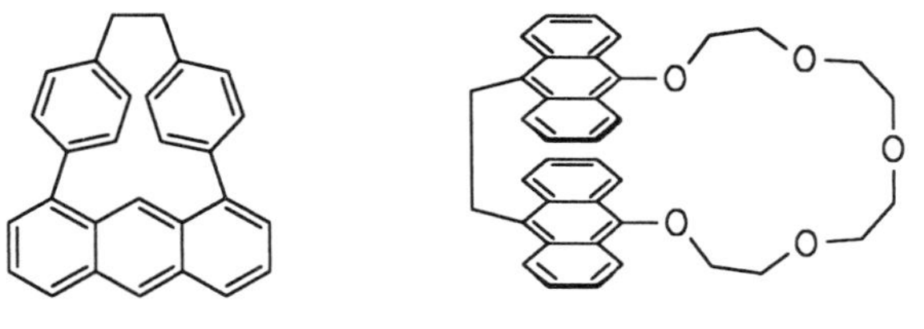

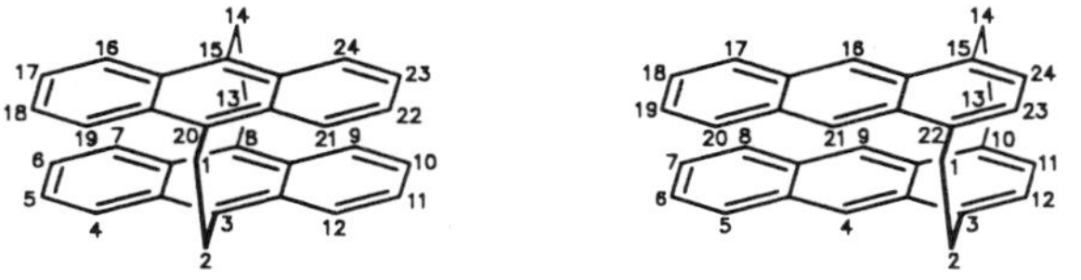

paracyclophanohelicene **31**; helicenophane **32**; and also dibenzo[*def,pgr*]tetra-
phenylene (**33**), with D_2 symmetry point group and a propeller-like geometry
similar to that of **30**.[16] Dihydro[7][7]circulene **35**, obtained from the dithia
[3.3]phane route by way of the intermediate **34**, can certainly be counted among
the 'exotic' examples.[17] The molecules **36** and **37a–c** are examples of multiply
interconnected [2.2]paracyclophanes and -metacyclophanes.[18,19] The structures

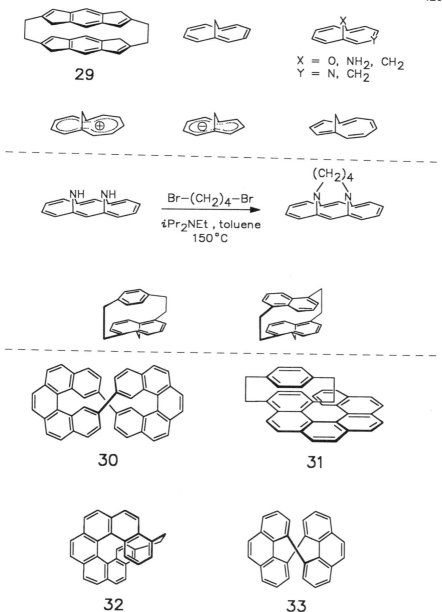

$X = O, NH_2, CH_2$
$Y = N, CH_2$

29

Br—$(CH_2)_4$—Br
———————————→
iPr$_2$NEt , toluene
150°C

$(CH_2)_4$

30

31

32

33

represented by **37**, unlike **36**, are chiral. The most stable isomer is **37a**, to which all other isomers are converted upon heating. **37a** exhibits a three-blade propeller symmetry, with all three [2.2]metacyclophane units in the *anti* orientation. It has been resolved by HPLC on (+)-poly(triphenylmethyl methacrylate) into its enantiomers, whose molecular rotations—M_{436}^{22} of (+)- and (−)-5235—are

34   **35**

considered to be large. **37a** fails to undergo racemization on prolonged heating at 310°C, at which temperature helicenes are fully racemized.

36

37a (all-*anti*)   **37b**   **37c**

Mixed, condensed multiple cyclophanes are illustrated by the quinones and hydrocarbons synthesized by de Meijere.[19a]

The salient property of these [2.2]paracyclophandienes is their reversible reduction to the tetra anion, and the radical anions show interesting electron correlation phenomena. The octaphenyl derivative, obtained in high yield from *in situ*-generated [2.2]paracyclophan-5,9-diyne and tetracyclone, can accept even more electrons. A strategy for obtaining polymeric webbed structures depends on construction of 'linear triple phanes' from p-benzoquinone and 5,6-dimethylene- or 5,6,9,10-tetramethylene[2.2]paracyclophane.

The readily prepared molecule **38** was found to possess a dish shape with corresponding basket-like cavity, as confirmed by X-ray crystallography.[20]

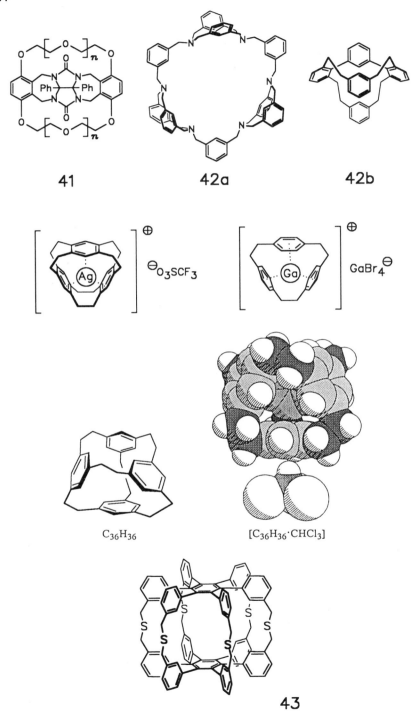

41

42a

42b

$C_{36}H_{36}$

$[C_{36}H_{36}\cdot CHCl_3]$

43

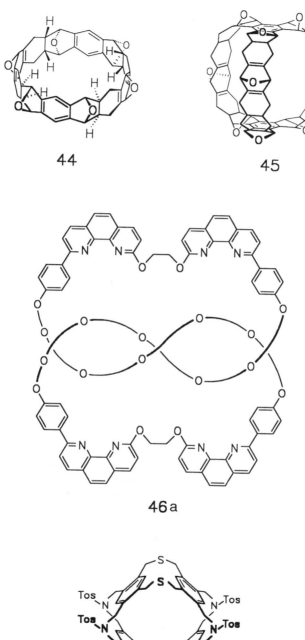

44

45

46a

46b

39 and **40** (page 423), examples of multicyclic phanes containing tetrathiafulvalene units, were prepared in connection with organic charge-transfer complexes and semiconductors.[21]

The crown compound **41** (page 424), prepared from a glycoluril, assumes a basket-like cavity shape and binds strongly with bis(ammonium) and alkali metal ions.[22]

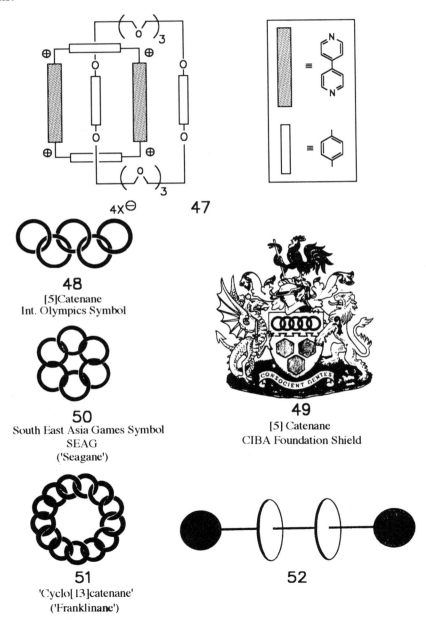

47

48
[5]Catenane
Int. Olympics Symbol

50
South East Asia Games Symbol
SEAG
('Seagane')

49
[5] Catenane
CIBA Foundation Shield

51
'Cyclo[13]catenane'
('Franklinane')

52

Examples of exotic but not easily prepared macrocycles are **42a**.[23a] and the 'cuppedophane' **42b**; even more extensively bridged multicycles are named 'cappedophanes' by Hart.[23b] The striking complex formation of 'deltaphane' and [2.2.2]paracyclophane with metal ions (page 424) is discussed in the textbook *Supramolecular Chemistry* (Wiley, Chichester, 1991). The heptacyclic hydrocarbon $C_{36}H_{36}$ ('spheriphane') was obtained by a tetrahedral clamping of four benzene rings into a spherical framework, with an intramolecular cavity diameter of 568 pm.[23c] The crystal structure analysis of this spherical phane reveals a helically chiral conformation for the 1:1 chloroform adduct, the $CHCl_3$ molecule being bound by a very short hydrogen π-bond to one of the benzene rings (p. 424).

The 'arenologue' of superphane **43** is of interest because of its chirality.[24] Particularly interesting, recently described exotic structures include kohnkene (**44**);[25] trinacrene (**45**)[26] prepared by Stoddart; the recent catenane (**46a**, 'trefoil knot') prepared by Sauvage;[27] and Stoddart's **47**.[28] The molecule **46b** was considered as part of a multi-step strategy to yield tube-shaped structures.[29]

The ultimate achievement in analogous molecular template synthesis would be the 'Olympic catenane' (**48**)[30] and similar, mechanically chained structures (**49–51**), as well as the first example of a 'molecular abacus' (**52**).[30,31]

References

1. J. L. Sessler, T. Murai and G. Hemmi, *Inorg. Chem.*, **28**, 3390 (1989); J. L. Sessler *et al.*, *J. Chem. Soc., Chem. Commun.*, 314 (1989); *J. Am. Chem. Soc.*, **110**, 5586 (1988); *J. Phys. Chem.*, **93**, 8111 (1989); J. L. Sessler and A. K. Burrell, *Top. Curr. Chem.*, **161**, 177 (1992).
   (a) J. L. Sessler, M. J. Cyr and B. G. Maiya, *Progress in Biomedical Optics*, **SPIE 1203**, 233 (1990).
2. H. Hopf, in *Cyclophanes*, Vol. II (P. M. Keehn and S. M. Rosenfeld, Eds.), Academic Press, New York, 1983; L. A. Paquette *et al.*, *J. Am. Chem. Soc.*, **114**, 2644 (1992); For cyclopropenyliophanes see R. Gleiter and M. Merger, *Tetrahedron Lett.*, **33**, 3473 (1992).
3. G. R. Newkome, V. K. Majestic and F. R. Fronczek, *Tetrahedron Lett.*, **22**, 3035, 3039 (1981).
   (a) J. Breitenbach, K. Rissanen, U. U. Wolf and F. Vögtle, *Chem. Ber.*, **124**, 2323 (1991).
4. M. Nakazaki, Y. Yamamoto and T. Toya, *J. Org. Chem.*, **45**, 2553 (1980); S. P. Adams and H. W. Whitlock, *J. Org. Chem.*, **46**, 3474 (1981).
5. B. P. Friedrichsen and H. W. Whitlock, *J. Am. Chem. Soc.*, **111**, 9132 (1989).
6. F. Diederich, Y. Rubin, C. B. Innobler, R. L. Whetten, K. E. Schriver, K. N. Houk and Y. Li, *Science*, **245**, 1088 (1989); Y. Li, Y. Rubin, F. Diederich and K. N. Houk, *J. Am. Chem. Soc.*, **112**, 1618 (1990); F. Diederich *et al.*, *J. Am. Chem. Soc.*, **113**, 495 (1991); J. Almlöf *et al.*, *J. Am. Chem. Soc.*, **113**, 1049 (1991).
7. (a) H. Misatome, J. Watanabe, K. Yamakawa and Y. Iitaka, *J. Am. Chem. Soc.*, **108**, 1333 (1986).
   (b) Review: Ch. Elschenbroich and A. Salzer, *Organometallchemie*, Teubner, Stuttgart, 1988; *Organometallics*, VCH Publishers, Weinheim, 1989.
   (c) M. Hisatome, J. Watanabe, Y. Kawajiri and K. Yamakawa, *Organometallics*, **9**, 497 (1990).

428

8. R. Gleiter, M. Karcher, M. L. Ziegler and B. Nuber, *Tetrahedron Lett.*, **28**, 195 (1987); R. Gleiter and M. Karcher, *Angew. Chem.*, **100**, 851 (1988); *Angew. Chem. Int. Ed. Engl.*, **27**, 840 (1988); E. Osawa, J. M. Rudzinski and Y.-M. Xun, *Struct. Chem.*, **1**, 333 (1990); R. Gleiter and D. Kratz, *Acc. Chem. Res.*, **26**, 311 (1993).
9. T. J. Katz, N. Acton and G. Martin, *J. Am. Chem. Soc.*, **91**, 2804 (1969).
10. F. Vögtle, N. Eisen, P. Mayenfels and F. Knoch, *Tetrahedron Lett.*, **27**, 695 (1986); N. Eisen and F. Vögtle, *Angew. Chem.*, **98**, 1029 (1986); *Angew. Chem. Int. Ed. Engl.*, **25**, 1026 (1986); F. Vögtle, N. Eisen, S. Franken, P. Büllesbach and H. Puff, *J. Org. Chem.*, **52**, 5560 (1987).
11. J. Nishimura, Y. Horikoshi, Y. Wada and H. Takahashi, *Tetrahedron Lett.*, **30**, 5439 (1989); H. Meier, E. Praß and K. Noller, *Chem. Ber.*, **121**, 1637 (1988); J. Nishimura et al., *J. Am. Chem. Soc.*, **112**, 1144 (1990); *Tetrahedron Lett.*, **32**, 2367 (1991).
12. Review: J. A. Reiss, in *Cyclophanes*, Vol. II (P. M. Keehn and S. M. Rosenfeld, Eds.), p. 443, Academic Press, New York, 1983.
13. P. Bickert, V. Boekelheide and K. Hafner, *Angew. Chem.*, **94**, 308 (1982); *Angew. Chem. Int. Ed. Engl.*, **21**, 304 (1982).
14. E. Vogel, M. Biskup, W. Pretzer and W. A. Böll, *Angew. Chem.*, **76**, 785 (1964); *Angew. Chem. Int. Ed. Engl.*, **3**, 642 (1964); E. Vogel, R. Feldmann and H. Düwel, *Tetrahedron Lett.*, 1941 (1970); M. Matsumoto, T. Otsubo, Y. Sakata and S. Misumi, *Tetrahedron Lett.*, 4425 (1977); cf. J. A. Marco, J. F. Sanz, E. Vogel, B. Schwartzkopff-Fischer, H. Schmikler and J. Lex, *Tetrahedron Lett.*, **31**, 999 (1990); T. Inaru et al., *ISNA-7*, July 1992, Poster no. 2.
15. F. Gerson, G. Gescheidt, U. Buser, E. Vogel, J. Lex, M. Zehnder and A. Riesen, *Angew. Chem.*, **101**, 938 (1989); *Angew. Chem. Int. Ed. Engl.*, **28**, 902 (1989); J. A. Marco and J. F. Sanz, *Tetrahedron Lett.*, **31**, 999 (1990).
16. Review: K. Meurer and F. Vögtle, *Top. Curr. Chem.*, **127**, 1 (1985).
17. K. Yamamoto et al., *J. Am. Chem. Soc.*, **110**, 3578 (1988); *ISNA-7*, Victoria 1992, Lecture no. 15; K. Yamamoto, Y. Saitho, D. Iwaki and H. Chikamatsu, *ISNA—International Symposium on Novel Aromatic Compounds*, Osaka, 1989.
18. M. Psiorz and H. Hopf, *Angew. Chem.*, **94**, 639 (1982); *Angew. Chem. Int. Ed. Engl.*, **21**, 640 (1982).
19. W. Kißener and F. Vögtle, *Angew. Chem.*, **97**, 227 (1985); *Angew. Chem. Int. Ed. Engl.*, **24**, 222 (1985).
    (a) O. Reiser, S. Reichow and A. de Meijere, *Angew. Chem.*, **99**, 1285 (1987); *Angew. Chem. Int. Ed. Engl.*, **26**, 1277 (1987); M. Stöbbe, O. Reiser, R. Näder and A. de Meijere, *Chem. Ber.*, **120**, 1667 (1987); A. de Meijere et al., in *Organometallics in Organic Synthesis* (H. Werner and G. Erker, Eds.), p. 255, Springer, Berlin, 1989; cf. H. Hopf and A. De Meijere, *Nachr. Chem. Tech. Lab.*, **38**, 319 (1990); A. de Meijere, J. Heinze, K. Meerholz, O. Reiser and B. König, *Angew. Chem.*, **102**, 1443 (1990); *Angew. Chem. Int. Ed. Engl.*, **29**, 1418 (1990); A. de Meijere et al., *Angew. Chem.*, **103**, 1350 (1991); *Angew. Chem. Int. Ed. Engl.*, **30**, 1361 (1991); A. de Meijere, F. Gerson et al., *J. Am. Chem. Soc.*, **112**, 6827 (1990); cf. A. de Meijere et al., *Synlett.*, **1991**, 20.
20. A. P. West Jr, D. van Engen and R. A. Pascal Jr, *J. Am. Chem. Soc.*, **111**, 6846 (1989).
21. B. Girmay, J. D. Kilbur, A. E. Underhill, K. S. Varma, M. B. Hursthouse, M. E. Harman, J. Becher and G. Bojesen, *J. Chem. Soc., Chem. Commun.*, 1406 (1989); K. Müllen et al., *Angew. Chem.*, **104**, 331 (1992); *Angew. Chem. Int. Ed. Engl.*, **31**, 309 (1992).
22. J. W. H. Smeets, R. P. Sijbesma, L. van Dalen, A. L. Spek, W. J. J. Smeets and R. J. M Nolte, *J. Org. Chem.*, **54**, 3710 (1989).
23. (a) H. Takemura et al., *Tetrahedron Lett.*, **29**, 1031 (1988).
    (b) T. K. Vinod and H. Hart, *J. Org. Chem.*, **55**, 881 (1990); *J. Am. Chem. Soc.*, **112**, 3250 (1990); *J. Org. Chem.*, **56**, 5630 (1991).

(c) F. Vögtle, J. Groß, Ch. Seel and M. Nieger, *Angew. Chem.*, **104**, 1112 (1992); *Angew. Chem. Int. Ed. Engl.*, **31**, 1069 (1992).

24. W. Kißener and F. Vögtle, *Angew. Chem.*, **97**, 782 (1985); *Angew. Chem. Int. Ed. Engl.*, **24**, 794 (1985).

25. F. H. Kohnke, A. M. Z. Slawin, J. F. Stoddart and D. J. Williams, *Angew. Chem.*, **99**, 941 (1987); *Angew. Chem. Int. Ed. Engl.*, **26**, 892 (1987); J. F. Stoddart, *Chem. Brit.*, **24**, 1203 (1988); F. H. Kohnke, J. P. Mathias and J. F. Stoddart, *Angew. Chem. Adv. Mater.*, **101**, 1129 (1989); *Angew. Chem. Int. Ed. Engl.*, **28**, 1103 (1989).

26. P. R. Ashton, N. S. Isaacs, F. H. Kohnke, G. Stagno d'Alcontres and J. F. Stoddart, *Angew. Chem.*, **101**, 1269 (1989); *Angew. Chem. Int. Ed. Engl.*, **28**, 1261 (1989); J. F. Stoddart *et al.*, *J. Am. Chem. Soc.*, **115**, 5422 (1993).

27. C. O. Dietrich-Buchecker and J. P. Sauvage, *Angew. Chem.*, **101**, 192 (1989); *Angew. Chem. Int. Ed. Engl.*, **28**, 189 (1989); V. Balzani, J.-P. Sauvage *et al.*, *J. Am. Chem. Soc.*, **113**, 4033 (1991); For a new different molecular knot see D. Walba, Q. Y. Zheng and K. Schilling, *J. Am. Chem. Soc.*, **114**, 6259 (1992); For rotaxanes see J.-P. Sauvage *et al.*, *J. Chem. Soc. Chem. Commun.*, 1131 (1992).

28. J. F. Stoddart *et al.*, *Angew. Chem.*, **101**, 1404 (1989); *Angew. Chem. Int. Ed. Engl.*, **28**, 1394 (1989); for cyclodextrin catenanes, see J. F. Stoddart *et al.*, *Angew. Chem.*, **105**, 944 (1993); *Angew. Chem. Int. Ed. Engl.*, **32**, 854 (1993).

29. F. Vögtle, A. Schröder and D. Karbach, *Angew. Chem.*, **103**, 582 (1991); *Angew. Chem. Int. Ed. Engl.*, **30**, 575 (1990); A. Schröder, D. Karbach, R. Güther and F. Vögtle, *Chem. Ber.*, **125**, 1881 (1992).

30. J. F. Stoddart, GDCh Lecture, Chemische Institute der Universität Bonn, 12 December, 1989.

31. For new rotaxanes, see J. A. Butcher, *Tetrahedron Lett.*, **33**, 6255 (1992); G. Werz and F. Wolff, *Angew. Chem. Int. Ed. Engl.*, **31**, 783 (1992).

11 Naturally Occurring Phanes

As can be seen from an overview of the phanes described in previous sections, they have been for the most part conceived and synthesized by chemists. Clearly the conception of new phane structures has been inspired or facilitated by consideration of natural products, in particular their ring structures. This does not imply, however, that there are no cyclophanes and other phanes to be found in nature; quite the contrary is the case. There exist a considerable number of bridged, aromatic compounds of biological origin, many of which show physiological activity.

It can not be the purpose in this chapter to present an exhaustive literature survey of all naturally occurring phanes.[1] Rather some of the earlier and more recent natural phane structures will be selected from the literature as examples. It is not always easy to locate specific natural phanes in the index of journals, for these substances are typically not classified under cyclophane chemistry but rather under natural product synthesis and physiological activity. Because, in fact, macrocyclic structures often are effective in providing strong physiological activity, basic cyclophane structures are continually of interest in structure modification for inducing pharmacological activity.

One of the simplest natural phanes is muscopyridine (6) which is available by synthesis.[2] In 1980 a report was published of the total synthesis of myricanol

a : R = H, X = H, OH
b : R = H, X = O
c : R = CH$_2$Ph, X = O
d : R = CH$_2$Ph, X = H, OAc
e : R = CH$_3$, X = O
f : R = CH$_3$, X = H, OAc

6   7

(7a) and myricanone (7b), which occur in the bark of the *Myrica nagi* tree.[3] The former shows insecticidal activity. From the X-ray crystal structure of 16-bromomyricanol it could be shown that the molecule is not planar and that the biphenyl axis is bent from linearity, the result being a relief of ring strain.

From the plant family Lythraceae over 40 alkaloids have been isolated, which are classified into four types A–D (A, lythranidine; B, lythrancine; C, lythrine; D, lagerine).[4] Some of these large rings have previously been synthesized.

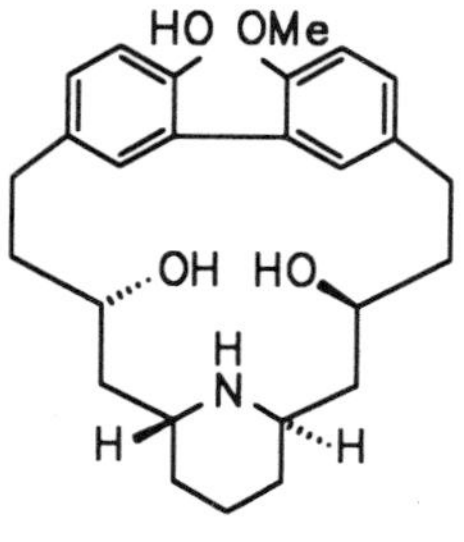

A

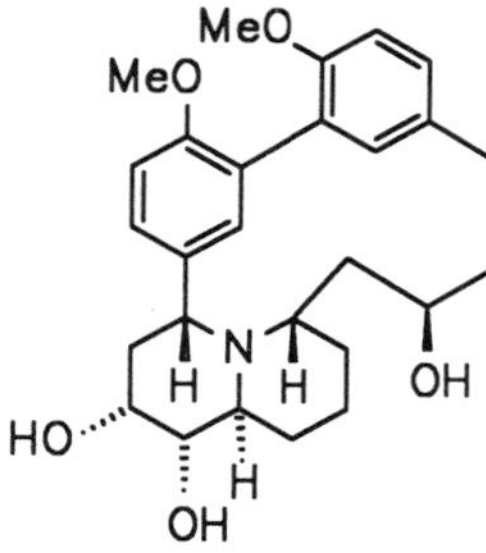

B

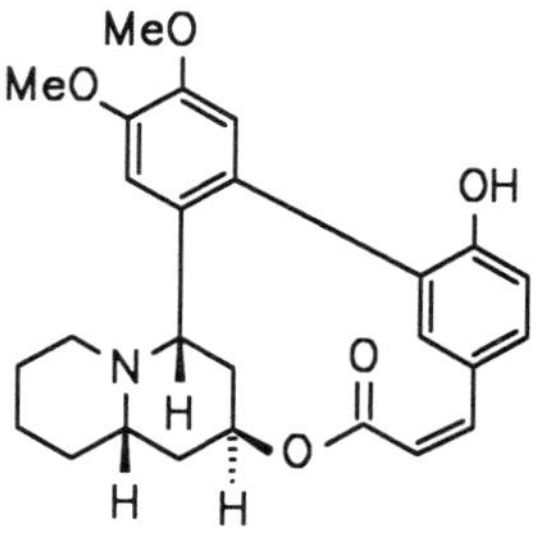

C

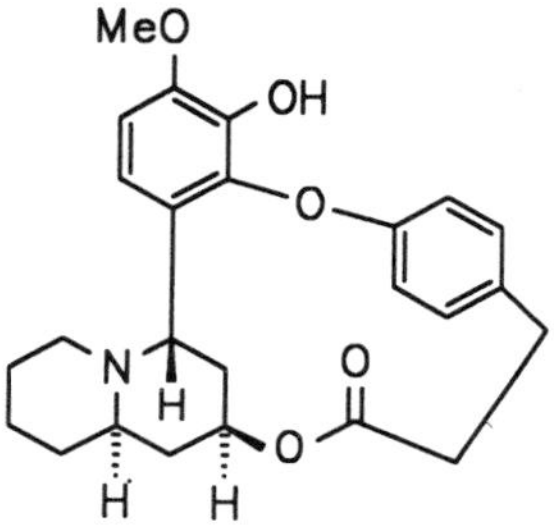

D

Riccardin B, which shows cytotoxic activity, features a diphenylethane unit. It was obtained by nickel-catalysed cyclization.[5]

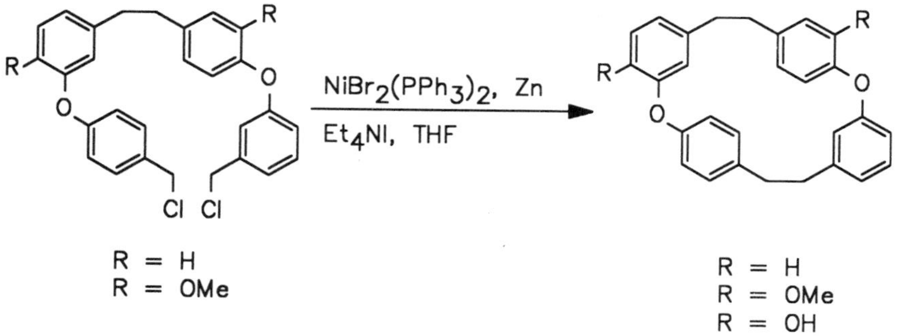

R = H
R = OMe

NiBr$_2$(PPh$_3$)$_2$, Zn
Et$_4$NI, THF

R = H
R = OMe
R = OH

Phane structures, often with partially hydrogenated arene units, occur in some important families of antibiotics: rifamycins, streptovaricins, tolypomycins, and geldanamycins. Only some of the well known, representative structures are shown here.[1a]

8

Rifamycin [1a]	X	Y	Z
B	H	OH	CH$_2$COOH
Y	OH	O=	CH$_2$COOH
L	H	OH	COCH$_2$OH

Rifamycin S [1e]

9

Antibiotics in the vancomycin group (14) owe their biological activity to their ability to form complexes with *N*-acyl derivatives of D-Ala-D-Ala, which leads to an interference in the biosynthesis of the bacterial cell wall. The phane ring structure holds the peptide backbone in a conformation such that it is oriented in a complementary fashion for substrate binding.[1e] The related antibiotic avoparcin (15) provides a similar binding site, whereas ristocetin (16) possesses a different phane structure. The sugar residues in 14–16 appear to play no major role in the binding sites. The phane ring in all three antibiotics 14–16 seems to control the conformation of the peptide binding site. They strongly bind complete dipeptide substrates selectively, even in polar solvents.[1e]

Several years ago macrocyclic oligopeptides and many-membered cyclophanes of the types 17 and 18 were synthesized, with a view to their effectiveness as vancomycin models and plausible active sites in vancomycin.[6]

(±)-*N*-Methylmaysenin (19), a member of the widely studied maytansenoids, is available in both enantiomeric forms from a total synthesis. In this way the natural, optically active *laevo*-form is obtained without resolution.[7] For purposes of comparison the structure of maytansin (20) is also shown. The individual synthetic steps, starting with commercially available tri-*O*-acetyl-D-glucal, can not be presented in detail. They proceed in high yield, and the various chiral centres are created with high stereochemical efficiency.

Tridentoquinone, a red mould pigment, possesses the macrocarbocyclic *ansa* structure 21, as determined by X-ray crystal structure analysis and chemical investigations. The total synthesis of the 17-membered cyclotripeptides OF 4949-III and OF 4949-IV was accomplished in 1989.[9] From this it was deduced that substituents at a substantial distance from the site of cyclization can exert

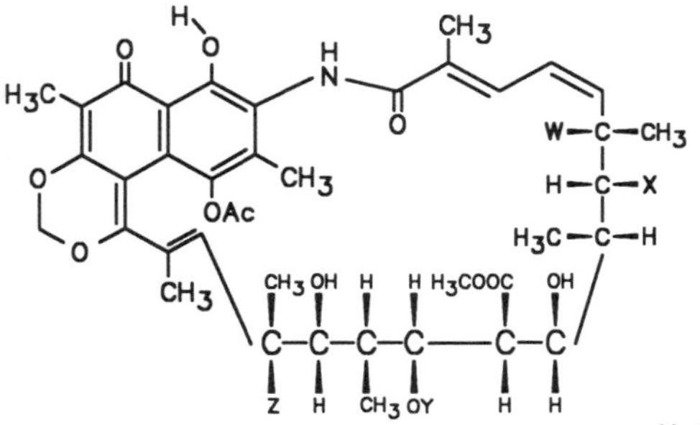

10

Streptovaricin [1a]

	W	X	Y	Z
A	OH	OH	Ac	OH
B	H	OH	Ac	OH
C	H	OH	H	OH
D	H	OH	H	H
E	H	O=	H	OH
G	OH	OH	H	OH

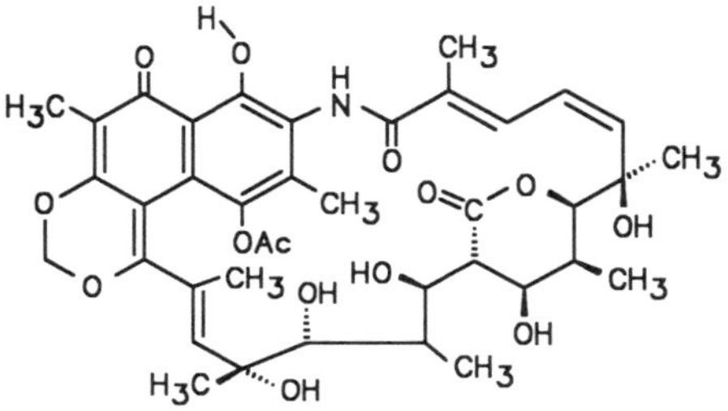

11

Streptovaricin F [1a]

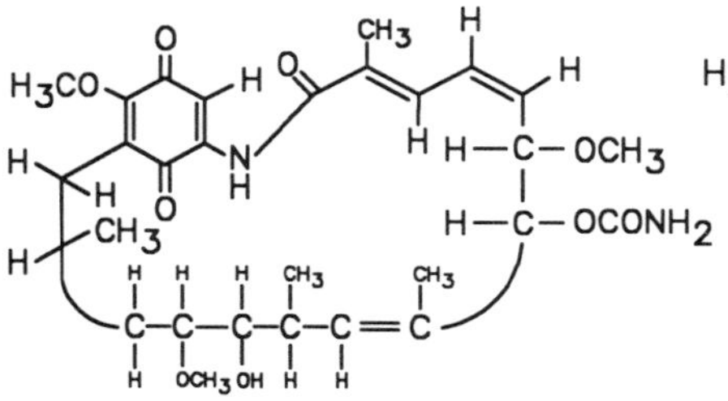

13

Geldanamycin

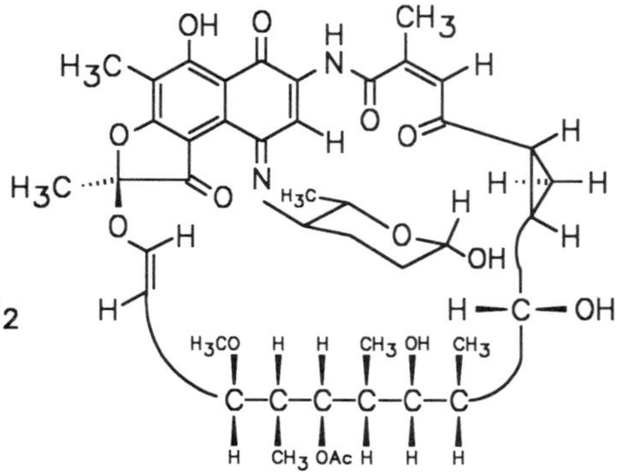

12

Tolypomycin Y

a strong influence on the rate of cyclization. These cyclotripeptides, isolated from *Penicillium rogulosum*, show strong aminopeptidase B inhibitor activity as well as antitumour activity. Analogues of the selective and highly active anti-tumour antibiotics bouvardin (**22**) and deoxybouvardin (**23**) are available synthetically.[10]

D. A. Evans and U. Schmidt, using pentafluorophenyl ester (PFPE-) cyclizations at varying positions, have succeeded in synthesizing OF 4949-III and moreover have constructed the necessary diphenyl ether-diamino acid by enantio- and diastereoselective synthesis.

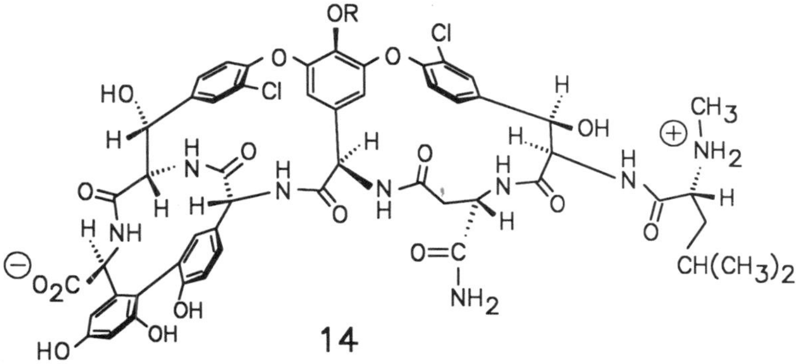

14

Vancomycin (R = sugar)

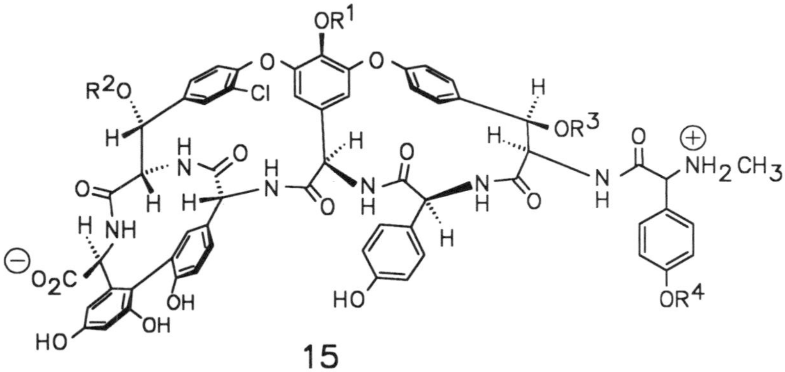

15

Avoparcin (R = sugar)

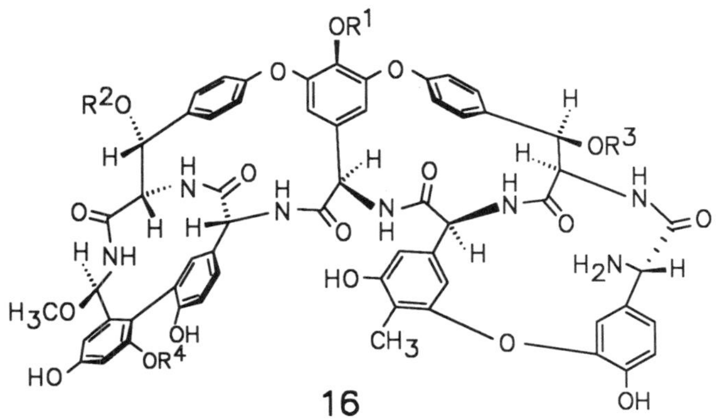

16

Ristocetin A (R = sugar)

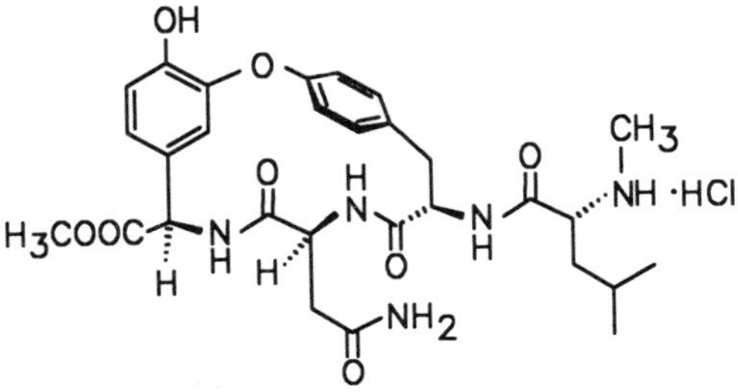

17

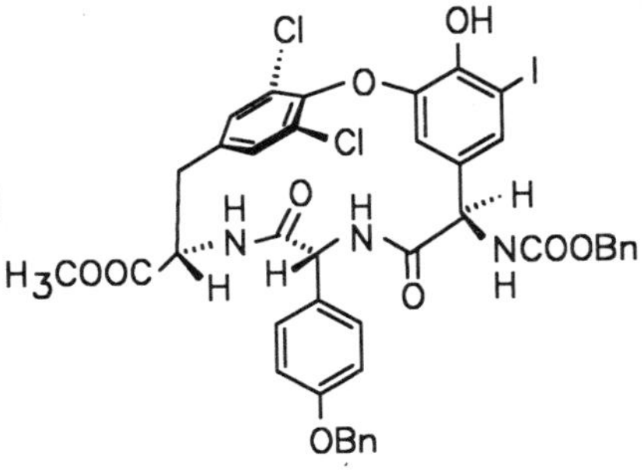

18

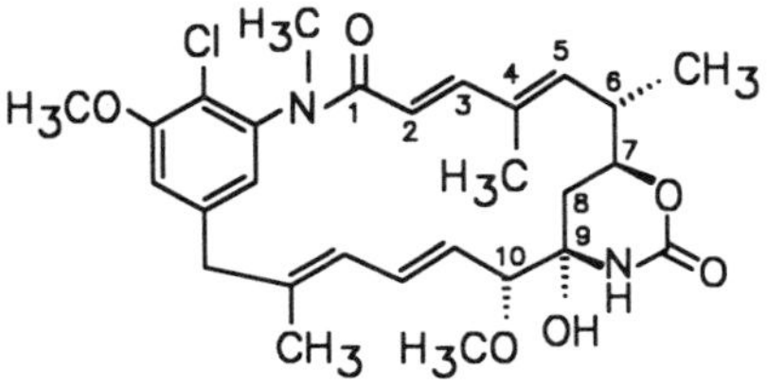

19

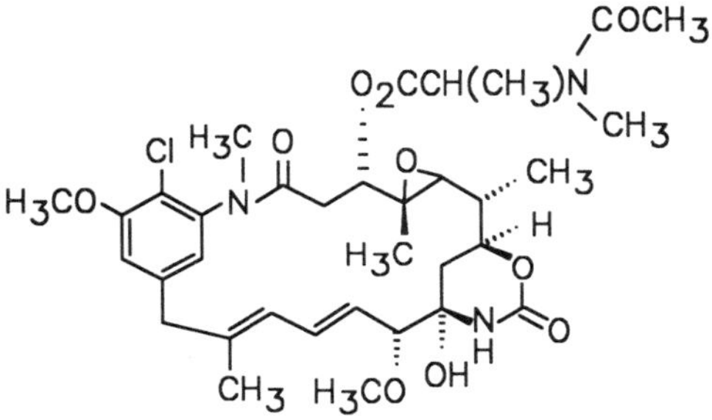

20

Maytansin

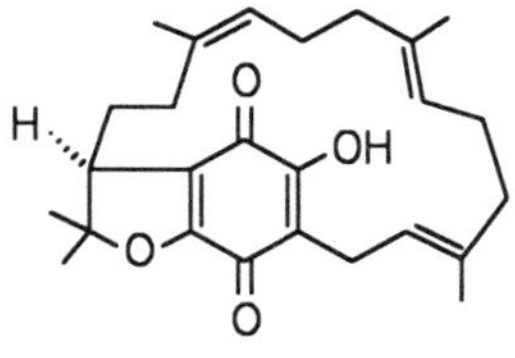

21

Cyclic peptides and peptolides with unusual structures are of particular interest because they often show highly active biological properties. The structures of *anti*-ACE (angiotensin converting enzyme) and renin cyclic peptides from *Lycii radicis* cortex (lyciumin A, for example) have recently been clarified by Japanese groups.[11]

When a peptide chain is cyclized, part of its flexibility is lost and the conformation is established for biologically important sections of the peptide. By studying the direction of attachment of active compounds at the receptor sites, important conclusions can be deduced about the structure of the receptor.[12] Many-membered cyclic peptides can be prepared in high yield (> 80%) by ring closure via catalytic hydrogenation of ω-(Z)-aminopentafluorophenyl ester, as

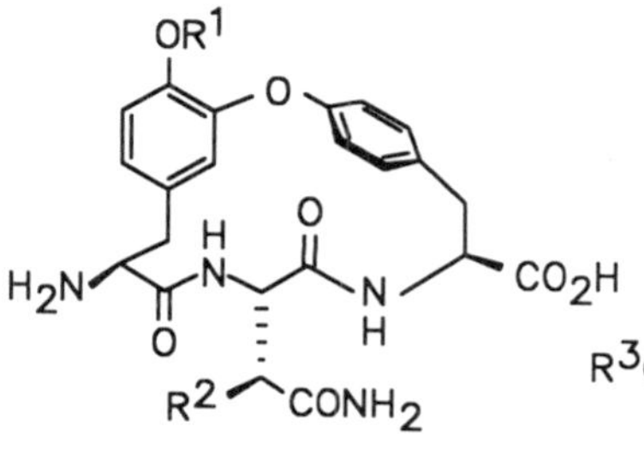

R^1	R^2	
CH$_3$	H	OF4949–III
H	H	OF4949–IV
CH$_3$	OH	OF4949–I
H	OH	OF4949–II

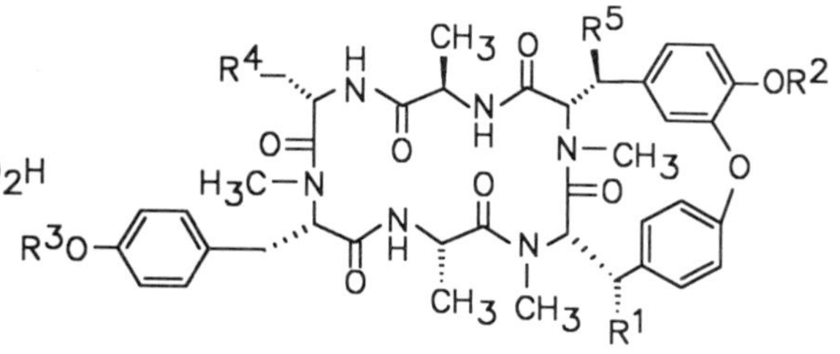

	R^1	R^2	R^3	R^4	R^5	
22:	OH	H	CH$_3$	H	H	Bouvardin
23:	H	H	CH$_3$	H	H	Deoxybouvardin, (RA–V)

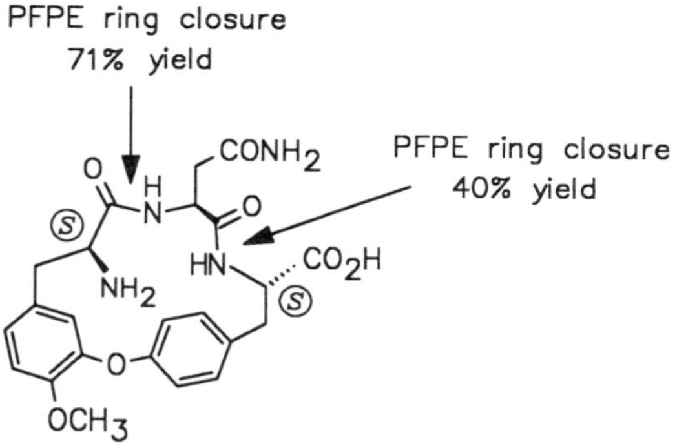

PFPE ring closure
71% yield

PFPE ring closure
40% yield

OF4949–III

is illustrated with the example of ziziphin A (**24**). The double bond was generated by oxidation of a phenylseleno ether. The cyclic peptide dolastatin (**25**), from the tunicate *Dolabella auricularia*,[1b] contains two thiazole rings.

Through synthesis of all 16 diastereomers by U. Schmidt and Shiori it was demonstrated that the structure originally proposed was wrong. Of particular interest in the synthesis is the fact that cyclization of the linear pentafluorophenyl

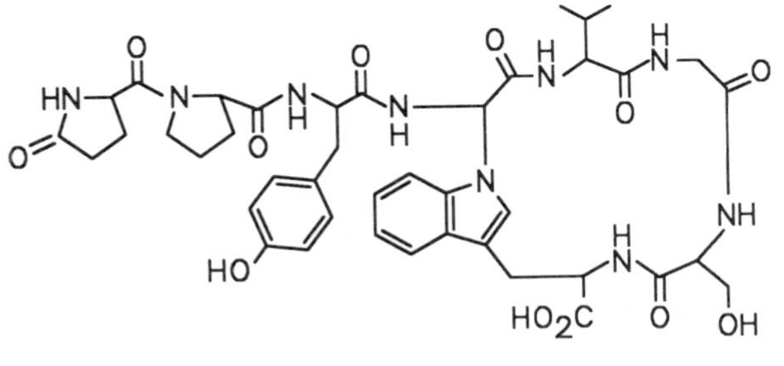

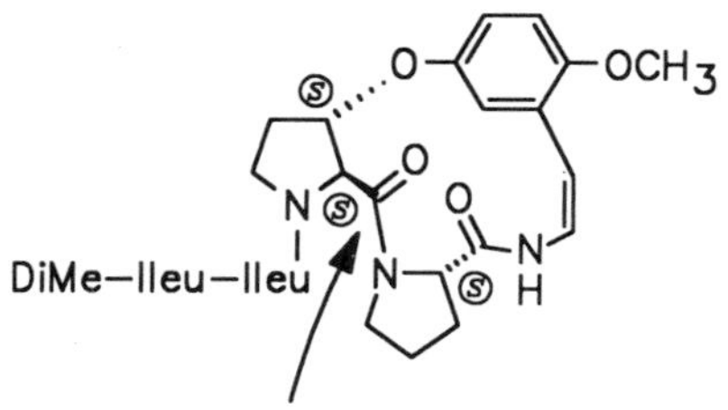

PFPE ring closure
80% yield

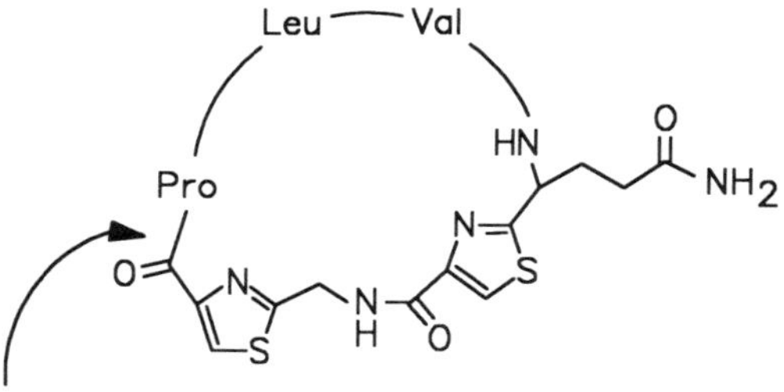

PFPE ring closure
95% yield

Ziziphin A

Dolastatin 3

24

25

ester in a phase-transfer system proceeded in nearly theoretical yield to afford the cyclic peptide. The strong cytostatic properties originally attributed to the compound could not be substantiated.

Cyclic peptides containing thiazole and oxazoline rings, such as the strongly cytostatic ulycyclamide (**26**), were isolated from the tunicate *Lissoclinum patella*. Ring closure once again was achieved by way of the Boc-amino PFP ester.[1b] The didemnins A and B (**27**), shown to have strong cytostatic properties, are

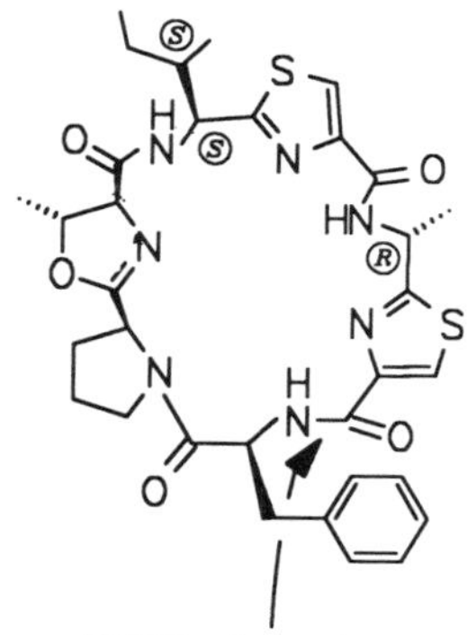

PFPE ring closure
20% yield

26

at the stage of clinical testing. Their molecular structure is based on X-ray crystal structure analysis. The total synthesis is designed for a large scale; in only a few minutes the ring can be formed in 70% yield by the PFPE method in a phase-transfer system (R = benzyloxycarbonyl or H).[1b]

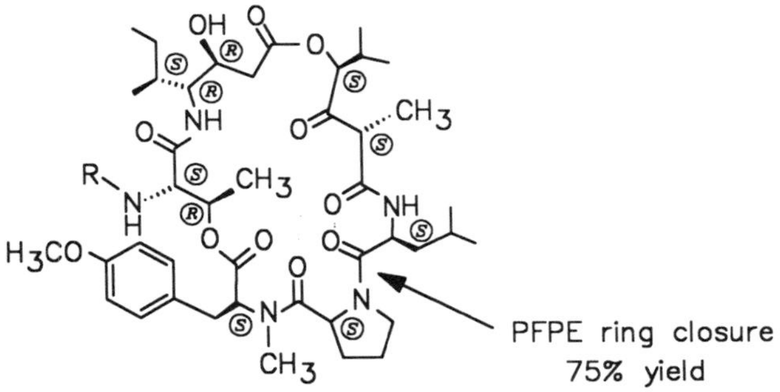

Didemnin A : R = H−(**R**)−MeLeu
Didemnin B : R = H−Lac−Pro−(**R**)−MeLeu

27

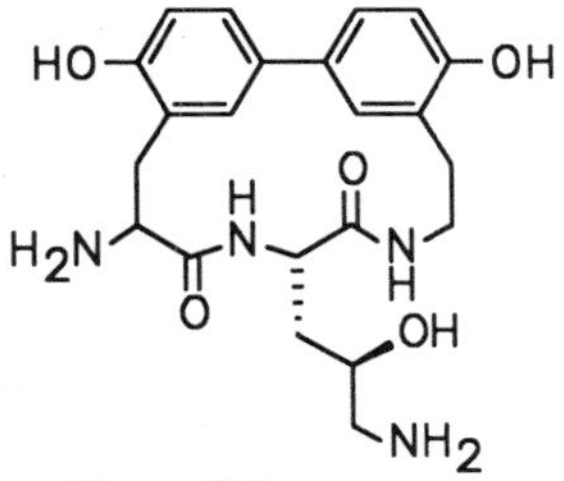

A : R = OH
B : R = H

28

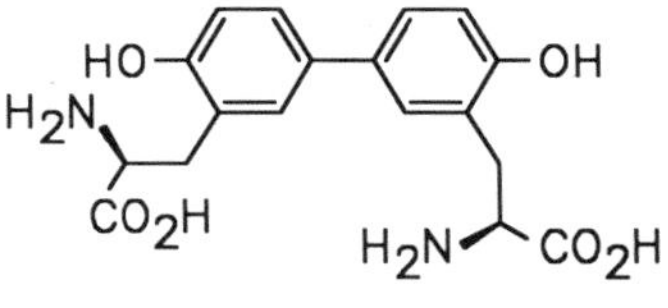

29

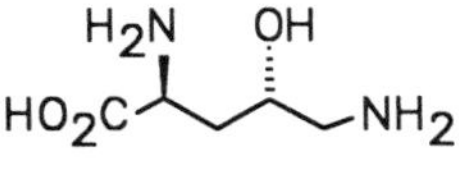

30

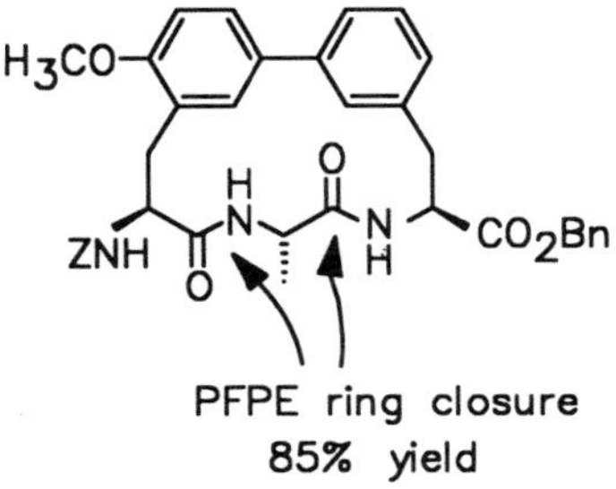

31

Cyclic peptides containing diphenyl or diphenyl ether units play a role in mould metabolites, as is evidenced from the continual isolation of several cyclic peptides with these units. The antibiotics of the vancomycin group (see above), already being used clinically, are members of this class. Both structural units arise from oxidative dimerization of hydroxyphenylglycine and tyrosine. The only cyclic peptides known to date with a single diphenyl unit are WS-43708A and B (28). They are highly effective antibacterial agents from *Streptomyces griseorubiginosus*. As *ansa* cyclic peptides, they contain the two unnatural amino acids diisotyrosine (29) and γ-hydroxyornithine (30).[1b] Synthesis of the *ansa* peptide 31 has just recently been accomplished, the ring closure of both amide positions being achieved in 85% yield by the PFPE phase-transfer process.[1b]

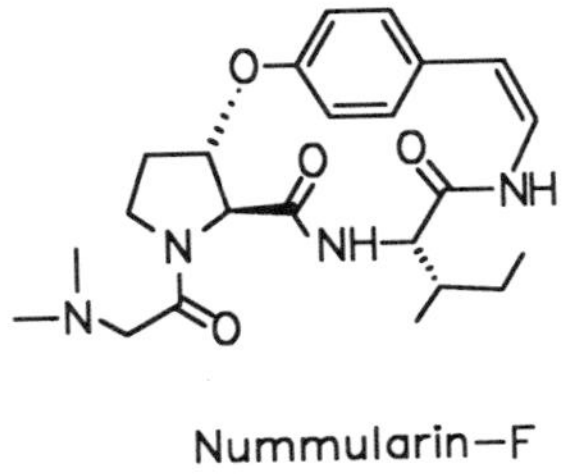

Nummularin−F

32

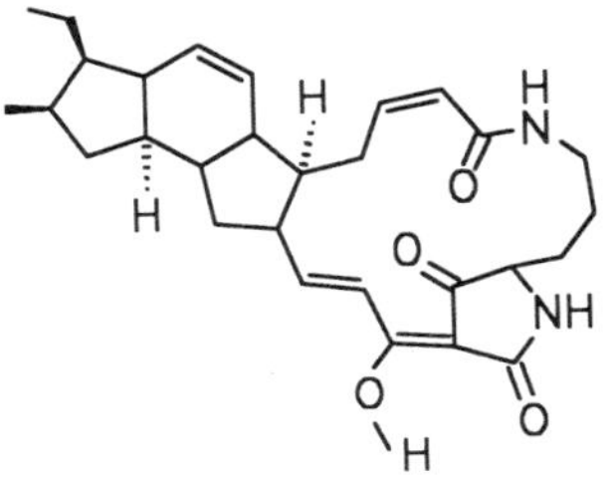

33

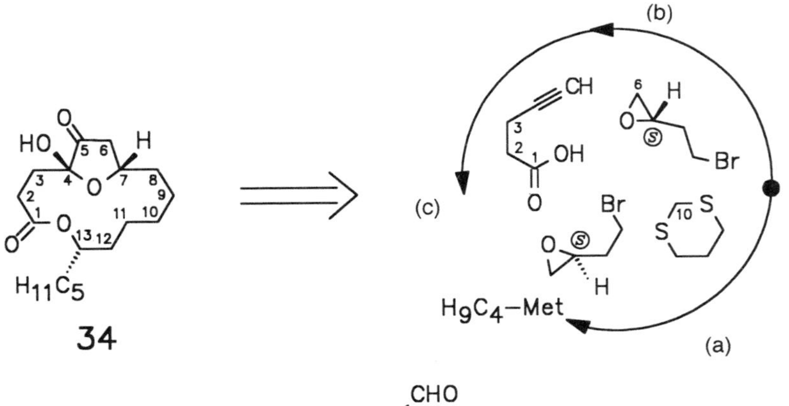

34

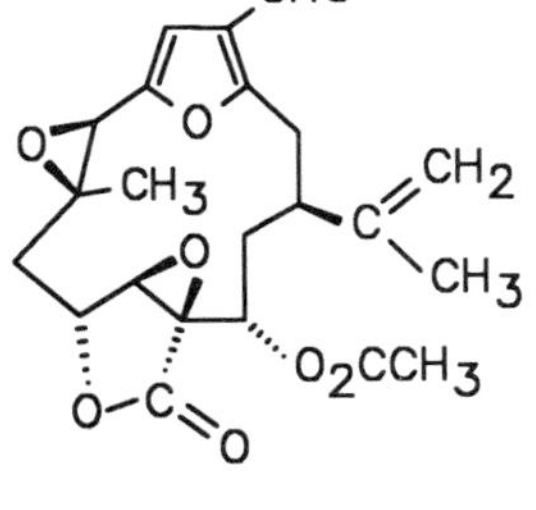

35

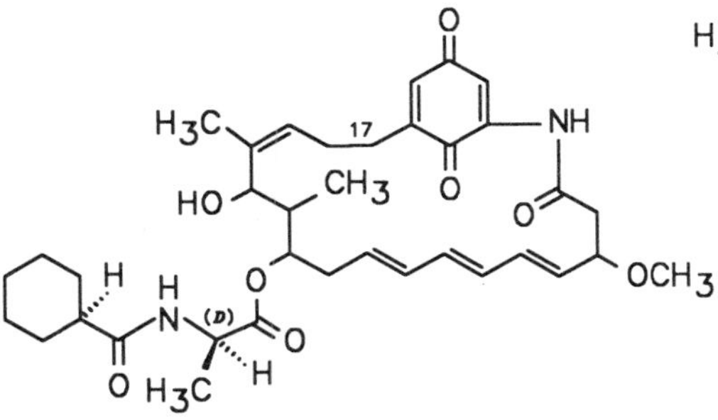

Ansatrienin A = Mycotrienin

36

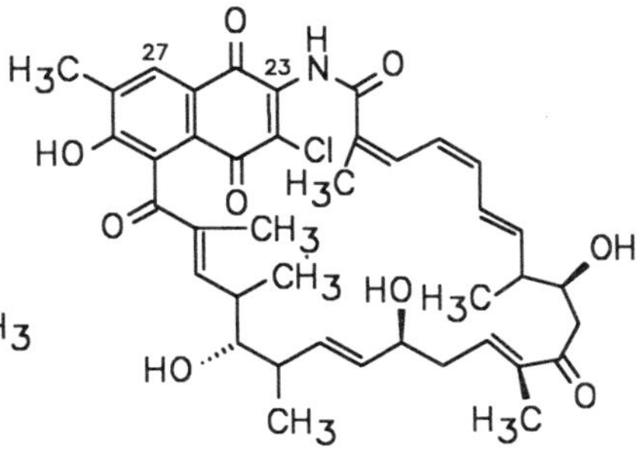

Naphthomycin A

37

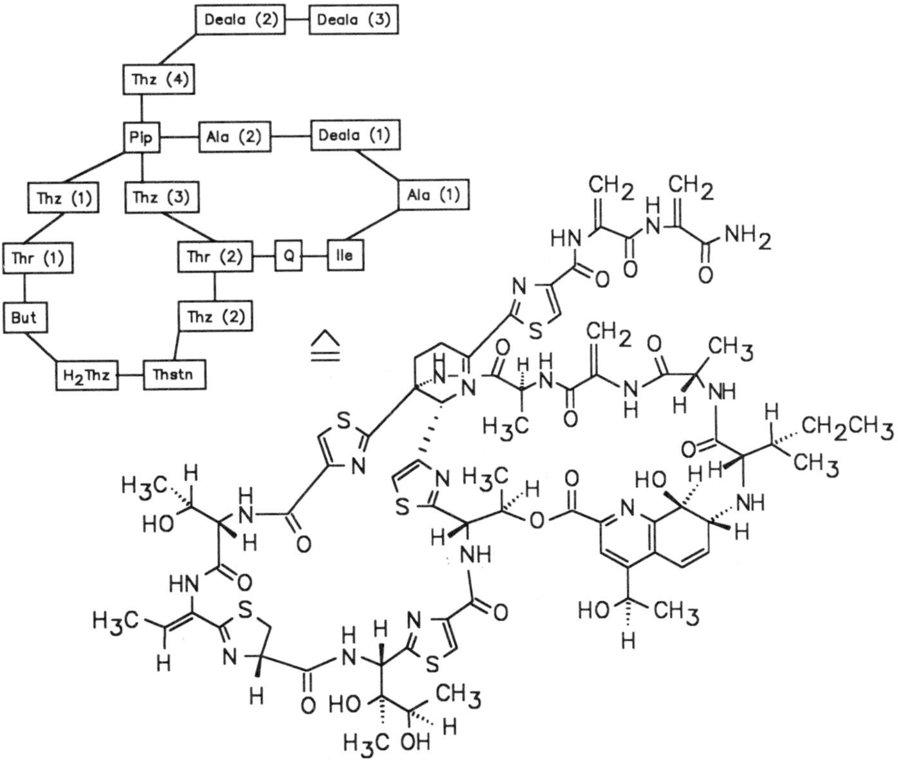

38

The highly strained 15-membered *para-ansa* cyclic peptide alkaloid 'nummularin F' (32) has been prepared in a similar way.[13]

A large number of natural macrocycles containing partially hydrogenated five- and six-membered rings could justifiably be designated as oligohydrophanes ('aliphanes'; see Chapter 9). This holds, for example, for ikarugamycin (33), for which Boeckmann *et al.* described an enantioselective and highly convergent synthesis.[14] Ikarugamycin, an antibiotic with widely varying biological activity, was isolated in 1972.[15]

Gloeosporon (34), extracted from the fungus *Colletotrichum gloeosporioides*, is a fungistatic. In 1987 Seebach reported the total synthesis of its tetra-hydrofuran derivative and established its absolute configuration. The key retrosynthetic analysis for 34 by the authors is outlined on page 440.[16] A contrasting example is lophotoxin (35), which, with a furan ring and two oxirane rings, is a genuine [10]furanophane.[17]

Finally, mention is made of the ansamycin antibiotics ansatrienin (36), naphthomycin A (37), both metabolites of *Streptomyces collinus*, and the complex sulphur-containing peptide antibiotics thiostrepton (38) and nosineptide (multhiomycin, 39). Both groups are effective against Gram-positive bacteria by virtue of the common features in their structures. They are bound to the ribosomal 50S subunit and thereby impede the functions of elongation factors Tu and G.[1c]

39

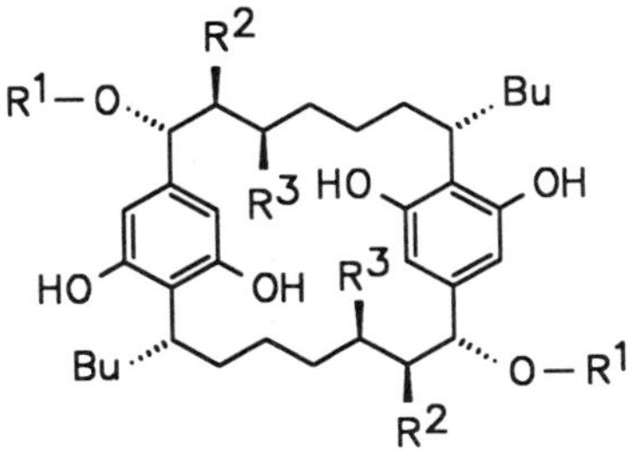

a: R^1, R^3 = H, R^2 = CH_3

b: R^1 = CH_3, R^3 = Cl, R^2 = H

40

Also included here are the recently isolated N,N'-dimethyl-2,12-dihydroxy-1, 10-diaza[3.3]paracyclophane (from citrus skin) and a [7.7]paracyclophane **40** (from algae).[18]

At the conclusion of this chapter, mention should be made of synthetic peptides resembling natural products (for example, **41** and **42**) reported by Feigel.[19] The macrocycle **41** in solution assumes a dish-like conformation, although both **41** and **42** show conformational flexibility.

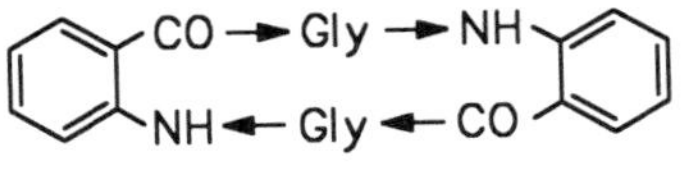

41

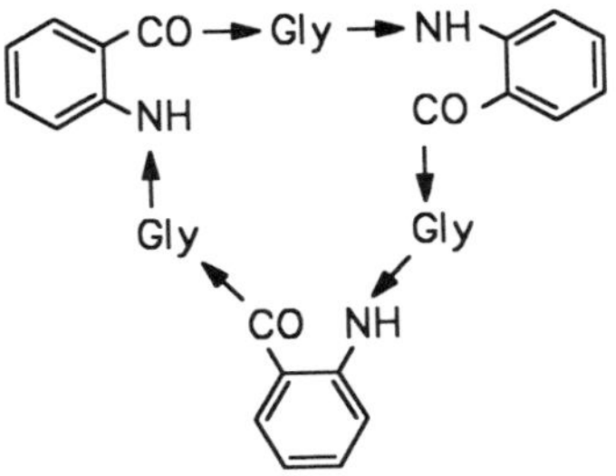

42

For recent information on naturally occurring phanes[20] and ansasteroids, the reader is referred to the original literature.

444

References

1. Reviews
   (a) K. L. Rinehart Jr, *Acc. Chem. Res.*, **5**, 57 (1972); K. L. Rinehart *et al.*, *Pure Appl. Chem.*, **62**, 1277 (1990); Rifamycin W: K. Tatsuza, *Tetrahedron Lett.*, **31**, 1585 (1990); L. Cellai *et al.*, *Helv. Chim. Acta*, **76**, 1459 (1993).
   (b) U. Schmidt, *Nachr. Chem. Tech. Lab.*, **37**, 1034 (1989). See also P. D. Edwards *et al.*, *J. Chem. Soc.*, *Perkin Trans 1*, 123 (1992).
   (c) H. G. Floss and J. M. Beale, *Angew. Chem.*, **101**, 147 (1989); *Angew. Chem. Int. Ed. Engl.*, **28**, 146 (1989). K. A. Reynolds *et al.*, *J. Am. Chem. Soc.*, **113**, 4339 (1991); Cyclothiazomycin: M. Aoki *et al.*, *Tetrahedron Lett.*, **32**, 221 (1991); H. G. Floss *et al.*, *J. Am. Chem. Soc.*, **115**, 5254 (1993).
   (d) S. M. Rosenfeld and K. A. Choe, in *Cyclophanes* (P. M. Keehn and S. M. Rosenfeld, Eds.), Vol. I, p. 351, Academic Press, New York, 1983.
   (e) I. O. Sutherland, in *Cyclophanes* (P. M. Keehn and S. M. Rosenfeld, Eds.) Vol. II, p. 602, Academic Press, New York, 1983; M. Born and Ch. Tamm, *Helv. Chim. Acta*, **73**, 2242 (1990).
2. K. Biemann, G. Büchi and B. H. Walker, *J. Am. Chem. Soc.*, **79**, 5558 (1957).
3. D. A. Whiting and A. F. Wood, *J. Chem. Soc.*, *Perkin Trans. 1*, 623 (1980).
4. K. Fuji, K. Ichikawa and E. Fujita, *Tetrahedron Lett.*, 361 (1979).
5. M. Iyoda, M. Sakaitani, H. Otsuka and M. Oda, *Tetrahedron Lett.*, **26**, 4777 (1985); see also H. D. Zinsmeister, H. Becker and T. Eicher, *Angew. Chem.*, **103**, 134 (1991); *Angew. Chem. Int. Ed. Engl.*, **30**, 130 (1991).
6. Y. Suzuki, S. Nishiyama and S. Yamamura, *Tetrahedron Lett.*, **30**, 6043 (1989).
7. E. J. Corey, L. O. Weigel, A. R. Chamberlin and B. Lipshutz, *J. Am. Chem. Soc.*, **102**, 1439 (1980).
8. H. Besl, H.-J. Hecht, P. Luger, V. Pasupathy and W. Steglich, *Chem. Ber.*, **108**, 3675 (1975).
9. D. L. Boger and D. Yohannes, *Tetrahedron Lett.*, **30**, 5061 (1989).
10. D. L. Boger and D. Yohannes, *Synlet*, 33 (1990); D. L. Boger and D. Yohannes, *J. Am. Chem. Soc.*, **113**, 1427 (1991).
11. S. Yahara, C. Shigeyama, T. Nohara, H. Okuda, K. Wakamatsu and T. Yasuhara, *Tetrahedron Lett.*, **30**, 6041 (1989).
12. H. Kessler, *Angew. Chem.*, **94**, 509 (1982); *Angew. Chem. Ind. Ed. Engl.*, **21**, 512 (1982); H. Kessler, A. G. Klein, R. Obermeier and M. Will, *Liebigs Ann. Chem.*, 269 (1989).
13. R. J. Heffner and M. M. Joullie, *Tetrahedron Lett.*, **30**, 7021 (1989).
14. R. K. Boeckman Jr, Ch. H. Weidner, R. B. Perni and J. J. Napier, *J. Am. Chem. Soc.*, **111**, 8036 (1989).
15. L. A. Paquette, D. Macdonald, L. G. Anderson and J. Wright, *J. Am. Chem. Soc.*, **111**, 8037 (1989).
16. G. Adam, R. Zibuck and D. Seebach, *J. Am. Chem. Soc.*, **109**, 6176 (1987).
17. W. Fenical *et al.*, *Science*, **212**, 1512 (1981); *J. Am. Chem. Soc.*, **104**, 6463 (1982).
18. Y. Matsubara, T. Mizuno, A. Sawabe, Y. Iizuka and K. Okamoto, *Nippon Nogei Kagaku Kaishi*, **63**, 1373 (1989); B. S. Moore *et al.*, *J. Am. Chem. Soc.*, **112**, 4061 (1990); *Tetrahedron*, **48**, 3001 (1992).
19. M. Feigel and G. Lugert, *Liebigs Ann. Chem.*, 1089 (1989); Overview on peptide conformation mimetics: G. Hölzemann, *Kontakte (Darmstadt)*, (1) 3 (1991).
20. Z. Dienes, M. Nógrádi, B. Vermes and M. Kajtár-Peredy, *Liebigs Ann. Chem.*, 1141 (1989); C. Bartolucci, L. Cellai, S. Cerrini, D. Lamba, A. L. Segre, V. Brizzi and M. Brufani, *Helv. Chim. Acta*, **73**, 185 (1990); W. F. Tinto, W. R. Chan, W. F. Reynolds and S. McLean, *Tetrahedron Lett.*, **31**, 465 (1990); S. Kadota, Y. Takamori, T. Kikuchi, A. Motegi and H. Ekimoto, *Tetrahedron Lett.*, **31**, 393 (1990). Naturally occurring aliphane with a cyclobutane ring: R. Hartmann, A. Sanmartin, O. Muñoz

and E. Breitmaier, *Angew. Chem.*, **102**, 441 (1990); *Angew. Chem. Int. Ed. Engl.*, **29**, 385 (1990); C. Bartolucci, L. Cellai, D. Lamba, A. L. Segre, V. Brizzi and M. Brufani, *Helv. Chim. Acta*, **73**, 185 (1990); J. Clardy and C. M. Ireland, *J. Am. Chem. Soc.*, **112**, 8080 (1990); A. M. Warshawsky and A. I. Meyers, *J. Am. Chem. Soc.*, **112**, 8090 (1990); K. Tatsuta *et al.*, *Tetrahedron*, **46**, 4629 (1990); D. H. Williams *et al.*, *J. Chem. Soc. Perkin Trans. I*, 2317 (1990); J. E. McMurry, R. G. Dushin, *J. Am. Chem. Soc.*, **112**, 6942 (1990); P. Hammam and G. Kretzschmar, *Tetrahedron*, **46**, 5603 (1990); M. S. Butler, T. K. Lim, R. J. Capon and L. S. Hammond, *Aust. J. Chem.*, **44**, 287 (1991); A. B. Smith, III, J. L. Wood, A. E. Gould, S. Omura and K. Komiyama, *Tetrahedron Lett.*, **32**, 1627 (1991); K. C. Nicolaou, G. Skokotas, S. Furaya, H. Suemune and D. C. Nicolaou, *Angew. Chem.*, **102**, 1066 (1990); *Angew. Chem. Int. Ed. Engl.*, **29**, 1064 (1990); N. Lindquist, W. Fenical, D. van Duyne and J. Clardy, *J. Am. Chem. Soc.*, **113**, 2303 (1990); H. D. Zinsmeister, H. Becker and T. Eicher, *Angew. Chem.*, **103**, 134 (1991); *Angew. Chem. Int. Ed. Engl.*, **30**, 130 (1991); A. B. Smith, III, J. L. Wood and S. Omura, *Tetrahedron Lett.*, **32**, 841 (1991); S.-K. Yoo, *Tetrahedron Lett.*, **33**, 2159 (1992).
21. E. Winterfeldt, *Chimia*, **47**, 39 (1993).

12 Molecular Recognition with Phanes as Host Molecules

In the broadest sense one understands the concept of *molecular recognition*[1] to mean the interaction between two molecules. In a narrower context a host molecule, which is characterized by a more concave topology, interacts with a complementary guest molecule which is normally convex. In Figure 1 the guest is symbolized as a hand with outstretched fingers, which selectively binds a *m*-toluidine derivative as guest in its palm.

In several previous sections we have been introduced to cyclophanes which show host properties towards guest molecules or ions. This is the case for the spherands, calixarenes, a number of porphyrins, some cyclotriveratrylenes (CTV) and cryptophanes derived from them. In addition, for some open-chain compounds with cyclophane-like stereochemistry, that is 'face-to-face' oriented benzene rings, and especially in the case of molecular tweezers (Chapter 9), molecular cavities and molecular 'palms' and their facility for complementary binding of guests were discussed.

In the following sections characteristic examples of molecular recognition with phanes as guest compounds will be selected. It should be noted that large rings devoid of aromatic units do not belong to the family of cyclophanes and are not included, even if they show host properties. Because host–guest chemistry was presented in detail in *Supramolecular Chemistry* (Wiley, Chichester, 1991), there is no need for an exhaustive treatment in this book. Even though all

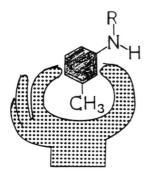

6th International Symposium

on

Molecular Recognition

and Inclusion

September 10. – 15. 1990
Berlin

Figure 1. Recognition and binding of an organic molecule (*m*-toluidine derivative) by a host stylized as a hand. Replication of the symbol of the 'International Symposium on Molecular Recognition and Inclusion', 1990[2]

crown compounds, complexones, and other complexing agents are not included in this section, nevertheless some of the most remarkable examples of molecular host–guest recognition will be described. This will illustrate the significance of the cyclophanes in this connection.[11] The reason is that desirable host compounds must not be either too rigid or too flexible. If they are too rigid, on the one hand, their kinetics of complexation and decomplexation are too slow; but if they are too flexible, perhaps because they lack rigid aromatic spacer units, there either is no cavity or it is occupied intramolecularly by portions of the ring itself. It is recognized that cyclophanes are the ideal source for good host molecules. By varying the flexible aliphatic portions and the rigid aromatic units of the phane, a wide variety of molecular cavities are available, which can be precisely designed as complementary to specific guest molecules and then synthesized.

For the sake of clarity, the molecular recognition with cyclophanes in the following sections is divided into three subsections according to the nature of the guest: complexation of ionic guests, of lipophilic guests, and of functionalized guests. The major interest from the point of view of cyclophane chemistry is the structural aspects, rather than quantification of the nature of the interaction.

12.1 IONS AS GUESTS

Crown ethers, which are broadly involved in complexation of alkaline and alkaline earth metal ions,[1] become phanes only when aromatic units such as

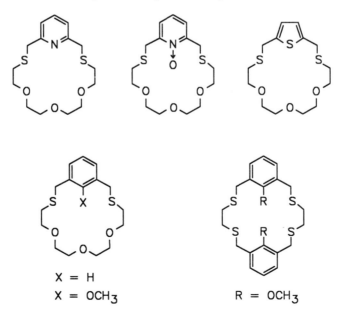

X = H
X = OCH$_3$

R = OCH$_3$

pyridine, thiophene, or substituted benzene rings are incorporated into the structure.[3] The first cryptands of this type and their alkali metal complexes

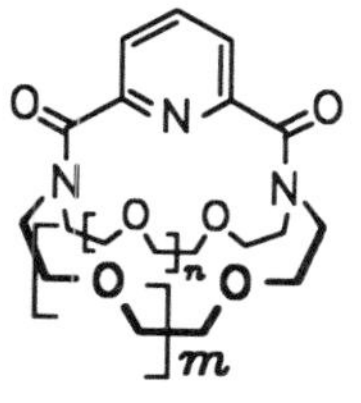

$n = 1, m = 2$

$n = m = 2$

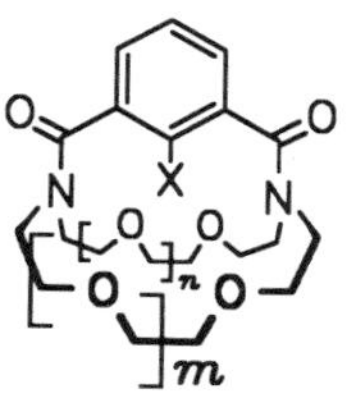

$n = 1, m = 2; X = H$

$n = m = 2 \quad ; X = NO_2$

were prepared by the same authors[4] and from this same research group came the first bipyridino-[5a] and phenanthrolino,[4] p-phenyleno[5b] and anthraquinono-[6] crowns.

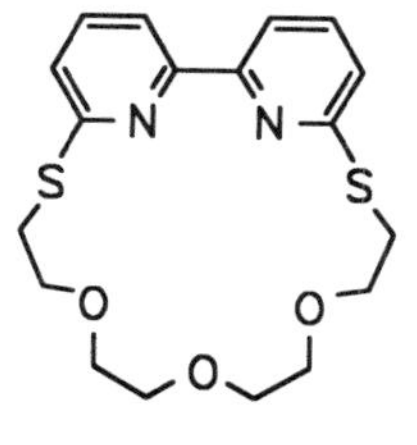

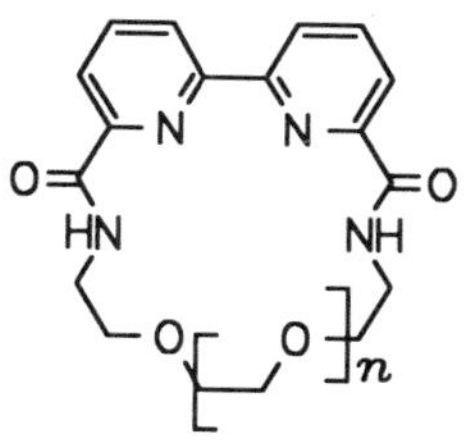

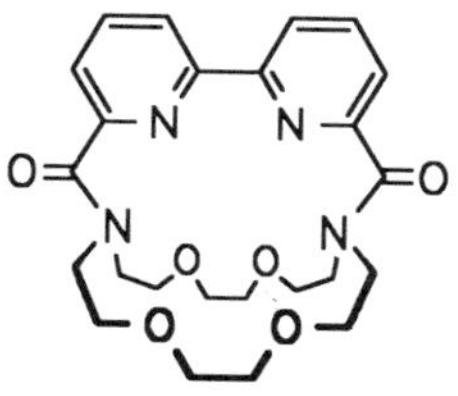

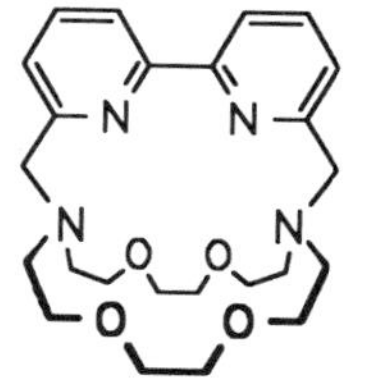

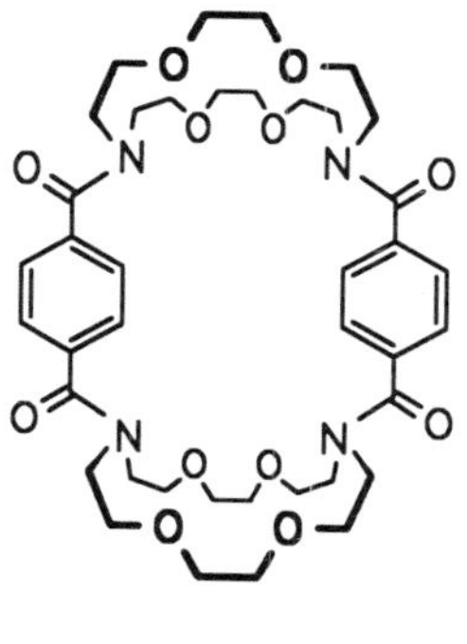

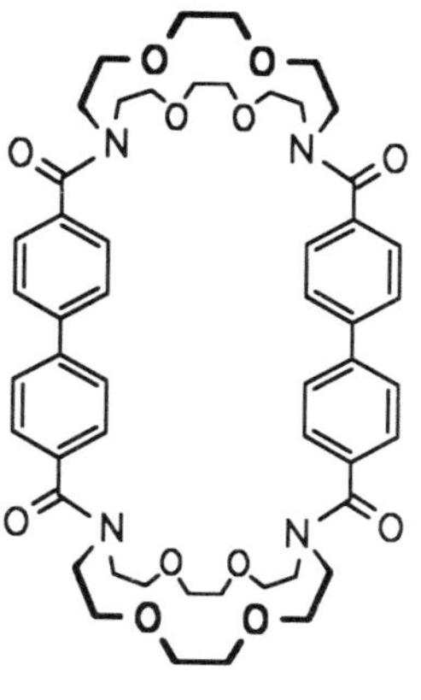

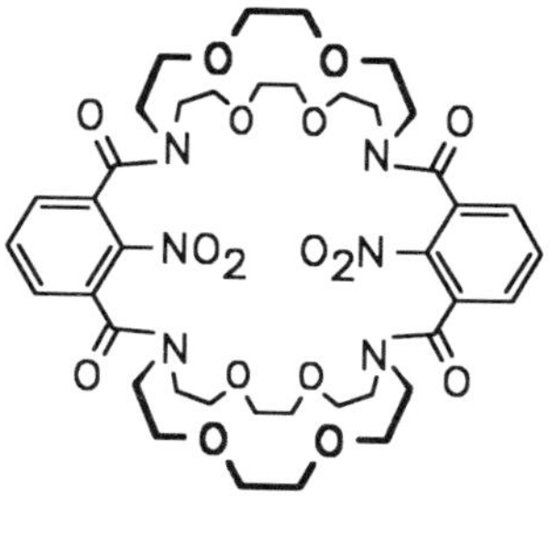

The cryptanal (6a), already mentioned, contains trigonal planar nitrogen atoms.[7] The X-ray crystal structure shows that, in the free ligand, one of the pyridine rings occupies the molecular cavity. Cryptanal forms complexes with $CoCl_2$ and $CuCl_2$, but their structures remain unknown. The orange, dipyrrino cryptand 6b, an acidic chromoionophore, also forms complexes (metal cryptaplexes)[7a] with various cations (Na^+, K^+, Cs^+, Ag^+, Co^{2+}, Ni^{2+}, Cu^+, Zn^{2+}).

6a

Sexipyridine **7** and its Na$^+$ complexes are still difficult to obtain.[8] Luminescent europium complexes **9** were obtained from the tris(bipyridine) ligand **8**.[1a,9]

That such cage-like compounds containing large cavities are not necessarily difficult to prepare is illustrated by the one-pot template synthesis of tris(bpy)

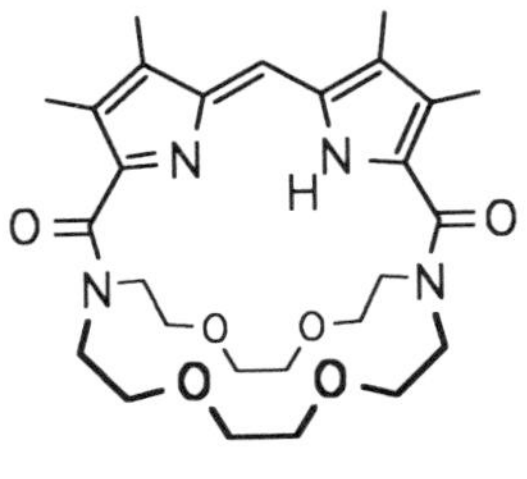

6b

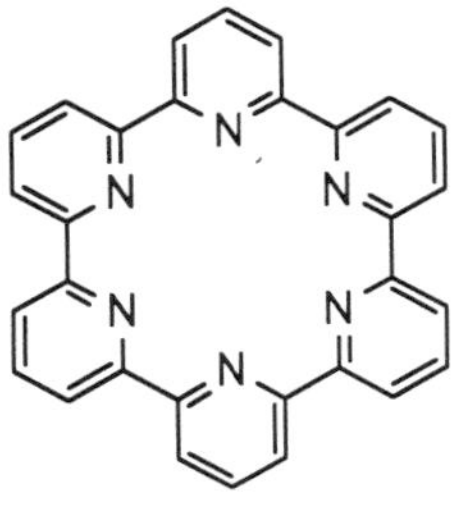

7

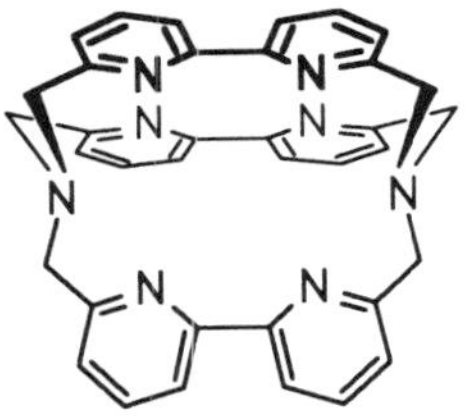

8 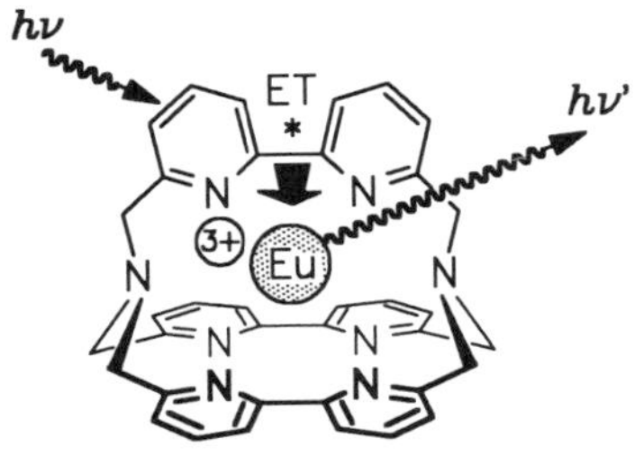   **9**

Na$^+$ complexes **10** and **11**.[1a,10,10a] The sodium is incorporated into the cryptand system in the course of the synthesis.

The Ru(bpy) cage complex **12** is distinctive not only because of its unusual photostability, but also because its photoexcited triplet state has a longer lifetime than the well known tris(bipy) Ru complex **13**. Thus **12** is the only luminescent cage complex of ruthenium known to date[11] and is therefore of interest in photophysical and photochemical investigations.

Similar cage-like bridged tris(catechol) complexes such as **14** were first synthesized by Vögtle et al.[12] The fact that Fe^{3+} ion fits into the cage was later confirmed through X-ray crystallography (Raymond).[13] The complexation constant for Fe(III) complexes of this type was determined to be extremely high, of the order of $> 10^{50}$ (see *Supramolecular Chemistry*, Wiley, Chichester, 1991). It should be emphasized that the bipyridine and catechol ligands in the complexes **12** and **14** are not strictly cryptands but possess large, internal,

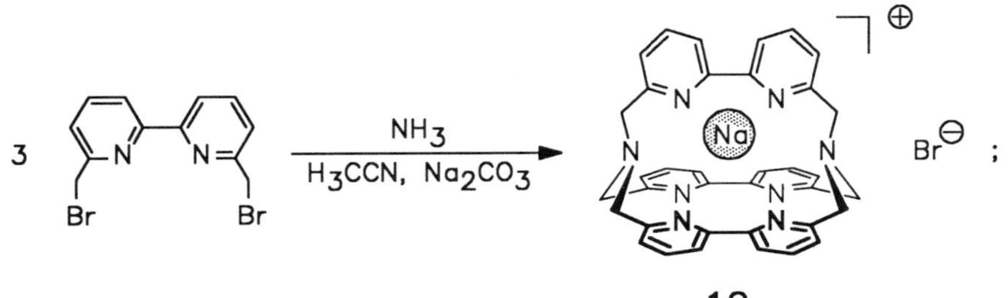

10

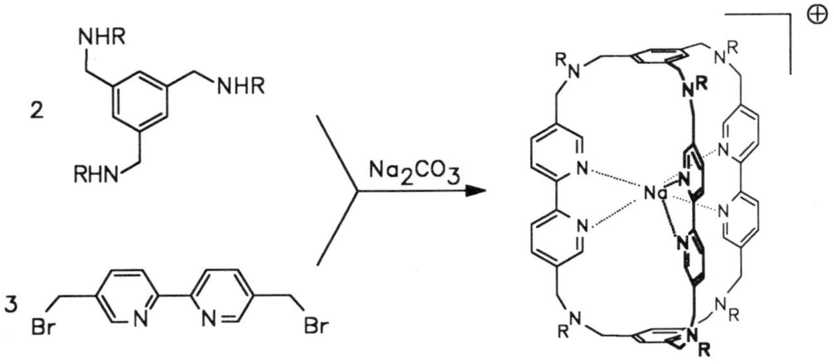

11

R = Benzyl
R = 1 − Naphthylmethyl

functionalized cavities without glycol units or donor centres that are separated by two C-atoms. These are, therefore, complex ligands of a new type, which go beyond crown ether/cryptands.

The cage-like bridged tris crowns 15 and 16[14] form multinuclear complexes with alkali metal cations; that is, there is not just a single cation complexed in the centre of the cavity of all three crown units. Upon complexation with KSCN, the anion is forced into a pocket of the cavity. A ligand particularly suited for large alkali metal cations is the recently synthesized catechol hexaether 17.[15] More nearly resembling a spherand than a cryptand, it assumes an unusual spherical donor geometry, in which the six aromatic methoxy donor oxygen atoms can be oriented in an octahedral arrangement. The cavity in 17 is exactly complementary to the voluminous Cs$^+$ ion (compare complex 18); and from the picrate method it can be shown that 17 is caesium-selective.[1d]

Catenands such as 19 and their complexes (catenates) were prepared by Sauvage[16] by template control from the Cu(I) complexes (for more details see reference 1d).

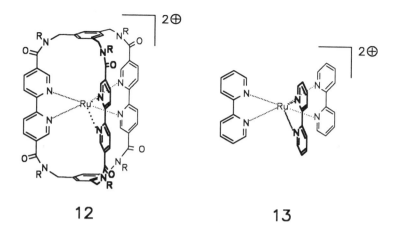

12

13

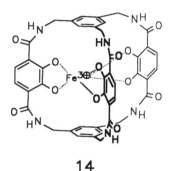

14

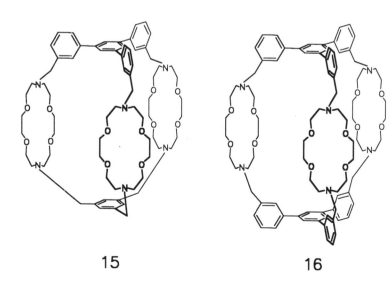

15

16

17   **18** : R = Benzyl

19 (a) $CH_2(CH_2OCH_2)_3CH_2$
    (b) $CH_2(CH_2OCH_2)_4CH_2$

From the point of view of biomimetic studies, multinuclear ligands—that is, those forming intramolecular complexes with several cations—are of special interest (for more details, see *Supramolecular Chemistry*, Wiley, Chichester, 1991). In the multinuclear complex **20**, which binds Ag$^+$ ion in the crown ether portion and Zn^{2+} ion in the porphyrin portion, the porphyrin group is photosensitive, and the 'coreceptor' bound Ag$^+$ ion is the electron acceptor site.[1a] This multinuclear orientation results in quenching of the excited singlet state of the Zn-porphyrin centre by an effective 'intracomplex electron transfer' and leads to a charge separation and to generation of a porphyrin cation with a long half-life.

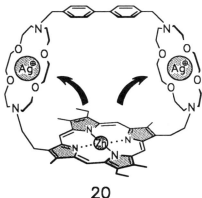

20

For information on spherands and calixarenes, see Section 7.2; on coloured ionophores (chromoionophores, fluorophores, and acerands), consult references 1d,g; on complexation of anions (mainly with non-cyclophane-like host compounds), consult references 1a,d. Further recent developments are found in the original literature.[17]

References to Section 12.1

1. Reviews with numerous additional references:
(a) J.-M. Lehn, *Angew. Chem.*, **100**, 91 (1988) (Nobel Lecture); *Angew. Chem. Int. Ed. Engl.*, **27**, 89 (1988).
(b) D. J. Cram, *Angew. Chem.*, **100**, 1041 (1988) (Nobel Lecture); *Angew. Chem. Int. Ed. Engl.*, **27**, 1009 (1988).
(c) H. M. Colquhoun, J. F. Stoddart and D. J. Williams, *Angew. Chem.*, **98**, 483 (1986); *Angew. Chem. Int. Ed. Engl.*, **25**, 487 (1986).
(d) F. Vögtle, *Supramolecular Chemistry*, Wiley, Chichester, 1993.
(e) E. Weber (Ed.), *Molecular Inclusion and Molecular Recognition—Clathrates II. Top. Curr. Chem.*, **149**, Springer, Berlin, 1988.
(f) F. Vögtle and E. Weber (Eds.), *Host Guest Complex Chemistry—Macrocycles*, Springer, Berlin, 1985.
(g) E. Weber, J. L. Toner, I. Goldberg, F. Vögtle, D. A. Laidler, J. F. Stoddart, R. A. Bartsch and C. L. Liotta, *Crown Ethers and Analogs* (S. Patai and Z. Rappoport, Eds.), Wiley, Chichester, 1989.
(h) B. S. Green, Y. Askani and D. Chipman (Eds.), *Chemical Approaches to Understanding Enzyme Catalysis, Biometric Chemistry and Transition-State Analogs*, Elsevier, Amsterdam, 1982.
(i) F. Diederich, *Cyclophanes. Monographs in Supramolecular Chemistry* (J. F. Stoddart, Ed.), The Royal Society, Cambridge, 1991; Cf. also: F. Diederich *et al.*, *J. Am. Chem. Soc.*, **114**, 1515 (1992).
(k) T. E. Edmonds (Ed.), *Chemical Sensors*, Chapman and Hall, New York, 1988.
(l) Symposium, *Supramolecular Organic Chemistry and Photochemistry*, Saarbrücken, August, 1989, Cf. *Nachr. Chem. Tech. Lab.*, **37**, 1157 (1989).
(m) B. Dietrich, P. Viout and J.-M. Lehn, *Macrocyclic Chemistry*, VCH, Weinheim, 1993.

456

2. W. Saenger and F. Vögtle, cf. circular, *6th International Symposium on Molecular Recognition and Inclusion*, Berlin, September, 1990.

3. F. Vögtle and E. Weber, *Angew. Chem.*, **86**, 126 (1974); *Angew. Chem. Ind. Ed. Engl.*, **13**, 149 (1974); see ref. 1b; see also boron esters as intra-annular groups: M. T. Reetz, C. M. Niemeyer and K. Harms, *Angew. Chem.*, **103**, 1515, 1517 (1991); *Angew. Chem. Int. Ed. Engl.*, **30**, 1472, 1474 (1991).

4. E. Buhleier and F. Vögtle, *Liebigs Ann. Chem.*, 1080 (1977).

5. (a) E. Buhleier, W. Wehner and F. Vögtle, *Chem. Ber.*, **111**, 200 (1978). Mention is also made of the recently described chiral cyclooctabipyridine ligands: X. Ch. Wang, Y. X. Cui, T. C. W. Mak and H. N. C. Wong, *J. Chem. Soc., Chem. Commun.*, 167 (1990).
(b) E. Buhleier, W. Wehner and F. Vögtle, *Chem. Ber.*, **112**, 546 (1979).

6. E. Buhleier and F. Vögtle, *Chem. Ber.*, **111**, 2729 (1978).

7. G. R. Newkome, V. K. Majestic and F. R. Fronczek, *Tetrahedron Lett.*, **22**, 3035 (1981); Schiff bases derived from 'tren': J. Hunter *et al.*, *J. Chem. Soc., Dalton Trans.*, 11 (1992); P. D. Beer *et al.*, *J. Chem. Soc., Chem. Commun.*, 602 (1992).
(a) Ch. Seel, M. Nieger and F. Vögtle, *Chem. Ber.*, **125**, 447 (1992).

8. J. L. Toner, *Tetrahedron Lett.*, **24**, 2707 (1983); G. R. Newkome and H. W. Lee, *J. Am. Chem. Soc.*, **105**, 5956 (1983); see also J. E. B. Ransohoff, H. A. Staab, *Tetrahedron Lett.*, **26**, 6179 (1985); T. W. Bell and A. Firestone, *J. Am. Chem. Soc.*, **108**, 8109 (1986); Torands: T. W. Bell *et al.*, *Angew. Chem.*, **104**, 319 (1992); *Angew. Chem. Int. Ed. Engl.*, **31**, 345 (1992); *Angew. Chem.*, **102**, 931 (1990); *Angew. Chem. Int. Ed. Engl.*, **31**, 348 (1992); For an ion-selective hydrazone-azine-tautomerization of a 14-membered macrocyclic ligand see: T. W. Bell and A. T. Papoulis, *Angew. Chem.*, **104**, 792 (1992); *Angew. Chem. Int. Ed. Engl.*, **31**, 749 (1992).

9. Cf. H. Dürr, K. Zengerle and H.-P. Trierweiler, *Z. Naturforsch.*, **43b**, 361 (1988); 'Sodiocryptatium' radical: J. M. Lehn *et al.*, *Angew. Chem.*, **103**, 884 (1991); *Angew. Chem. Int. Ed. Engl.*, **30**, 838 (1991).

10. Cf. F. Ebmeyer and F. Vögtle, *Chem. Ber.*, **122**, 1725 (1989).
(a) For a similar tren-based tris-silver complex see: J. de Mendoza, F. Vögtle, P.-M. Windscheif, K. Rissanen, J. M. Lehn *et al.*, *Angew. Chem.*, **103**, 1365 (1991); *Angew. Chem. Int. Ed. Engl.*, **30**, 1331 (1991).

11. F. Ebmeyer and F. Vögtle, *Angew. Chem.*, **101**, 95 (1989); *Angew. Chem. Int. Ed. Engl.*, **28**, 79 (1989); L. de Cola, F. Barigelletti, V. Balzani, P. Belser, A. v. Zelewsky, F. Vögtle, F. Ebmeyer and S. Grammenudi, *J. Am. Chem. Soc.*, **110**, 7210 (1988); F. Barigelletti, L. de Cola, V. Balzani, P. Belser, A. v. Zelewsky, F. Vögtle, F. Ebmeyer and S. Grammenudi, *J. Am. Chem. Soc.*, **111**, 4662 (1989); P. Belser, *Chimia*, **44**, 226 (1990).

12 P. Stutte, W. Kiggen and F. Vögtle, *Tetrahedron*, **43**, 2065 (1987); cf. F. Vögtle, Ch. Seel, F. Ebmeyer, A. Wallon, M. Bauer, W. M. Müller and U. Werner, *Kemia-Kemi*, **16**, 796 (1989).

13. (a) Th. J. McMurry, S. J. Rodgers and K. N. Raymond, *J. Am. Chem. Soc.*, **109**, 3451 (1987).
(b) Th. J. McMurry, M. W. Hosseini, Th. M. Garrett, F. E. Hahn, Z. E. Reyes and K. N. Raymond, *J. Am. Chem. Soc.*, **109**, 7196 (1987).

14. F. Vögtle, A. Wallon, W. M. Müller, U. Werner and M. Nieger, *J. Chem. Soc., Chem. Commun.*, 158 (1990); A. Wallon, U. Werner, W. M. Müller, M. Nieger and F. Vögtle, *Chem. Ber.*, **123**, 859 (1990).

15. J. Peter-Katalinic, F. Ebmeyer, Ch. Seel and F. Vögtle, *Chem. Ber.*, **122**, 2391 (1989).

16. A.-M. Albrecht-Gary, C. O. Dietrich-Buchecker, Z. Saad and J.-P. Sauvage, *J. Am. Chem. Soc.*, **110**, 1467 (1988); J.-P. Sauvage *et al.*, *J. Am. Chem. Soc.*, **112**, 8002 (1990).

17. A. Nakano, Q. Xie, J. V. Mallen, L. Echegoyen and G. W. Gokel, *J. Am. Chem. Soc.*, **112**, 1287 (1990).

12.2 LIPOPHILIC GUESTS

Recognition processes at the molecular level are understood to mean the ability of tailor-made host molecules to encompass or attach *selectively* to smaller, sterically complementary guests (molecules, ions) in solutuion—by analogy to biological receptors and enzymes.[1] Just as the masking of cations can lead to a change in their analytical properties, such as solubility, for example, so the properties of guest molecules are altered if they reside in the cavity or in a niche of a host.

In this section particular attention will be paid to functionless, lipophilic guest molecules.[1a-o] Exolipophilic guests such as benzene, naphthalene, and other aromatic and aliphatic compounds (chloroform, halothane, etc.) must be held in a molecular cavity or at least in a molecular niche, a situation similar to the well known inclusion of a guest in the conical cyclodextrin cavity (for more details see reference 1d). In this case the complexation does not depend on a specific binding of the guest in the host cavity; rather the guest molecule fits reasonably well in the host cavity. In order to strengthen the host–guest association, one imitates the cyclodextrins. That is, one uses a hydrophilic solvent (water, alcohol) and an exohydrophilic host, in order to attract the lipophilic guest, such as an oil drop, into the endolipophilic interior of the host. The advantage of this process is that the lipophilic surface of the guest and the inner surface of the host become smaller than they were before complexation. Said another way: the hydrogen-bonding structure of the solvent is not disrupted by *two* lipophilic 'layers', for these are closely associated and the water structure is less distorted.[11]

For the reasons above, non-functionalized cyclophanes are particularly suitable for complexation of lipophilic, neutral molecules.[1] The challenge rests in preparing relatively large host cyclophanes with cavities which are water- or methanol-soluble.

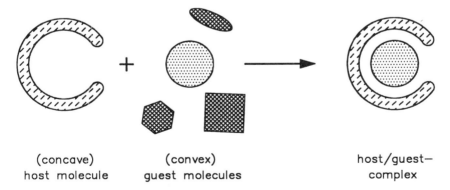

| (concave) | (convex) | host/guest— |
| host molecule | guest molecules | complex |

Figure 1. Selective complexation (molecular recognition) of a molecular guest (in this case circular) by a topologically complementary (concave) organic host molecule (schematic). Gratitude is expressed to Dr K. Wagemann, Dechema, Frankfurt, for suggesting this depiction

The beginning of this area of host–guest chemistry involved the early work on tetraaza macrocycles by Stetter and Roos,[2] which was repeated by Vögtle and Rossa. Their complexes with benzene and other aromatics were shown to be clathrates from the X-ray crystal structure analysis carried out collaboratively with Hilgenfeld and Saenger in 1982.[3] At about this same time investigations

$$n = 2$$
$$n = 3$$
$$n = 4$$

I

were undertaken by several research groups (Tabushi, Murakami, Koga, Whitlock) to synthesize water-soluble, cyclophane-like macrocycles and to study their tendency to serve as host molecules in aqueous solution.[1] One of the first stoichiometric complexes actually bound to a lipophilic guest in solution was the chloroform complex of a macromonocyclic 30-membered hexalactam (**6**).[4] The X-ray crystal structure of the 1:1 complex **7** showed a threefold axis for the host and guest, directed along the C–H bond of the $HCCl_3$. It also revealed how the chloroform is nested in the niche of the cavity (see **8**).

$$R = CH_2Ph$$

6

The breakthrough in this field must be attributed to work of Koga *et al.* In 1980, following the lead of Stetter, they were able for the first time to provide unequivocal evidence for the stoichiometric inclusion of a non-polar guest molecule (durene) in the cavity of a water-soluble cyclophane (**9**), not only in aqueous solution but also in the solid state by X-ray crystal structure analysis.[1k] The complex binding constant for 2,7-naphthalenediol as guest in HCl solution was about 10^3 l/mol. Furthermore, the guest molecule is held in the cavity in a specific, highly favourable orientation, such that the aromatic protons of durene are pointing toward the 'corners' formed by the diphenylmethane units. A crystalline 1:1 complex with durene as guest, obtained from HCl solution, has been clearly characterized as a cavity inclusion complex.[1k]

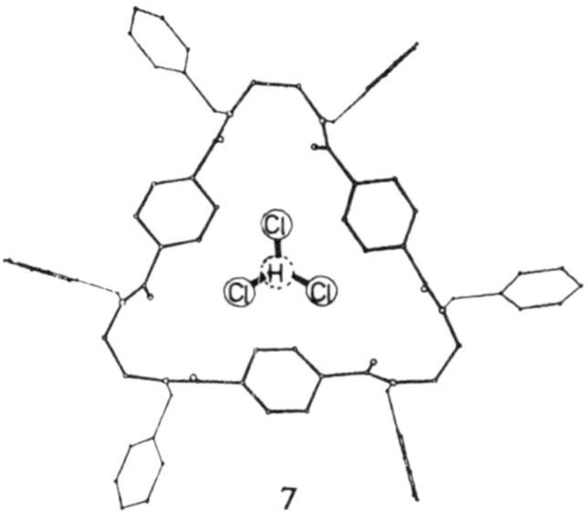

7

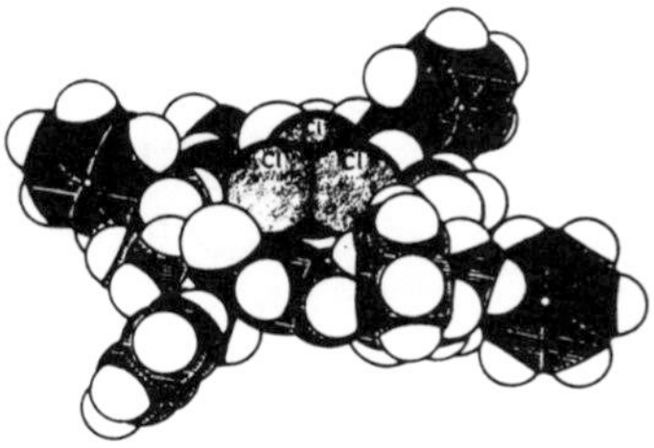

8

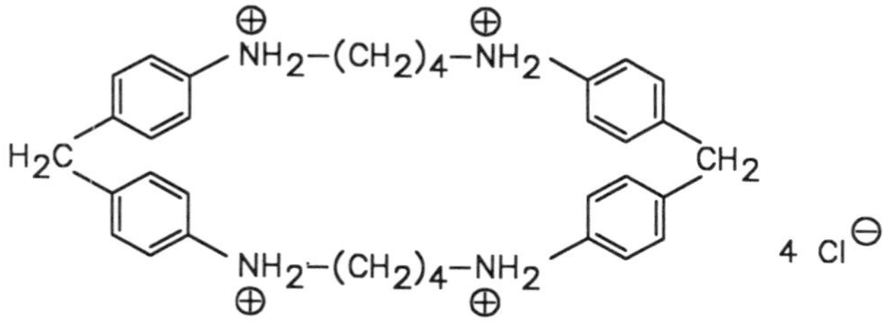

9

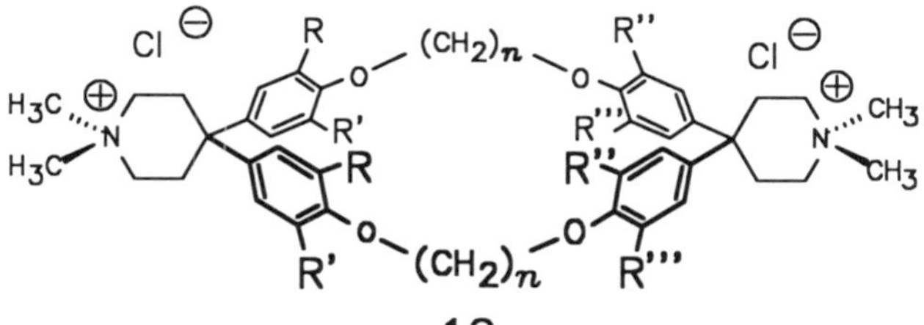

10

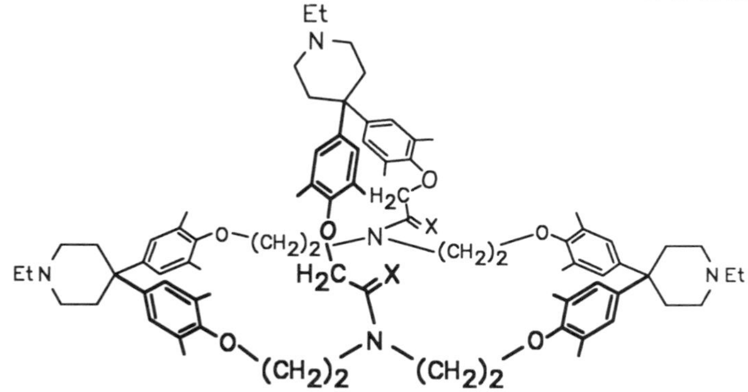

X = O
X = 2H

11

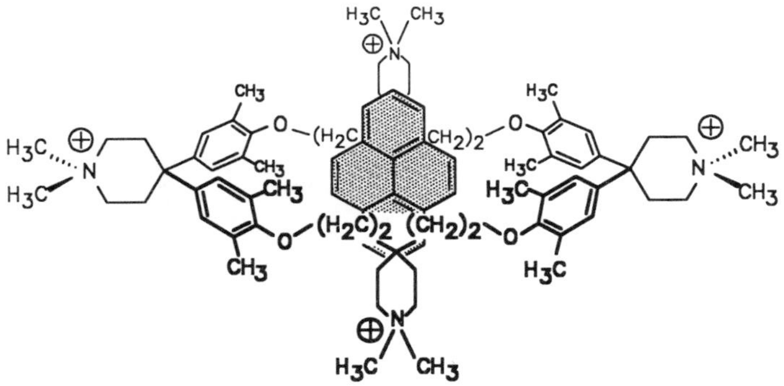

12

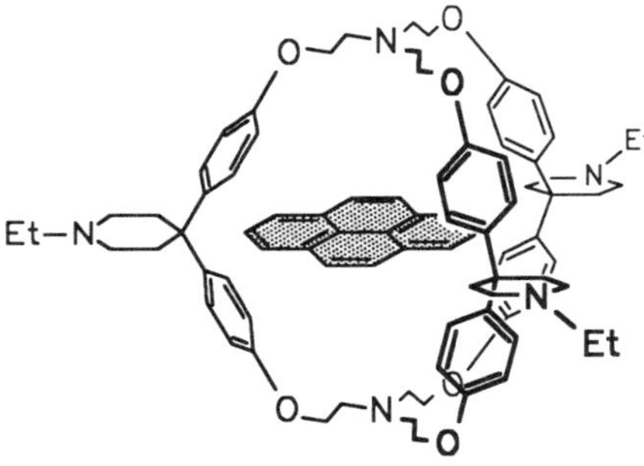

13

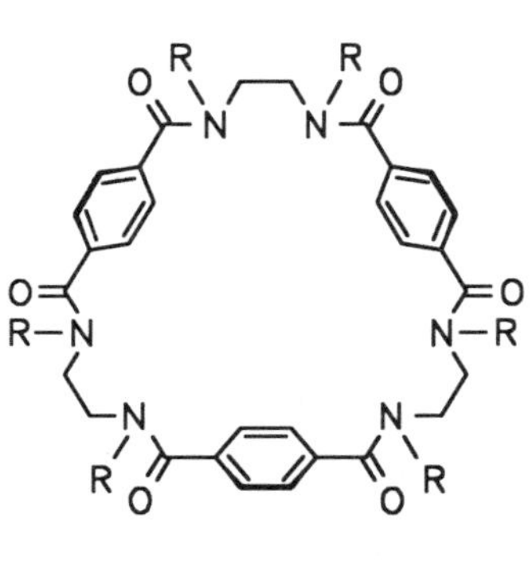

14

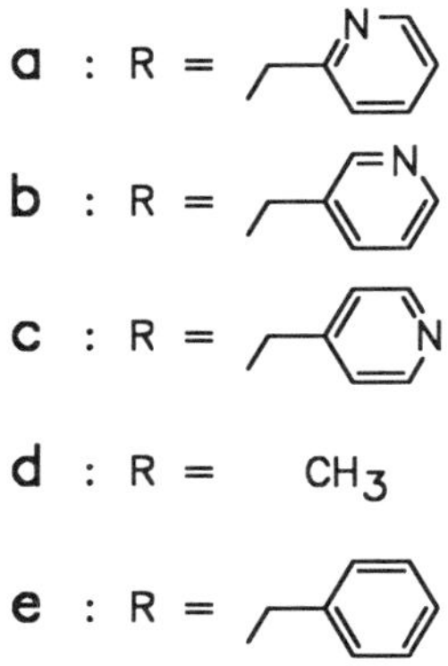

Figure 2. Sketch of two favourable guest orientations (guest: pyrene) in the host cavity, as determined by ^{1}H-NMR spectroscopy in aqueous solution[1i]

By 'removal' of the ammonium functions, which contribute to the water solubility, Diederich could achieve even higher complexation constants; for example, with macromonocycles of type **10** and macrobicycles such as **11**.[li] Two favourable orientations of the guest within the cavity (see **12** and **13**) are assumed for the complexes obtained with pyrene as guest in aqueous solution.[li]

Host compounds of type **14a–c** form complexes with assorted guests such as 2,7-dihydroxynaphthalene in aqueous acid solution, as can be detected by the strong high-field shift of the guest protons.[1d] The reference substances **14d,e** are water-insoluble.

The two isomeric host molecules **15** and **16**, synthesized by Vögtle *et al.*, show differing guest selectivity. While the sterically undistorted cavity of the *out/out* isomer **15** accepts spherical adamantane as guest in aqueous acid solution, the *in/out* isomer **16** is not capable of doing so. On the other hand, planar guest molecules such as naphthalene are complexed by both host isomers.[1d,m] Dougherty observed binding of anthracene and pyrene by the macrocyclic ligand **17**.[5]

15   16

17

The easily soluble macrobicycles **18–21** were obtained in a one-step cyclization reaction. Arenes as large as benzo[*ghi*]perylene, benzo[*a*]pyrene, and triphenylene are selectively complexed by **19** in aqueous acid solution (see complex **22** from **19** and triphenylene). Because protonated **19** in water

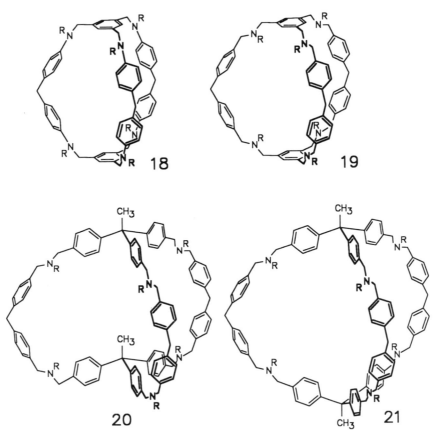

solubilizes phenathrene effectively but anthracene to only a small extent, it is possible by solid–liquid extraction to separate phenanthrene from anthracene. Partially hydrogenated arenes can also be separated from the corresponding arenes by extraction methods: dihydrophenanthrene from phenanthrene and triphenylene from dodecahydrotriphenylene, for example.[6,6a] Some of the highly successful types of guest molecules are listed alongside structure **22**.

The previously described cryptophanes of Collet (see **23**) form inclusion complexes of the type **24** (schematic) with small, aliphatic, lipophilic molecules such as CH_2Cl_2, $CHCl_3$, and CCl_4, and with other appropriately sized guest molecules. This can be demonstrated by the strong high-field ^{1}H-NMR shifts.[7] The chiral host **25**, described by Wilcox,[8] is capable of complexing benzene derivatives in aqueous solution. Its cavity binding sites consist of a diphenyl-methane unit and a variation of Tröger's base.[9]

Hünig has reported a new variation of cationic host molecules of the quat-type **26** with π-acceptor properties.[10] These onium hosts are easily obtained

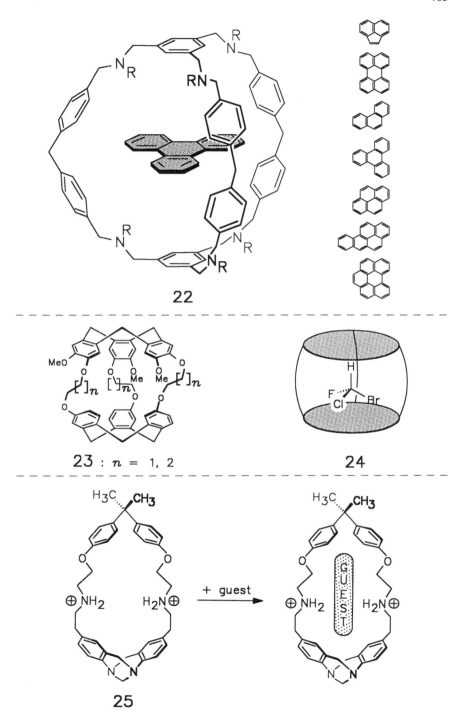

22

23 : *n* = 1, 2

24

25

26 : n = 1

by quaternization. In the course of their synthesis, intentionally added guests—aromatic compounds varying in size from benzene to pyrene—exert a catalytic effect ('molecular template effect'). The dominating binding interactions here are assumed to be (a) dispersion forces, including charge-transfer interactions and (b) electronic 'edge-to-face' interactions of the 'T-type'.[10]

Although the aesthetically pleasing macropolycyclic tertiary amine **27** ('cubaphane') is indeed rigid, upon protonation it becomes water-soluble and capable of internally accepting hydrophobic guests (Murakami[11]). In this way it shows a remarkable capability for distinguishing between isomeric naphthalenedisulphonate guests, which indicates a preferential orientation of host and guest in the complexation. In this host, as with many analogous macropolycyclic ammonium hosts, effective hydrophobic interactions may indirectly be the driving force in host–guest complexation.

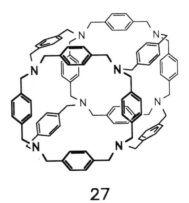

27

According to Cram the nature of guest binding in 'carcerand' (**28**) is of a spectacularly new type, what he terms a new 'phase state' (vacuum + guest in varying proportions).[12] A guest molecule of dimethyl sulphoxide in the host molecule **28** can not be driven out even during a 12-hour heating period in dimethylformamide. Cyclobutadiene was imprisoned inside such a cavity.

In the case of host molecules like **30** (which correspond roughly to the two 'halves' of the carcerands), Cram was able to demonstrate recently that such organic molecules with high 'structural recognition' dimerize ('velcraplexes'), provided they possess a large lipophilic surface containing two preorganized

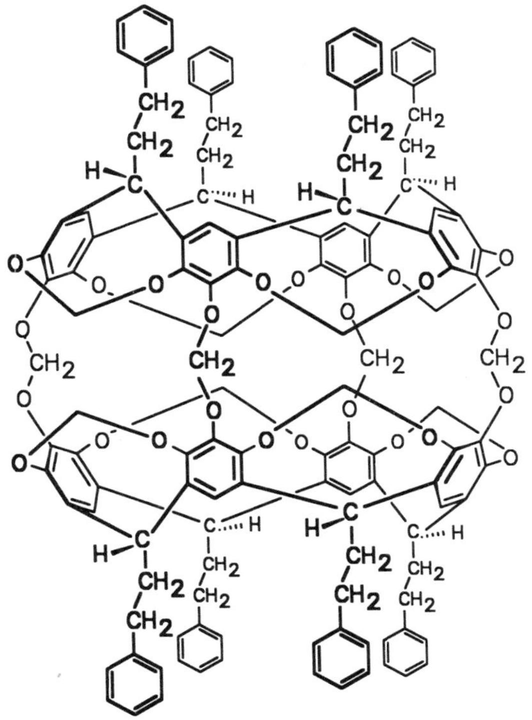

28

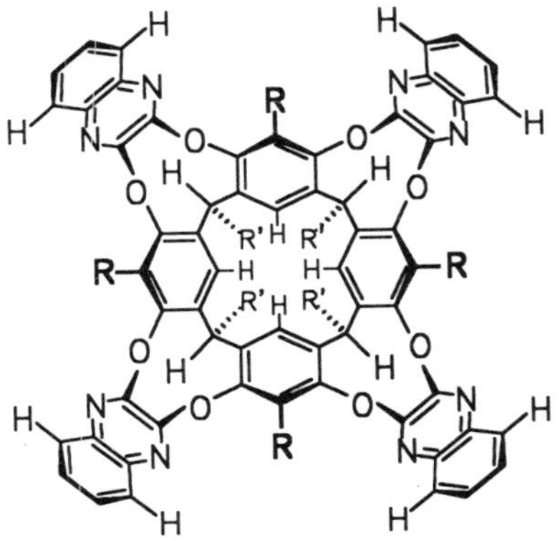

29 : R = H

30 : R = CH$_3$

31 : R = C$_2$H$_5$

R' = C$_5$H$_{11}$

and complementary host and guest regions.[13] Whereas **30** exists in CDCl$_3$ solely as the dimer, **29** and **31** remain as monomers. In the latter molecule the complementarity is lacking. In this unique dimer formation, the first of its kind, the usual binding forces—polar–polar and polar–dipole forces, metal complexation, hydrogen bonding, and hydrophobic interactions—apparently play no role.

466

References to Section 12.2

1. Reviews with numerous additional references:
   (a) J.-M. Lehn, *Angew. Chem.*, **100**, 91 (1988) (Nobel Lecture); *Angew. Chem. Int. Ed. Engl.*, **27**, 89 (1988).
   (b) D. J. Cram, *Angew. Chem.*, **100**, 1041 (1988) (Nobel Lecture); *Angew. Chem. Int. Ed. Engl.*, **27**, 1009 (1988).
   (c) H. M. Colquhoun, J. F. Stoddart and D. J. Williams, *Angew. Chem.*, **98**, 483 (1986); *Angew. Chem. Int. Ed. Engl.*, **25**, 487 (1986).
   (d) F. Vögtle, Supramolecular Chemistry, Wiley, Chichester, 1993.
   (e) *Molecular Inclusion and Molecular Recognition—Clathrates II* (E. Weber, Ed.), *Top. Curr. Chem.*, **149** (1988), Springer, Berlin.
   (f) *Host Guest Complex Chemistry—Macrocycles* (F. Vögtle and E. Weber, Eds.), Springer, Berlin, 1985.
   (g) E. Weber, J. L. Toner, I. Goldberg, F. Vögtle, D. A. Laidler, J. F. Stoddart, R. A. Bartsch and C. L. Liotta, *Crown Ethers and Analogs* (S. Patai and Z. Rappoport, Eds.), Wiley, Chichester, 1989.
   (h) *Chemical Approaches to Understanding Enzyme Catalysis. Biomimetic Chemistry and Transition-State Analogs* (B. S. Green, Y. Askani and D. Chipman, Eds.), Elsevier, Amsterdam, 1982.
   (i) F. Diederich, *Cyclophanes, Monographs in Supramolecular Chemistry* (J. F. Stoddart, Ed.), The Royal Society of Chemistry, Cambridge, 1991; F. Diederich, *Angew. Chem.*, **100**, 372 (1988); *Angew. Chem. Int. Ed. Engl.*, **27**, 362 (1988).
   (k) K. Odashima, K. Koga, in *Cyclophanes, II* (P. M. Keehn and S. R. Rosenfeld, Eds.), p. 629, Academic Press, New York, 1983.
   (l) I. Tabushi and K. Yamamura, in *Cyclophanes, I* (F. Vögtle, Ed.), *Top. Curr. Chem.*, **113**, 145 (1983).
   (m) F. Vögtle and W. M. Müller, *J. Incl. Phenom.*, **2**, 369 (1984); F. Vögtle, W. M. Müller and W. H. Watson, *Top. Curr. Chem.*, **125**, 131 (1984); F. Vögtle, H.-G. Löhr, J. Franke and D. Worsch, *Angew. Chem.*, **97**, 721 (1985); *Angew. Chem. Int. Ed. Engl.*, **24**, 727 (1985).
   (n) F. Vögtle, Ch. Seel, F. Ebmeyer, A. Wallon, M. Bauer, W. M. Müller and U. Werner, *Kemia-Kemi*, **16**, 796 (1989).
   (o) H.R. Christen and F. Vögtle, *Organische Chemie. Von den Grundlagen zur Forschung*, Vol. II, pp. 535–562, Salle-Sauerländer, Frankfurt a.M./Aarau, 1990.
2. H. Stetter and E.-E. Roos, *Chem. Ber.*, **88**, 1390 (1955).
3. R. Hilgenfeld and W. Saenger, *Angew. Chem.*, **94**, 788 (1982); *Angew. Chem. Int. Ed. Engl.*, **21**, 781 (1982) Suppl. 1690; F. Vögtle, W. M. Müller and L. Rossa, unpublished results.
4. F. Behm, W. Simon, W. M. Müller and F. Vögtle, *Helv. Chim. Acta*, **68**, 940 (1985); F. Vögtle, H. Puff, E. Friedrichs and W. M. Müller, *J. Chem. Soc., Chem. Commun.*, 1398 (1982).
5. T. J. Shepodd, M. A. Petti and D. A. Dougherty, *J. Am. Chem. Soc.*, **110**, 1983 (1988).
6. F. Vögtle, W. M. Müller, U. Werner and H.-W. Losensky, *Angew. Chem.*, **99**, 930 (1987); *Angew. Chem. Int. Ed. Engl.*, **26**, 901 (1987); A. Wallon, J. Peter-Katalinic, U. Werner, W. M. Müller and F. Vögtle, *Chem. Ber.*, **123**, 375 (1990).
   (a) Cf. J.-M. Lehn *et al.*, *J. Chem. Soc., Chem. Commun.*, 540 (1993).
7. For a review see reference 1d.
8. C. S. Wilcox and M. D. Cowart, *Tetrahedron Lett.*, **27**, 5563 (1986).
9. Review of Tröger bases: F. Vögtle, *Reizvolle Moleküle der Organischen Chemie*, Teubner, Stuttgart, 1989; *Fascinating Molecules in Organic Chemistry*, Wiley, Chichester, 1992.
10. M. Bühner, W. Geuder, W.-K. Gries, S. Hünig, M. Koch and Th. Poll, *Angew. Chem.*, **100**, 1611 (1988); *Angew. Chem. Int. Ed. Engl.*, **27**, 1553 (1988); contains information on earlier reports.

11. Y. Murakami, J. Kikuchi, T. Ohno and T. Hirayama, *Chem. Lett.*, 881 (1989); Y. Murakami, J. Kikuchi, T. Ohno, T. Hirayama and H. Nishimura, *Chem. Lett.*, 1199 (1989); Y. Murakami *et al.*, *J. Chem. Soc., Chem. Commun.*, 950 (1991).
12. J. C. Sherman and D. J. Cram, *J. Am. Chem. Soc.*, **111**, 4527 (1989); M. L. C. Quan and D. J. Cram, *J. Am. Chem. Soc.*, **113**, 2754 (1991); D. J. Cram *et al.*, *J. Am. Chem. Soc.*, **113**, 2194, 2790 (1991); *J. Org. Chem.*, **57**, 40 (1992); Cyclobutadiene inside a carcerand: *Angew. Chem.*, **103**, 1084 (1991); *Angew. Chem. Int. Ed. Engl.*, **30**, 1024 (1991).
13. D. J. Cram *et al.*, *J. Am. Chem. Soc.*, **112**, 1254, 1255, 1659 (1990).

12.3 FUNCTIONALIZED MOLECULES AS GUESTS

In the discussion of 'molecular tweezers'[1] in Chapter 9, it was pointed out that guest molecules are also bound even if they are not fully enclosed but when complementary functional groups in the host can provide directed binding such as hydrogen bonding. In order to bind selectively guest molecules containing convergent functional groups (FG) in the niche of a host, it is imperative to have a host cavity or host niche with complementary functional groups. If the guest bears phenolic OH groups, for example, then the host niche must contain basic functions such as NH_2 groups or pyridine N-atoms at the binding sites so that the guest can become attached. In Figure 1 such a binding of complementary functionalized host and guest molecules is represented schematically.[2-6]

Guest molecules may contain ammonium groups, in which case the ammonium compounds form complexes with crown ethers which are stabilized by hydrogen bonding and dipole–dipole interactions.[2-9] Collet has described a 'speleand', containing both a crown ether and a cyclotriveratrylene unit, which binds with methylammonium compounds to give a 'speleate' (**6**).[2] Speleands are defined by Lehn as macropolycyclic structures containing both apolar structural components and polar bonding units. Larger substrates do not fit into the narrow cavity of the speleands leading to complex **6**; the result is that

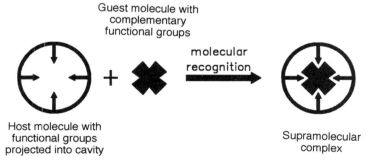

Figure 1. Binding or attachment of a guest molecule in the cavity or niche of a host. The functional groups in the host and guest must be so constituted that they mutually interact

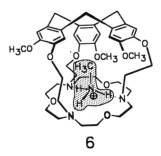

6

the speleand is selective for methylammonium ions.[7b] If one assembles two crown ether units separated by two spacer groups, the result is cylindrical cavity structures or cylindrical 'ditopic coreceptors'.[3,7] With terminally functional diammonium ions they form $^+H_3N-[CH_2]_n-NH_3{}^+$-cryptates of type **7**. From X-ray crystal structures of these supramolecules it can be shown that the substrate

7

$$\boxed{R} = \text{—}\bigcirc\text{—} \; ; \; \bigcirc\bigcirc \; ; \; \text{—}\bigcirc\text{—}\bigcirc\text{—} \; ; \; \text{—}\bigcirc\text{—}\bigcirc\text{—}\bigcirc\text{—}$$

resides in the centre of the cavity of the receptor molecule, such that it is anchored to the macrocyclic binding sites through the two NH_3^+ groups. As the length of the bridge R in **7** is altered, the complexation selectivity shifts in favour of the substrate with the corresponding length. NMR relaxation measurements have shown that the optimal host–guest partners in a receptor–substrate pair undergo similar molecular motion. Thus, complementarity in supramolecular species implies that not only the steric properties but also the dynamic behaviour must be matched.[7b]

In supramolecular complexes such as **8** the possibility exists for multi-recognition by metalloreceptors.[3] Metalloreceptors are 'heterotopic coreceptors', which can bind metal ions and also organic molecules with substrate-specific structural units. For the binding of metal ions porphyrin and α,α'-bipyridyine (bpy) groups are introduced into the macrocyclic coreceptor, which also contains binding positions for anchoring NH_3^+ groups. Through simultaneous binding of metal and diammonium ions, such receptors

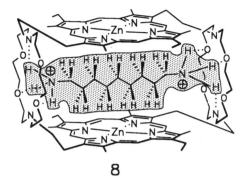

8

form supramolecules such as **8**, with which it is possible to study the interactions and reactions between simultaneously bound organic and inorganic species. They also have application as model bioinorganic systems.

Host cavity and guest are required not only to be matched according to size and shape ('lock and key' principle), but there must be complementary functional groups between host and guest in order to ensure strong host–guest binding.[1] To date there are relatively few examples of complexation of organic guest molecules in which multiple hydrogen bonding is utilized in non-aqueous media such that the guest does not compete with the host for hydrogen bonding sites. It it well known that crown ethers complex with phenols, amines, polar methyl compounds such as nitromethane and acetonitrile, urea and other neutral molecules through loose hydrogen bonding with the upper and lower sides of the crown ether.[7b,11] With these as models, Stoddart,[5] among others,

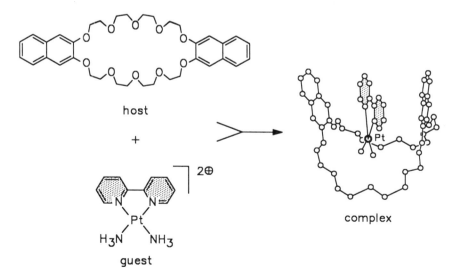

Figure 2. Structure of the 1:1 adduct of $[Pt(bpy)(NH_3)_2]^{2+}$ and dinaphthaleno[30]-crown-10 as determined by X-ray crystallography[5]

was able to construct macrobicyclic and macropolycyclic oligoethers as receptors for transition metal ammine complexes ('coordination in the second sphere'). In this way it was possible to bind the complex $[Pt(bpy)(NH_3)_2]^{2+}$ with dinaphthaleno[30]crown-10, whereupon a charge-transfer interaction between the bipyridine (as acceptor) and the naphthalene rings (as π-donors) created a host niche, in which the guest complex could almost perfectly fit.[5]

An even stronger binding between both ammine ligands of a *cis*-diammine transition metal complex through hydrogen bonding can be achieved with macrobicyclic polyethers containing aromatic rings: that is, multiply bridged cyclophanes of type **9**.[5] Then each crown ether ring can bind one of the ammine ligands.

9

Reinhoudt was able to tailor a benzo crown compound of type **10** which not only binds Ba^{2+} but also binds simultaneously a UO_2^{2+} cation and a molecule of urea (Figure 3).[12] The U–O(urea) distance is 237 pm; U–N(urea) 255–259 pm; U–O(uranyl) 178 pm; and N(urea)–O(crown) 294–314 pm. Binding of urea and

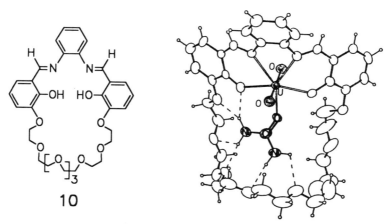

10

Figure 3. Crown host **10** (left) and a drawing of the complex of **10** with UO_2^{2+} and urea[12]

other physiologically important compounds has significant ramifications for medicine (blood exchange, artificial kidneys).

Aoyama has investigated the important molecular recognition of glycerol, D-glucose, D-ribose, riboflavin, vitamin B_{12}, and haemin with the aid of resorcinol-aldehyde cyclooligomers of type **11** and **12**.[13] The substances, all insoluble in CCl_4 and benzene with the exception of D-glucose, can be selectively dissolved with the help of the host **11a** through hydrogen bonding. By contrast, no complexation is observed with the lipophilic host **12a**.

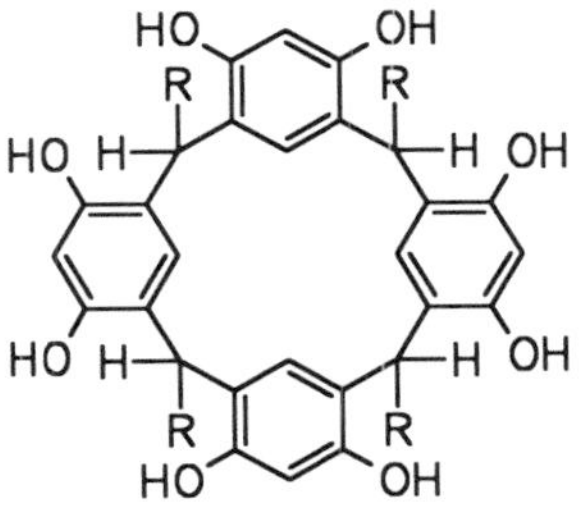

a : R = $(CH_2)_{10}CH_3$

b : R = CH_3

c : R = C_6H_5

11

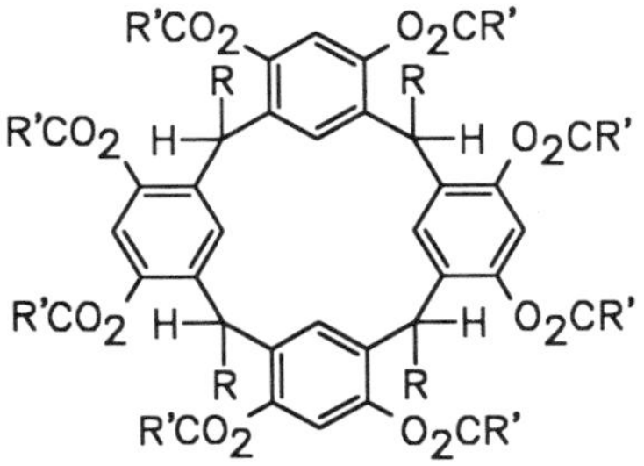

a : R = $(CH_2)_{10}CH_3$, R' = CH_3

b : R = CH_3, R' = CH_3, C_2H_5

c : R = C_6H_5, R' C_3H_7

12

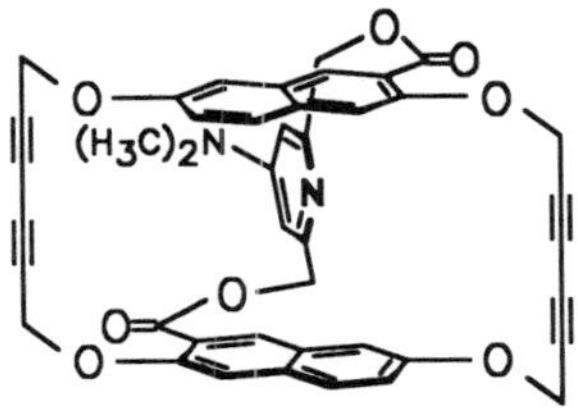

13

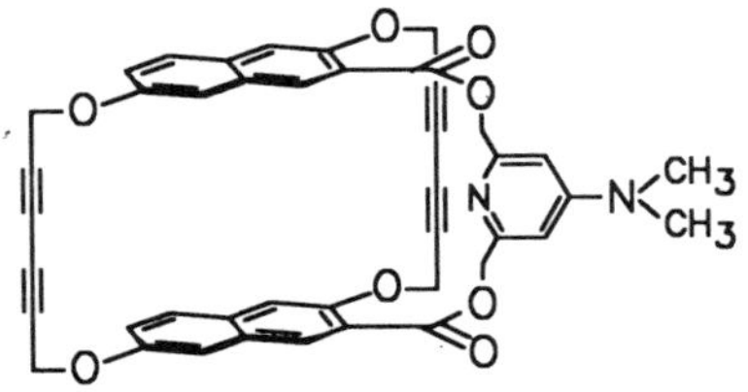

p-nitro-phenol

14

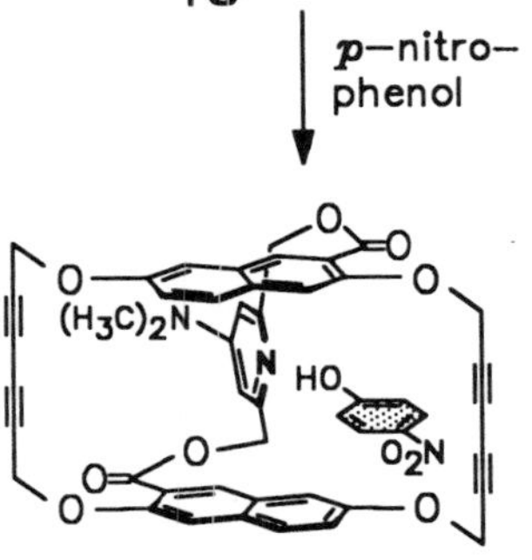

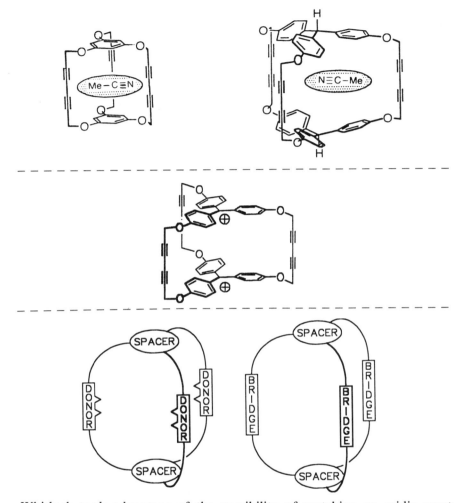

Whitlock took advantage of the possibility of matching an acidic guest molecule with a proton acceptor functional group in the host molecule.[14] As shown in host molecules **13** and **14**, a guest such as *p*-nitrophenol can bind partially to the pyridine nitrogen atom through hydrogen bonding. Because of the rigid conformations of **13** and **14**, the guest molecule must insert itself into their cavities in order to form a hydrogen bond and become associated with the pyridine nitrogen. This insertion into the cavity is facilitated by the accompanying π-interaction between the *p*-nitrophenol guest and the oxa-substituted naphthalene rings, which serve as π-donors. The result is that the isomer **13** binds *p*-nitrophenol with an association constant $K_{assoc} = 3000\ \text{l/mol}$, while phenol, remarkably enough, is scarcely bound ($K_{assoc} = 20\ \text{l/mol}$). It is of interest that the isomer **14** has an even higher association constant with *p*-nitrophenol of $13\,700\ \text{l/mol}$. Isomer **14** is also vastly superior for binding of

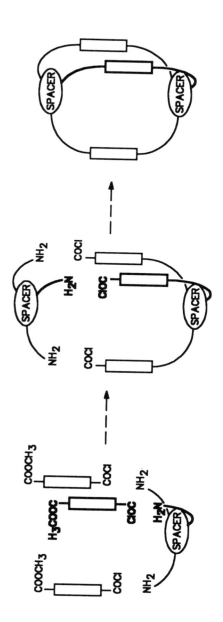

benzoic acid ($K_{assoc} = 5700 \, 1/mol$). This example illustrates clearly that the guest must display a very exact, well designed, complementary acidity corresponding to the basicity of the host molecule [p-dimethylaminopyridine (DMAP) unit]. Indeed p-nitrophenol is strongly bound because its acidity affords favourable hydrogen bonds. More strongly acidic guests form no hydrogen bonds but rather transfer their proton fully to the pyridine, with generation of an ionic structure.[14a]

Compounds bridged analogously but bearing benzene and triphenylmethane units were found by X-ray analysis to enclose acetonitrile.[14b] The orange triphenylmethyl dication, designated as a concave dye, has recently been synthesized (page 472).[14c]

Vögtle *et al.* have developed an approach for the synthesis of functionalized cavities of widely varying size within the main cavity.[2,15] The resulting receptor molecules possess the following topology. One anchor group of variable size, serving as 'spacer', holds three donor-bearing (or donor-free) bridges together in a cage. The synthesis proceeds according to the scheme on page 473. In this way two series of intra-annularly functionalized cavity compounds were obtained, the one (15–18) with acidic OH groups, the other (19–22) with basic bipyridine units. The tris(bipyridine) hosts showed an especially high guest

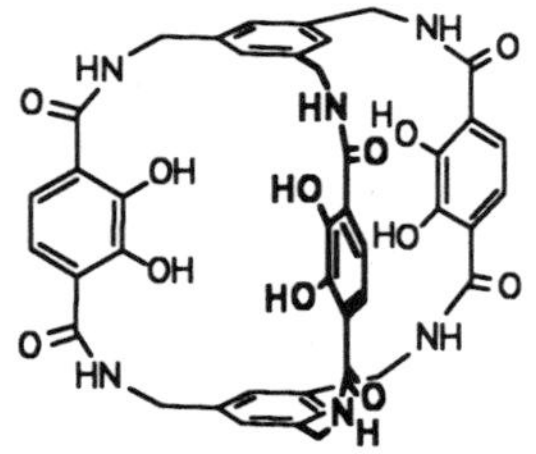

15

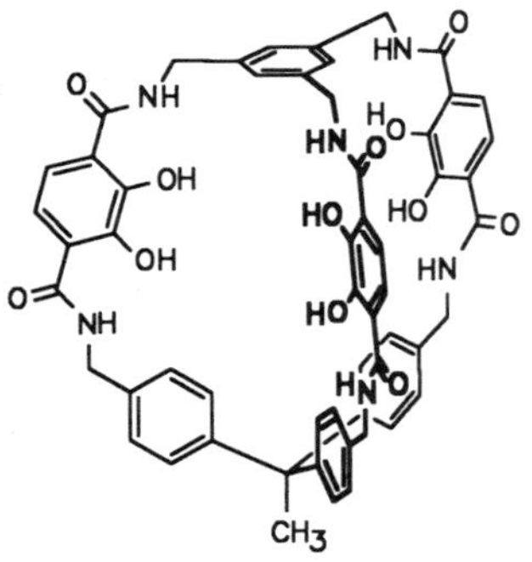

16

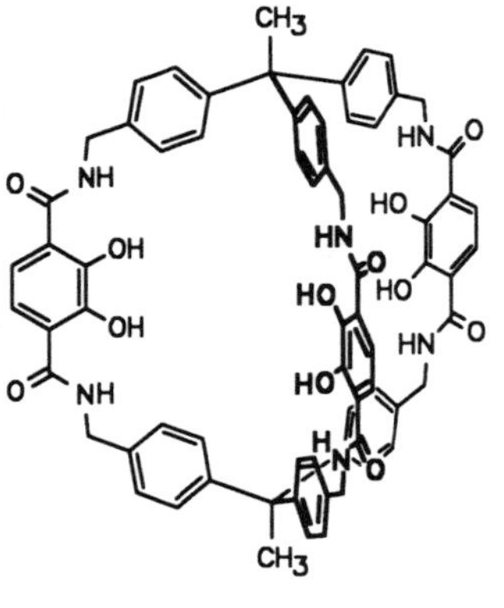

17

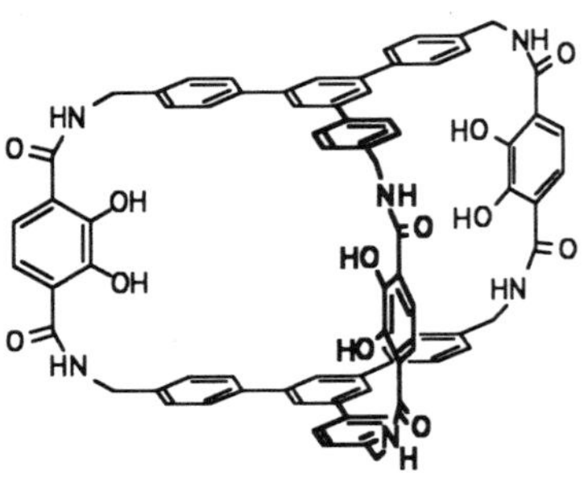

18

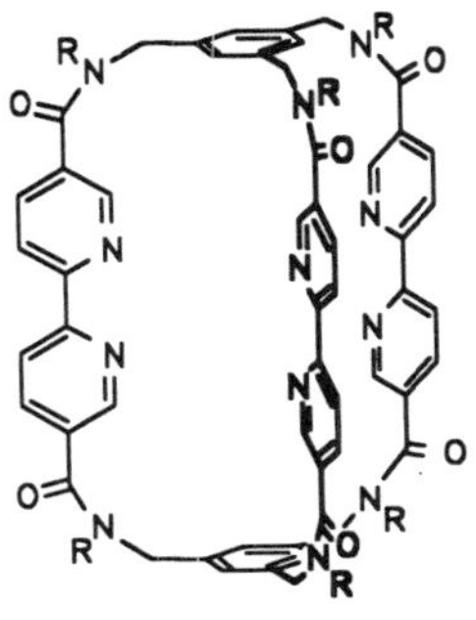

19

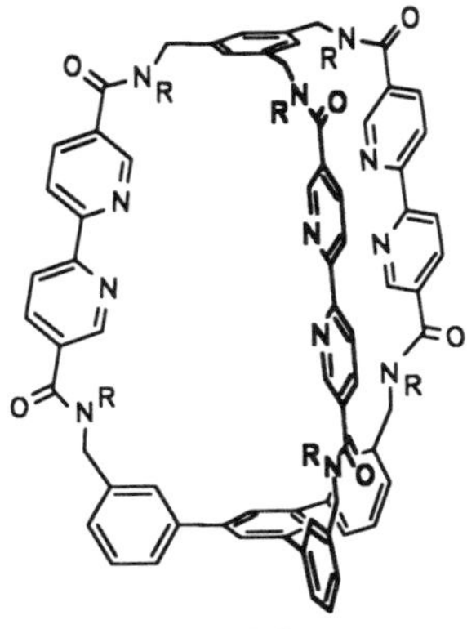

20

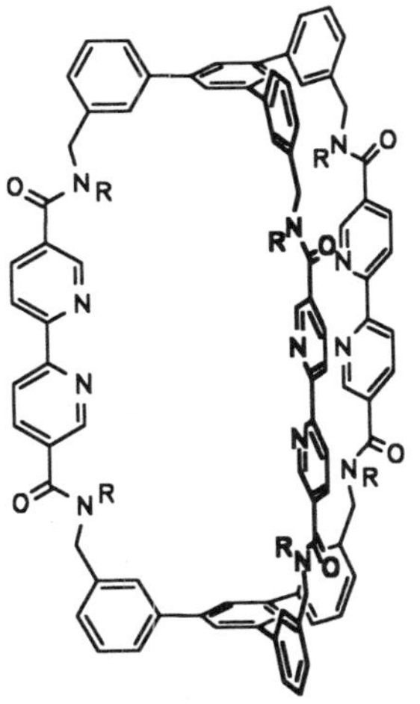

21

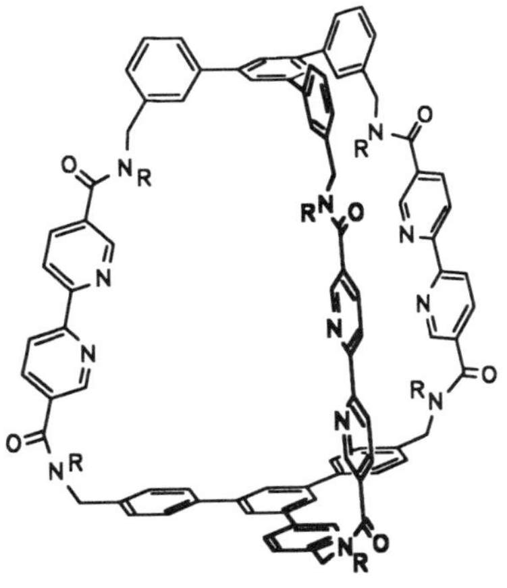

22

selectivity; the host compound **21** in particular, with a medium-sized cavity, is well suited to complex phloroglucinol (1,3,5-trihydroxybenzene), the result being the complex with structure **23**.[15] Complexation is readily observed: a sample of crystalline phloroglucinol is suspended in chloroform, in which it is not fully soluble. When the host compound **21** is added, the phloroglucinol goes into solution. From determination of complexation constants, it was established that complex stability is of the order of those for the Whitlock complexes mentioned earlier. Complexation is selective. Only the host compounds **20** and **21** from among the group **19–22** are capable of complexing strongly with phloroglucinol. This implies that if the cavity is too small as in **19** or too large as in **22**, the host–guest complementarity is not sufficient for complexation. Only phloroglucinol and one of its three isomers (1,2,4-trihydroxybenzene) are complexed by **20** and **21**; no complexation is observed with 1,2,3-trihydroxybenzene. The trihydroxybenzoic acids, although stronger

23

acids than the trihydroxybenzenes, are not complexed to any significant extent, which is also the case for 1,3,5-benzenetricarboxylic acid. The *endo*-acidic catechol host **16** has been shown to complex selectively and to transport nucleobases as guests.[15d]

The rigid *endo*-preorganization of the nitrogen donor atoms of three phenanthroline host units allows the complexation of three 2,9-disubstituted 1,10-phenanthroline guests *via* three Cu$^+$ ions:[15c]

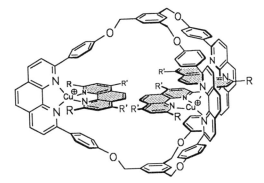

The ligands **24** and **25**, which are bridged with two spacer groups by the incorporation of three crown ether units, show interesting binding properties.[16] Contrary to expectation, none of the tridentate, trifunctional guest compounds such as phloroglucinol can be bound by these hosts. Rather, a strong host–guest interaction occurs with, for example, 2,7-dihydroxynaphthalene, whose two OH groups are apparently able to bind with two of the crown ether units in the cavity.

The energy requirement for a conformation in which three phenolic OH groups and three crown ethers undergo threefold binding is apparently too high. Complexation of dihydroxynaphthalene can be detected unambiguously by the high-field shifts in the host and guest protons.

24   **25**

It should be noted here that C. Still has reported an interesting enantioselective binding of simple guests containing amide structures by the C_2-symmetric host molecules **26a,b**.[17] **26b**, for example, complexes carboxylic acids in C_6D_6, a

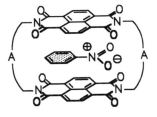

a: R = H
b: R = CH_2Ph

26

conclusion based upon the fact that the NMR spectra of both host and guest are changed significantly. For instance, the NH protons of the *N*-methylacetamides

27: A = $(CH_2)_8$

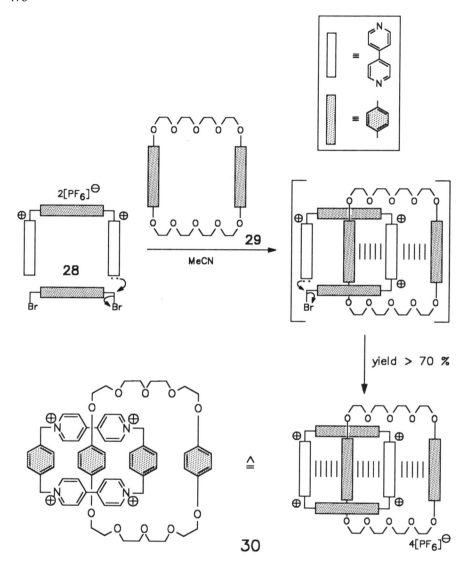

in both host and guest are shifted downfield by more than 1 p.p.m., and the acetyl CH_3 groups in the guest move 0.5 p.p.m. upfield. The enantiomers of chiral amides give rise to varying shifts upon complexation with the host **26b**.

In this final section on molecular recognition with phanes, apart from the binding of *p*-benzoquinone[17a] reference is made to a particular form of complexation which we encountered in Chapter 10 with crown ether–catenanes described by Stoddart: that is, binding through donor–acceptor interactions. Designed 'receptors' of the intercaland type, which contain intercalating units in a macrocyclic system, are of interest not only for binding small

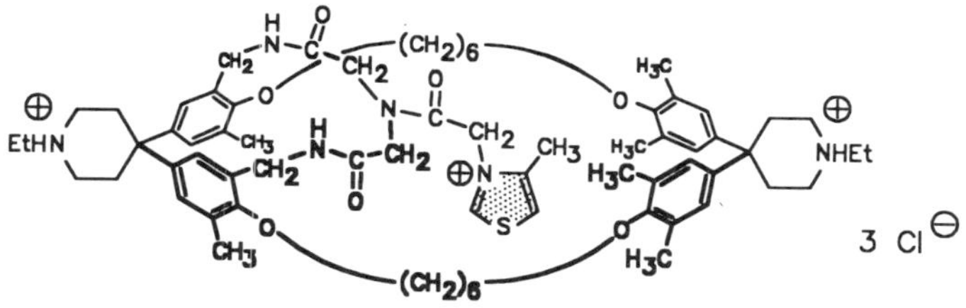

31

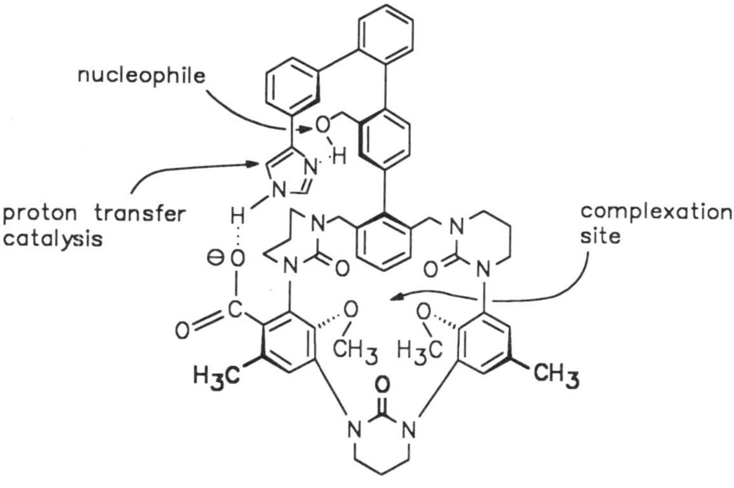

nucleophile

proton transfer
catalysis

complexation
site

32

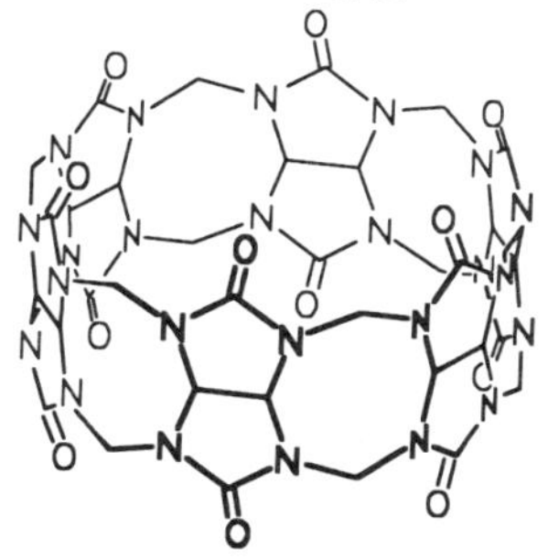

33

molecules but because they show selective interactions with nucleic acids. In the supramolecule **27** a nitrobenzene molecule has been inserted as guest between two planar structural (intercalands) units of the host. Such receptor molecules are especially well suited for recognition of planar substrates (page 477).[3]

480

In the case of the successful synthesis by Stoddart of donor–acceptor stacked molecules from viologen or hydroquinone units, which have interesting electrochemical properties, the alkylation cyclization of the bipyridinium precursor **28** leads to a [2]catenane **30** in the presence of the hydroquinone crown ethers **29**.[18] The high yield of >70% can only be explained through intervention of host–guest association through the ring component (shaded portion). In a catenane of type **30** one of the rings can 'slide' along until it encounters the 'next' π-complex formation with a second benzene ring (page 478).

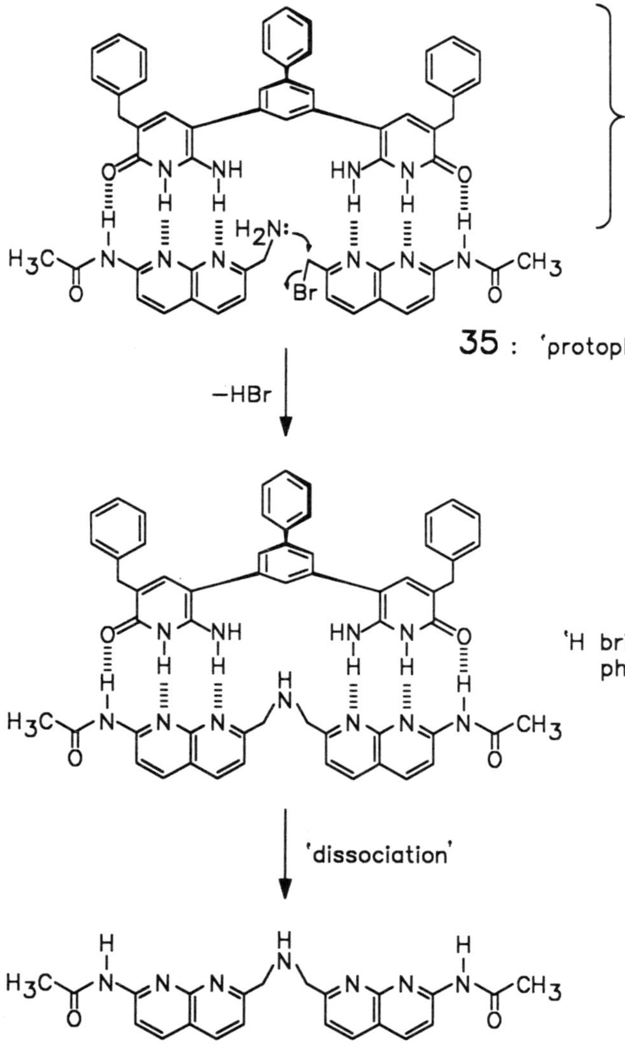

34

35 : 'protophane'

−HBr

'H bridged phane'

'dissociation'

36

The catalysis of the benzoin condensation and of ester hydrolysis ('transacylase mimic') by 'catalytic cyclophanes' such as **31**[19] and **32**[20] should also be mentioned (page 479).

Even hosts to which their own catalytic functional groups have been synthetically attached can exert an effect on the rate of certain reactions. Mock observed a 100 000-fold increase in rate for a cycloaddition in the presence of the synthetic torus host cucurbituril (**33**),[21] which is one of the 'aliphanes' (Section 9.2) (page 479).

As a final note, mention should be made of two further catalysts with promise for the future, in which only loosely held 'phanes' are formed as intermediates through hydrogen bonding. Kelly was able to align two reaction partners onto an open-chain 'matrix host' (**34**);[22] The result was a molecular complex **35** bound by hydrogen bonds, which facilitated the bimolecular reaction of the substrates (see arrows), with formation of the product **36**.

Rebek was successful in constructing a self-replicating system[23] by taking advantage of DNA-like pairing of molecules[22a] (see **37**) with a 'hydrogen-bonding cyclized' 'phane' **38** ('H-bonding phane').

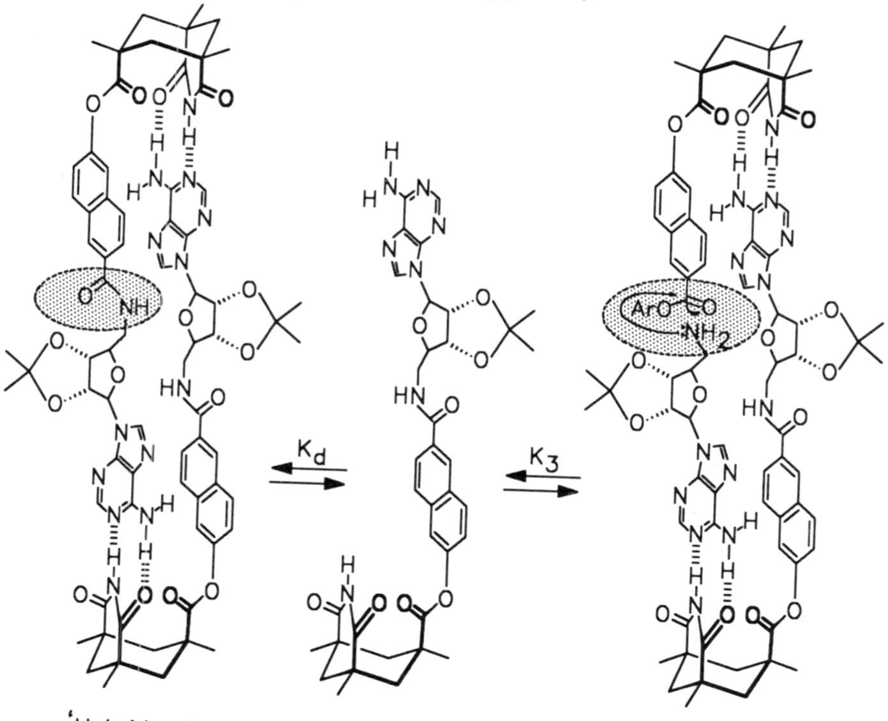

'H bridged phane'

38

'protophane'

37

References to Section 12.3

1. (a) E. Weber and F. Vögtle, *Chemie in uns. Zeit*, **23**, 210 (1989).
   (b) J. Rebek Jr, *Top. Curr. Chem.*, **149**, 189 (1988); *J. Incl. Phenom.*, **7**, 7 (1989); J. Rebek Jr, *et al.*, *J. Am. Chem. Soc.*, **111**, 1082 (1989).
2. (a) F. Vögtle, *Supramolecular Chemistry*, Wiley, Chichester, 1993.
   (b) H.-J. Schneider and I. Theis, *Angew. Chem.*, **101**, 757 (1989); *Angew. Chem. Int. Ed. Engl.*, **28**, 753 (1989); H.-J. Schneider and Th. Blatter, *Angew. Chem.*, **100**, 1211 (1988); *Angew. Chem. Int. Ed. Engl.*, **27**, 1163 (1988); includes information on earlier work. H.-J. Schneider and R. Busch, *Angew. Chem.*, **96**, 910 (1984); *Angew. Chem. Int. Ed. Engl.*, **23**, 911 (1984); H.-J. Schneider and A. Junker, *Chem. Ber.*, **119**, 2815 (1986); H.-J. Schneider, D. Güttes and U. Schneider, *J. Am. Chem. Soc.*, **110**, 6449 (1988).
3. J.-M. Lehn, *Angew. Chem.*, **100**, 91 (1988) (Nobel Lecture); *Angew. Chem. Int. Ed. Engl.*, **27**, 89 (1988).
4. D. J. Cram, *Angew. Chem.*, **100**, 1041 (1988) (Nobel Lecture); *Angew. Chem. Int. Ed. Engl.*, **27**, 1009 (1988).
5. H. M. Colquhoun, J. F. Stoddart and D. J. Williams, *Angew. Chem.*, **98**, 483 (1986); *Angew. Chem. Int. Ed. Engl.*, **25**, 487 (1986); see also J. F. Stoddart *et al.*, *Angew. Chem.*, **101**, 1402, 1404 (1989); *Angew. Chem. Int. Ed. Engl.*, **28**, 1261, 1394 (1989).
6. H. R. Christen and F. Vögtle, *Organische Chemie. Von den Grundlagen zur Forschung*, Vol. II, pp. 535–562, Salle-Sauerländer, Frankfurt a.M./Aarau, 1990.
7. (a) I. O. Sutherland, in *Cyclophanes, Vol. II* (P. M. Keehn and S. M. Rosenfeld, Eds.), p. 679, Academic Press, New York, 1983.
   (b) D. Gehin, P. A. Kollman and G. Wipff, *J. Am. Chem. Soc.*, **111**, 3011 (1989); cf. P. D. J. Grootenhuis and P. A. Kollman, *J. Am. Chem. Soc.*, **111**, 4046 (1989).
8. *Host Guest Complex Chemistry—Macrocycles* (F. Vögtle and E. Weber, Eds.), Springer, Berlin, 1985.
9. E. Weber, J. L. Toner, J. Goldberg, F. Vögtle, D. A. Laidler, J. F. Stoddart, R. A. Bartsch and C. L. Liotta, *Crown Ethers and Analogs* (S. Patai and Z. Rappoport, Eds.), Wiley, Chichester, 1989.
10. J. Canceill, A. Collet, J. Gabard, F. Kotzyba-Hibert and J.-M. Lehn, *Helv. Chim. Acta*, **65**, 1894 (1982).
11. (a) F. Vögtle, W. M. Müller and W. H. Watson, *Top. Curr. Chem.*, **125**, 131 (1984).
    (b) F. Vögtle and W. M. Müller, *J. Incl. Phenom.*, **1**, 369 (1984).
    (c) W. H. Watson, J. Galloy, F. Vögtle and W. M. Müller, *Acta Crystallogr.*, **C40**, 200 (1984); M. R. Caira, W. H. Watson, F. Vögtle and W. M. Müller, *Acta Crystallogr.*, **C40**, 491 (1984); W. H. Watson, J. Galloy, D. A. Grossie, F. Vögtle and W. M. Müller, *J. Org. Chem.*, **49**, 347 (1984).
12. C. J. van Staveren, D. E. Fenton, D. N. Reinhoudt, J. van Eerden and S. Harkema, *J. Am. Chem. Soc.*, **109**, 3456 (1987); F. C. J. M. van Veggel, M. Bos, S. Harkema, W. Verboom and D. N. Reinhoudt, *Angew. Chem.*, **101**, 800 (1989); *Angew. Chem. Int. Ed. Engl.*, **28**, 746 (1989).
13. Y. Aoyama, Y. Tanaka, H. Toi and H. Ogoshi, *J. Am. Chem. Soc.*, **110**, 634 (1988); Y. Aoyama, Y. Tanaka and S. Sugahara, *J. Am. Chem. Soc.*, **111**, 5397 (1989); For binding of amino acids see Y. Aoyama *et al.*, *J. Am. Chem. Soc.*, **110**, 4076 (1988); Complexation of sugars: T. Kunitake *et al.*, *J. Am. Chem. Soc.*, **113**, 444 (1991).
14. R. E. Sheridan and H. W. Whitlock Jr, *J. Am. Chem. Soc.*, **110**, 4071 (1988); *J. Am. Chem. Soc.*, **108**, 7120 (1986); K. M. Neder and H. W. Whitlock Jr, *J. Am. Chem. Soc.*, **112**, 9412 (1990).
    (a) Concave reagents with intra-annular pyridine nitrogen: U. Lüning and F. Schillinger, *Chem. Ber.*, **123**, 2073 (1990); Concave benzoic acids: U. Lüning, C. Wangnick, K. Peters and H. G. von Schnering, *Chem. Ber.*, **124**, 397 (1991).

483

(b) R. Berscheid, M. Nieger and F. Vögtle, *J. Chem. Soc., Chem. Commun.*, **19**, 1364 (1991).

(c) R. Berscheid and F. Vögtle, *Synthesis*, **1992**, 58.

15. (a) F. Ebmeyer and F. Vögtle, *Angew. Chem.*, **101**, 95 (1989); *Angew. Chem. Int. Ed. Engl.*, **28**, 79 (1989).

(b) F. Vögtle, Ch. Seel, F. Ebmeyer, A. Wallon, M. Bauer, W. M. Müller and U. Werner, *Kemia-Kemi*, **16**, 796 (1989).

(c) F. Vögtle, I. Lüer, V. Balzani and N. Armaroli, *Angew. Chem.*, **103**, 1367 (1991); *Angew. Chem. Int. Ed. Engl.*, **30**, 1333 (1991).

(d) Ch. Seel and F. Vögtle, *Angew. Chem.*, **103**, 433 (1991); *Angew. Chem. Int. Ed. Engl.*, **30**, 442 (1991); Ch. Seel and F. Vögtle, in S. R. Cooper (Ed.), *Crown Compounds*, p. 191, VCH, Weinheim, 1992; cf. also R. Hoss and F. Vögtle, *J. Chem. Soc., Chem. Commun.*, **21**, 1584 (1992); *Chem. Ber.*, **126**, 1003 (1993).

16. F. Vögtle, A. Wallon, W. M. Müller, U. Werner and M. Nieger, *J. Chem. Soc., Chem. Commun.*, 158 (1990).

17. P. E. J. Sanderson, J. D. Kilburn and W. C. Still, *J. Am. Chem. Soc.*, **111**, 8314 (1989); W. C. Still *et al.*, *J. Am. Chem. Soc.*, **113**, 5111 (1991); S. S. Yoon, W. C. Still, *J. Am. Chem. Soc.*, **115**, 823 (1993); overview on peptide receptors: H. J. Schneider, *Angew. Chem.*, **105**, 890 (1993); *Angew. Chem. Int. Ed. Engl.*, **32**, 848 (1993).

(a) Ch. A. Hunter, *J. Chem. Soc., Chem. Commun.*, 749 (1991).

18. J. F. Stoddart *et al.*, *Angew. Chem.*, **101**, 1404 (1989); *Angew. Chem. Int. Ed. Engl.*, **28**, 1396 (1989); *Angew. Chem.*, **103**, 1055, 1058 (1991); *Angew. Chem. Int. Ed. Engl.*, **30**, 1036, 1039 (1991); *J. Am. Chem. Soc.*, **113**, 4335 (1991); For a new type of catenanes see: C. A. Hunter, *J. Am. Chem. Soc.*, **114**, 5303 (1992) and R. Hoss, St. Meier and F. Vögtle, *Angew. Chem.*, **104**, 1628 (1992); *Angew. Chem. Int. Ed. Engl.*, **31**, 1619 (1992).

19. F. Diederich and H.-D. Lutter, *J. Am. Chem. Soc.*, **111**, 8438 (1989); cf. R. Breslow and E. Kool, *Tetrahedron Lett.*, **29**, 1635 (1988); D. Hilvert and R. Breslow, *Bioorg. Chem.*, **12**, 206 (1984); R. Breslow, *Science*, **218**, 532 (1982); *Acc. Chem. Res.*, **13**, 170 (1980); I. Chao and F. Diederich, *Rec. Trav. Chim. Pays Bas*, **112**, 335 (1993).

20. D. J. Cram. P. Y.-S. Lam and S. P. Ho, *J. Am. Chem. Soc.*, **108**, 839 (1986); D. J. Cram and P. Y.-S. Lam, *Tetrahedron*, **42**, 1607 (1986); cf. S. Kumar and H.-J. Schneider, *J. Chem. Soc., Perkin Trans. 2*, 245 (1989).

21. W. L. Mock *et al.*, *J. Org. Chem.*, **54**, 5302 (1989).

22. T. R. Kelly, C. Zhao and G. J. Bridger, *J. Am. Chem. Soc.*, **111**, 3744 (1989); T. R. Kelly, G. H. Bridger and C. Zhao, *J. Am. Chem. Soc.*, **112**, 8024 (1990).

(a) A. D. Hamilton *et al.*, *J. Am. Chem. Soc.*, **115**, 5314 (1993).

23. T. Tjivikua, P. Ballester and J. Rebek Jr, *J. Am. Chem. Soc.*, **112**, 1249 (1990); Review: J. Rebek Jr, *Angew. Chem.*, **102**, 261 (1990); *Angew. Chem. Int. Ed. Engl.*, **29**, 245 (1990).

24. Gigantocycles: F. M. Menger *et al.*, *Angew. Chem.*, **104**, 1542 (1992); *Angew. Chem. Int. Ed. Engl.*, **31**, 1492 (1992).

Concluding Remarks and Future Outlook

This journey into cyclophane chemistry should demonstrate how widely this originally isolated branch of organic chemistry has developed up to the present and what varied results and problems have been generated. Especially to be highlighted are the, for the most part, clearly defined rigid and dynamic stereochemistry, transannular effects, steric and electronic interactions, and—last but not least—the use of diverse phane frameworks for host–guest interactions and as skeletal structures for molecular organization, assembly and replication.

Ring compounds have fascinated mankind for a long time, and rings in general are an undeniable aspect of human culture and art, as was illustrated in the introductory section. Molecules containing rings bear information which open-chain compounds can scarcely convey at the same high level; so it is to be expected that organic and inorganic rings (with outer *and* inner surfaces) will remain the basis for research and application.

The future of cyclophane chemistry will in all likelihood include the following areas:

- Novel molecular and supramolecular structures and resulting unique properties
- Large molecules extended in three dimensions, which provide large cavities and expansive outer surfaces
- Extremely distorted aromatic rings
- Highly selective complexation chemistry, host–guest chemistry, and supramolecular chemistry (sensors)
- Transition metal complexes (metallocenophanes; metallaphanes)
- Donor–acceptor complexes, including electrically conducting substances
- Molecular recognition (e.g. switches, light-induced energy transfer, photo-induced electron transfer between donor–acceptors, molecular motors or elements of motors)
- Specifically designed transannular electronic and steric effects
- Unusual transannular reactions
- Structure/chiroptical relationships in new optically active phanes
- Development of new synthetic methods for strained phanes
- Phanes as models for intercalations
- Interactions between sterically proximate functional groups within a phane
- Exotic phane structures (unusual molecular skeletons, but also phanes with convergent functional groups)

486

- Even larger macro(poly)cyclic ring systems of nano scale dimensions ('gigantocycles', see page 483, ref. 24); unique ring types
- Polycycloarenes (beyond 'Kohnkene')
- Three-dimensional giant structures beyond 'trinacrene'
- Three-dimensional structures with larger cavities than in the 'carcerands'
- Molecular loops; novel catenane types
- Soluble, highly defined giant molecules, in which the enhanced solubility is introduced by incorporation of multiple functional groups such as carboxylic acids, sulphonic acids, ammonium ion, or oligoethylene glycol ether units.
- Micelle- and membrane-forming phane structures
- Extension of porphyrinophanes and phthalocyaninophanes (bioinorganic models)
- Dyestuff phanes with transannular electronic colour effects (concave dyes)
- Extension of calixarenes and cyclotriveratrylenes as standards for variable cavities and as readily available standard skeletal structures.
- Use of phane cavities for chromatographic stationary phases for highly selective separation of guest substances, including chiral resins for resolution of enantiomers by MPLC and HPLC
- Phanes as physiological agents in pharmaceutical chemistry, for special dyestuffs, polymers. . .
- Phanes as skeletal structures for non-linear optics (NLO), molecular computers, and other 'high-chem' applications.

With this enumeration of the future perspectives, we come to the conclusion of this overview of the phanes.

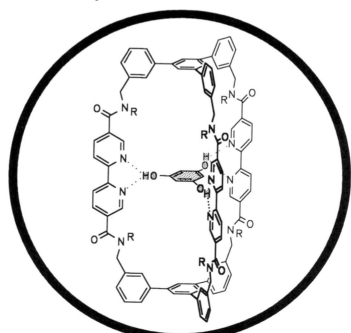

The formula shows the molecular recognition of 1,3,5-trihydroxybenzene (phloroglucinol) in the large cavity of a tris(bipyridine) host molecule bearing complementary functional groups. For more detail, see Section 12.3 (computer drawing, Acad-Programme, F. Vögtle, Ch. Seel, F. Ebmeyer).

Author Index

Subject Index

500

Indexes compiled by G. Jones